AF519297

WASTEWATER Treatment Technologies and ENVIRONMENT

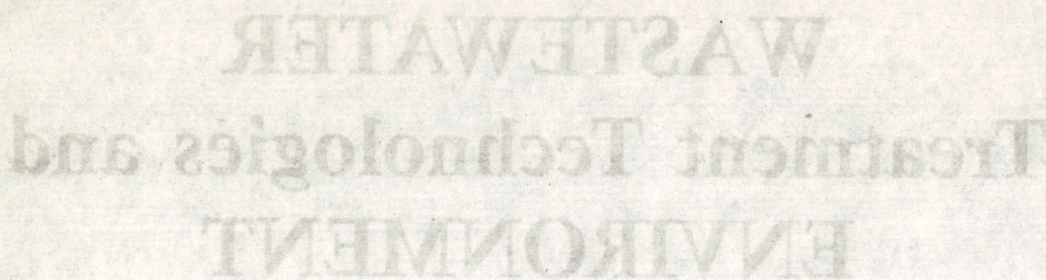

WASTEWATER
Treatment Technologies and ENVIRONMENT

— Vol. 2 —

— Editors —
Dr. S. N. Kaul
Dr. Lidia Szpyrkowicz
Professor (Dr.) Arvind Kumar

— Assocaite Editors —
N. N. Rao
D. M. Dharmadhikari
National Environmental Engineering Research Institute
Nagpur - 440 020

2004
DAYA PUBLISHING HOUSE
Delhi - 110 035

ISBN 81-7035-342-4 (Set)
ISBN 81-7035-343-2 (Vol. 1)
ISBN 81-7035-344-0 (Vol. 2)

Published by : **Daya Publishing House**
1123/74, Deva Ram Park
Tri Nagar, Delhi - 110 035
Phone : 27383999
Fax : (011) 27382329
e-mail : dayabooks@vsnl.com
website : www.dayabooks.com

Showroom : 4762-63/23, Ansari Road, Darya Ganj
New Delhi - 110 002
Phone: 23245578, 23244987

Laser Typesetting : **Classic Computer Services**
Delhi - 110 035

Printed at : **Chawla Offset Printers**
Delhi - 110 052

PRINTED IN INDIA

Preface

The past few years have seen the emergence of a growing, widespread desire in this country, and indeed everywhere, that positive actions be taken to restore the quality of our environment, and to protect it from the degrading effects of all forms of pollution - air, noise, solid waste, and water. Since pollution is a direct or indirect consequence of waste, if there is no waste, there can be no pollution, and the seemingly idealistic demand for "zero discharge" can be construed as a demand for zero waste. However, as long as there is waste, we can only attempt to abate the consequent pollution by converting it to a less noxious form. In those instances, in which a particular type of pollution has been recognized, three major questions usually arise:

1. How serious is the pollution?
2. Is the technology to abate it available? and
3. Do the costs of abatement justify the degree of abatement achieved ?

The principal intention of this book on *Wastewater Treatment Technologies and Environment* is to help in formulating useful answers to the second and third of these questions, and to examine their costs in the light of the real level of benefits afforded.

The traditional approach of applying tried-and-true solutions to specific pollution problems has been a major factor contributing to the success of environmental engineering, and in large measure has accounted for the establishment of a methodology of pollution control. However, realizing the already great complexity of current environmental problems, and understanding that, as time goes on, these issues will become even more complex and interrelated; render it imperative that intelligent planning of pollution abatement systems must be undertaken. Prerequisite to such planning is an understanding of the performance, potential, and limitations of the various methods of pollution abatement available for environmental engineering. We believe that the unification of the concepts and engineering methodology found in this book is a logical step in the evolution of environmental engineering, particularly for wastewater treatment.

The treatment of various engineering systems presented will show how an engineering formulation of the subject flows naturally from the fundamental principles and theory of chemistry, physics and mathematics. This emphasis on fundamental science is based on the recognition that engineering practice has of necessity in recent years become more firmly based of scientific principles, rather than depending largely on our empirical accumulation of facts, as was earlier the case. It was not intended, though, to neglect empiricism when such data lead quickly to the most economic design; certain engineering systems are not readily amenable to fundamental scientific analysis, and in these instances we have resorted to less science in favour of more art and empiricism.

An engineer must understand science within a context of applications. It is, therefore, necessary to equip oneself with promising potential technologies. The book on *Wastewater Treatment Technologies and Environment* is the outcome of scientific collaboration between National Environmental Engineering Research Institute (NEERI, India); and Department of Environmental Sciences, University of Venice, Venice, Italy; under 'Bilateral Programme of CSIR, India and CNR, Italy Collaborative Project'.

The editors are pleased to acknowledge the encouragement and support received from their colleagues at NEERI and Department of Environmental Sciences, University of Venice, Venice, Italy, during the conceptual stages. The editors wish to thank the contributing authors for their time and effort, and for having borne patiently our numerous queries and comments. Last but not the least, the editors appreciate the meticulous efforts rendered by Mr. G. B. R. Sharma, Mr. C. R. Bakre and Mr. P. A. Ingle in the preparation and production of this book.

One of the authors, Professor (Dr) Arvind Kumar wishes to express his sincere gratitude to Professor Indu Dhan, Hon'ble Vice Chancellor and Professor Binodanand Jha, Hon'ble Pro-Vice Chancellor of S. K. University, Dumka who have been the source of constant inspiration. He is also thankful to his wife, Professor Kumari Bimla and sons, Kumar Pallav Ramachandran and Kumar Prasun Ramakrishnan whose natural smiles extended him relief from this tiresome endeavour.

It is my sincere duty to express my heartiest thankfulness to Shri Anil Mittal, Proprietor, Daya Publishing House, Delhi for taking keen interest in publishing of this book.

Dr. S. N. Kaul

Dr. Lidia Szpyrkowicz

Professor (Dr.) Arvind Kumar

Contents

— VOLUME 1 —

42

Removal of Hexavalent Chromium from Industrial Effluents Using Low Cost Adsorbents

V. PATANJALI & P.K. MISHRA

Department of Chemical Engineering and Technology Institute of Technology, Banaras Hindu University, Varanasi

ABSTRACT

Removal of hexavalent chromium from synthetically prepared chromium bearing wastewater was investigated using coconut shell carbon, rice husk, wood charcoal and sawdust carbon as adsorbents. The effect of pH, concentration, temperature and particle size were studied in batch and continuous processes. The removal was in general most effective at low pH and low initial hexavalent chromium concentrations. Batch adsorption kinetics has been suitably described by Lagergren equation. The applicability of the Langmuir isotherm for the present system has been tested for different temperatures. Thermodynamic parameters indicate the endothermic nature of hexavalent chromium adsorption on sawdust carbon, where as the others indicate exothermic nature of the reaction. It was found that the break through point depends on the flow rate and the height of the bed, during continuous bed experiments.

Introduction

Water pollution by toxic heavy metals like chromium has become a major concern. Hexavalent chromium, a waste from industries like textile, fertilizer, tanning and electroplating, adversely affects the flora and fauna and can cause severe health problems, if more than permissible limits. Chromium having many acute affects on all biological systems has to be removed from water. In the present study an attempt has been made to remove the hexavalent chromium using low cost adsorbents through adsorption technique.

The adsorption is a surface phenomenon. The increase in concentration if the two phases are in mutual contact, relative to the concentration in the bulk liquid of the phase is called adsorption. The material being concentrated or adsorbed is called adsorbate and the adsorbing phase is the adsorbent. The process of adsorption is mainly due to the forces of imbalance at the solid or liquid surface. Migration of a substance to the surface or boundary result in reduction of the above force acting at the surface. Now it has become a widely used wastewater treatment operation.

Efficiency of adsorption process mainly depends on nature of adsorbent, nature of adsorbate, pH, concentration, temperature, foreign ions, time and agitation. Various workers (1-9) have reported encouraging results regarding removal of Cr (VI) using low cost adsorbents. As coconut shell, rice husk, wood charcoal and sawdust are easily available and very economic they have been used for the present work.

Experimental Setup

Materials

Various adsorbents are * and prepared as per following details:

Coconut Shell Carbon

Coconut shell was obtained locally in Varanasi, U.P., treated with distilled water and dried in oven at 50 °c. Coconut shell carbon was prepared by reacting 10 gm. of coconut shell with 10 ml. Of concentrated sulphuric acid and was heated in a oven at 150°C for 12 hours. The carbonized material was washed with distilled water to remove excess acid and then dried in oven at 50°C.

Wood Charcoal

Wood charcoal was obtained locally in Varanasi, U.P. It has been washed with water and dried in oven at 50 °c.

Rice Husk

Rice husk was obtained from a rice mill at Varanasi, U.P. It has been washed with water and dried in oven at 50 °c.

Sawdust Carbon

Sawdust was obtained from wood mill in Varanasi, U.P. It has been washed with water and dried in oven at 50 °c.Sawdust carbon was prepared by reacting 10 gm. of sawdust with 10 ml. of concentrated sulphuric acid and was heated in a oven at 150°C for 12 hours. The carbonized material was then washed with distilled water to remove excess acid and dried in a oven at 50°C.

Physico-Chemical Properties of Adsorbents

As the removal of pollutants from wastewater by an absorbent is significantly affected by physico-chemical nature of adsorbent, therefore the adsorbents were characterized before hand for determining their relevant physical properties and chemical composition.

1. *Physical properties:* bulk densities of the adsorbents were determined with the help of densitometry and particle size of all sizes of adsorbents were determined with the help of sieve analysis, whereas, surface area of adsorbents were determined by BET method, using

low temperature N_2 gas adsorption. All the adsorbents were sieved through 177, 252 and 353 micromesh. The physico-chemical properties of adsorbents are listed in Table 1.

2. *Proximate analysis:* proximate analysis of the adsorbents to be utilized in experimental investigation was done as per standard procedure for fixed carbon and volatile matter. Proximate analyses of adsorbents are shown in Table 2.
3. Infra red analysis of adsorbent was recorded using potassium bromide pellet and a Perkin Elmer Spectrophotometer for determining the nature of the bonding present on the surface of the adsorbent particles.
4. *X-ray diffraction analysis:* X-ray diffractions of adsorbents were obtained by a Philips diffraction unit using a copper target for determining the nature of component phases present in adsorbent particles.

Table 1: Physical Properties of the Adsorbents

Physical Properties	*Saw Dust Carbon*	*Rice Husk*	*Coconut Shell Carbon*	*Wood Charcoal*
Surface area (m^2/g)	228	13	36	22
Bulk density (kg/m^3x1000)	0.4325	0.5745	0.6082	0.7825
Particle size (Mm)	177,252,353	177,252,353	177,252,353	177,252,353

Table 2: Proximate Analysis

	Saw Dust Carbon	*Rice Husk*	*Coconut Shell Carbon*	*Wood Charcoal*
Ash content	9.01	2.10	10	10.25
Volatile matter	7.65	4.50	6.54	5.40
Moisture	1.80	2.20	2.20	2.10
Fixed carbon	65	5.40	45.50	30.16

The hexavalent chromium compound exhibits intense and characteristic adsorption of visible radiant energy under the optimum conditions of hydrogen ion concentrations. A linear plot of absorbance versus concentration passing through the origin conforms that the Lamberts-Beers equation is applicable at lower concentrations and isobathic wavelength. Accordingly direct visible spectrophotometer was employed for determination of hexavalent chromium concentration with Systronic-20, make UV spectrophotometer.

Standard stock-solution of adsorbate was made by accurately weighed amount of adsorbate in distilled water and absorbances were noted at 540nm, and calibration curve was plotted. For maintaining constant pH throughout the experimental work, solutions of sulphuric acid and sodium hydroxide were used. Further, an intermediate stock-solution and its subsequent dilutions were made to obtain standard working solutions of known concentrations.

Methods

Both batch as well as continuous studies was performed to test the suitability of these low cost adsorbents.

Batch Tests

Batch adsorption experiments were conducted by taking 0.25 gm of sawdust carbon and 1gm of each of other adsorbents of known particle size in 50ml of aqueous solution of the adsorbate namely sodium dichromate of desired concentration, pH and temperature in a 250 ml glass bottle. This bottle is shaken in a shaking thermostat at a desired agitation speed, the extent of adsorption was determined at different intervals till the attainment of equilibrium and the suspension was taken out as soon as equilibrium is achieved, it was then centrifuged and the supernatant liquid was analyzed spectrophotometrically, to find the residual adsorbate concentration. The adsorbate concentration was adjusted using NaOH and H_2SO_4 solutions.

Adsorbate concentration was maintained at 50mgl^{-1}in all sets of experiment except in those runs, which meant for investigating the effect of concentration. For adsorption isotherm, chromium solutions of different concentrations were agitated for known amount of adsorbent till equilibrium was achieved. In order to study the effect of temperature, pH, adsorbate concentration, speed of agitation, and particle size, chromium solution of same concentration were contacted with a known amount of each of the adsorbents, till the equilibrium was attained and residual concentration of chromium was subsequently determined. The temperature was maintained at 30° C in all runs, except those meant for investigating the effect of temperature.

Continuous Experiments

Continuous column studies were carried out in 20mm diameter columns. The study was conducted to find out how long the bed will last before regeneration becomes necessary. The experiments were conducted to determine break through curves for varying bed heights and to find corresponding break through times. The effects of flow rates were evaluated on these parameters. Hexavalent chromium solution was allowed to pass through the columns having chosen adsorbents; uniform flow rate of solution was maintained through constant head glass tank (Fig. 1).

The hexavalent chromium solution coming out of the column was collected at different intervals of time and residual concentration was determined. Similar procedure was repeated at different bed heights and flow rates. The parameters like initial concentration, pH and temperature were kept constant during the course of operation.

Results and Discussion

Results of Infrared Spectroscopy and X-ray Analysis

Infrared spectroscopy and x-ray analysis helped in the identification of various forms of components and functional groups present in the constituents of adsorbents.

Coconut shell carbon showed the presence of hydroxyl group, amine, alkene, aromatic ring, silica oxide calcium oxide, magnesium oxide, iron oxide and aluminium oxide.

Sawdust carbon showed the presence of hydroxyl group, amine, alkene, carbon-carbon double bond, aromatic ring, silica oxide, iron oxide, and aluminium oxide, which help in identification of various forms of components and functional group in the adsorbent.

The presence of hydroxyl group, amine, alkene, carbon-carbon double bond, aromatic ring, silica oxide, magnesium oxide, iron oxide, and aluminium oxide was observed in wood charcoal.

Hydroxyl group, alkene group, ketonic group, aromatic ring, silica oxide, iron oxide, aluminium oxide and calcium oxide were found to be present in rice husk.

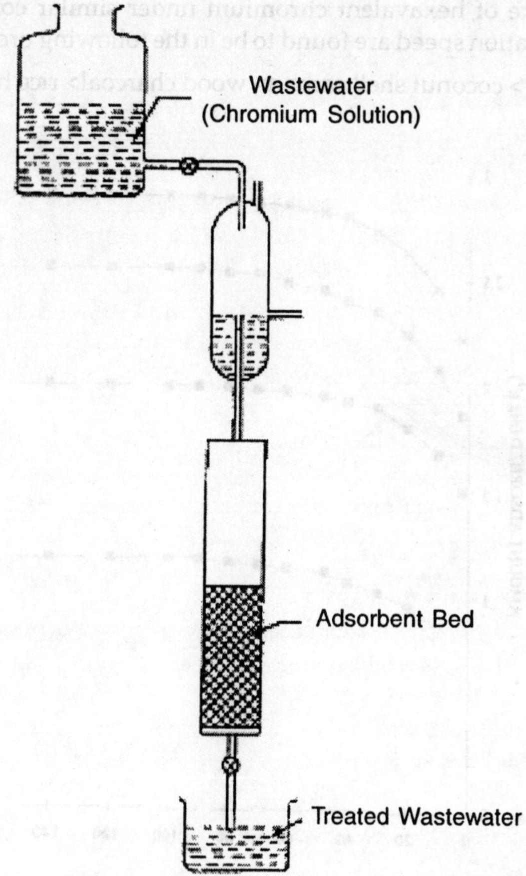

Fig. 1: Experimental Setup for Continuous Tests

Effect of Initial Concentration

Effect of initial chromium concentration on removal efficiency of adsorbent and time to reach equilibrium were studied over a wide range of chromium concentrations (25-100 mgL^{-1}), with all other conditions kept constant and the pH was maintained at the respective optimum values. The result for coconut shell carbon is shown in Fig. 2, similar results are observed for others. It is observed that the uptake of hexavalent chromium is rapid in initial stages and becomes slower in latter stages. The time required to attain equilibrium for various adsorbents *viz.* coconut shell carbon, sawdust carbon, rice husk and wood charcoal was found to be 90, 50, 90 and 60 minutes respectively. Initial hexavalent chromium concentration was found to have no significant effect on the equilibrium time. The results also indicate that the percentage removal increases with decrease in initial concentration and the hexavalent chromium uptake per unit mass of sawdust carbon was found to increase with increase in initial concentration. The efficiency of coconut shell carbon, sawdust carbon, rice husk and wood charcoal on the uptake of hexavalent chromium under similar conditions of temperature, pH, concentration and agitation speed are found to be in the following order:

Sawdust carbon> > coconut shell carbon> wood charcoal> rice husk.

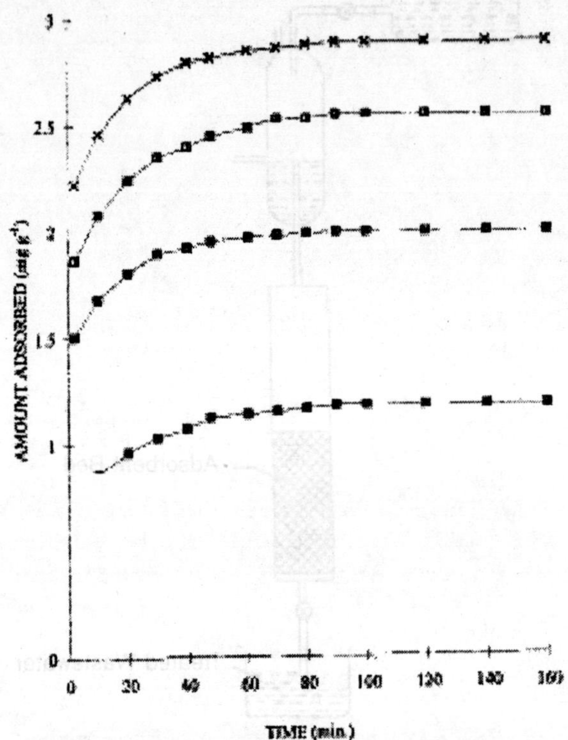

Fig. 2: The Effect of Variation of Time on Adsorption of Chromium (IV) on Coconut Shell Carbon at Different Initial Concentrations

Effect of Temperature

The effect of temperature on the removal of hexavalent chromium for coconut shell carbon is shown in Fig. 3, rice husk and wood charcoal shower similar results but that for saw dust is shown in Fig. 4, at three different temperatures *viz.* 30, 45 and 60°C under identical conditions and the results show that:

1. At initial stages of adsorption, the uptake of hexavalent chromium is rapid at each temperature and becomes slow at later stages and finally attains saturation state.
2. The equilibrium time is independent of temperature.
3. The adsorption capacity of coconut shell carbon, rice husk and wood charcoal for hexavalent chromium decreases with increasing temperature, thus indicating exothermic nature of adsorption and that of sawdust carbon increases with increasing temperature, thus indicating endothermic nature of adsorption.

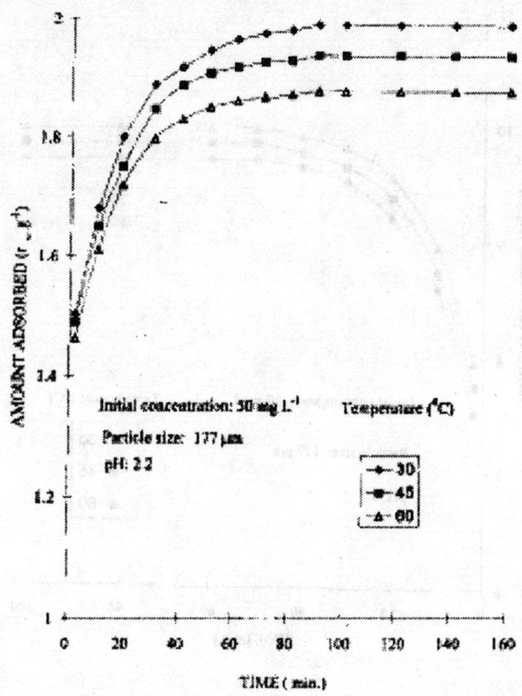

Fig. 3: The Effect of Variation of Time on the Adsorption of Chromium (IV) on Coconut Shell Carbon at Different Temperatures

Adsorption Isotherm

The plots of c_e/q_e versus c_e (Fig. 5 shows for rice husk, similar results were obtained for others.) show a linear relationship, which conforms the applicability of Langmuir equation for hexavalent chromium.

Effect of pH

The plots of the effect of pH change on the adsorption of hexavalent chromium for wood charcoal (Fig. 6, similar results were obtained for others.), with all other conditions kept identical, are initially curved due to rapid removal and finally become flat with lapse of time. The plot of pH versus removal for optimum pH for coconut shell carbon (Fig. 7, similar results were obtained for others.) shows clearly that adsorption increases with decrease in pH. The value of pH for maximum adsorption for coconut shell carbon, sawdust carbon, rice husk and wood charcoal are 2.2, 2.2, 1.8 and 2.0 respectively.

Effect of Particle Size

In the present work, effect of particle size of the adsorbents was investigated using three particle sizes (177, 252 and 353 Mm) under identical conditions of initial concentration, adsorbent dose and temperature. The pH was maintained at the respective optimum value. The result for saw dust carbon (Fig. 8, similar results were obtained for others.) show that the removal of hexavalent chromium increases with decrease in particle size of adsorbents.

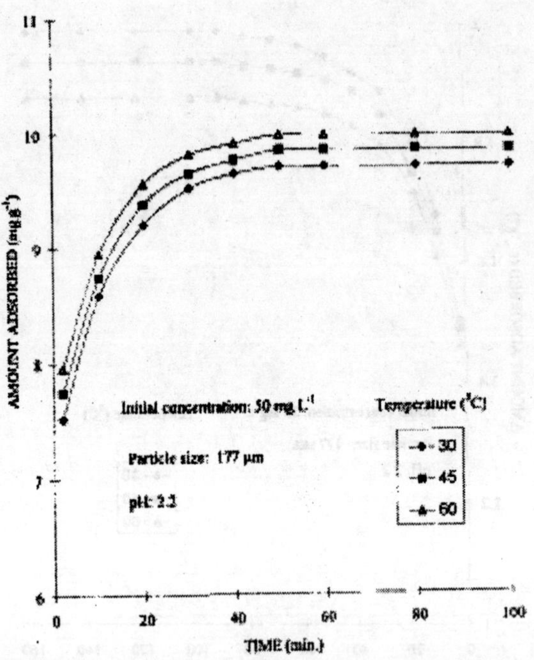

Fig. 4: The Effect of Variation of Time on the Adsorption of Chromium (IV) on Saw Dust Carbon at Different Temperatures

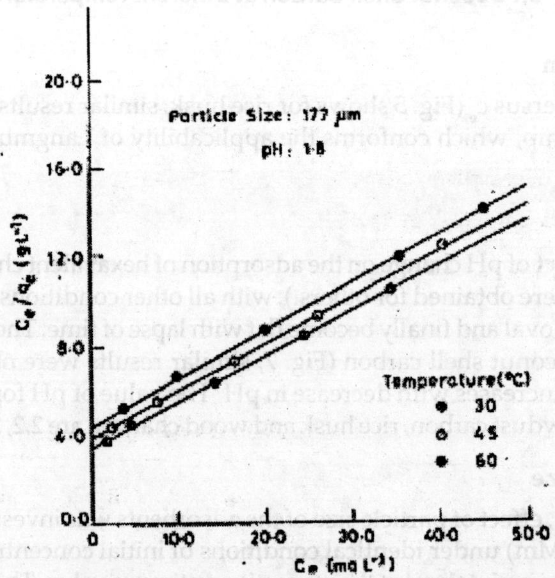

Fig. 5: Langmuir Isotherm for the Adsorption of Chromium (IV) on Rice Husk at Different Temperatures

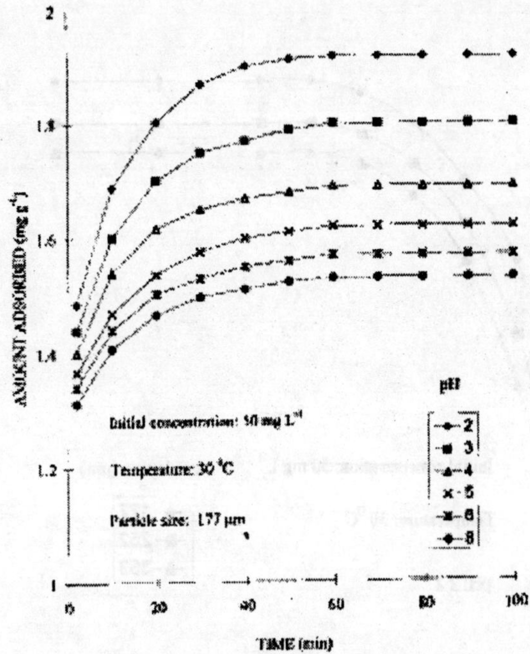

Fig. 6: The Effect of Variation of Time on the Adsorption of Chromium (IV) on Wood Charcoal at Different pH Values

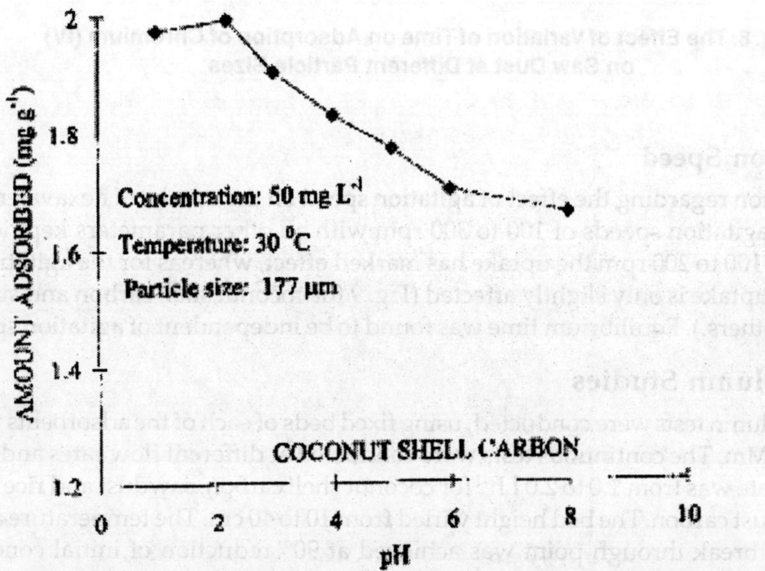

Fig. 7: Amount Adsorbed at Equilibrium Condition as a Function of pH

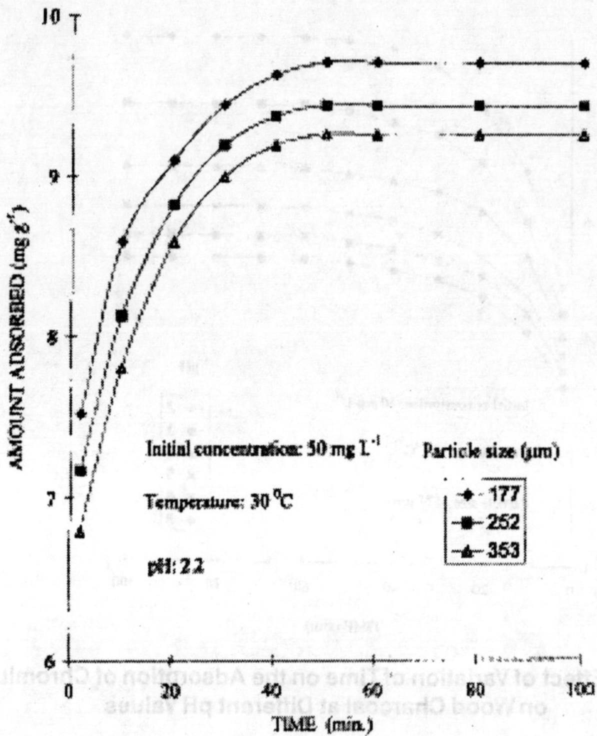

Fig. 8: The Effect of Variation of Time on Adsorption of Chromium (IV) on Saw Dust at Different Particle Sizes

Effect of Agitation Speed

The investigation regarding the effect of agitation speed on the uptake of hexavalent chromium has been done for agitation speeds of 100 to 300 rpm with all other parameters kept identical. For agitation speeds of 100 to 200 rpm the uptake has marked effect, whereas for the agitation speeds of 200 to 300 rpm the uptake is only slightly affected (Fig. 9 for coconut shell carbon and similar results were obtained for others.). Equilibrium time was found to be independent of agitation speed.

Continuous Column Studies

Continuous column tests were conducted, using fixed beds of each of the adsorbents with average particle size of 500 Mm. The continuous tests were conducted at different flow rates and bed heights. The range of flow rate was from 1.0 to 2.0 Lh^{-1} for coconut shell carbon, sawdust and rice husk and 1.0 to 3.0 Lh^{-1} for sawdust carbon. The bed height varied from 10 to 40 cm. The temperature and pH were kept constant. The break through point was achieved at 90%reduction of initial concentration of hexavalent chromium, while the inlet concentration of hexavalent chromium taken was 25 mg L^{-1} in the experiments. Break through curve for saw dust carbon is shown in Fig. 10 (similar results were obtained for others). It is clear from these plots that as bed height increases saturation time also

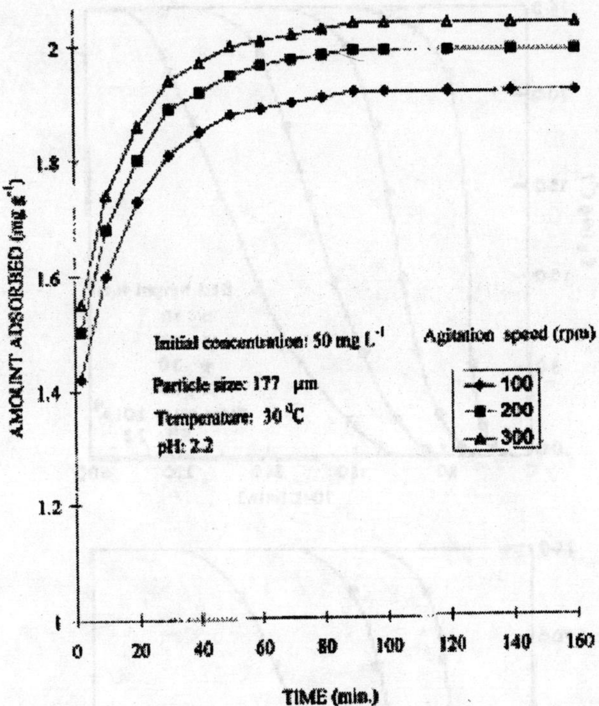

Fig. 9: The Effect of Variation of Time on Adsorption of Chromium (IV) on Coconut Shell Carbon at Different Agitation Speed

increases, which indicates the use of adsorbent bed for longer period before regeneration is required. The plots also show that as flow rate increases saturation time decreases.

Conclusions

Following conclusions were drawn on the basis of the above investigations:

1. Physico chemical analysis of the adsorbents by X ray diffraction analysis show the presence of aluminium oxide, silicon oxide, iron oxide, magnesium oxide and carbon in coconut shell carbon, the analysis of sawdust carbon shows the presence of aluminium oxide, iron oxide, silicon oxide and carbon, rice husk was found to contain aluminium oxide, silicon oxide and calcium oxide and in wood charcoal silicon oxide, magnesium oxide, iron oxide, aluminium oxide and carbon. The presence of these compounds is also proved by infrared spectrum.
2. The amount of hexavalent chromium adsorbed increases with increase in initial concentration.
3. The effect of temperature using coconut shell carbon, rice husk and wood charcoal as adsorbents indicates the exothermic nature while in case of sawdust carbon it shows endothermic nature of adsorption.

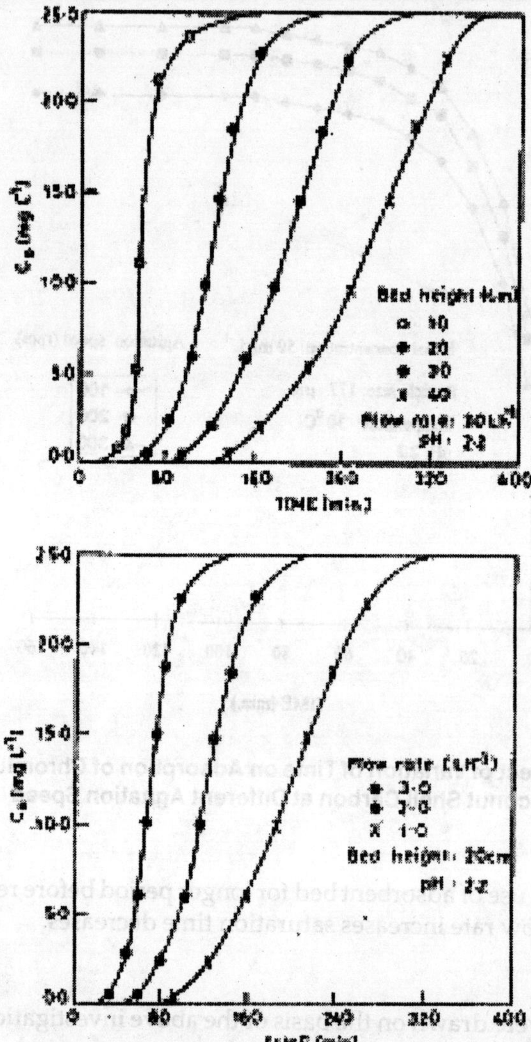

Fig. 10: Break Through Curves for the Adsorpton of Chromium (IV) on Fixed Beds of Saw Dust Carbon

4. The chromium uptake capacity was found to be high at lower pH range and the optimum pH was found to be 2.2 for coconut shell carbon, sawdust carbon and wood charcoal and 1.8 for rice husk.
5. The removal of hexavalent chromium increases with decrease in particle size.
6. Agitation speed has marked effect on uptake of hexavalent chromium between 100 to 200 rpm, whereas other agitation speeds greater than 200 were found to have minor effects on agitation speed.

7. The adsorption equilibrium data were found to fit well in isotherm model proposed by Langmuir.
8. The adsorption through fixed beds shows that the breakthrough point depends upon the flow rate and bed height.

REFERENCES

1. Ali and Deo, 1992, With rice straw on Cr (VI), batch process.
2. Bansal and Sharma, 1992, With rice husk ash on Cr (VI), batch process ci=50mgl^{-1}, pH=2.0, temp=30°c, dose 50gml^{-1}.
3. Bhargava *et al.*, 1989, With sawdust on Cr (VI), Pb, Cd, and 10 t0 50mgL^{-1} concentration.
4. Deo and Ali, 1992, With rice straw on Cr (VI), c_i=5mgL^{-1}, dose 5 gml^{-1}.
5. Manju and Anirudhan, 1997, With coconut fiber pitch carbon on Cr (VI) batch process pH =2.0, temp=30°C.
6. Rai and Kumar, 1999, With brick kiln ash, fly ash on Cr (VI) batch process c_i=50mgL^{-1}, temp=30°c.
7. Sharifov *et al.*, 1998, With sawdust on Cr (VI), treated with 7% H_2SO_4.
8. Srinivasan *et al.*, 1998, With rice husk on Cr (VI), column process.
9. Zarras and Mahmoud, 1995, With sawdust on Cr (VI), batch process.

43

New Polymeric Membranes for Dye Wastewater Treatment: Transportation and Compaction Characteristics

A.N. VAIDYA & A.S. BAL

Scientist, National Environmental Research Institute, Nehru Marg, Nagpur

Introduction

The application of membrane technologies in industrial wastewater, water reuse, and product recovery and recycle are the result of the increasing environmental regulations enforced on industry. Membranes are also used by industries in the purification of water, intermediates, and products. Membrane separation processes are the key component behind the ability of industries to achieve zero discharge goals (Johns, F.J.). Membrane is the most vital element in membrane separation processes and polymeric membranes that are widely employed in wastewater treatment are cellulose acetate, polyamides. Polysulphones. Other polymers that are used in wastewater treatment include polypropylene, nylon, polyacrylonitrile, polyvinyl alcohol, and polycarbonate. In addition, ceramic and metallic membranes are sometimes used. Cellulose acetate and polyamide membranes exhibit excellent flux and rejection characteristics but are prone to bacterial and chemical attack, especially chlorine that irreversibly damages the membrane. Cellulose acetate membranes cannot sustain temperatures more than 38°C and pH outside the range 4 to 7. Polysulphone membranes exhibit excellent resistance to bacterial and chemical attack, however yield low throughputs and are costly.

Efficacy of the membrane technology entirely depends upon the selection of membrane material with appropriate properties to with stand wastewater environment. Ideally a membrane that is used for wastewater treatment should exhibit high throughput, high rejection of the desired pollutant, excellent resistance to bacterial and chemical attack, and good pH tolerance. Even though membrane technology has been commercialized since long time, research endeavors through out the world are introducing newer and newer membrane materials in an effort to minimize the problems associated with the existing membrane materials. Extensive research efforts at National Environmental Engineering Research Institute (NEERI) have resulted in the development of polymethyl methacrylate (PMMA) as a new useful membrane material for the application in wastewater treatment. Polymethyl methacrylate has excellent resistance to bacterial attack and it can tolerate a wide pH range. Dilute acids or organic chemicals do not affect it. Its glass transition temperature is 108°C and therefore it can sustain higher temperatures. The hydrodynamic and interfacial parameters of PMMA are quite favorable for its use as a membrane material (Sourirajan). Textile industries traditionally generate a large quantity of wastewater, which is normally discharged after expensive treatments. Compared to conventional methods of treating textile effluents, pressure driven membrane processes enable recovery of valuable waste components such as dyes and sizing agents (Caetano A.T. *et al.*). This paper describes preparation of PMMA membranes for the treatment of dye wastewater, its transportation characteristics, and its compaction characteristics.

Materials and Methods

Membrane Preparation

Membranes were prepared by standard phase inversion method (Cheryen). Membrane casting solution was prepared by dissolving PMMA in a mixture of N-N dimethyl formamide, methanol, and organic solvent, preferably substituted ether that served as a flux modifier. The weight percentage of PMMA in the casting solution was in the range of 30 to 40 %. The homogenous casting solution was allowed to stand overnight to expel the entrapped air. Membranes were cast under the casting conditions detailed in Table 1. Flat membrane sheets were cast on a clean glass plate with the help of an applicator similar to one used in thin layer chromatography. After a lapse of desired evaporation time flat membrane sheet was immersed in ice-cold gelation water for 1 hour and stored under water. Membranes were marked properly according to batch numbers. The results and discussions in this paper pertain to the most useful membranes of high flux-high rejection type that are cast from solutions containing polymer in the range of 32 to 35 % (w/w). The PMMA was procured from M/S NOCIL, Thane, Mumbai and its average molecular weight was in the range of 80000 to 120000, as measured by viscosity method and acetone as a solvent. All the organic solvents were GR grade and were procured from E Merc Company.

Table 1: Casting Conditions for PMMA Membranes

1.	Casting temperature	15°C
2.	Evaporation time	60 seconds
3.	As cast thickness	200 microns
4.	Gelation media	ice cold distilled water
5.	Gelation time	1 hour
6.	Pre-treatment	not required

Membrane Testing Assembly

Flat membranes were tested in a testing cell made up of stainless steel. The cell had one pressure chamber and one permeate chamber. Pressure chamber was 6.5 cm in diameter and provided a flow channel having a width of 6.5 cm and height of 1 cm. Membrane testing cell was fixed into membrane testing assembly that consisted a 50 lit/hr capacity high pressure triplex plunger pump, feed and by pass lines, pressure regulator and permeate collection system. Schematic of the assembly is shown in Figure 1.

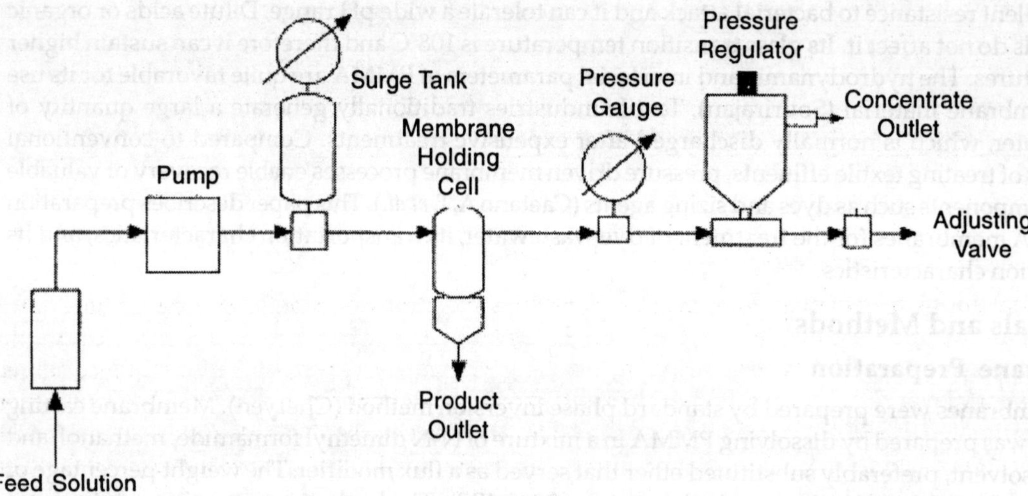

Fig. 1: Flow Diagram of the Apparatus for Testing the Ultrafiltration Flat Membranes

Membrane Testing

A sheet of desired size was cut from a membrane and was placed into testing cell with proper support. Assembly was then securely connected to triplex pump and pressure regulator. Except for compaction studies, membranes were pre compacted under a pressure of 18 atm. for two hours to obtain consistent flux. Compaction studies were conducted with pure water. Treatment studies were conducted with aqueous dye solution of commasie blue dye. The properties of this dye are detailed in Table 2. During the experimental runs, feed either pure water or dye solution was passed over membrane at desired pressure and feed and permeate samples were collected after half an hour so that consistent results were obtained. The concentration of dye was initially analyzed by colorimetric methods however; dilution errors involved were significant due to high concentration of dye in feed solution.

Therefore concentration of dye in feed and permeate was estimated by measuring chemical oxygen demand (COD), as dye was the only component that exerted COD. Analytical methods prescribed in Standard Methods were used. Permeation rates were measured as lit/hr and then converted to $m^3/m^2/day$ utilizing appropriate membrane area.

Table 2: Properties of Dye Used in UF Experiments

1.	Commercial Name	Ultrazone Blue (Commasie Blue)
2.	Colour	Blue
3.	Type of Dye	Direct Dye
4.	Molecular weight	830.9
5.	Reactive Group	(SO_3)-
6.	Procured from	textile industry
7.	Color index	42655
8.	Chemical Oxygen Demand (COD) Exerted by 1 % Dye Solution in Water	1200 mg/L (after 100 times dilution)
9.	Solubility	Easily soluble in water
10.	Wavelength of the filter required for colorometry	550 nm

Results and Discussions

Compaction Characteristics

Initial throughput of membrane gradually declines and ultimately stabilizes at some value that is lower than the initial value, the phenomenon that is known as membrane compaction. Under the applied pressure, the physical structure of membrane gets compacted increasing the resistance offered by the spongy layer of the membrane, which results in reduced throughput or flux. Extent of compaction can be estimated by stability factor that is denoted as ' m'. Relationship between flux and time for the quantification of compaction due to applied pressure is assumed to be exponential (Belfort G.). The relation can be written as $J_t = J_1 (t)^m$ or $\log (J_t/J_1) = m.\log t$ where J_t is flux at any time, J_1 is reference flux usually flux at 1 hr, and 'm' is stability factor. PMMA membranes 'A', 'B', and 'C' were tested for compaction using pure water as feed to avoid the effects of concentration polarization. Casting solution compositions for these membranes are given in Table 3. Pure water flux was measured at regular time intervals. Variations in pure water flux with time for all the membranes 'A', 'B', and 'C' are presented in Figure 2, 3 & 4 respectively. The general trend indicated that for all the three membranes, flux decline was quite sharp initially and gradually stabilized over a period of time. The log values of J_t/J_1 were plotted against log values of time.

Table 3: Casting Solution Composition for PMMA Membranes

Membrane Type	*Polymer (Weight %)*	*Solvent (Weight %)*	*Additive + Modifier (Weight %)*
A	32	38	30
B	32	40	28
C	32	42	26

The trend was almost linear and value of stability factor 'm' was calculated from the slope of best-fit line for all the membranes as shown in Figures 5, 6, & 7 respectively. It was observed that the stability factor was minimum for membrane 'B', while it was relatively higher for membranes 'A' and 'C'. Ideally the value of 'm' should be zero, which indicates constant flux. However for practical membranes values less than 0.1 are quite satisfactory. In case of PMMA membranes values of 'm' were around 0.2, which are quite satisfactory considering higher throughput of the membrane. It could also

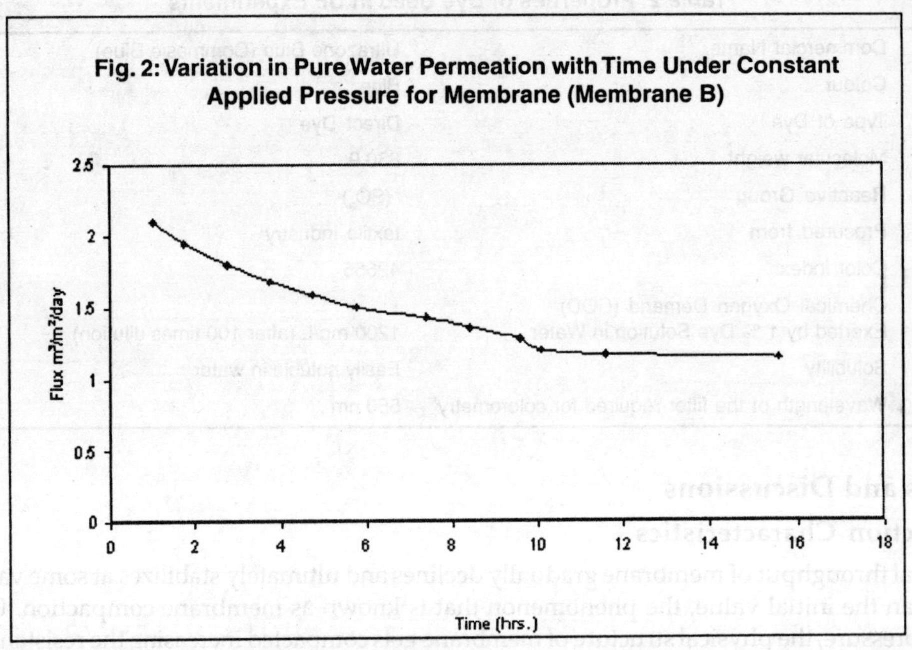

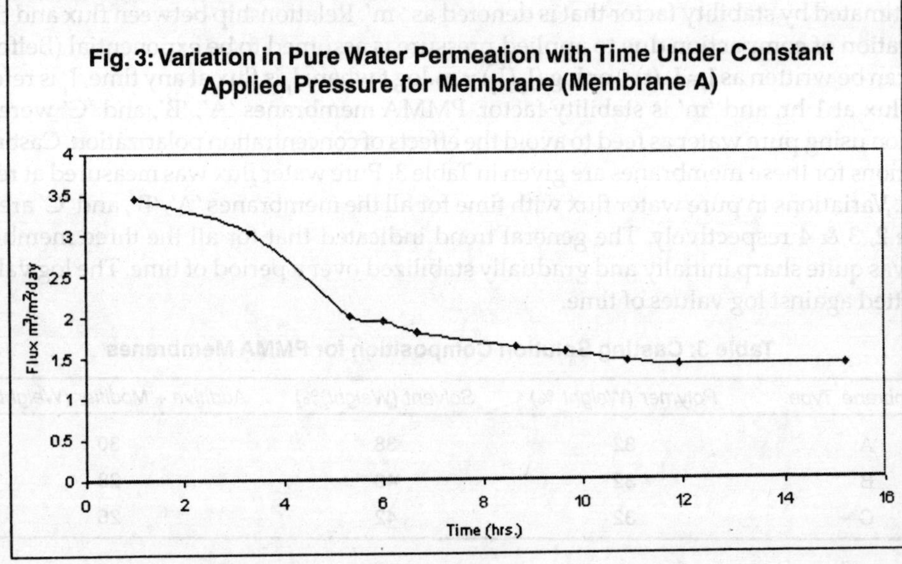

be seen that the values of 'm' changed with variation in solvent content of the membrane casting solution and membranes exhibit rapid flux stabilization at optimum solvent content which is 40 % (w/w) in case of PMMA membranes. The solvent content of the membrane casting solution plays an important role in the formation of membrane structure during evaporation and gelation steps in membrane preparation by interrupting polymer-polymer interaction. Solvent content more or less than optimum value results into an uneven porous sub layer that controls the flux through membrane.

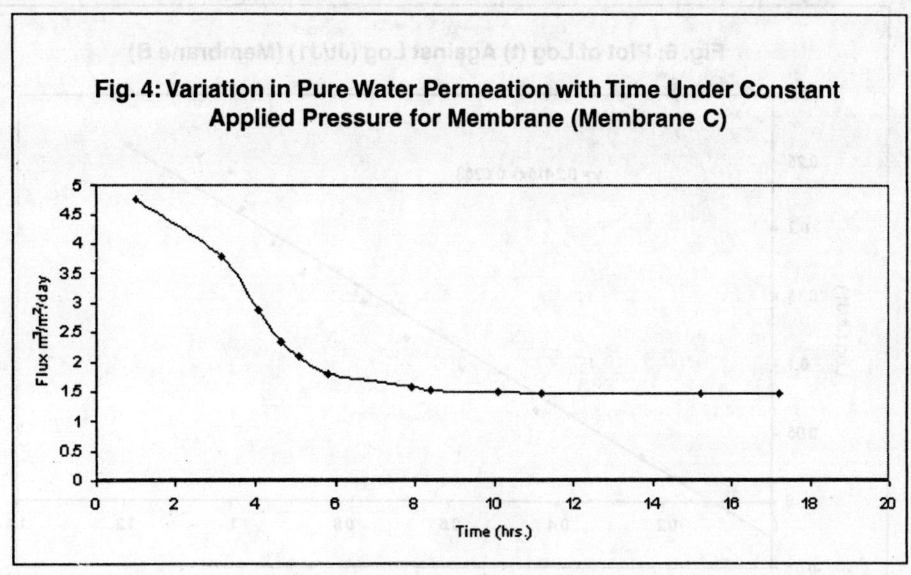

Fig. 4: Variation in Pure Water Permeation with Time Under Constant Applied Pressure for Membrane (Membrane C)

Obviously such uneven porous sub layer easily collapses under applied pressure resulting into rapid flux decline. Scanning Electron Microscopic photographs of membrane cross-section before and after compaction are shown in Figure 8.

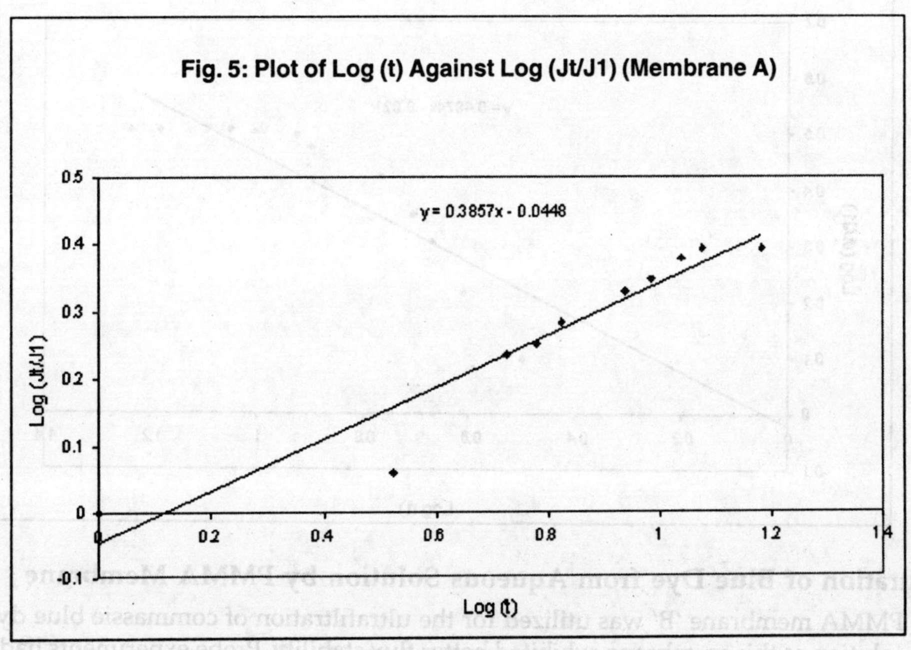

Fig. 5: Plot of Log (t) Against Log (Jt/J1) (Membrane A)

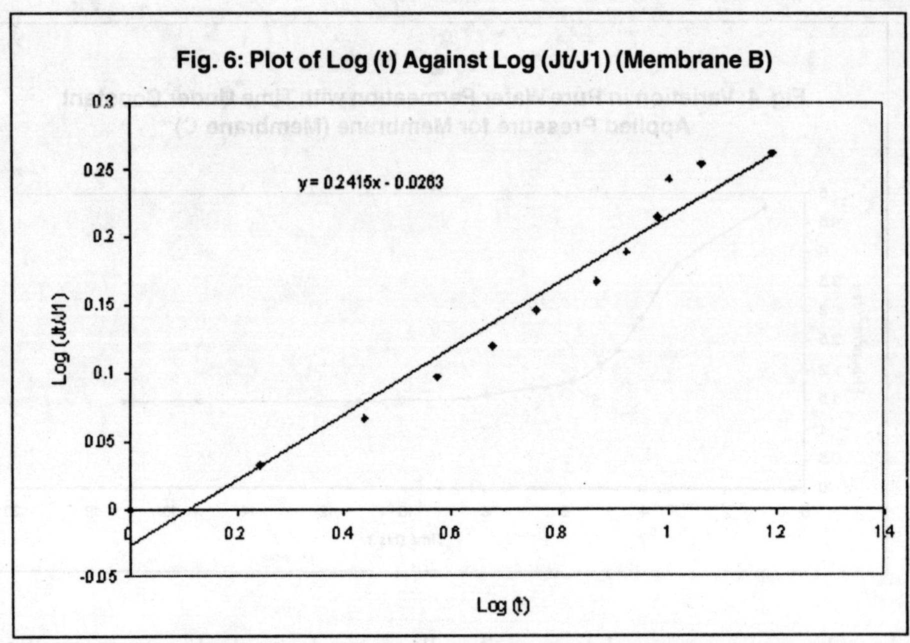

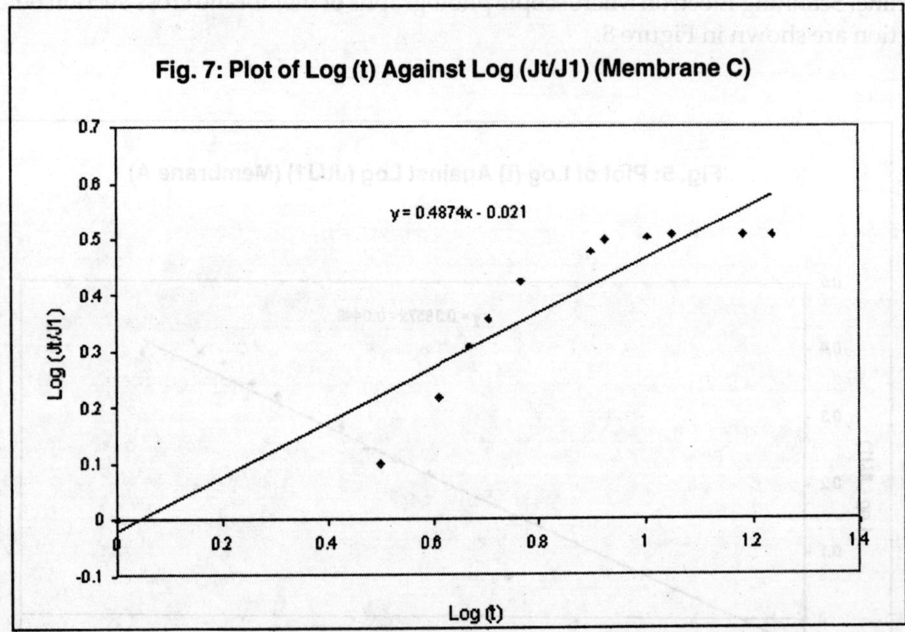

Ultrafiltration of Blue Dye from Aqueous Solution by PMMA Membrane

The PMMA membrane 'B' was utilized for the ultrafiltration of commassie blue dye from its aqueous solution as this membrane exhibited better flux stability. Probe experiments had indicated that these membranes showed excellent rejection properties with respect to comasie blue dye. Aqueous

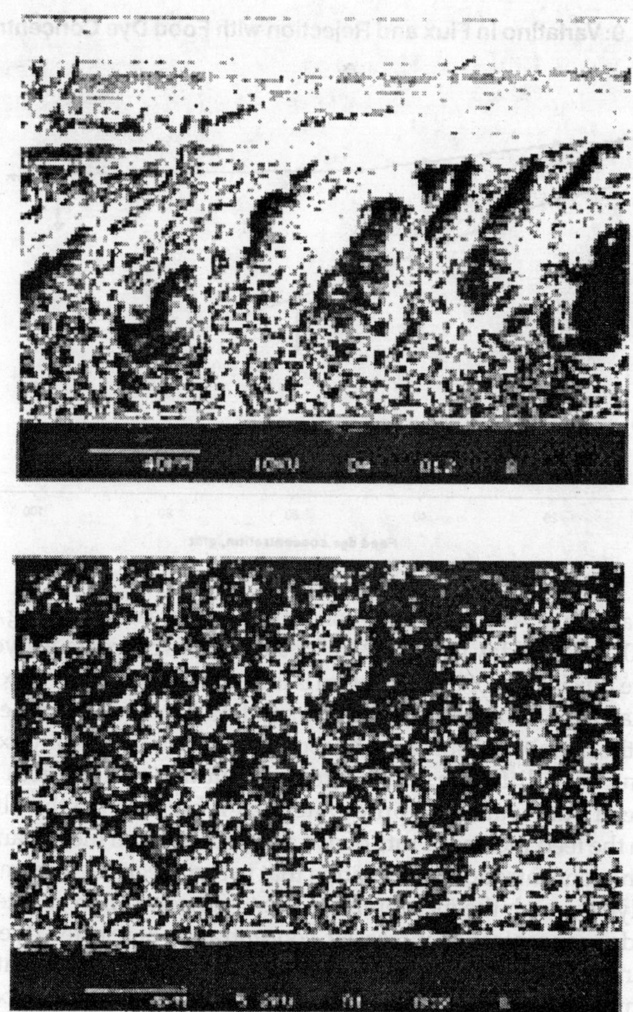

Fig. 8: Scanning Electron Microscope Photographs of PMMA Membrane Before and After Compressions

dye solutions with different dye concentrations were passed over PMMA membrane 'B' at constant feed rate of 30 lit/hr. Feed dye concentrations of 5,10,20,40,50, and100 g/lit were used for this purpose. At each feed dye concentration experimental runs were conducted for at least 8 hours and stable flux values were recorded.

Each time permeate was analyzed for COD and percentage dye rejection was estimated. Before the experiments with dye solutions, pure water flux through the membrane was measured. The variation in membrane flux and percentage dye rejection with varying feed dye concentration is shown in Figure 9. It was observed that membrane flux decreased with increasing feed dye concentration, so also percentage dye rejection. It was also observed that membrane flux in ultrafiltration of dye solution

Fig. 9: Variatino in Flux and Rejection with Food Dye Concentration

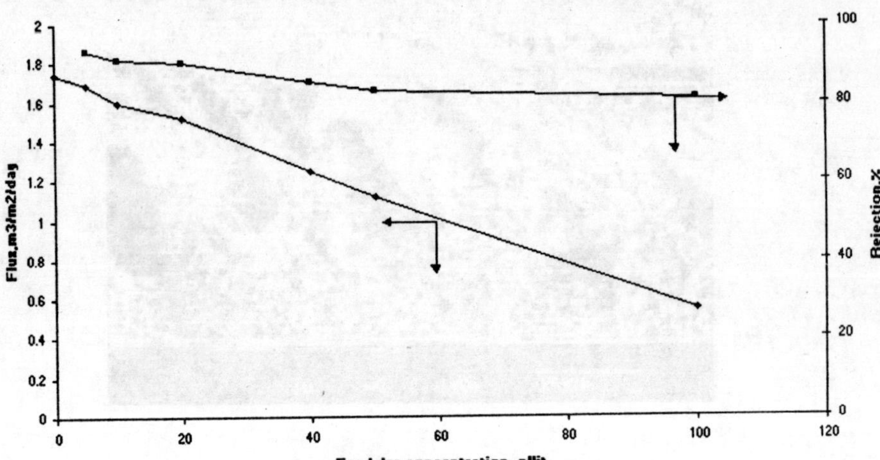

was less than pure water flux. As the feed dye concentration increased from 5g/lit to100g/lit, permeate flux reduced from 2.01 m^3/m^2/day to 0.70 m^3/m^2/day whereas percentage dye rejection reduced from 93% to 81%. Permeate flux in case of ultrafiltration is lower than pure water flux due to the phenomenon of concentration polarization. When aqueous dye solution is passed over membrane, dye molecules get accumulated on the membrane surface and concentration of dye locally exceeds than the bulk dye concentration causing back diffusion of dye into bulk solution that reduces flux. This phenomenon does not exist in case of pure water and pure water flux is higher than the ultrafiltration flux. As the dye concentration in the feed increased rate of dye accumulation on membrane surface also increased thereby enhancing the concentration polarization that in turn decreased membrane flux. On the other hand, transport of dye through the membrane is due to concentration gradient that exists across the membrane. As feed dye concentration increased, concentration of rejected dye near membrane surface increased resulting in higher concentration gradient across the membrane that enhanced the transport of dye through the membrane decreasing percentage dye rejection. These experiments showed that PMMA membranes exhibit usual ultra filtration trends. The membranes are useful for the separation of commasie blue dye and feed dye concentrations up to 20 g/lit can be readily handled at permeate rates around 1.0 m^3/m^2/day that are practically acceptable. The percentage dye rejections are usually more than 85% indicating the potential of PMMA membranes for the treatment of dye wastewaters.

Conclusions

Polymethyl methacrylate is useful membrane material that has satisfactory resistance to bacterial and chemical attack, the properties that are essential for application of membrane technologies in wastewater treatment. The PMMA membranes can be effectively applied for the separation of blue dye from aqueous solutions and are potentially useful in the treatment of dye wastewater from textile industry. The PMMA membranes yield reasonably well permeate flux and dye rejection is more than 85%. Manipulating the solvent content in the membrane casting solution can control the membrane compaction and flux stability. Further studies on separation of other dyes from multiple dye solutions are in progress.

Acknowledgement

Authors are thankful to the Director, NEERI, for kindly permitting the publication of this research paper. Authors also sincerely thank Dr. S.N. Kaul, Director Level Scientist, NEERI, Nagpur for his keen interest in this work.

REFERENCES

1. Belfort G., 1984, "Synthetic Membrane Processes; Fundamentals and Water Applications, Edited by Belfort G. Academic Press Inc.
2. Caetano, A.T., Muntau, H. and Depinho, M.N., 1992, Proceddings of Euromembrane 92, Paris, October 5-8.
3. Cheryan, M., 1986, Ultrafiltration Handbook, Technomic, Lancaster, PA.
4. Johns, F.J., 2000, "Membrane Technologies for Industrial and Municipal Waste", Water Environment Federation, pp. 2-16
5. Sourirajan, S., 1991, Lectures on Membrane Separations, Indian Membrane Society, Vadodara.

44

Treatment of Synthetically Prepared Pharmaceutical Waste by Pervaporation Technique

P.K. MISHRA & GAURAV BOSE

Department of Chemical Engineering and Technology, Institute of Technology
Banaras Hindu University, Varanasi

ABSTRACT

The pharmaceutical waste is rich of highly toxic compounds, which are degraded with great difficulty. The cost of converting into allowable range for discharge in sewers and receiving range by conventional methods is very high. The degradation of pharmaceutical waste can be carried out by biological treatment or incineration. Solvent streams containing higher concentration of organics and impurities cannot be discharged to biological waste treatment plant because of high BOD/COD loading or potential toxicity.

Primary investigations have been made by using Pervaporation technique for the removal of organic constituents from synthetically prepared pharmaceutical wastes. Investigations are carried out to determine separation characteristics of cellulose nitrate, Poly vinyl alcohol, Poly tetraflouro ethylene for separation of binary mixtures *viz.* Water-Ethanol and Water-Acetic acid.Based on these studies, suitability of membranes for separation of organic contaminants from synthetically prepared pharmaceutical wastewater using pervaporation process are discussed.

In second stage the most suitable membrane out of these is used to study the separation performance of pervaporation technique for removal of organic contaminants. Effect of salt concentration and pressure on separation were also investigated.

It was found that cellulose nitrate is not suitable for the water-ethanol system & water acetic acid system at high concentration as well as at low concentration. PVA is good for low water concentration, but our aim is to get organics from the effluents, where water percent is high and concentration of organics is low. PTFE membranes are found to be suitable for water-ethanol and water-acetic acid system, so it has been selected this membrane for treatment of pharmaceutical waste. Salt concentration adversely affects the permeate flux. Flux as well as selectivity increases with increasing vacuum.

Introduction

Pharmaceutical aqueous waste streams are generally characterized by the presence of moderate to high (20-40 wt%) concentrations of solvents (ethanol, methanol, ethyl acetate, acetic acid), inorganic salts (sodium chloride) and organics (MW 200-600) (1). Pharmaceutical waste originates principally from the spent liquors from fermentation process with the addition of floor washing and laboratory waste. Liquid waste from pharmaceutical plants producing antibiotics and biological can be classified as:

1. Strong fermentation beers, small in volume but having 4000 to 8000 ppm BOD.
2. Inorganic solids: Waste slurry with little BOD.
3. Washing of floor and equipment from 600 to 2500 ppm.
4. Chemical waste: Solution of solvents, which exert a substantial BOD when diluted with other waste.
5. Barometric Condenser water: Resulting from solids and volatile gases being mixed with condenser water, causing 60 to 100 ppm BOD (2).

Typical composition of pharmaceutical wastewater is given in Table 1.

Table 1: Composition of Typical Pharmaceutical Waste (1)

Compound	*Approximate Wt %*
Water	64
Ethanol	24
Methanol	2
Sodium Chloride	4
Acetic Acid & Methyl Acetate	6

The effect of pharmaceutical wastes on municipal sewers and receiving waters can be varied. Some wastes have a clear acid or alkaline reaction. Other effluents contain compounds having strong and unpleasant odors. Various toxic substances, such as cyanides, heavy metals harm the growth of biological life in stream and disturb the biological processes in sewage treatment plats. The most unpleasant aspect of antibiotics wastes is the disagreeable odor imparted to the receiving water by organics solvents. The microorganism in the stream can be adversely affected if the antibiotics are allowed to mix into wastes (3).

The degradation of pharmaceutical waste can be carried out by biological treatment or incineration. Solvent streams containing higher concentration of organics and impurities cannot be discharged to biological waste treatment plant because of high BOD/COD loading or potential toxicity (4).

Pervaporation is one of the most popular areas of current membrane research, which can be successfully employed for removal of organics wastes from water. The concept of Pervaporation for separation is not new. The phenomenon of Pervaporation was first observed by Kober in 1917, who originated the term in a publication reporting selective permeation of water from aqueous solution of albumin and toluene through cellulose nitrate films. The usefulness of Pervaporation for separation and concentration was recognized in 1935 by Farber. However, the first quantitative work on Pervaporation was published by Heisler *et al.*, in 1956 for the separation of water/ethanol mixture using cellulose membrane.

One of the major advantages of using Pervaporation for the waste treatment in that pharmaceutical process are batch process (low volume high value product based processes) and hence the quantities of waste generated are small in volume. As a result, the membrane area requirement is generally less. Also Pervaporation systems are usually modular in nature and thus highly flexible (1).

Keeping all these advantages in mind, primary investigation has been made by using Pervaporation technique for the removal of organic constituents from synthetically prepared pharmaceutical wastes.

To restrain the number of constituents in our study we have omitted methyl acetate and considered rest as the constituent of pharmaceutical effluents.

Investigations are made to determine separation characteristics of cellulose nitrate, Poly vinyl alcohol, Poly tetraflouro ethylene for separation of binary mixture. *e.g.*

1. Water-Ethanol
2. Water-Acetic acid

Based on these studies, suitability of membrane for pervaporation process for separation of organic contaminants from synthetically prepared pharmaceutical wastewater are discussed.

In second stage suitable membranes are used to study the separation performance of pervaporation technique for removal of organic contaminants. Effect of salt concentration, pressure on separation are studied.

Effect of following parameters on separation have been studied in present investigations.

1. Permeate Flux
2. Selectivity
3. Transmembrane Pressure
4. Salt Concentration
5. Extent of COD reduction

Experimental

Experimental Setup

The setup used for the experiment is shown in Figure 1. The experimental setup consisted of a feed tank, a stop valve, a tubular pipe made of Perspex, condenser, collector, peristaltic pump and a vacuum pump. The feed tank was made of a small steel vessel that could hold approximately around 2.5 litres of the feed solution. It was placed at a height of 61" above the ground level so that the feed solution can pass through the membrane arrangement easily with the help of gravity. The feed solution was prepared according to the experimental run and fed in this steel vessel. From these, the solution

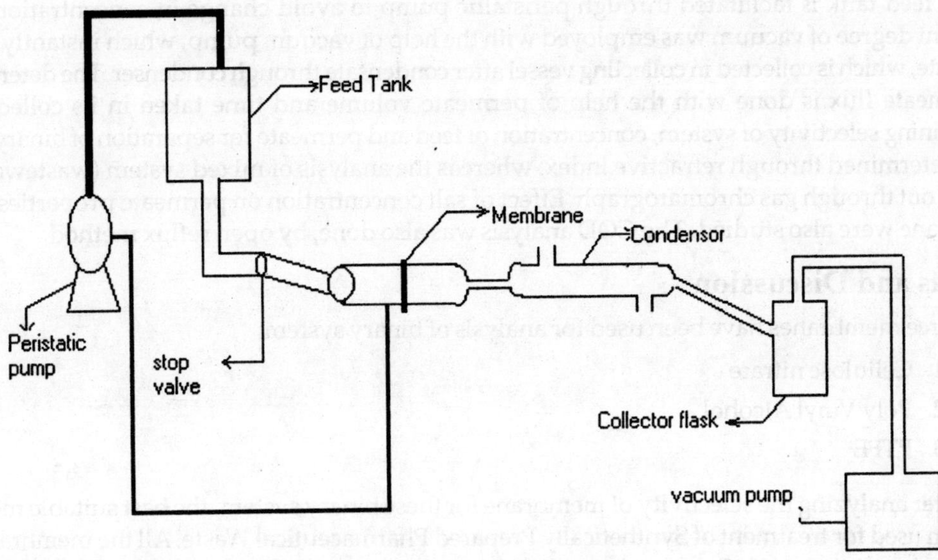

Fig. 1: Experimental Setup

slid down through the glass stop valve till it reached the process vessel. The glass stop valve was placed at 54 inches above the ground level. A 20-inch long mild steel pipe acted as an intermediary between the stop valve and the process vessel. The process vessel was fabricated by joining together two half pieces of Perspex tube, one have 8 inches long and another have measuring 4 inches long, with the help of flanges. The tube was 3 inches in diameter. The cellulose nitrate (CN), polyvinyl alcohol (PVA) and Posolytetraflouroethylene (PTFE) membrane, which were 47 mm in diameter, were placed between the flanges. One side of the membrane remained in contact with the feed solution, whereas the other side of the membrane remained dry due to the presence of vacuum, which immediately pulled away the liquid/vapor coming out of the membrane. Then, the vapors pass through a straight tube condenser, which is 18 inches long and finally get collected in a collecting vessel, which was placed at a height of 33 inches above the ground level. Peristaltic pump, which was placed at a height of 50 inches above the ground level, used for recycling the feed solution.

Analytical Methods

1. For binary systems, refractive index property has been used for determining the concentration of various constituents of permeate and feed.
2. For complex systems, gas chromatograph has been used for elemental analysis in feed as well as permeates.
3. COD analysis has been made to account COD reduction due to separation. This was done by the open reflux method.

Experimental Procedure

Synthetically prepared wastewater (2 litre in volume) was placed in feed tank, which due to gravity flows towards membrane and comes in to its contact. As permeate flux is very low, recycle of feed to feed tank is facilitated through peristaltic pump to avoid change in concentration of feed. Different degree of vacuum was employed with the help of vacuum pump, which instantly removes permeate, which is collected in collecting vessel after condensate through condenser. The determination of permeate flux is done with the help of permeate volume and time taken in its collection. For determining selectivity of system, concentration of feed and permeate for separation of binary mixture were determined through refractive index, whereas the analysis of mixed system (wastewater) was carried out through gas chromatograph. Effect of salt concentration on permeate properties through membrane were also studied. The COD analysis was also done, by open reflux method.

Results and Discussion

Three membranes have been used for analysis of binary system:

1. Cellulose nitrate
2. Poly Vinyl Alcohol
3. PTFE

After analyzing the selectivity of membrane for these binary system, the best suitable membrane has been used for treatment of Synthetically Prepared Pharmaceutical Waste. All the membranes used were having diameter of 47 mm and pore size 0.1 Mm. operations were carried out for binary system at 30 mm Hg vacuum and for synthetically prepared pharmaceutical waste, separation characteristics were determined at different vacuum.

Binary System

Water-Ethanol System

Figure 2 shows the standard curve of refractive index of water-ethanol system. This has been used to determine the permeate concentrations. Plot of flux vs. feed concentrations in shown in Figure 3. Different characteristics have been observed for different membrane. For cellulose nitrate membrane maximum flux is achieved and for other two *viz.* PVA and PTFE lesser flux have obtained. For CN membrane it has been observed that initially flux is increasing as ethanol concentration in the feed is increasing but at ethanol concentration of 30% flux attains a constant value of 29 kg/m^2. hr. In case of PVA membrane flux initially decreases sharply with increase in ethanol concentration but latter on it is nearly constant. For PTFE membrane flux increases with increase in ethanol concentration. This particular curve does not clearly indicate which one is suitable for water ethanol system. So selectively of these membranes for ethanol in ethanol-water system is shown in Figure 4. It is clear from the plot that PTFE membrane is highly selective for ethanol, at all feed concentration. CN membrane is not found suitable for ethanol, whereas for PVA membrane, selectivity slightly increases with increase in ethanol concentration in the feed, In case of PVA membrane water is preferentially permeating component through the membrane. But we are interested for lower concentration of ethanol in water as it is the concentration range present in pharmaceutical wastes. So we can say that PTFE is suitable for removal of ethanol from pharmaceutical wastes.

Water-Acetic Acid System

The standard graph of water-acetic acid system is shown in Figure 5. This graph has been used to determine the permeate concentration. Variation of flux vs. feed concentration for different membranes is shown in Figure 5. For CN membrane it has been observed that flux increases as the concentrations of acetic acid in feed increases. In case of PVA membrane, flux decrease with increase in concentrations of acetic acid. It can be seen that lowest flux of 6.2 kg/m^2.hr has been obtained at 90% acetic acid

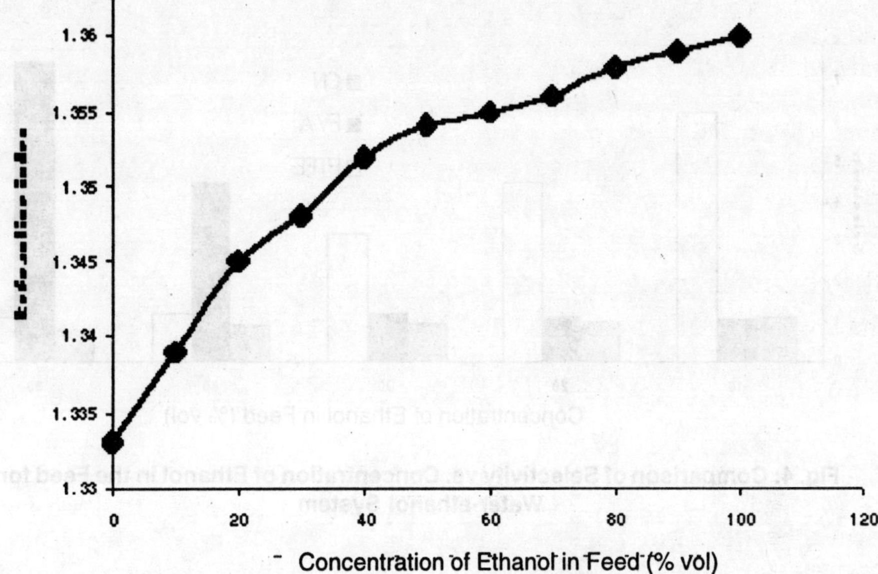

Fig. 2: Standard Graph for Water-ethanol System

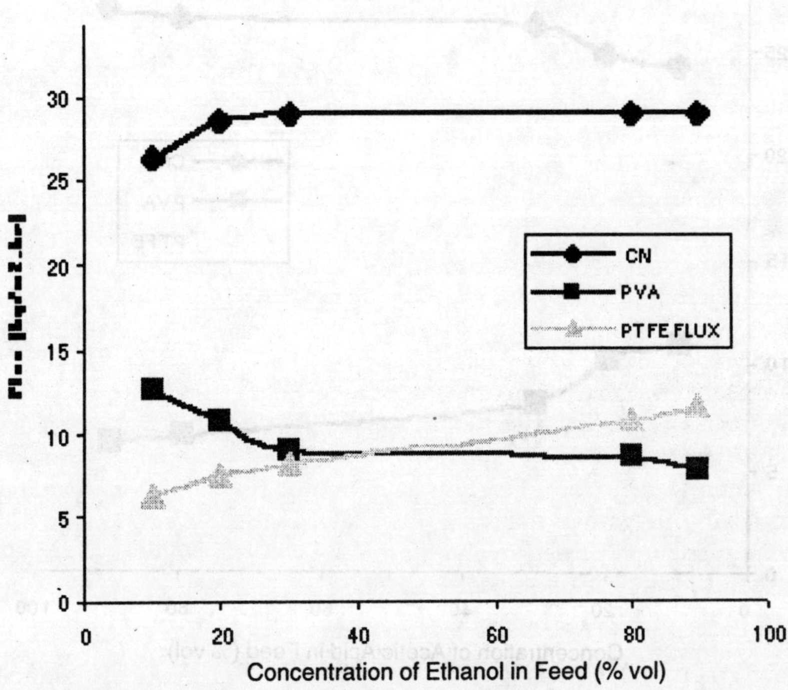

Fig. 3: Plot of Flux vs. Concentration of Ethanol in the Feed for Water-ethanol System

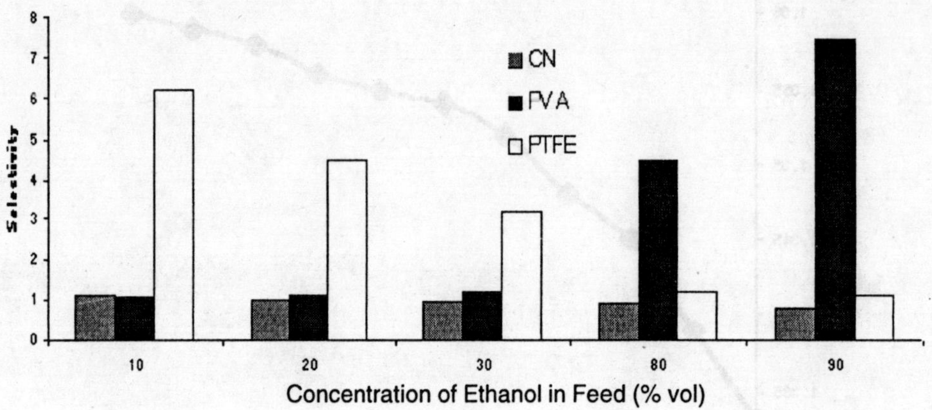

Fig. 4: Comparison of Selectivity vs. Concentration of Ethanol in the Feed for Water-ethanol System

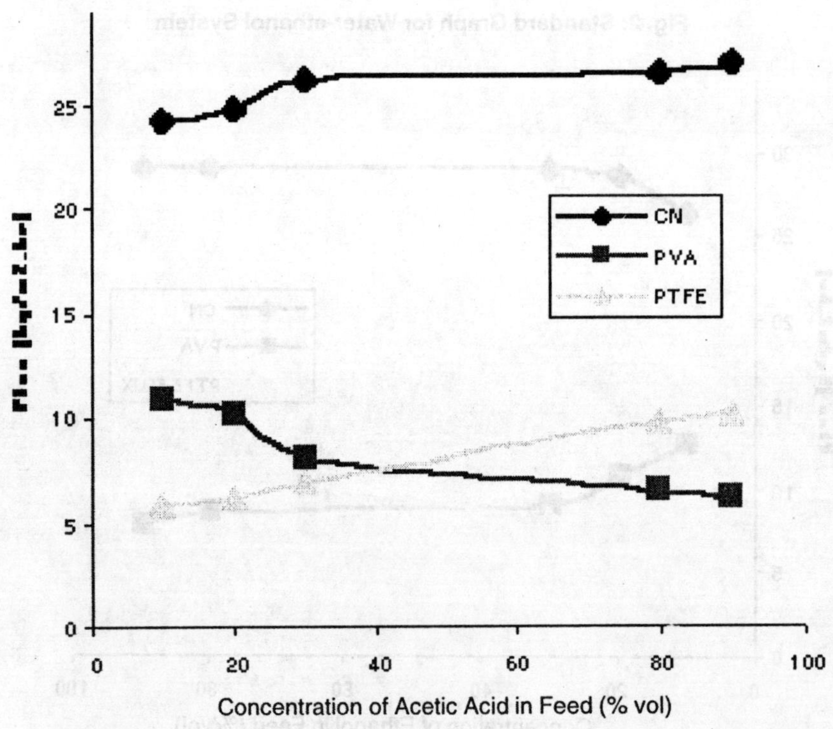

Fig. 5: Plot of Flux vs. Concentration of Acetic Acid in the Feed for Water-acetic Acid System

concentration in the feed. For PTFE membrane, flux increases with acetic acid concentration in the feed. Plot of selectivity vs. feed concentration for acetic acid system is shown in Figure 6. It can be seen from the graph that PTFE membrane is suitable for low as well as higher concentration of acetic acid in the feed. PVA membrane in suitable for low concentration of water, so we can infer from selectivity curve that PTFE membrane is suitable for removal of acetic acid from pharmaceutical wastes.

COD Analysis of Binary System

Figure 7 shows the COD of feed and permeate through PTFE membrane for binary system *viz.* water-acetic acid, water-methanol and water-ethanol, respectively. Considerable increase in COD has been observed in permeate, indicating increase of organic compound in it. As flux through pervaporation membrane is always low so it should be preferred that component having lesser amount in feed should permeate through membrane, which will encourage the process. So it is clear that organic compound *viz.* ethanol, methanol and acetic acid in being removed from feed and hence PTFE

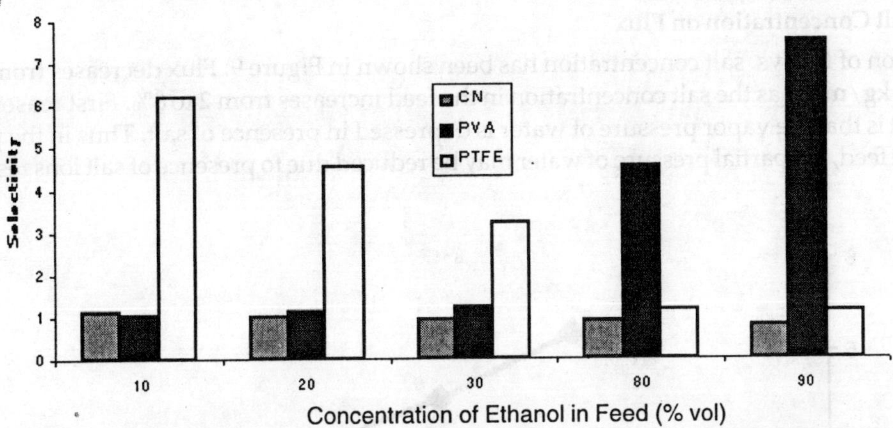

Fig. 6: Comparison of Selectivity vs. Concentration of Ethanol in the Feed for Water-ethanol System

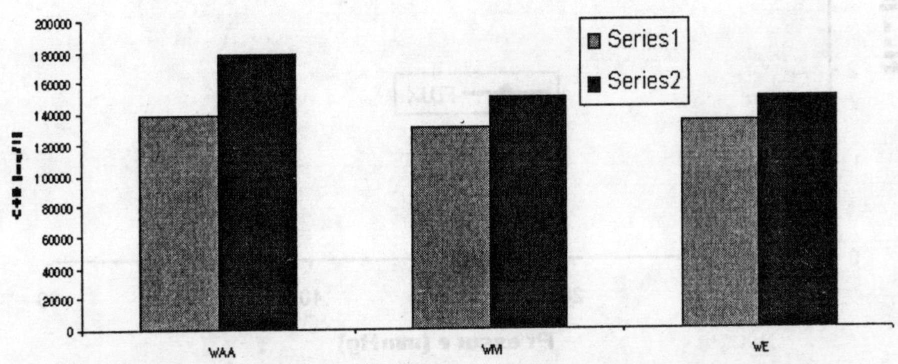

Fig. 7: Comparison of COD for Different System Using PTFE Membrane

membrane is suitable for treatment of pharmaceutical waste, which is having these organic contaminants in lesser amount *i.e.* in the range of 10-20%.

Synthetically Prepared Pharmaceutical Waste

To test the suitability of PTFE membrane for pharmaceutical waste investigations have been made to study the effect of the vacuum on flux (Figure 9), effect of salt concentration on flux (Figure 9), effect of vacuum on selectivity of various organic components through PTFE membrane and COD reduction of feed through these membrane (Figure 7 & 11).

Effect of Vacuum on Flux Through PTFE Membrane

Maximum flux for synthetically prepared pharmaceutical wastes can be observed when the vacuum is maximum (20 mm Hg). Increasing the vacuum can cause the increase the driving force (pressure difference between the feed and permeate side) for separation of pharmaceutical wastes. The flux for synthetically prepared pharmaceutical waste system increase from the 3.2 kg/m².hr to 5.2 kg/m².hr as the vacuum in the permeate is varied from 50 to 20 mm Hg.

Effect of Salt Concentration on Flux

Variation of flux vs. salt concentration has been shown in Figure 9. Flux decreases from 4.5 kg/m².hr to 1.6 kg/m².hr as the salt concentration in the feed increases from 2 to 8%. First reason for this drop in flux is that the vapor pressure of water is depressed in presence of salt. Thus in the presence of salt in the feed, the partial pressure of water may be reduced due to presence of salt ions resulting in

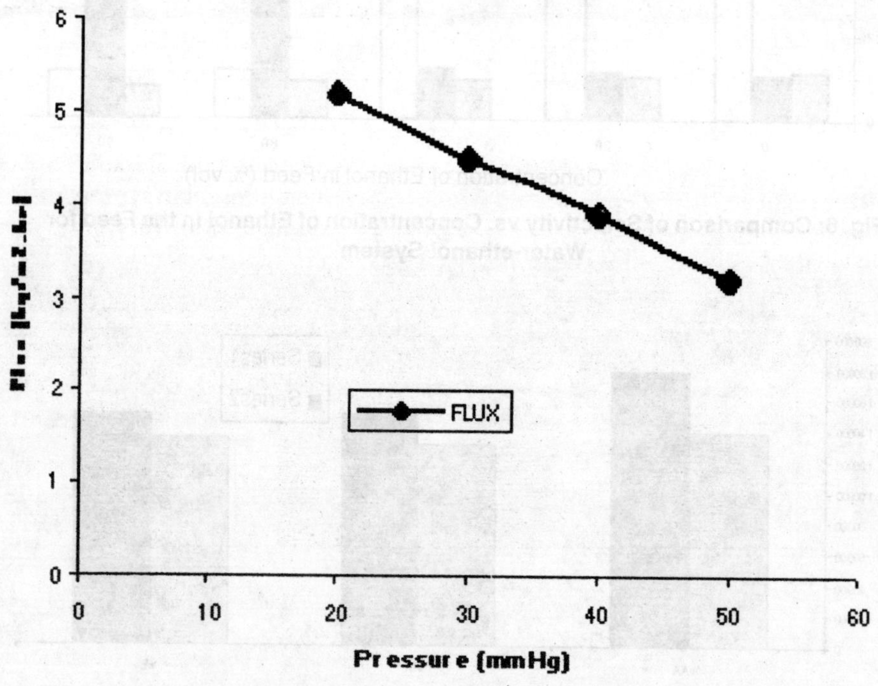

Fig. 8: Effect of Pressure on Permeate Flux

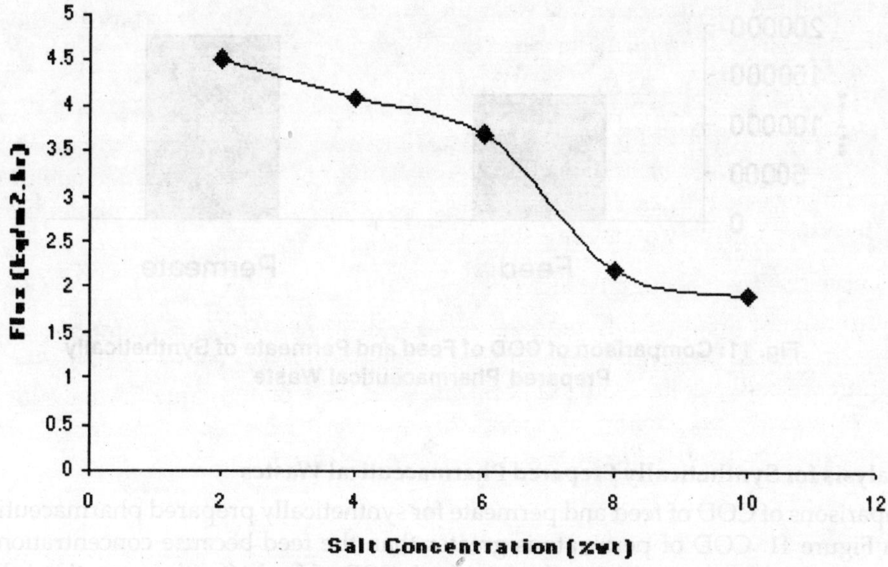

Fig. 9: Effect of Salt Concentration on Flux for Synthetically Prepared Pharmaceutical Waste

a lower driving force. Another possible reason for drop in flux at high salt concentration could be physical blocking of pores on the permeate side by micro-salt crystallites. The salt ions could drag with water into the membrane and precipitate on the permeate side causing the drop in permeate flux.

Effect of Vacuum on Selectivity

Effect of vacuum on selectivity for synthetically prepared pharmaceutical waste has been shown in Figure 10. Selectivity is increasing for all the organic component present in synthetically prepared pharmaceutical waste with the increase in vacuum in the permeate side. Very good selectivity is observed when in the permeate side is lowest (20 mm Hg).

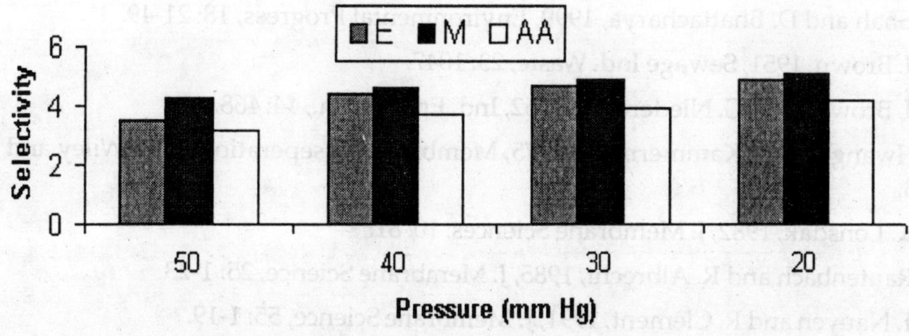

Fig. 10: Effect of Pressure on Selectivity

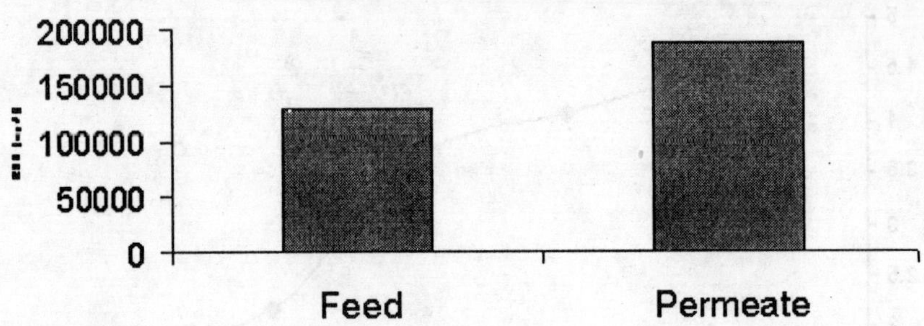

Fig. 11: Comparison of COD of Feed and Permeate of Synthetically Prepared Pharmaceutical Waste

COD Analysis for Synthetically Prepared Pharmaceutical Wastes

Comparisons of COD of feed and permeate for synthetically prepared pharmaceutical waste is shown in Figure 11. COD of permeate is greater than the feed because concentration of organic components in the permeate is higher than the feed. COD of feed after the operation is less than the permeate due to the permeation of organic components through the PTFE membrane.

Conclusion

1. Cellulose nitrate is not suitable for the water-ethanol system and water acetic acid system at high concentration as well as low concentration.
2. PVA is good for low water concentration, but our aim is to get organics from the effluents, where water percent is high and concentration of organics is low.
3. PTFE membranes are suitable for water-ethanol and water-acetic acid system, so it has been selected for treatment for pharmaceutical waste.
4. Salt concentration adversely affects the permeate flux.
5. Flux as well as selectivity increases with increasing vacuum.

REFERENCES

1. D. Shah and D. Bhattacharya, 1999, Environmental Progress, 18: 21-49.
2. J.M. Brown, 1951, Sewage Ind. Waste, 23: 1047.
3. J.M. Brown and J.G. Niedercorn, 1952, Ind. Eng. Chem., 44: 468.
4. S. Hwang and K. Kammermyer, 1975, Membranes in seperation, John Wiley and Sons, Inc., 99-123.
5. H.K. Lonsdak, 1982, J. Membrane Sciences, 10: 81.
6. R. Rautenbach and R. Albrecht, 1985, J. Membrane Science, 25: 1-23.
7. Q.T. Nauyen and R. Clement, 1991, J. Membrane Science, 55: 1-19.
8. Ulrich Sandev and Pavel Souknr, 1990, J. Membrane Science, 51: 169-179.

45

Characterization of Cellulose Nitrate Membrane for Separation of Various Aqueous Organic Mixtures Through Pervaporation

P.K. MISHRA, V. PATANJALI, B.N. RAI & R.R. NARAYANAN

Department of Chemical Engineering and Technology, Institute of Technology
Banaras Hindu University, Varanasi

ABSTRACT

Selectivity of cellulose nitrate membrane for the separation of some organic substances from their water mixtures has been studied using pervaporation. It was seen that cellulose nitrate membrane could be used for the separation of 1,4-dioxane, propane 1,2-diol, MEK, glycerol at low concentrations and diethanolamine from their binary systems with water. The membrane could be used for the wastewater treatment of streams containing 1,4-dioxane, propane 1,2-diol, MEK, glycerol, ethyl formate, methyl acetate, formic acid, acetone, ethyl acetate, n-butanol and diethanolamine. It was found to be useless for the water binary mixtures of dimethyl formamide, benzyl alcohol, aniline, acetic anhydride, trichloroacetic acid, furfural, cyclohexanol, tetrahydrofurfuryl alcohol, iso propanal, ethylene glycol, monoethanolamine, carbitol and ethanol.

Keywords: Pervaporation, Aqueous Organic Mixture and Cellulose Nitrate Membrane.

Introduction

Among the different membrane technologies, Pervaporation is one of the most promising technologies for environmental applications. Pervaporation is a pressure-driven separation process in which a liquid mixture is in direct contact with one side of the membrane and the permeated product called the pervaporate is removed from the other side in a vapor state. It is different from the other membrane process as it involves a phase change requiring heat of vaporization. Maintaining down stream partial pressure lower than saturation pressure at that temperature brings pervaporation. This is usually done by creating a vaccum or by sweeping a inert carrier gas (1,2). There are two kinds of pervaporation *viz.* hydrophobic pervaporation and organophilic pervaporation (3).

Experimental

The set up used for the experiment is shown in Fig.1. The experimental setup consisted of a feed tank, a stop valve, a tubular pipe made of Perspex, condenser, collector and a vacuum pump arranged in series. The feed tank was made up of a small steel vessel that could hold approximately 1.2 liters of the feed solution. It was placed at a height above the ground so that feed solution can pass through to the membrane arrangement easily with the help of gravity. The feed solution was prepared according to the experimental run and fed in this steel vessel. From here the solution slid down through the glass stop valve to reach the process vessel. A 33-inch long mild steel pipe acted as an intermediary between the stop valve and the process vessel. The process vessel was fabricated by joining together two half pieces of Perspex tube, each measuring 6-inches long, with the help of flanges. The tube was 1.5 inches in diameter. The cellulose nitrate membrane that was 47 mm in diameter was placed between the flanges. One side of the membrane remained in constant contact. One side of the feed solution was in constant contact with the feed solution; where as the other side of the membrane remained dry due to the presence of vacuum, which immediately pulled away liquid/ vapor coming out of the membrane. Then the vapors pass through a straight tube condenser, which is 23-inches long and finally get collected in a collecting vessel. The vacuum pump, which had a capacity of 100 l/min, was placed on the ground.

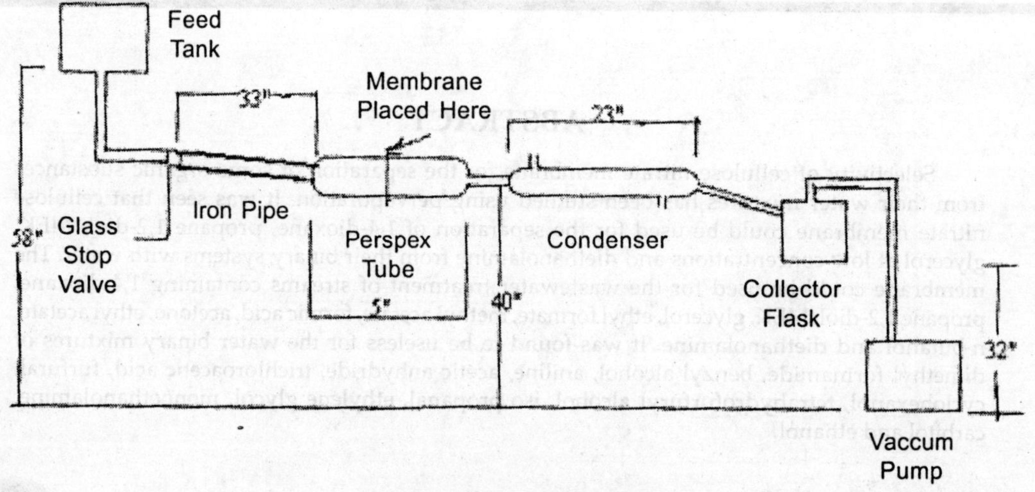

Fig. 1: Experimental Setup

Results and Discussion

The reason for choosing the cellulose nitrate is that it is a hydrophilic membrane, whose cost is quite low. Cellulose nitrate being a new membrane is easily available and also has a long shelf life. It could be operated at a temperature of 145°C and had a pore size as low as 0.25 Mm. The diameter of the membrane was 47-mm. Time lag (measured in seconds) refers to the time difference between the opening of the feed valve and the collection of permeates in the flask. The concentration of the feed and the permeate solutions was found out using a standard plot between refractive indices versus mixture compositions. In all the plots of permeate concentration versus feed concentration series 1 indicated the permeate concentration whereas series 2 indicated the feed concentration.

System 1: 1,4-dioxane + Water

The first system taken up was 1,4-dioxane water system. The plot of permeate concentration versus feed concentration is shown in Fig. 2. No separation was indicated at low concentrations and separation starts to take place only at high at concentrations more than 15-18%. The ideal range for this system is 20 to 40% range of organic and cannot be used to remove beyond a stage of 10%. Since the membrane gets dissolve in the system at concentration greater than 60% organic the upper limit for the use of membrane is also fixed. The pervaporation selectivity ranges from 1 to 0.833 (Fig. 3) reaching a minimum of 0.75 at concentration of dioxane ranging from 20 to 40%. The minimum the pervaporation selectivity, the better since membrane used is a hydrophilic one. So cellulose nitrate is used for separation only in high concentration ranges for this system. On increasing the temperature (100°C) the permeate concentration is found to decrease.

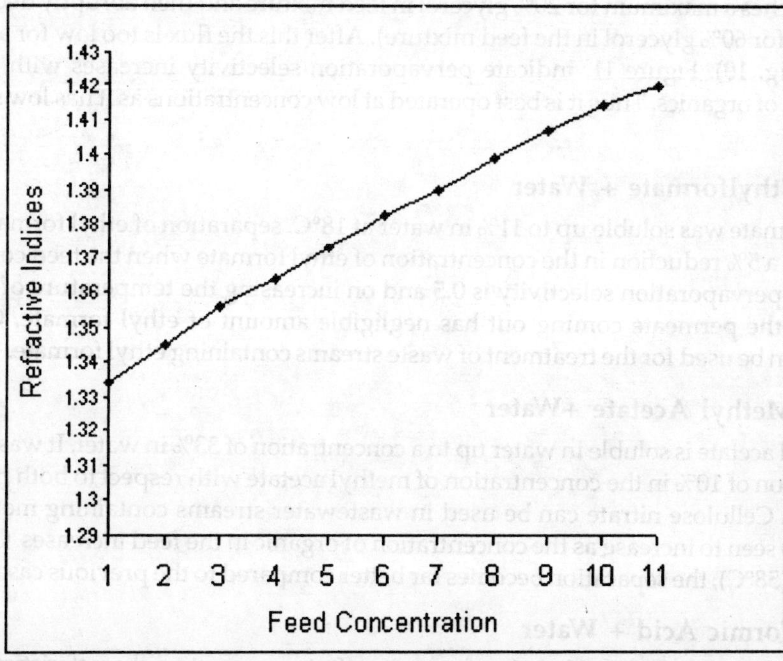

Fig. 2: Standard Graph for Dioxane-water System Feed Concentrations Indicated in Numbers with "1" Indicating % Dioxane and "2" Indicating 100% Dioxane in Increasing Steps of 10%

System 2: Propane 1,2-diol + Water

The plot of permeate concentration versus feed concentration is shown in Fig. 4. for this system. From this figure it can be concluded that as feed concentration increase permeate concentration increases. At low concentration the separation takes place effectively so cellulose membrane can be used for wastewater treatment with propane 1,2-diol as contaminant. As the concentration increases, the flux initially increases reaches a maximum value and finally drops (Fig. 5). At low organic feed concentrations the selectivity is low but at higher organic feed concentrations, the selectivity ranges in 0.75 to 0.87 (Fig. 6). Increase in feed temperature (100°C) cause the concentration of organic to increase.

System 3: Methyl Ethyl Ketone +Water

MEK is soluble only up to 33% in water therefore; the membrane has been used only up to this level for study. In Figure 7, it is observed that the permeate concentration of the organic phase is very low and a good separation factor is observed. Thus, MEK is separated as a retentate to a very large extent. This also implies clearly from figure 8, facilitating the use of cellulose nitrate membrane for streams containing MEK along with water. At high temperatures of feed (80°C), a marginal decrease in permeate concentration of MEK is found.

System 4: Glycerol + Water

The glycerol water system indicates (Fig. 9) that the separation of glycerol is limited using cellulose nitrate membrane. Taking both the feed and permeate into consideration, there is a reduction of only approximately 5%in the concentration of organic. Thus the desire amount of separation can be achieved by using a number of membranes placed in series. Another interesting fact is that the flux initially increases, reaches a maximum for 20% glycerol in feed mixture and then abruptly decreases to a value as low as 5.3 (for 60% glycerol in the feed mixture). After this the flux is too low for any separation to take place (Fig. 10). Figure 11. indicate pervaporation selectivity increases with increase in feed concentration of organics. Thus it is best operated at low concentrations as it has low selectivity in this region.

System 5: Ethylformate + Water

Ethyl formate was soluble up to 11% in water at 18°C. separation of ethyl formate is observed in this case with a 5% reduction in the concentration of ethyl formate when the feed contains 10% ethyl formate. The pervaporation selectivity is 0.5 and on increasing the temperature of feed to 55°C,we observe that the permeate coming out has negligible amount of ethyl formate. Cellulose nitrate membrane can be used for the treatment of waste streams containing ethyl formate.

System 6: Methyl Acetate +Water

Methyl acetate is soluble in water up to a concentration of 33% in water. It was found that there was a reduction of 10% in the concentration of methyl acetate with respect to both permeate and the feed streams. Cellulose nitrate can be used in wastewater streams containing methyl acetate. The overall is also seen to increase as the concentration of organic in the feed increases. On increasing the temperature (58°C), the separation becomes far better compared to the previous case.

System 7: Formic Acid + Water

In formic acid water system there is an opposite effect on separation characteristics. The membrane, which is hydrophilic in nature, allows formic acid to pass through for, feed concentrations up to a level of 20%in the feed. Thus the retentate contains more water than permeate up to this level of feed

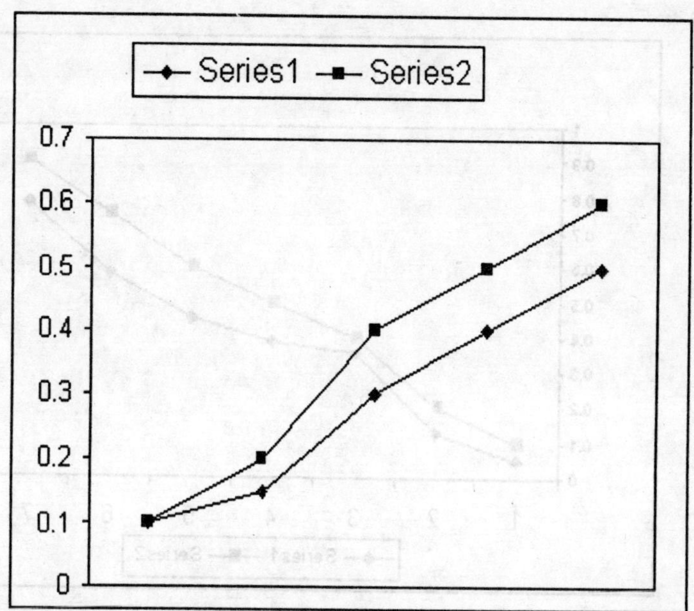

Fig. 3: Plot of Permeate Concentration vs. Feed Concentration Series 1 Indicating Feed Concentration While Series 2 Indicating Peremate Concentration

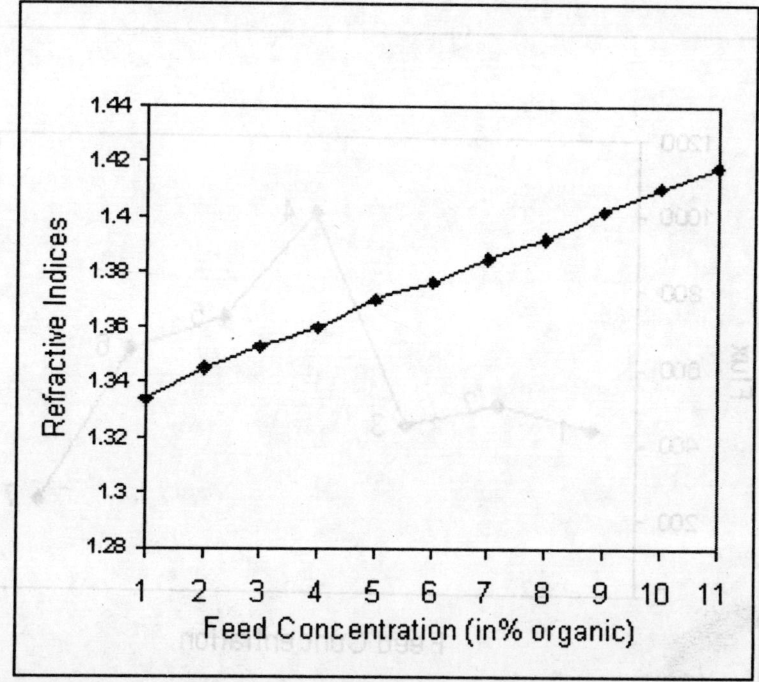

Fig. 4: Standard Graph for Propane 1,2 Diol System Feed Concentration is Indicated in Number with 1" Indicating % Propane and "2" Indicating 100% Propane 1,2 Diol in Increasing Steps of 10%

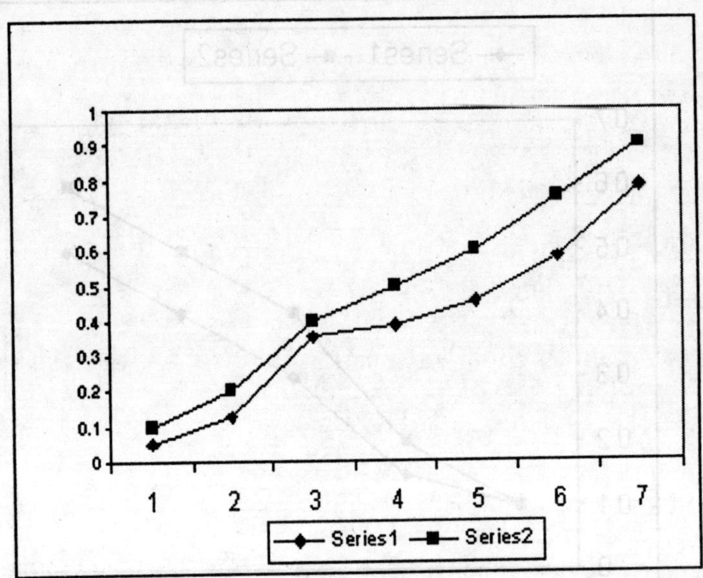

Fig. 5: Plot of Permeate Concentration vs. Feed Concentration Series 1 Indicating Feed Concentration While Series 2 Indicating Peremate Concentration

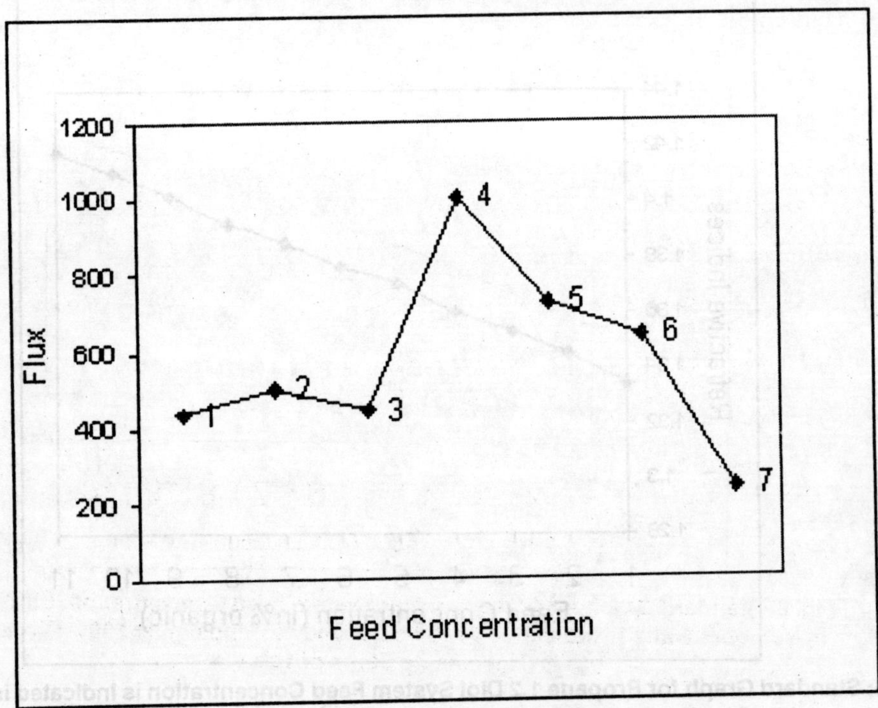

Fig. 6: Flux vs. Feed Concentration

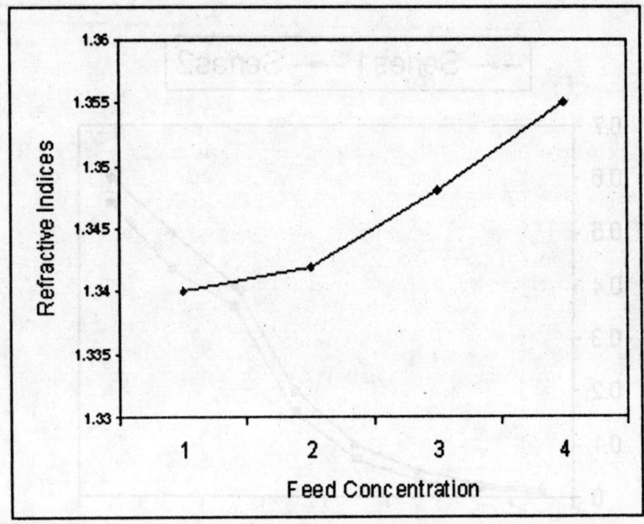

Fig. 7: Standard Graph for MEK-water System Feed Concentration is Indicated in Numbers with "1" Indicating 0% MEK and "4" Indicating 100% MEK in Increasing Steps of 10%

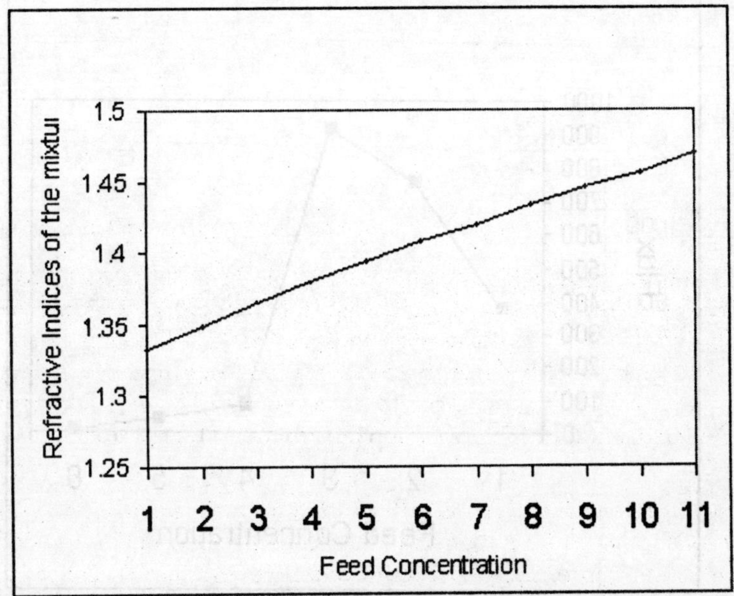

Fig. 8: Standard Graph for Glycerol-water System Feed Concentration is Indicated in Numbers with "1" Indicating 0% Glycerol and "11" Indicating 100% Glycerol in Increasing Steps of 10%

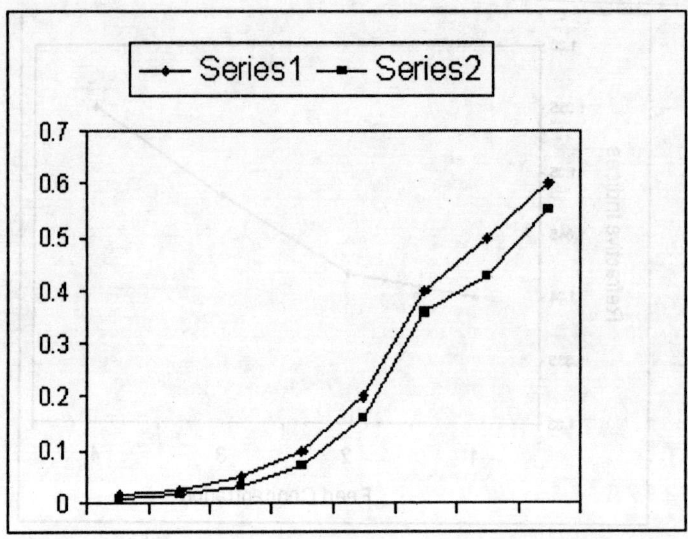

Fig. 9: Plot of Permeate Concentration vs. Feed Concentration Series "1" Indicating Feed Concentration While Series "2" Indicating Permeate Concentration

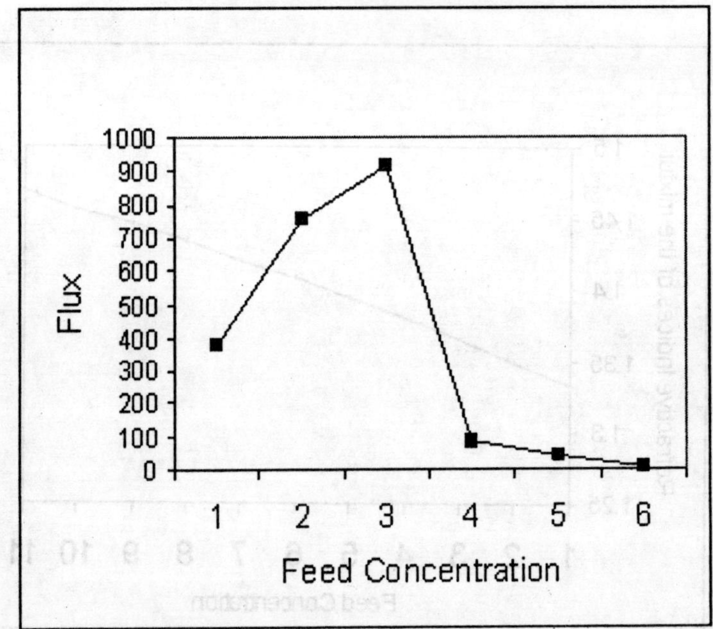

Fig. 10: Plot of Flux vs. Feed Concentration. Feed Concentration is Indicated in Numbers with "1" Indicating 10% Glycerol and "6" Indicating 100% Glycerol in Increasing Steps of 10%

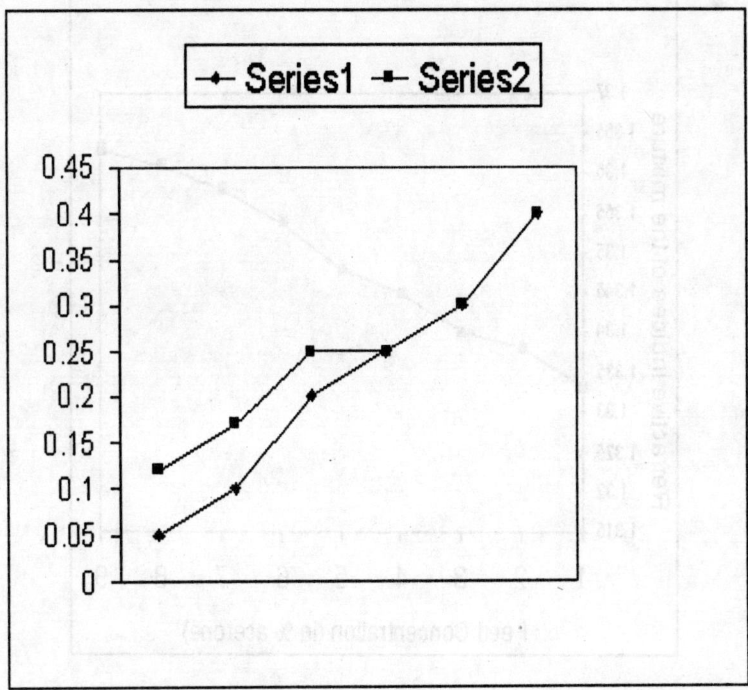

Fig. 11: Plot of Permeate Concentration vs. Feed Concentration Series "1" Indicating Feed Concentration while Series "2" Indicating Permeate Concentration

solution. From a level of 25% formic acid in feed onwards, separation is achieved. The pervaporation selectivity (Fig. 12) also flattens and becomes 1.0 in this case.

System 8: Acetone + Water

It was found (Fig. 13) that separation of acetone occurs appreciably and the acetone is separated in the retentate stream up to a feed concentration of 25%. After that, the membrane selectivity increases (Fig. 14.), thus allowing acetone to pass through. There is a significant reduction of acetone through the permeate compared to that of the feed. The membrane dissolves for feed mixtures with more than 50% acetone so becomes useless. Hence cellulose nitrate membrane can be use for treatment of wastewater streams with low concentrations.

System 9: Diethyl formamide + Water

The suitability of cellulose nitrate was tested and not much success is met with the help of the cellulose nitrate membrane. This was also found true for other systems such as benzyl alcohol, aniline, acetic anhydride,trichloro acetic acid, furfural, cyclohexanol, tetra hydro furfuryl alcohol, iso propanol, ethylene glycol, ethanol, triethanolamine and carbitol where we don't have any separation taking place. Ethyl acetate is one system, where the solubility of the material is around 8.5 ml in 100 ml of water. We find that appreciable amount of separation takes place on considering a 5% and 7% solution of ethyl acetate in water.

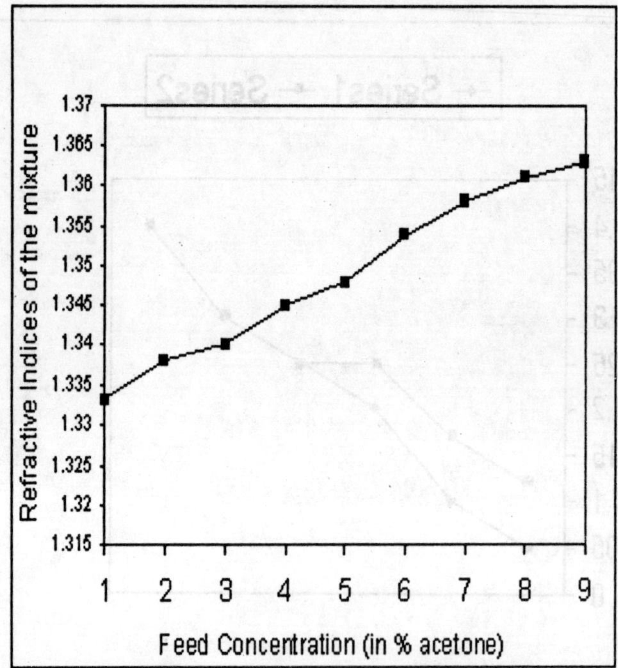

Fig. 12: Standard Graph for Acetone Water System Feed Concentration is Indicated in Number with "1" Indicating 10% Acetone and "9" Indicating 80% Acetone in Increasing Steps of 10%

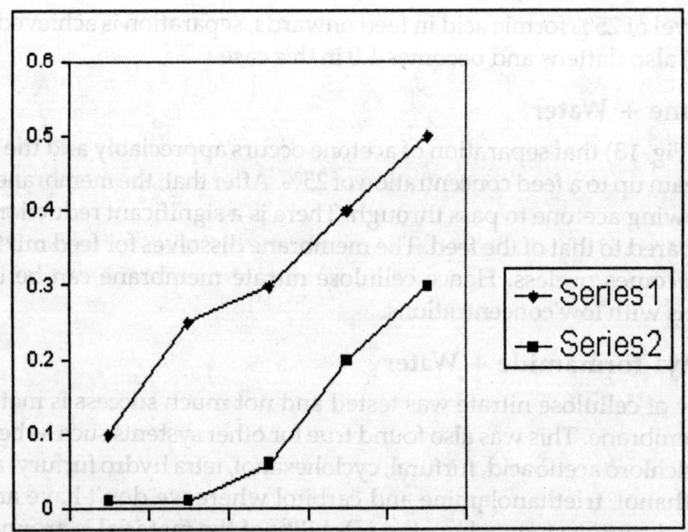

Fig. 13: Plot of Permeate Concentration vs. Feed Concentration Series "1" Indicating Feed Concentration while Series "2" Indicating Permeate Concentration

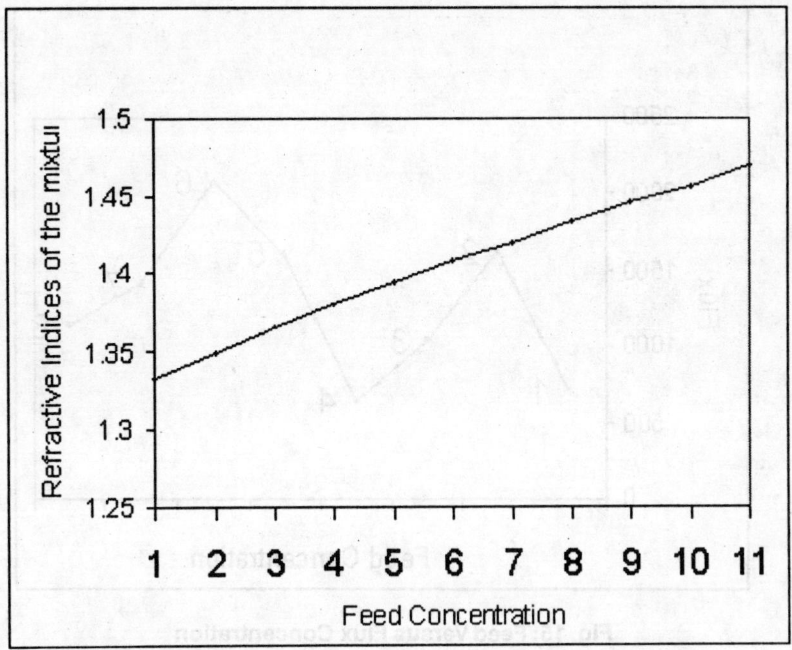

Fig. 14: Standard Graph for Diethanol Amine-water System Feed Concentration is Indicated in Numbers with "1" Indicating 10% Dietanolamine and "11" Indicating 100% Diethanolamine in Increasing Steps of 10%

System 10: n-Butanol + Water

It was found that there was a marginal decrease in the refractive index of the permeate solution as compared to that of the feed solution. n-butanol is soluble in water only to an extent of 8% in water. For a waste stream containing n-butanol it is better to have a series of membrane so that the treatment proceeds progressively and a water rich permeate is obtained at the far end of the pervaporation process.

System 11: Diethanolamine + Water

The variation of flux in this case is quite unpredictable, in the sense that the flux follows no major pattern. A major increase is observed in the range of 20%diethanolamine to 50%and after that there is a gradual decrease in the flux for higher concentration of the organic in the mixture. The mixture becomes viscous as the concentration of the diethanol amine increases with the result that the flux decreases. Figure 15 compares the curves of both the permeate concentration and the feed concentration, it can be observed that there is a constant difference between the permeate concentration and the feed concentration with the result that both the graphs are parallel to each other. The lines are slightly wide apart initially and as the feed concentration increases the lines get closer to each other. This shows that cellulose nitrate membrane is much more effective for the treatment of dilute streams of diethanol. It is also observed that at low concentrations the membrane is highly effective. Figure 16 shows the comparison of pervaporation selectivity with the feed concentration. The figure indicates that the pervaporation selectivity is also lower compared to high concentration organic in the organic water mixture.

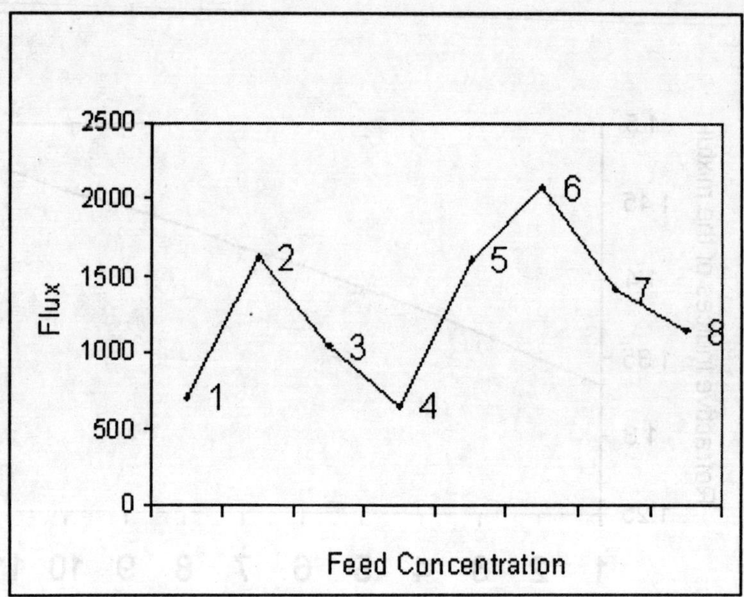

Fig. 15: Feed Versus Flux Concentration

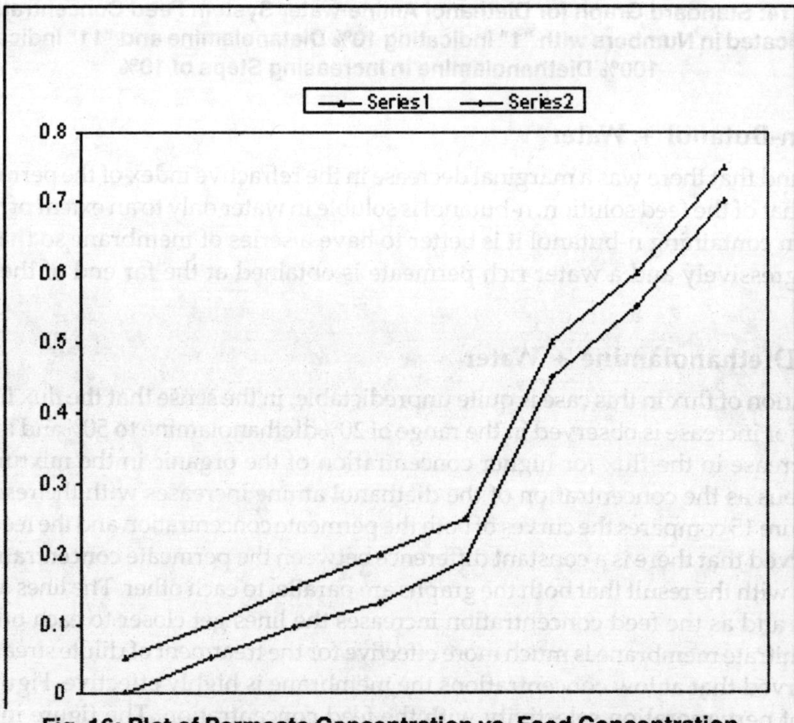

Fig. 16: Plot of Permeate Concentration vs. Feed Concentration Series "1" Indicating Feed Concentration while Series "2" Indicating Permeate Concentration

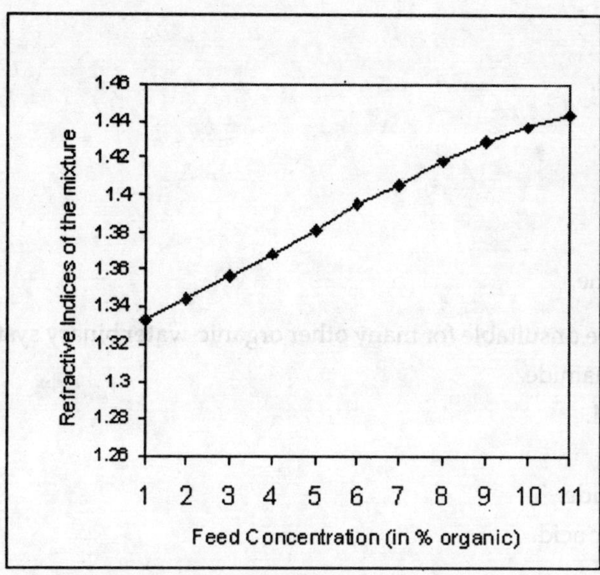

Fig. 17: Standard Graph for Monoethylamine-water System Feed Concentration is Indicated in Number with "1" Indicating 10% Monoethylamine and "11" Indicating 100% Monoethylamine in Increasing Steps of 10%

System 12: Monoethanolamine + Water

Favorable results were not observed in this case. Figure 17 shows that the permeate concentration is the same as that of the feed concentration. Hence the pervaporation selectivity is 1 in all cases for this system.

Conclusion

The following conclusions may be drawn considering cellulose nitrate as a membrane used for wastewater treatment for the wide range of materials discussed by pervaporation process.

1. It is found to be best when used for the following organic binary systems with water in the field of separation.

   1,4-Dioxane

   Propane 1,2-diol

   Methyl ethyl ketone

   Glycerol at low concentrations

   Diethanolamine

2. Cellulose nitrate membrane may be used in the treatment of wastewater for the following organic-water systems.

   1,4-Dioxane.

   Propane 1,2-diol.

   Methyl ethyl ketone.

Glycerol.

Ethyl formate.

Methyl acetate.

Formic acid.

Acetone.

Ethyl acetate.

n-butanol.

Diethanolamine.

3. It is found to be unsuitable for many other organic-water binary systems.

Dimethyl formamide.

Benzyl alcohol.

Aniline.

Acetic anhydride.

Trichloroacetic acid.

Furfural.

Cyclohexanal.

Tetrahydrofufuryl alcohol.

Isopropanal.

Ethylene glycol.

Monoethanolamine.

Trietanolamine.

Carbitol.

Ethanol.

It is found to be unsuitable for many other binary systems.

REFERENCES

1. International union of pure and applied chemistry recommendations, 1996, J. Membrane. Science 120: 149.
2. H.K. Lonsdak, 1982, J. Membrane Science, 10: 81.
3. Frank Lipnizki, Stephan Hausmanns, Po-kiong Ten, Robert W. 1999, Field and Gunter Laufenberg, Chemical Engineering Journal, 73: 113-29.

46

Enhanced Photocatalytic Degradation of Monocrotophos in Aqueous Solution Over Titania Hybridised with β-Zeolite

M.V. SHANKAR, K.K. CHERALATHAN, BANUMATHI ARABINDOO, M. PALANICHAMY & V. MURUGESAN

Department of Chemistry, Anna University, Chennai - 600 025

ABSTRACT

Photocatalytic degradation of monocrotophos has been carried out in a slurry type batch reactor. The reaction variables were optimised to obtain maximum degradation efficiency. The degradation rate of monocrotophos is significantly higher for acidic solutions than for alkaline solutions. It is also observed that the high concentration environment of the pollutant is formed around the hybridised TiO_2, enhancing the mineralisation rate of monocrotophos. Hence an increase in relative photonic efficiency is obtained.

Keywords: Monocrotophos; Hybrid photocatalyst; Relative photonic efficiency,

Introduction

In recent years, photocatalytic degradation attracts increasing attention as cleaner and greener technology for removal of toxic organic and inorganic pollutants in water and wastewater (1-3). It is postulated that the photodegradation reaction is initiated with attack of the target substrate by photogenerated active oxygen species such as $O_2^{\bullet-}$ and $^{\bullet}OH$ (4). However, the development of a practical photocatalytic system has not yet been achieved successfully, because many operational parameters are to be considered. Further the success of photocatalytic degradation process in wastewater treatment depends on saving energy and minimisation of the process cost. In order to achieve these two targets several attempts have been made to increase the photoefficiency of the photocatalyst either by noble metal deposition or by ion doping. In photocatalytic degradation of organic pollutants, which are essentially total oxidation reactions, such modifications did not enhance the photocatalytic activity of titania and they also favoured electron-hole recombination, which is detrimental for the photoefficiency. A third way to possibly increase the photocatalytic efficiency of titania is by adding an inorganic co-adsorbent to the catalyst, such as silica, alumina, zeolite or clay (5-8). Takata *et al.* (4) concluded that optimum adsorption strength of the co-adsorbent is needed to improve titania's photoactivity.

Previous results obtained in our laboratory clearly showed the importance of adsorption characteristics of the supported system to obtain high degradation rates of pollutants. It has been observed that the dye molecules are mainly adsorbed on the adsorbent sites to make high concentration environment around the loaded titania resulting in the enhanced photodegradation efficiency (9). H.Yoneyama and T.Torimoto (10) observed lower degradation rate with titania loaded on carbon due to high adsorption constant of the activated carbon. This led to the slow diffusion of adsorbed pollutant. Takeda *et al.* (4) studied different adsorbents for the improvement of photodecomposition rate and found out that the supports having moderate adsorbability form high concentration environment around the loaded titania and allow a fairly easy supply from the support to the loaded titania.

In the present study attempt has been made to hybridise titania with B-Zeolite to improve the photonic efficiency in the degradation of monocrotophos present in wastewater. Monocrotophos, (Figure 1) a foliar organophosphorus insecticide and acaricide mainly used for cotton, is active against wide range of insects and potentially toxic in aquatic environment. β-Zeolite is a crystalline aluminosilicate with uniform three dimensional pore system, possessing high surface area and good adsorption capability. The electric field present in the zeolite framework is capable of effectively separating the photogenerated electron/hole to enhance the photonic efficiency of titania. Based on this fact degradation studies were carried out in a slurry type batch reactor using naked titania and titania hybridised with β-Zeolite. The effects of concentration of monocrotophos, titania loading, pH and irradiation time on the degradation rate of monocrotophos were studied to optimise the experimental parameters.

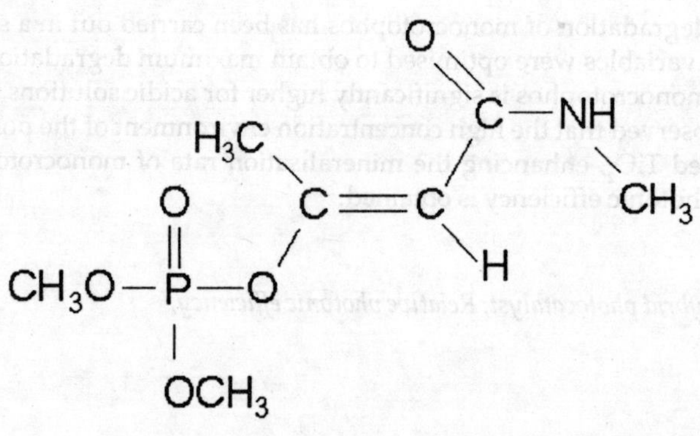

Fig. 1: Structure of Monocrotophos

Experimental

Materials

The photocatalyst, TiO_2 (Degussa P-25) with surface area 50 m^2/g containing mainly anatase form (ca.70%) was obtained from Degussa AG, Germany and used in this study. Na-form of zeolite β (Si/Al = 15), zeolite Y (Si/Al = 3) and ZSM-5 (Si/Al = 50) were obtained from Sud-Chemie India Ltd., Baroda and were converted into H-form by ion exchange with 1M ammonium nitrate solution for 24 h at 80°C (repeated three times) with subsequent calcination at 550°C in air for 6 h. The resultant H-form of zeolites was denoted as HB, HY and HZSM-5. The hybridised catalyst was obtained by grinding TiO_2 and HB-zeolite in 1: 1 ratio using pestle and mortar.

Characterisation

The BET surface area analysis of HB, HY and HZSM-5 zeolites was carried using a Quantochrome Autosorb 1 sorption analyser. Shimadzu HPLC system comprising of an isocratic pump and UV-Visible detector adjusted at 296 nm for the detection of monocrotophos and equipped with reverse-phase column (length 250 mm internal diameter 4.6 mm) CLC-ODS (M) (Shim-pack) was used for the degradation studies. The mobile phase was methanol and triple distilled water (60/40 v/v) and the flow rate was 1 ml/min. Euroglas TOC 1200 instrument was used for Total Organic Carbon (TOC) analysis.

Photoreactor and Light Source

The cylindrical photochemical reactor having dimensions 30 cm × 3 cm (height × diameter) provided with a water circulation arrangement in order to maintain the temperature in the range of 25-30°C was used in all the experiments. The cover of the reactor had ports for sampling gas purger and outlet. Irradiation was carried out using 8 × 8 W low-pressure mercury arch lamp built into a lamp housing with polished anodised aluminium reflectors and placed 6.5 cm away from the reactor. The lamps emit predominantly UV radiation at a wavelength of 254 nm. The reactor set up was covered with aluminium foil followed by cloth to prevent UV leakage.

Methods

Photocatalytic degradation was performed with aqueous solution of monocrotophos in batch reactor. Aqueous solutions of monocrotophos of different concentrations (10-75 mg/l) were prepared. To 100 ml of monocrotophos solution, 100 mg of titania was added and the resultant slurry was stirred with a magnetic stirrer for a few minutes for equilibration and irradiated with UV lamp. At specific time intervals suitable aliquots of the sample were withdrawn and analysed after centrifugation followed by filtration with Gelman GHP acrodisc 0.45 μm.

Results and Discussion

Adsorption Studies

Initially three different zeolites namely HB, HY and HZSM-5 were taken and the extent of adsorption of the monocrotophos in aqueous medium (200 mg/100 ml) was investigated using HPLC. The results showed that HB zeolite adsorbs higher concentration of monocrotophos. Hence HB zeolite was selected as an adsorbent. Further 100 mg TiO_2/100 mg HB – Zeolite (1: 1 ratio) shows good adsorption (Table 1). The lower adsorption property of HY-zeolite and HZSM-5 compared to HB may be due to their higher hydrophilic nature and lower surface area respectively. HB-zeolite is considered for the better adsorption of anionic form of monocrotophos over the acidic sites of the HB-zeolite. Secondly effective separation of electron/hole recombination is expected.

Table 1: Adsorption of Monocrotophos Over Different Catalysts

Type of Catalyst	*Amount of Monocrotophos After 12 h Adsorption (mg/l)*	*Amount of Monocrotophos Adsorbed (%)*
HB	21.0	58.0
HY	40.3	19.4
HZSM-5	43.1	13.8
TiO_2 (P-25)	43.5	13.0
TiO_2 + HB (1: 1)	29.2	41.6

Initial concentration of monocrotophos = 50 mg/l
Amount of catalyst = 200 mg
Volume of monocrotophos = 100 ml

Effect of Initial Monocrotophos Concentration

The effect of initial concentration of monocrotophos on the rate of degradation was studied by varying the initial concentration from 10 to 75 mg/l (pH 3.9-3.1) with constant catalyst loading (100 mg) and the results are reported in Table 2. Figure 2 shows the relationship between normalised monocrotophos concentration and irradiation time observed during the photocatalytic degradation of monocrotophos and it follows first order. It is obvious that the rate of degradation increases upto 50 mg/l and then the degradation decreases drastically. The rate of degradation is related to the formation

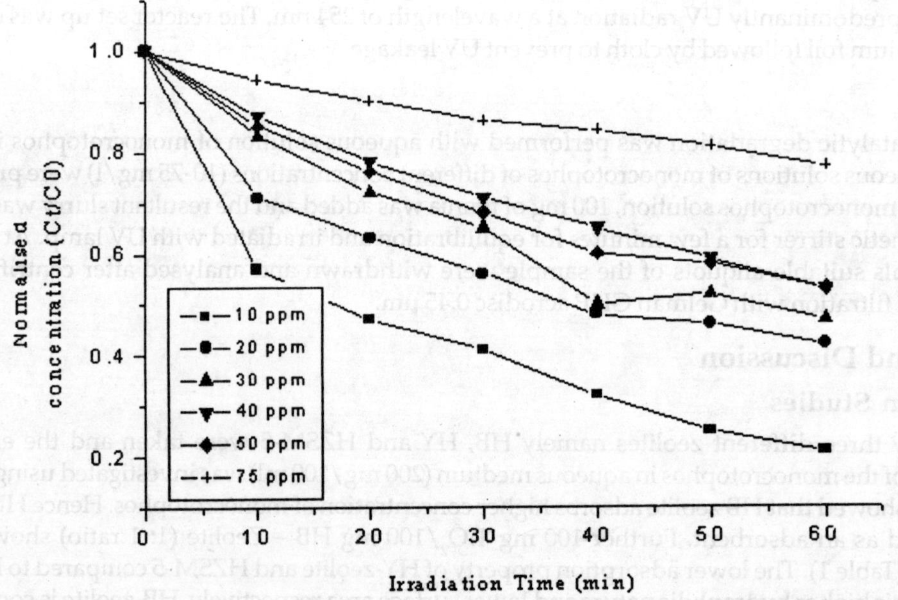

Fig. 2: Effect of Initial Concentration on Monocrotophos Degradation

of $^{\bullet}OH$ radical which is the critical species in the degradation process (11). The equilibrium adsorption of reactants on the catalyst surface and rate of reaction of $^{\bullet}OH$ radicals with other chemicals are also significant in the rate of degradation. Hence k can be expressed as follows:

$$k = k_0 \, P_{\bullet OH} \, P_{monocrotophos} \quad (1)$$

where,

k is the overall rate constant, k_0 is the reaction rate constant, $P_{\bullet OH}$ is the probability of generation of $^{\bullet}OH$ radicals on the catalyst surface and $P_{monocrotophos}$ is the probability of $^{\bullet}OH$ radicals reacting with monocrotophos molecules. The reaction rate constant k_0 is independent of the initial concentration of the substrate but $P_{\bullet OH}$ and $P_{monocrotophos}$ depend on the substrate concentration implicitly. $^{\bullet}OH$ are formed by the reaction of holes with adsorbed OH^- and H_2O. If we assume that the adsorbed OH^- positions are replaced by monocrotophos ions (MCP^-) which are generated from the dissociation of monocorotophos molecules:

$$(MCP\text{-}H) \longrightarrow (MCP^-) + H^+ \quad (2)$$

then the generation of $^{\bullet}OH$ radicals will be reduced since there are only a fewer active sites available for the generation of $^{\bullet}OH$ radicals. Therefore, $P_{\bullet OH}$ is adversely affected by the catalyst site, coverage fraction of monocrotophos and its oxidation intermediates.

Table 2: Effect of Initial Concentration on Degradation Rate of Monocrotophos

Monocrotophos Concentration (mg/l)	*k ($\times 10^3 s^{-1}$)*	*R ($\times 10^3$ mg $l^{-1}s^{-1}$)*
10	23.9	3.983
20	13.3	4.43
30	12.1	6.05
40	10.0	6.66
50	10.3	8.58
75	4.1	5.125

Titania = 100 mg/100 ml
pH = 3.1-3.9

Effect of Catalyst Loading

Experiments were carried out to assess the optimum catalyst loading by varying the amount of catalyst from 100 mg to 250 mg/l for the monocrotophos solution 50 mg/l (Table 3, Figure 3). The results demonstrate that the rate of photodegradation increases linearly with catalyst loading upto 200 mg/l. Above this loading, increase in turbidity of the solution reduces the light transmission through the solution, while below this level, it is assumed that the catalyst surface and absorption of light by the catalyst are the limiting factors. Many authors (12-15) have investigated the reaction rate as a function of catalyst loading under different experimental conditions. Wei and Wan (16) have reported that the catalyst amount has both positive and negative impact on the photodecomposition rate. Furthermore, rate may decrease as a result of loss in surface area available for light-harvesting occasioned by agglomeration (particle-particle interactions) at high solid concentration. These effects cause additional complications in the design of large-scale multiphase photoreactors compared to conventional slurry system.

Table 3: Effect of Titania Loading on Degradation Rate of Monocrotophos

Titania Loading (mg/l)	*k (x10^{3}s^{-1})*	*R (x 10^3 mg l^{-1}s^{-1})*
100	10.3	8.58
150	10.6	8.83
200	11.1	9.25
250	10.6	8.83

Monocrotophos = 50 mg/l
Volume of monocrotophos =100 ml
pH 3.4

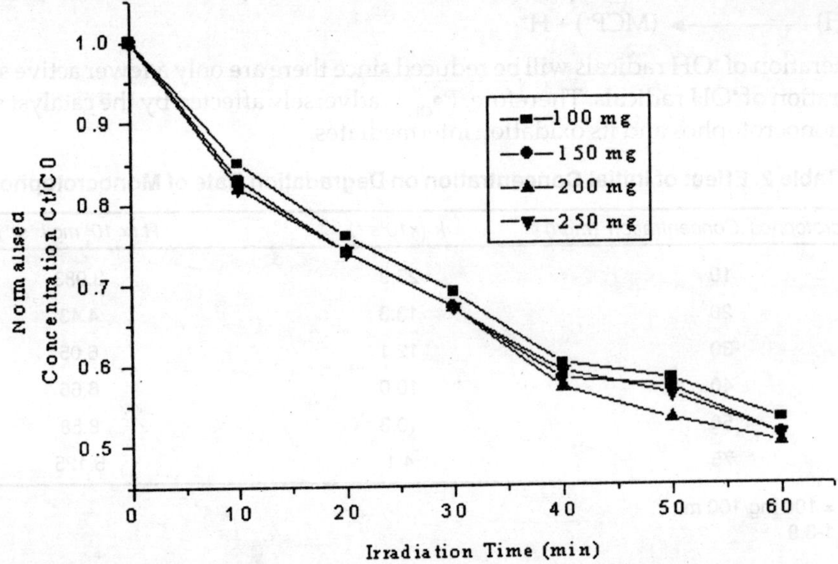

Fig. 3: Effect of TiO_2 Loading on Monocrotophos Degradation

Effect of pH

The amphoteric behaviour of titania influences the surface charge of the photocatalyst. The role of pH on the rate of photocatalytic degradation was studied in the pH range 2-8 at constant monocrotophos concentration 50 mg/l and 200 mg titania. The results are presented in Table 4 and illustrated in Figure 4. It is observed that the rate of degradation increases with increase in pH exhibiting maximum at pH4. Bahnemann *et al.* (17) also observed that acid-base property of the metal oxide surfaces can have considerable implications upon their photocatalytic activity. The effect of pH can be explained on the basis of zero point of charge of titania. The adsorption of H_2O molecules at surface metal sites is followed by the dissociation of OH^-groups leading to coverage (4-10 nm^{-2}) with chemically equivalent metal hydroxyl groups (M-OH). Due to the amphoteric behaviour of titania the following two equilibria are considered.

$$Ti\text{-}OH \rightleftharpoons Ti\text{-}O^- + H^+_{aq} \quad (3)$$

$$Ti\text{-}OH + H^+_{aq} \rightleftharpoons Ti\text{-}OH_2^+ \quad (4)$$

Table 4: Effect of pH on Degradation Rate of Monocrotophos

pH	*k* $(x10^3 s^{-1})$	*R* $(x\ 10^3\ mg\ l^{-1} s^{-1})$
2	10.4	8.66
4	11.2	9.33
6	11.1	9.25
8	10	8.33

Concentration of monocrotophos = 50mg/L
Titania = 200 mg
Volume of monocrotophos = 100ml

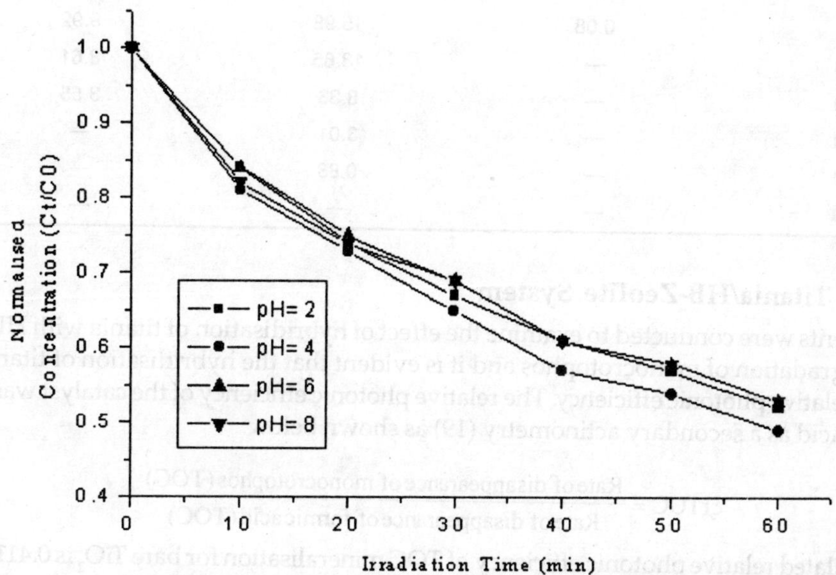

Fig. 4: Effect of pH on Monocrotophos Degradation

The zero point of charge (pzc) pH_{zpc} for titania (Degussa P25) is 6.8, below this pH it is positively charged and above this pH it is negatively charged. At low pH range, electrostatic interactions between the positive catalyst surface and monocrotophos anions lead to strong adsorption of the latter on the metal oxide support. Zhang *et al.* (18) reported that TiO_2 is the most active photocatalyst for the degradation of eosin in the acidic medium particularly at pH 4.

Effect of Irradiation Time

The relationship between photodegradation efficiency of the monocrotophos and irradiation time is presented in Table. 5. It is clearly seen that the rate of degradation increases with increase in irradiation time. The light induced degradation proceeds with slow kinetics after one hour irradiation. This reveals the formation of intermediates and its competitiveness with monocrotophos molecules in the photocatalytic degradation process. The slow kinetics of monocrotophos degradation after certain time limit is due to the difficulty in converting the N-atoms and P-atoms in monocrotophos into oxidised nitrogen and phosphorous compounds. Another reason for this slow kinetics is the slow reaction of short chain aliphatics with $^{\bullet}OH$ radicals. Under the given experimental conditions 300 min irradiation is required for complete degradation. The formation of $^{\bullet}OH$ and $O_2^{-\bullet}$ increases with increase in irradiation time and hence the monocrotophos degrades in the course of time.

Table 5: Effect of Irradiation Time on Monocrotophos Degradation and TOC Mineralisation

Irradiation Time (min)	*Monocrotophos (mg/l)*	*TOC of Bare TiO_2 (mg/l)*	*TOC of TiO_2/ HB zeolite (mg/l)*
0	50	25.19	25.19
30	11.44	22.18	12.46
60	7.72	17.92	10.53
90	0.08	15.98	9.92
120	—	13.65	8.61
180	—	8.33	3.55
240	—	3.01	—
300	—	0.88	—
330	—	—	—

Titania and Titania/HB-Zeolite System

Experiments were conducted to examine the effect of hybridisation of titania with HB-zeolite on the rate of degradation of monocrotophos and it is evident that the hybridisation of titania with HB enhances its relative photonic efficiency. The relative photonic efficiency of the catalyst was calculated using formic acid as a secondary actinometry (19) as shown below:

$$\xi rTOC = \frac{\text{Rate of disappearance of monocrotophos (TOC)}}{\text{Rate of disappearance of formic acid (TOC)}}$$

The calculated relative photonic efficiency of TOC mineralisation for bare TiO_2 is 0.411 and that of TiO_2/HB-zeolite system is 0.593.

Zhang *et al.* (18) observed the relative photonic efficiencies (measured by TOC mineralisation) of the two pesticides namely Propamocarb and Lufenuron to be 0.48 and 0.73 respectively. They conducted the experiments using titania in suspension, All the efficiencies are less than unity, indicating that the initial photocatalytic oxidative degradations of the monocrotophos, at the selected initial concentration is less efficient than for standard substance.

This enhancement of relative photonic efficiency is due to two reasons:

1. The higher adsorption property of TiO_2/HB-zeolite system produces a high pollutant concentration environment around the TiO_2 particles which enhances the possibility of attack of photogenerated active oxygen species over the pollutant molecules.
2. When titania is illuminated with light < 388nm it produces electrons and holes. Unless a suitable electron scavenger is used the electron and hole will recombine within a few

nanoseconds. If this happens the holes will no longer be available for oxidation reactions. The strong electric field present in the zeolite frame work effectively separates the electrons and holes produced during photoexcitation of titania and thereby it hinders electron and hole recombination which in turn improves the relative photonic efficiency.

Conclusion

It may be concluded from the results of the present investigation that adsorbents can be used as supports for TiO_2 to enhance the photocatalytic activity. Since monocrotophos and its intermediates are adsorbed well over hybridised titania they are completely mineralised. An acidic pH is favourable (below *zpc* of titania) for better adsorption of monocrotophos anions. Under optimal conditions HPLC and TOC data indicate complete mineralisation of the monocrotophos in 300 min with bare TiO_2 while hybridised TiO_2 takes less than 210 min for complete mineralisation of the monocrotophos. Further the amount of TiO_2 in the hybridised TiO_2 is only 100 mg which has enhanced the relative photonic efficiency by 44%. Hybridised titania makes the treatment process cost effective as zeolite is less expensive than TiO_2.

Acknowledgement

The authors gratefully acknowledge the financial support by the All India Council for Technical Education, Govt. of India, New Delhi for this project. One of the authors, M.V.Shankar is grateful for the award of Senior Research Fellowship by CSIR, New Delhi.

REFERENCES

1. R.W. Matthews, 1991, Wat. Res., 24: 1169.
2. C.Nasr, K.Vinodgopal, L.Fisher, S.Hotchandani, A.K. Chattopadhyay and P.V.Kamat, 1996, J. Phy. Chem., 100: 8436.
3. B. Neppolain, H.C. Choi, S. Sakthivel, Banumathi Arabindoo and V. Murugesan, 2002, J. Haz. Mat. B89 9: 303.
4. N. Takeda, M. Ohtani, T. Torimoto, S. Kuwabata and H. Yoneyama, 1997, J. Phys. Chem. B, 2644.
5. C. Minero, F. Catozzo, E. Pelizzetti, 1992, Langmuir, 8: 489.
6. J. Matos, J. Laine and J.M. Herrmann, 1998, Appl.Catal. B, 281
7. Y. Xu, W. Zheng and W. Liu, 1999, J. Photochem. Photobiol. A, 122: 57.
8. C. Anderson and A.J. Bard, 1995, J.Phys.Chem, 99: 3882
9. S. Saklthivel, M.V. Shankar, M. Palanichamy, B. Arabindoo and V. Murugesan, 2002, J. Photochem. Photobiol. A, 5990: 1.
10. H.Yoneyama, T.Torimoto, 2000, Catal. Today 58: 133.
11. M.V. Shankar, B. Neppolian, S. Sakthivel, Banumathi Arabindoo, M. Palanichamy & V. Murugesan, 2001, Indian J. Engg. & Mat. Sci., 8: 104.
12. C. Kormann, D.W. Bahnemann, and M.R. Hoffmann, 1991, Environ. Sci. & Technol., 25: 494.
13. Z. Mengyue, C. Shifu, T. Yaowu, 1995, J. Chem. Technol. Biotechnol., 64: 339.
14. L. Zhang, C.Y. Liu, X.M. Ren, 1995, J. Photochem. Photobiol., A85: 239.

15. S. Sakthivel, B. Neppolian, M.V. Shankar, B. Arabindoo, M. Palanichamy and V. Murugesan, 2002, Solar Energy Materials and Solar Cells., in press.
16. T.Y. Wei and C.C. Wan, Ind. Eng. Chem. Res. 30: 1293.
17. D.W. Bahnemann, 1991, In: E. Pelizzetti and M. Schiavello (Eds.) Photochemical conversion and storage of solar energy (Kleuwer Academic Publishers, Netherland), 251.
18. F. Zhang, J. Zhao, T. Shen, H. Hidaka, E. Pelizzetti, N. Serpone, 1998, Appl. Catal. B 15: 147.
19. N. Serpone and A. Salinaro, 1999, Pure & Appl. Chem., 71: 303.

47

Utility of Column Lysimeters in Wastewater Renovation Studies

P. NEMA, ARVIND KUMAR, C.S.P. OJHA, P. KHANNA & A. CHELANI

National Environmental Engineering Research Institute (NEERI), Nagpur – 440 020

ABSTRACT

Land treatment of wastewater is a potential and promising alternative in the conventional wastewater treatment technologies. The Soil Aquifer Treatment (SAT), essentially a land treatment system for wastewater is a simple and low cost technology and has been adopted in many countries. SAT systems are however site specific and require detailed hydrogeological investigations to confirm suitability of soil matrix at their proposed locations. These investigations are usually expensive and time consuming. In certain situations, field soils and their behavior can be simulated under controlled conditions by column lysimeter experiments which can provide reliable data to evolve operational parameters for a field scale SAT system. This paper presents a case study wherein column lysimeter tests are shown to have utility for design and operation of a full scale SAT system under Indian conditions.

Keywords: Wastewater renovation, column lysimeters, field scale SAT system, organic loading, SS loading.

Introduction

Water scarcity problem all over the world has warranted judicious use of available water resources. There is also a growing realization that new technologies be developed to conserve the limited water resources and renovate and reuse the wastewater in an environmentally safe and sustainable manner. Thus, renovation of wastewater and its reuse need to be incorporated as a practical supplemental strategy in any water resource planning, development and management programme (Bartone, 1991). Experience has shown that the municipal wastewater could be an alternate potential source of water because of its assured round the year availability, stable composition, and easy amenability to treatment to any desired degree depending on intended use. A number of conventional treatment methods are available for treatment of wastewater which usually aim at disposal of the treated waste in the environment without much consideration for conservation and reuse. Moreover, these methods are cost-prohibitive and require skilled operation and maintenance which is usually a major constraint in setting up of adequate treatment facilities.

Soil Aquifer Treatment (SAT) is essentially a land treatment system for wastewater which involves physical, chemical and biological processes occurring in the soil zone (Loehr *et al.*, 1985) These processes are similar to those taking place in a conventional wastewater treatment plant except that these occur in an unconfined reactor filled with soil particles whereas separate tanks (reactors) filled with liquid or slurry perform these tasks to achieve complete waste treatment consisting of sedimentation, biological treatment, and sludge treatment including disposal. Biological treatment occurs in the upper soil layer as the applied organics are stabilized by the microorganisms in the soil. In SAT system, energy is needed to transport the wastewater for application on to the infiltration basins and for recovery of treated wastewater. In conventional systems, on the other hand, energy is needed for transport, mixing, aeration, and handling of side streams (return sludge, effluent recirculation) and residuals (scum and digested sludge). Therefore, SAT systems can be less energy intensive than conventional systems.

SAT systems are site-specific and require detailed hydrogeological investigations to confirm the suitability of soil matrix at their proposed locations. These investigations are usually expensive and time-consuming. In certain situations, field soils and their behaviour can be simulated under controlled conditions by column lysimeter tests which can provide reliable data to understand the removal of various pollutants in the wastewater. Lysimeter studies provide a convenient and inexpensive means of assessing the fundamental behaviour of soils under varying conditions of input sewage and may thus prove a valuable tool in estimating the performance of large scale SAT systems (Orlob, 1956).

Several studies have been carried out using columns filled with soil to study the removal of various pollutants in the wastewater vis-a-vis change in soil characteristics. Treated effluents, mostly secondary and tertiary have been used as influent to SAT system. Lance *et al.* (1980) compared the renovation of primary and secondary effluents using high rate filtration in the laboratory columns packed with soil. It was observed that the infiltration rates for primary effluent were only marginally lower (15%) in comparison to secondary effluent though the SS concentrations in the two effluents were 51-181 mg/L and 10 mg/L respectively. Rice (1974) studied the phenomenon of clogging in soil columns filled with secondary effluent and observed that the reduction of infiltration rates were attributable to physical clogging due to high SS concentration. However, biological clogging occurred when long flooding period with low SS concentration was used. The biological clogging caused a decrease in hydraulic conductivity due to entrapped gases, probably nitrogen, which accumulated in the soil pores as a result of denitrification or depletion of oxygen in soil air by microbial activity.

Despite the recognition of lysimeter experiments in SAT studies, there is no evidence of utilizing the results of lysimeter studies in planning and design of field scale SAT systems.

The Present Study

From the review of the previous work on SAT technology for its wastewater renovation efficacy using column lysimeters, the present study was formulated with the following objectives:

1. To study the functional behaviour of soil column with reference to removal of various pollutants
2. To determine upper limits of cumulative input loading for lysimeter experiments
3. To assess the utility of column lysimeter studies in regulating the input load to a field scale SAT system

Experimental Setup

The experimental programme for the present study consisted of two column lysimeters and a field SAT system. Primary settled municipal wastewater was used directly in these experiments with soil from the Sabarmati riverbed, Ahmedabad. (Nema *et al.*, 2001).

Two column lysimeters (Lys-I and Lys-II) were fabricated and installed in the compound of the Vasana Sewage Treatment Plant (VSTP) of Ahmedabad Municipal Corporation. The experimental set up as shown in Fig. 1 comprised of the following components:

1. Two column lysimeters of mild steel (MS) with a total column height of 3 m and a diameter of 0.75 m were fabricated. The columns were assembled in 4 sections of 0.75 m each and flange-jointed and welded to a steel base plate. Each MS column section was provided with sampling ports for collection of effluent samples as well as manometer tubes for head loss measurement.
2. A special arrangement was made at each sampling port by providing a 50-mm diameter half-cut MS pipe filled with gravel across the diameter of the column. This was done with a view to collecting a sample representative of passage through the entire overlying sand column as also to minimise the wall effects. The bottom-most 10 cm of each column lysimeter was filled with gravel and a sampling port provided there. The columns were filled with the sand representative of the sandy layer in the top 1m of the riverbed. The representative sand was obtained by mixing the sand collected from 5 locations spread across the riverbed. The depth of the sand column in Lysimeter I and Lysimeter II was 259 cm and 257 cm respectively. To ensure constant surface heading on top of the sand surface, an overflow port was provided at 16 and 19 cm above the sand top in the two columns respectively. To ensure a uniform distribution of influent over the entire sand surface area in each column, a specially designed sprinkler system was fitted at the delivery end of the influent pipe discharging on top of the sand.
3. To obtain primary settled sewage (as influent to the column lysimeters), a MS settling tank was fabricated and installed in the vicinity of the two columns lysimeters. Raw sewage was pumped from the main inlet chamber at the VSTP and applied after settling for about 2 hrs on to the soil column surface at a constant rate.

The infrastructure for the pilot SAT system in the Sabarmati riverbed consisted of two primary settling basins, two infiltration basins, one production well located in the center (for recovery of renovated wastewater) and four observation wells at four corners of each basin for monitoring of SAT processes. The production well was fitted with a submersible pump of suitable capacity. A 150 mm

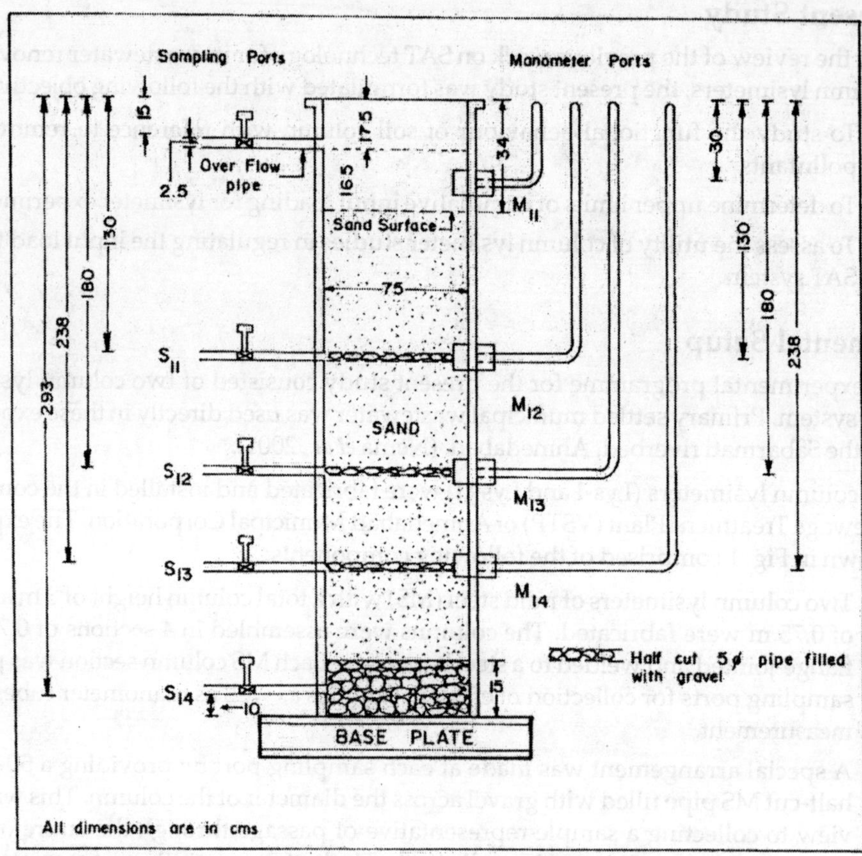

Fig. 1: Schematic of Column Lysimeter Experimental Setup

diameter PVC pipe was provided for taking settled sewage to infiltration basin under gravity. Rectangular notches and control valves were also installed at appropriate locations to measure/regulate the inflow and outflow rates (Fig. 2).

Experimental Programme

Column Lysimeter

Initially, experiments were carried out to determine the hydraulic characteristics of the surface soils filled in the column lysimeters from the riverbed. These experiments comprised of determining grain size distribution, hydraulic conductivity, porosity and dispersivity of the soil in the lysimeter. Plain tap water was used for hydraulic conductivity test whereas sodium chloride (NaCl) was used as a tracer for determination of porosity and dispersivity.

Based on the above experimental data, the average values of these parameters *viz.* hydraulic conductivity, porosity and dispersivity were determined as 9.2 m/day, 0.32 and 0.00019 m^2/min respectively which were taken as a representative values for the near surface soil matrix underlying the river bed at the pilot plant.

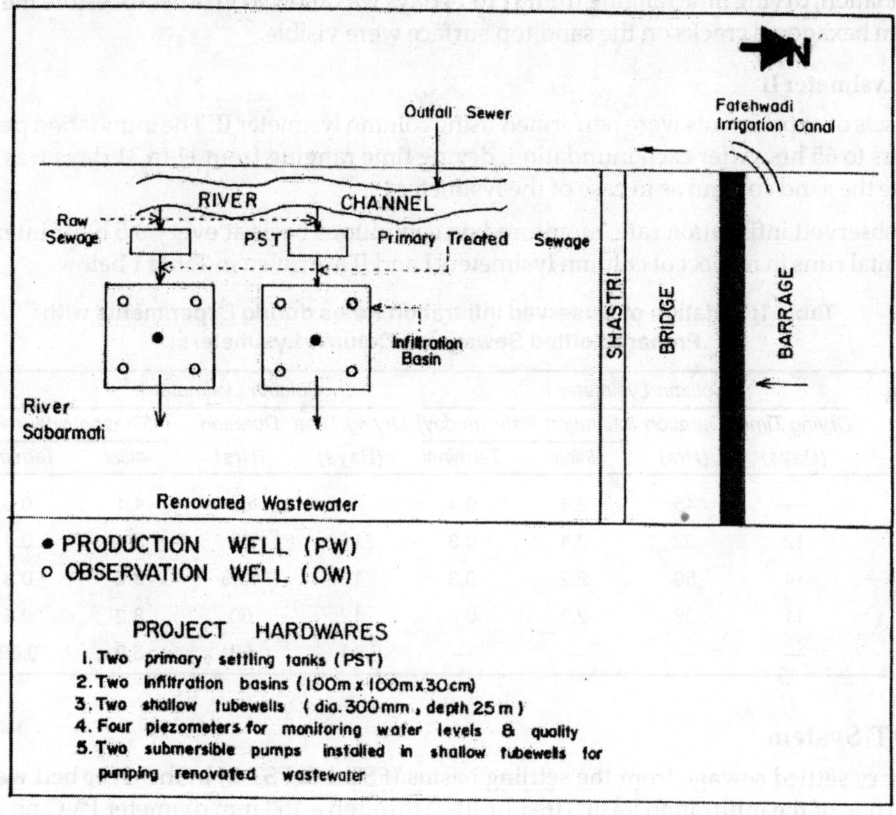

Fig. 2: Schematic Layout of the Pilot SAT System in Sabarmati River Bed

The second phase of experiments was carried out with primary settled sewage as the influent for assessment of the efficiency of the soil matrix in retarding/removal of various wastewater pollutants. Raw sewage was pumped from the inlet chamber of the VSTP and settled in the MS primary settling tank. A detention time of about 2 hrs was provided. The settled sewage was applied on top of the soil surface in the column through a pump at a constant input rate. A constant head of water column was ensured by wasting the excess flow from the overflow port located above the sand top surface. Outflow was monitored continuously at the port (S_{14}, Fig. 2) and samples of influent and effluent were collected simultaneously for laboratory analyses for relevant parameters. The length of experimental run was governed by following considerations:

1. The breakthrough of the influent wastewater quality was observed in the effluent quality.
2. The filtration rate (*i.e.* outflow rate) declined to 5-10% of the initial rate due to either clogging or entrapment of air in the soil column.

Column Lysimeter I

Four sets of experiments were performed on column lysimeter I. The inundation period varied from 32 hrs to 50 hrs and the experiments were terminated when considerable decline in the inflow rate and an abrupt increase in the headloss indicating clogging of the sand bed was observed. Between each inundation, drying time ranging from 11 to 14 days was allowed in order to restore the infiltration rates when hexagonal cracks on the sand top surface were visible.

Column Lysimeter II

Five sets of experiments were performed using column lysimeter II. The inundation period varied from 43 hrs to 65 hrs. After each inundation, drying time ranging from 11 to 31 days was allowed to rejuvenate the sand column as in case of the lysimeter I.

The observed infiltration rate, monitored on continuous basis at every 0.5 hour interval for the experimental runs in respect of column lysimeters I and II are given in Table 1 below.

Table 1: Variation of Observed Infiltration Rates during Experiments with Primary Settled Sewage on Column Lysimeters

Exp	*Column Lysimeter I*				*Column Lysimeter II*			
	Drying Time (Days)	*Duration (Hrs)*	*Infiltration Rate (m/day)*		*Drying Time (Days)*	*Duration (Hrs)*	*Infiltration rate (m/day)*	
			Initial	*Terminal*			*Initial*	*Terminal*
1	—	45	2.4	0.4	—	65	4.1	0.6
2	12	32	2.4	0.3	11	62	3.4	0.4
3	14	50	2.2	0.3	12	43.5	3.6	0.3
4	11	38	2.3	0.3	12	60	3.2	0.3
5	—	—	—	—	31	53	3.9	0.09

Pilot SAT System

Primary settled sewage from the settling basins (PST-1 & PST-2) in the river bed was taken by gravity to one of the infiltration basin (IB-1 or IB-2) through a 150 mm diameter PVC pipe provided with a control valve. At any given time, one of the two infiltration basins was kept under flooding and the other left for drying. Initially, the duration of the flooding and drying cycle was maintained at 10-14 days each and then reduced to about one week as the summer approached its peak (April-May) and evaporation rate increased. Further, the duration of flooding and drying cycle was adjusted to ensure: (a) complete flooding of an infiltration basin to a maximum liquid depth of 0.3 m, and (b) complete drying of the other basin until prominent shrinkage cracks developed in the biological mat over the basin bed.

The flow rates of the influent and pumped renovated wastewater were monitored on regular basis. Samples of influent (primary settled sewage) and effluent (renovated wastewater) were collected periodically. Laboratory analyses were carried out for (i) organic pollutants, (ii) nutrients and (iii) bacteria. The experiments were carried out in two phases during the post monsoon period of 1995 (44 days) and 1996 (138 days).

During flooding, a water column of 16.5 cm and 19 cm was maintained on the sand surface top of column lysimeter I and II respectively.

Results and Discussions

As mentioned earlier, one of the objectives of the study was to perform lysimeter experiments with a loading rate significantly much different from that used in field SAT system. The idea was to study the response of lysimeter under a given cumulative input loading. It was presumed that if such input loadings are applied to SAT systems, their process performance would be within the desirable limits.

Only two parameters namely BOD and SS have been considered for evaluation of lysimeter as well as field scale SAT systems. The performance data of column lysimeter I & II and pilot SAT system in respect of BOD and SS removal are shown in Table 2.

Table 2: System Performance: Column Lysimeter and SAT

Parameter	*Column Lysimeter I*		*Column Lysimeter II*		*SAT*	
	Influent	*Effluent*	*Influent*	*Effluent*	*Influent**	*Effluent***
BOD, mg/L	77-103 (88)	12-39 (26)	65-118 (93)	22-35 (28)	52-200 (120)	2-24 (8)
SS, mg/L	28-57 (48)	14-17 (15)	21-58 (40)	15-24 (20)	72-249 (149)	4-28 (12)

* Settled Sewage; ** Production Well; Note: Values in parenthesis are average values

Effluents Quality vs Loadings

The knowledge of average percent removals of pollutants may not be adequate to understand the system's response under various loading patterns. Thus, certain correlations, if available, between the effluent quality and the loading, can be achieved for a given SAT system, and it is expected that this will help a better understanding of the potential of the system.

Plots between cumulative mass loading and the effluent quality in terms of BOD and SS were attempted for column lysimeters. The objective of this exercise was to visualise any improvement in variation trend between these parameters as this approach followed the filtration theory wherein the filtrate quality is influenced by the deposition of impurities versus time. From these plots (Fig. 3), as expected deterioration in effluent quality with increasing cumulative loading was clearly discernible. Such plots were also developed between effluent quality and cumulative hydraulic loading with a view to observe the relative influence of these two parameters on the removal of pollutants (Fig. 4).

Similar plots correlating cumulative mass loading and the effluent quality in terms of BOD an SS were prepared for SAT system. In this case also, the effluent quality showed deterioration with increasing loading (Fig. 5). As in case of lysimeter, a correlation between effluent quality and cumulative hydraulic loading was also attempted (Fig. 6).

However, the trend observed was more or less similar between effluent quality and cumulative mass & hydraulic loading rates in both the cases.

Evaluation of Upper Loading Limits

One of specific objectives of the column lysimeter studies was to determine the upper limits of hydraulic and mass loadings on the pilot SAT system to obtain the desired effluent quality and ascertain if the loadings obtained from column lysimeter experiments are useful for the operation of the pilot SAT system. The experimental data was, therefore, analysed to observe simultaneous performance of the column lysimeters and pilot SAT system with regard to effluent quality in terms of BOD and SS. The data relating to different cumulative loading rates obtained for the desired effluent quality in terms of BOD and SS for SAT system and column lysimeters is presented in Table 3. It is interesting to see that for each experiment, the cumulative mass as well as hydraulic loadings for lysimeters are higher than those for pilot SAT system. Table 4a and 4b depict the ratio of cumulative mass loadings and hydraulic loadings between the column lysimeters and pilot SAT system. Thus, lysimeter tests can be used to plan SAT loadings without having any adverse effect on the operation of SAT systems.

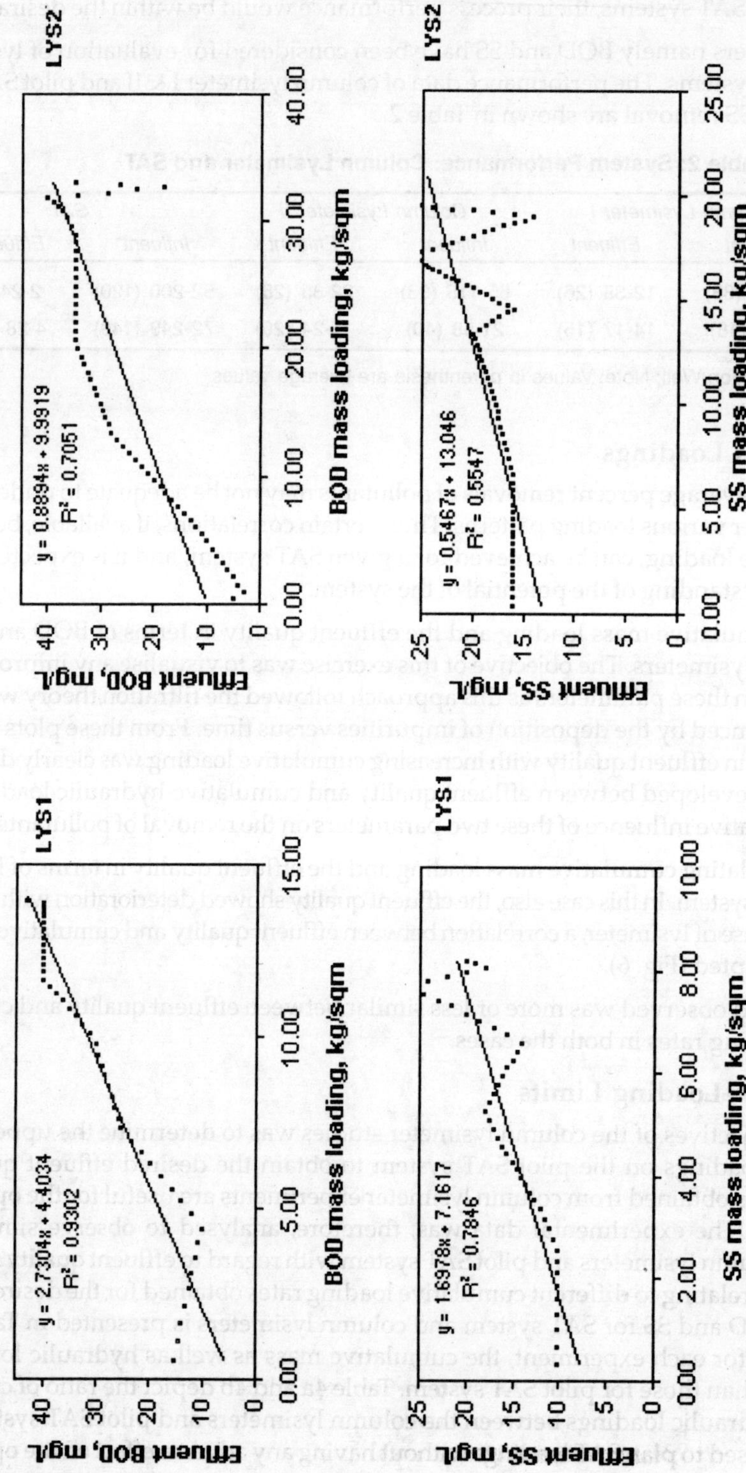

Fig 3: Variation Between Cumulative Mass Loading and Effluent Quality (Lysimeter): Typical Plots

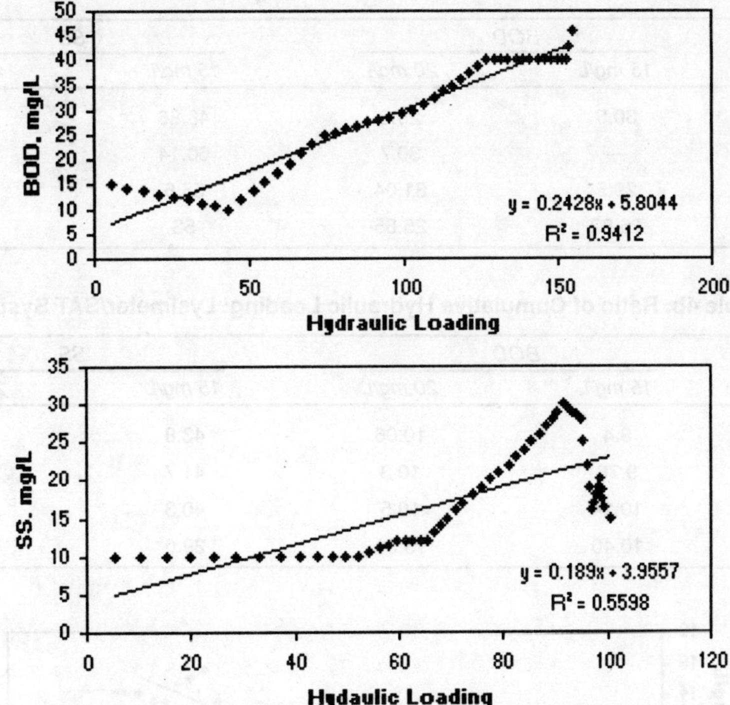

Fig. 4: Variation Between Cumulative Hydraulic Loading and Effluent Quality (Lysimeter) Typical Plots

Table 3: Cumulative Mass Loading Rates (kg/sqm) for Desired Effluent Quality

Sl.No.	*BOD*		*SS*	
	15 mg/L	*20 mg/L*	*15 mg/L*	*20 mg/L*
SAT System				
1.	0.174	0.25	0.079	0.129
2.	0.17	0.22	0.069	0.109
3.	0.19	0.23	0.08	0.119
4.	0.211	0.26	0.09	0.139
Column Lysimeter				
1.	5.23	6.40	3.86	5.12
2.	—	6.77	4.15	7.05
3.	5.61	7.14	4.45	7.19
4.	5.66	6.67	4.77	4.33

Table 4a: Ratio of Cumulative Mass Loading: Lysimeter/SAT System

Sl.No.	*BOD*		*SS*	
	15 mg/L	*20 mg/L*	*15 mg/L*	*20 mg/L*
1.	30.5	25.6	48.86	39.68
2.	—	30.7	60.14	64.67
3.	29.5	31.04	55.6	60.42
4.	26.82	25.65	53	31.15

Table 4b: Ratio of Cumulative Hydraulic Loading: Lysimeter/SAT System

Sl.No.	*BOD*		*SS*	
	15 mg/L	*20 mg/L*	*15 mg/L*	*20 mg/L*
1.	9.4	10.06	43.8	37.9
2.	9.76	10.3	41.7	37.9
3.	10.14	10.5	40.3	36.9
4.	10.40	13.31	39.0	36.2

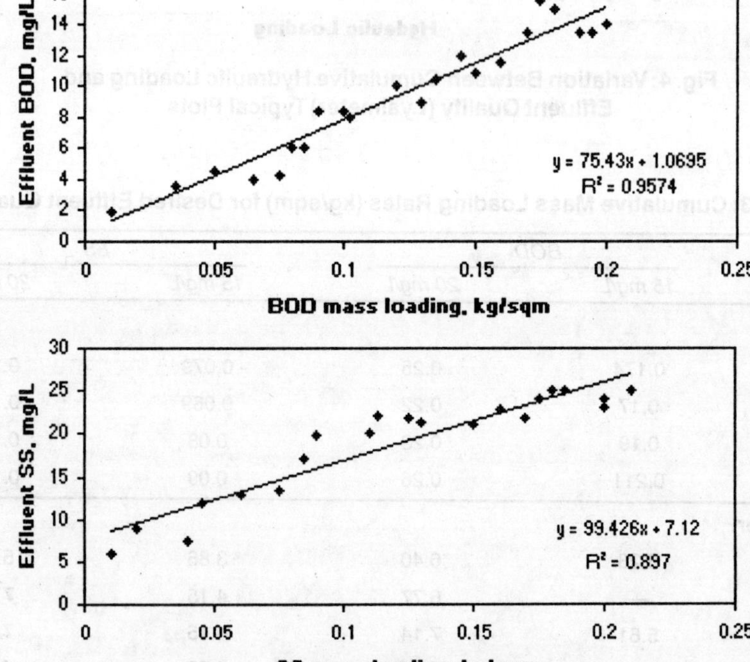

Fig. 5: Variation Between Cumulative Mass Loading and Effluent Quality (SAT) Typical Plots

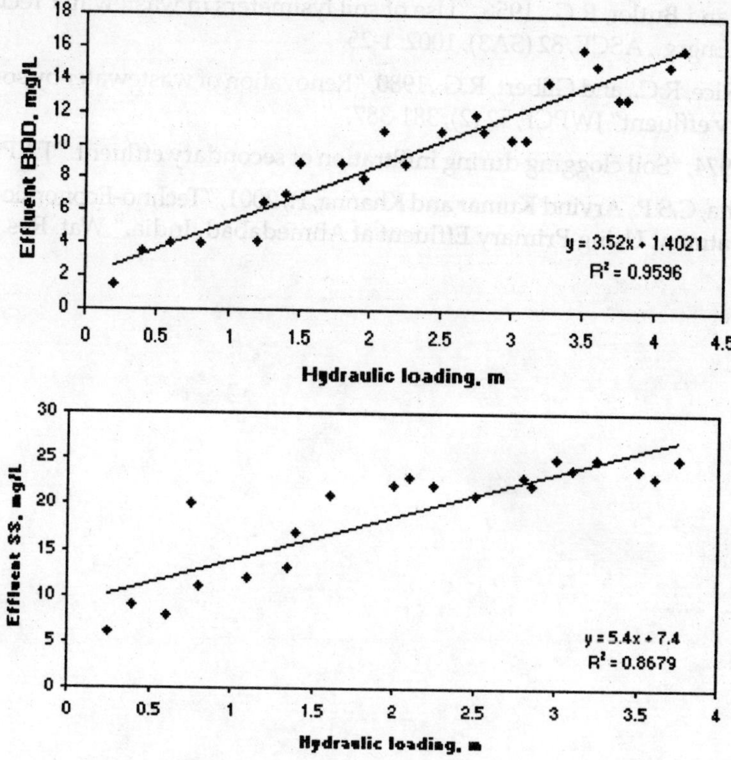

Fig. 6: Variation Between Hydraulic Loading and Effluent Quality (SAT) Typical Plots

Conclusions

The necessity of performing lysimeter experiments is evident from the results of the experiments described above. Here, the issue is regarding the loading to which the lysimeters should be subjected. Maintaining the same loading rate in lysimeters as that of pilot SAT system will eventually result in very longer runs of lysimeters and this may help in precisely quantifying the run of pilot SAT system under given loading. However, this may not be a desirable option in preference to the lysimeter runs of shorter duration subjected to higher loadings than that of SAT. For this purpose only, lysimeter runs were subjected to higher loading rates. From the data obtained in this study, it can be concluded that the upper loadings based on lysimeter runs can safely be assumed as the upper loadings for designing a field scale SAT system. The development of a biofilm which is usually observed in case of pilot SAT system may further reinforce the validity of this assumption.

REFERENCES

1. Bartone, C.R., 1991, "International perspective on water resources management and wastewater reuse appropriate technologies." Water Sci Technol., 23, 2039-2047.
2. Loeh, R.C., and Michael, R.O., 1985, "Land treatment of wastes: concepts and general design." J. Envir. Engrg., ASCE, 111 (2), 141-160.

3. Orlob, G.T., and Butler, R.G., 1956, "Use of soil lysimeters in wastewater reclamation studies." Proc. J. San. Engrg., ASCE, 82 (SA3), 1002: 1-25.

4. Lance, J.C., Rice, R.C., and Gilbert, R.G., 1980, "Renovation of wastewater by soil columns flooded with primary effluent" JWPCF, 52 (2), 381-387.

5. Rice, R.C., 1974, "Soil clogging during infiltration of secondary effluent." JWPCF, 46 (4), 708-716.

6. Nema, P. Ojha, C.S.P., Arvind Kumar and Khanna, P., 2001, "Techno-Economic Evaluation of Soil-Aquifer Treatment Using Primary Effluent at Ahmedabad, India." Wat. Res. Vol. 35, No. 9, pp. 2179-2190.

48

An Innovative Method of Ferritization for Wastewater Treatment

D.M. DHARMADHIKARI, S.K. MANDAVGANE & S.S. DARA

National Environmental Engineering Research Institute, EAI Division, Nehru Marg, Nagpur - 20

ABSTRACT

The paper deals with the studies on retrieval of heavy metal ions (Co^{+2}, Zn^{+2}, Ni^{+2} and Mn^{+2}) in solution by ferritization and its potential application in wastewater treatment. The optimum procedure for ferritization of heavy metal ions in solution has been evolved with respect to pH, concentration of Fe^{+2}, temperature, rate of aeration and time of oxidation. The recommended procedure consists of controlled aeration of the solution containing heavy metal ions and ferrous ions at pH 9.5 to 10.5 at about 50°C until the black, granular soft ferrite of respective metal separates out.

The room temperature ferritization phenomenon has been observed for the first time in this work. Room temperature ferritization process has been invented by ageing the mixed metal hydroxide (pH 10-11) at room temperature for a period of 2-3 days, so that the ferrite crystallizes and can be separated. The metal ferrites recovered have been characterized by x-ray diffractrometry.

The laboratory scope experiments conducted with synthetic heavy metal solutions as well as wastewater from industries showed that heavy metal ions could be effectively removed from solution to sub-ppm levels. This promising technique has potential application in removal of heavy metal ions from industrial wastewater and using the recovered soft ferrites as microwave absorbers, catalysts, antennae rods and have wide applications in electronic and telecommunication industries.

The studies on leaching tests of ferrite sludge show that in the pH range 4-10, the concentration of leachable metals is extremely low, showing the stability of the ferrite material.

Keywords: Heavy Metals, Recycle and Reuse, Soft Ferrites, Chemical Process, X-Ray Diffraction, Leachate

Introduction

Quality criterion of water is assessed on the basis of total dissolved matter present in it. Heavy metals are one of the prime pollutants affecting the quality of water and are most hazardous among the determinants of global significance, because of their non-degradable nature and property to affect all kinds of ecological systems, even when present in very small quantities.

Exposure to heavy metal toxicity can result from every facet of national activity such as agriculture, mining, transport, energy, industry, metal coating and finishing industries, tanneries, smelting and refining of non-ferrous metals, ferrous industries and battery manufacturing. The industrial wastewater is generally industry specific and vary widely both in quality and characteristics. Meal processing, finishing and plating industries are obvious sources of heavy metal wastes, and the nature of waste directly reflects the particular combination of metals and manufacturing process as utilized by a given plant. Due to the toxic and hazardous nature of heavy metals even in trace levels, the abatement is of prime importance in implementing the environmental act.

The available literature reveals that the electroplating and pickling industries seem to contribute most of the metal bearing industrial waste. The volume of the heavy metal waste is significant and will have to be recycled within manufacturing processes. In 1986, EPA-F006 – classified waste alone (National Report, US, EPA, 1991), which represents the metal sludge generated only from electroplating operations, accounted for 11.2 million tons or 1.5% of the 747 million tons of waste generated that year. Ground water and surface waters contaminated with low levels of toxic heavy metal ions also represent a significant environmental problem. A prominent example (Tech. Report, 1991) is the Berkeley pit, filled with 75.7 million m^3 of acidic mine leachate effluent contaminated with heavy metal ions, including arsenic (0.8 ppm), cadmium (1.5 ppm) and nickel (1.1 ppm).

Literature survey reveals that several workers have developed major chemical processes technologies for removal of heavy metal ions from wastewater (Brooks, 1991; Peters *et al.*, 1985; Krishnan, 1993). Many of these processes generate other forms of hazardous wastes (such as metal hydroxide sludges/carbonate sludges), which also require highly regulated and costly disposal. Presently, chemical and textile industries are using methods involving chemical oxidation and reduction, precipitation as hydroxides, sulphides, carbonates and xanthates, chemical displacement, evaporation, adsorption, ultra-filtration, electro-dialysis and reverse osmosis.

The present work is aimed for recycling and reuse of heavy metals by improving the technique of ferritization. An attempt has also been made to reduce the ratio of Fe^{+2} ions to heavy metal ions, temperature, pressure and contact time, so as to develop a simpler and more convenient method for retrieval of heavy metals (Zn^{+2}, Co^{+2}, Ni^{+2} and Mn^{+2}) from wastewaters in the form of environmentally stable ferrites having a general formula AB_2O_4. Where, A = Cation to be retrieved and B = Fe^{+2} ions.

It is the need of the time to remove the heavy metals very effectively and economically and at the same time to recover them in some or the other form while simultaneously achieving pollution control and reuse of the waste. It is precisely in this context that the process to retrieve the heavy metals in solution in the form of corresponding ferrites has been developed. Ferrites are environmentally stable compounds.

Material and Methods

Solutions of metal ions namely Co^{+2}, Zn^{+2}, Ni^{+2} and Mn^{+2} were prepared from Analar grade salts of the respective metals. Wastewaters containing these metals were also sampled from Galvanising Plant, Electroplating Industry, Manganese Mines and Copper Mines. Aeration of the solution was done with the help of small vacuum pump. The rate of air flow used was measured by a Perspex rotameter (filflab) calibrated with a gap kit meter.

Experimental Procedure

The solution containing known concentrations of the heavy metal ions, namely Zn^{+2}, Co^{+2}, Ni^{+2} and Mn^{+2}, were mixed in separate experiments with appropriate quantities of Fe^{+2} ions (Ferrous Sulphate) in separate beakers. The pH of each of the solutions was adjusted, between 9.5 and 10.5 as the case may be, by adding sodium hydroxide solution and the contents of the beaker were heated at about 60°C while simultaneously passing air into the solution at an appropriate flow rate until the black magnetic ferrite was formed. The ferrite formed in each case was separated by filtration. The ferrites formed were dried, powdered and subjected to X-ray diffraction for identification of the structure.

For the room temperature ageing experiment, the solutions containing the required amounts of Fe^{2+} and the heavy metal ions were brought to pH 10 and the contents were allowed to stand for about 7-8 days with occasional manual stirring until the ferrites were formed.

Mechanism for Ferrite Formation

When an equivalent amount of alkali is added to an aqueous solution containing Fe^{2-} and divalent non-ferrous metal ions, the following reaction takes place (Okuda *et al.*, 1975).

$$x\,M^{2+} + Fe^{2+}{}_{3-x} + 6OH \rightarrow M_x\,Fe_{3-x}\,(OH)_6$$

When the solution containing this mixed hydroxide is oxidized under specific conditions, phenomenon such as redissolution and complex formation take place. Eventually, a black spinel compound (ferrite) separates out as follows:

$$M_x\,Fe_{3-x}\,(OH)_6 + \tfrac{1}{2}O_2 \rightarrow M_x\,Fe_{3-x}\,O_4 + 3\,H_2O$$

The following mechanism has also been suggested for ferrite formation under the above conditions (Mandaokar *et al.*, 1994).

$$M^{2+} + 2Fe^{3+} + 8OH \rightarrow M\,(OH)_2 + 2\,Fe\,(OH)_3$$

$$M(OH)_2 + 2\,Fe(OH)_3 \xrightarrow[50^\circ C]{\text{Aeration}} \underset{\text{Metal Ferrite}}{M\,Fe_2\,O_4} + 4H_2O$$

Characterization of Ferrites

X-ray diffraction studies were carried out using CuK_a radiations on Philips diffractometer (Philips, model 1710). The X-ray patterns of the collected ferrites were indexed with standard procedures of X-ray crystallography and the data was compared with that of pure ferrites from ASTM data.

Quantification of Heavy Metals in Effluents and Influents

The concentration of heavy metals in filtrate and in ferrite phase (residue) were analysed using Atomic Absorption Spectrophotometery (GBC, model 904 with deuterium background).

Results and Discussion

The result on treatability studies of metal ion concentration in the range of 5 mg/L – 1000 mg/L are shown in figures 1–4 as:

1. Concentration of Fe (total) in filtrate and ferrite phase as a function pH are shown for Zn^{+2}, Co^{+2}, Ni^{+2} and Mn^{+2} are shown in Fig. 1.
2. Optimization of dose of $FeSO_4$ is shown in Fig. 2.
3. Optimum concentration of Fe^{+2} required for ferrite formation is shown in Fig. 3.
4. Optimization of Fe^{+2}/M^{+2} ratio is shown in Fig. 4.

Time taken for ferrite formation as a function of M^{+2} ion concentration is shown in Fig. 5.

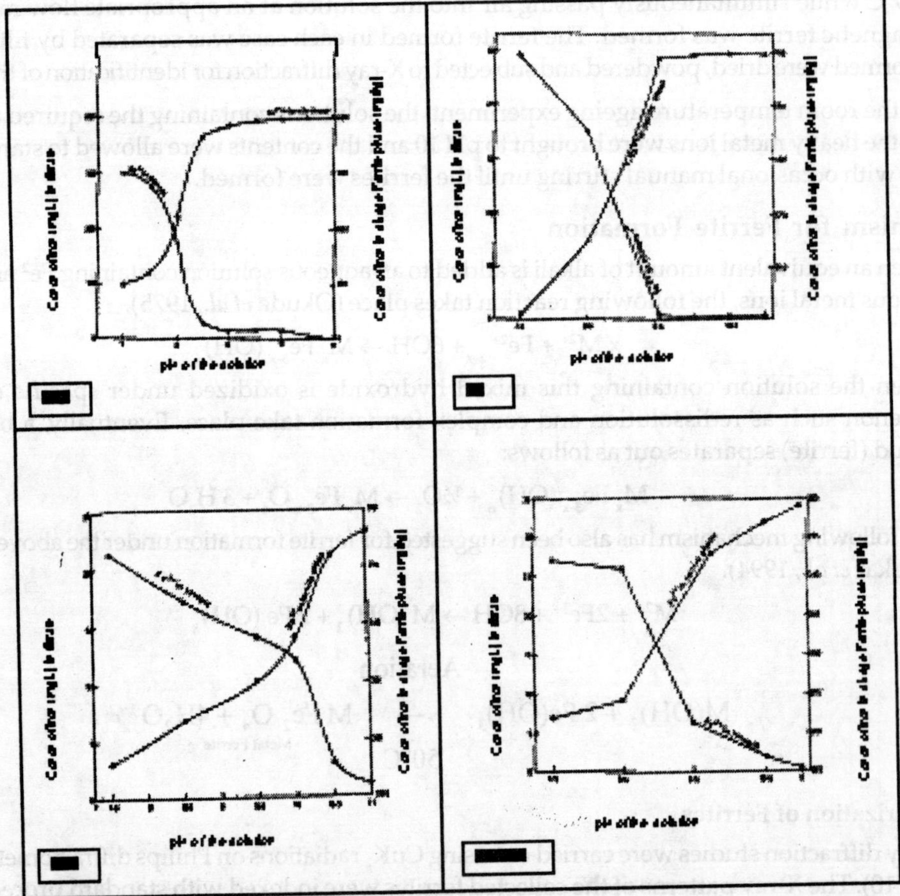

Fig. 1: Concentration of Fe^{+2} Ions Present in the Ferrite Phase and Filtrate After the Formation of (Cobalt, Zinc, Nickel and Manganese) Ferrite at Different pH

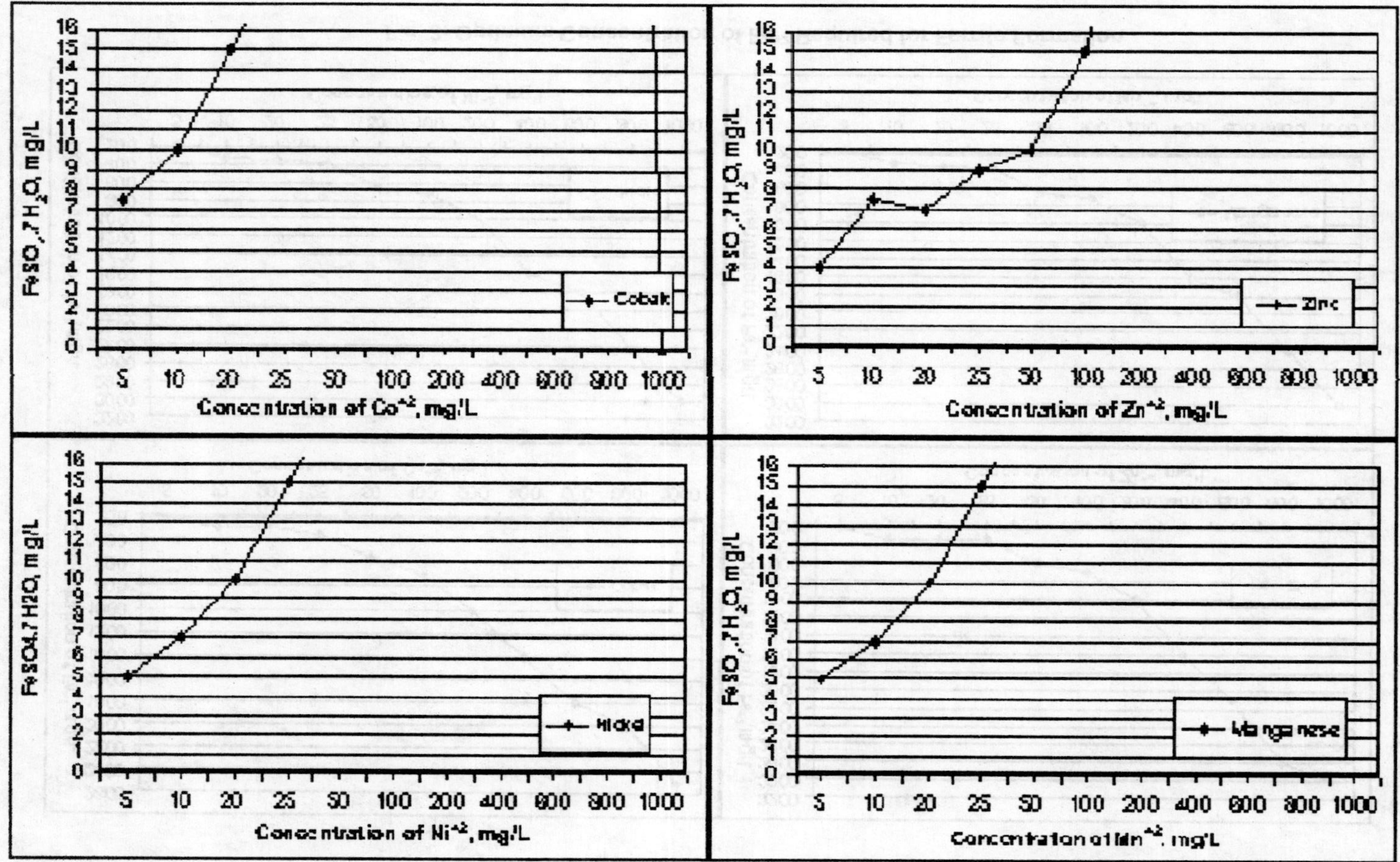

Fig. 2: Dose of FeS O_4.7 H_2O Required for Removal of Heavy Metals from Solution

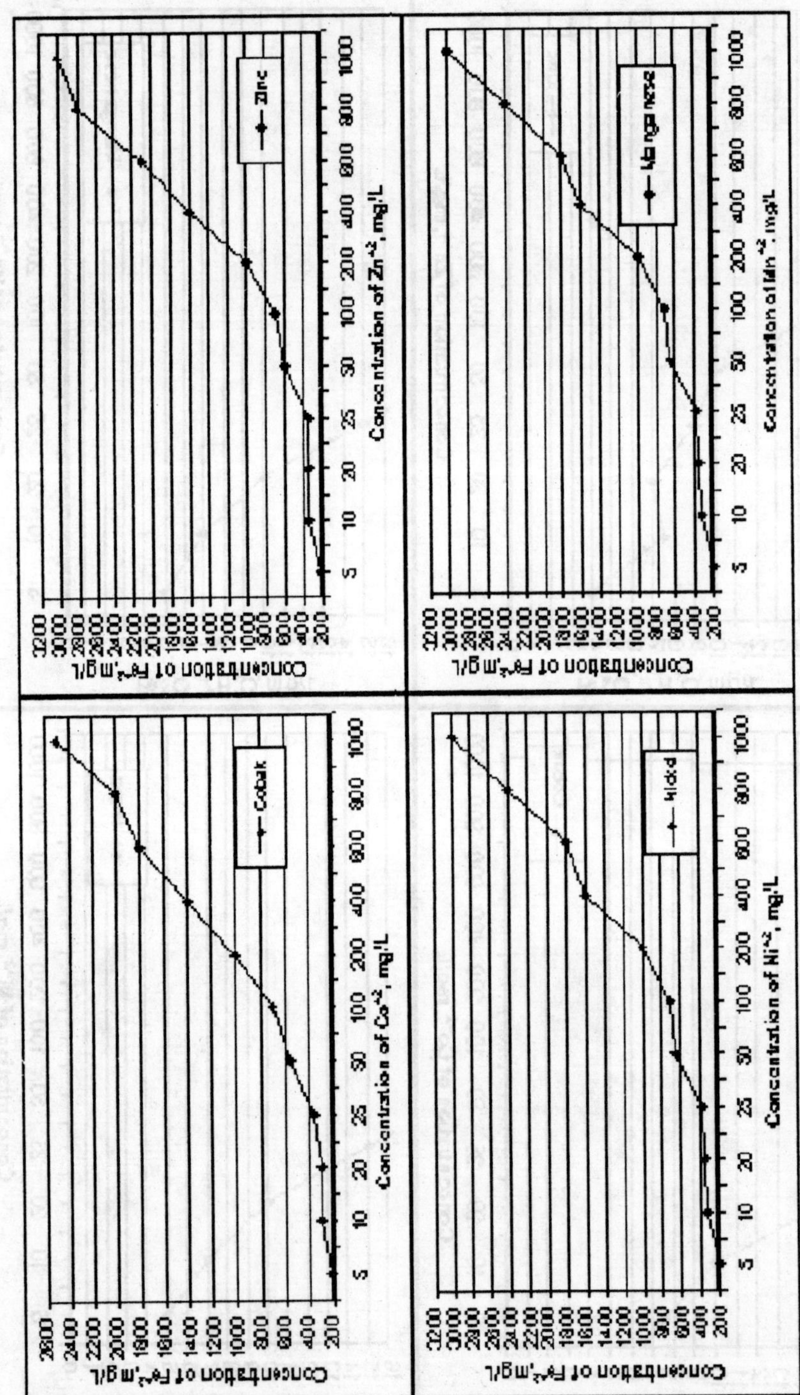

Fig. 3: Optimum Concentration of Fe^{+2} Required for Ferrite Formation

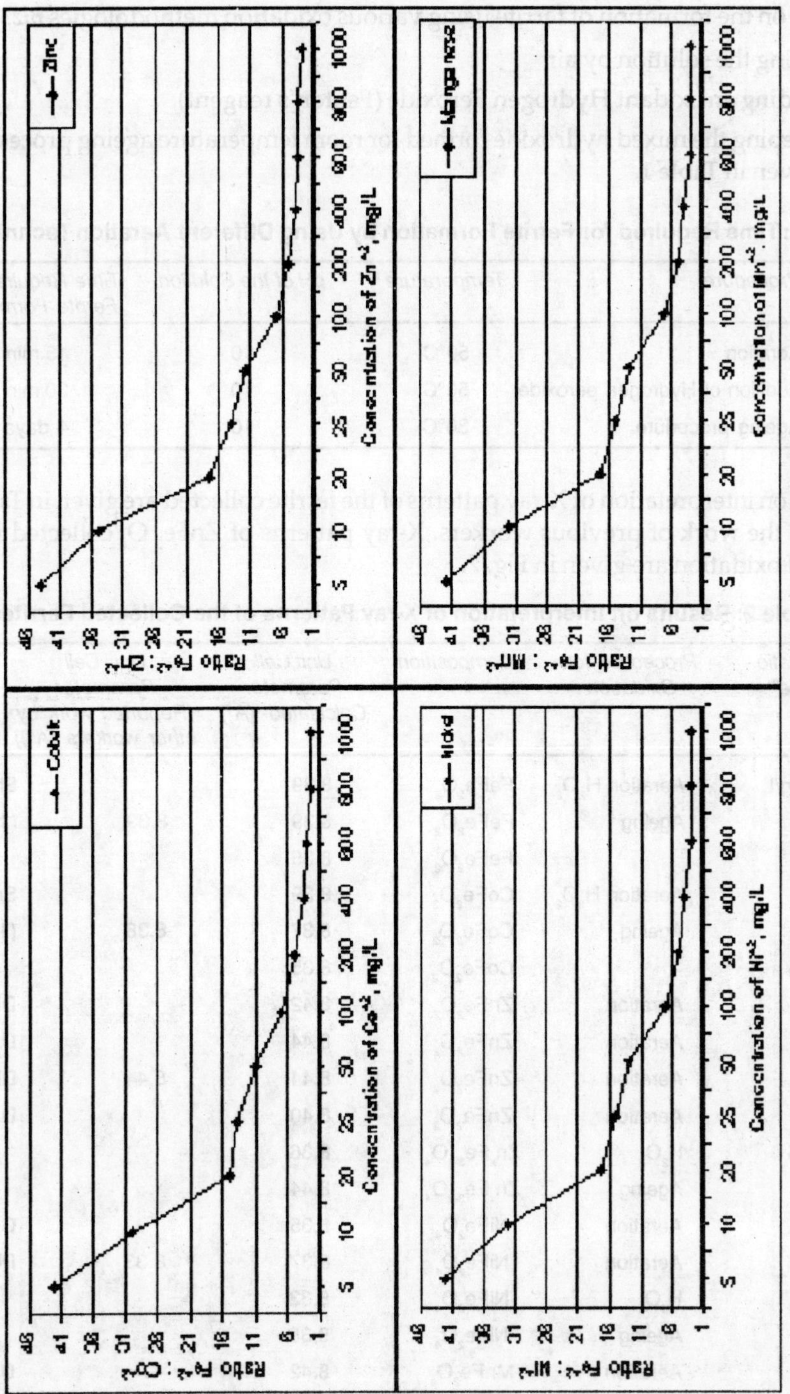

Fig. 4: Ration of Fe^{+2} : M^{+2} Ions Required for Retrieval of M^{+2} Ions as Ferrite

The results on optimization of pH for the process are shown in Fig. 6.

The results on the formation of ferrite using various oxidation methodologies *viz.*

1. Aerating the solution by air
2. By adding an oxidant Hydrogen Peroxide (Fenton's reagent)
3. By keeping the mixed hydroxide formed for room temperature ageing process for few days are given in Table 1.

Table 1: Time Required for Ferrite Formation by Using Different Aeration Techniques

Sl. No.	*Procedure*	*Temperature*	*pH of the Solution*	*Time Required for Ferrite Formation*
1.	Aeration	50°C	10	15 min
2.	Addition of Hydrogen peroxide	50°C	10	30 min
3.	Ageing procedure	30°C	10	4 days

The results on interpretation of X-ray patterns of the ferrite collected are given in Table 2 and also compared with the work of previous workers. X-ray patterns of $ZnFe_2\ O_4$ collected in solution by various ways of oxidation are given in Fig. 7.

Table 2: Results on Interpretation of X-ray Patterns of the Collected Ferrites

Sl. No.	*Mole Ratio M^{+2}: Fe^{2+}*	*Procedure for Oxidation*	*Composition*	*Unit Cell Parameter Calculated (A°)*	*Until Cell Parameter (Reported work by other workers (A°)]*	*Reference*
1.	4000 mg/L	Aeration H_2O_2	$FeFe_2O_2$	8.39		Sharma, V.N.
	Fe^{2+}	Ageing	$FeFe_2O_2$	8.39	8.39	(1965)
			$FeFe_2O_2$	8.35		
2.	1: 3	Aeration H_2O_2	$CoFe_2O_2$	8.39		Smit, J. *et al.*
	1: 7	Ageing	$CoFe_2O_2$	8.37	8.38	(1954)
	1: 35		$CoFe_2O_2$	8.35		
3.	1: 3	Aeration	$ZnFe_2O_2$	8.42		Dharmadhikari,
	1: 4	Aeration	$ZnFe_2O_2$	8.44		D.M. *et al.* (1986);
	1: 6	Aeration	$ZnFe_2O_2$	8.44	8.44	Dharmadhikari,
	1: 7	Aeration	$ZnFe_2O_2$	8.40		D.M., *et al.* (1978)
	1: 15	H_2O_2	$Zn_xFe_{3-x}O_4$	8.36		
	1: 3	Ageing	$Zn_xFe_{3-x}O_4$	8.44		
4.	1: 3	Aeration	$NiFe_2O_4$	8.35		Gorter, E.W.
	1: 4	Aeration	$NiFe_2O_4$	8.37	8.33	Philips (1954)
	1: 7	H_2O_2	$NiFe_2O_4$	8.33		
	1: 7	Ageing	$NiFe_2O_4$	8.38		
5.	1: 3	Aeration	$MnFe_2O_4$	8.42		Dharmadhikari.
	1: 15	H_2O_2	$Mn_xFe_{3-x}O_4$	8.35	8.49	D.M., (1978)
	1: 30	Ageing	$Mn_xFe_{3-x}O_4$	8.33		

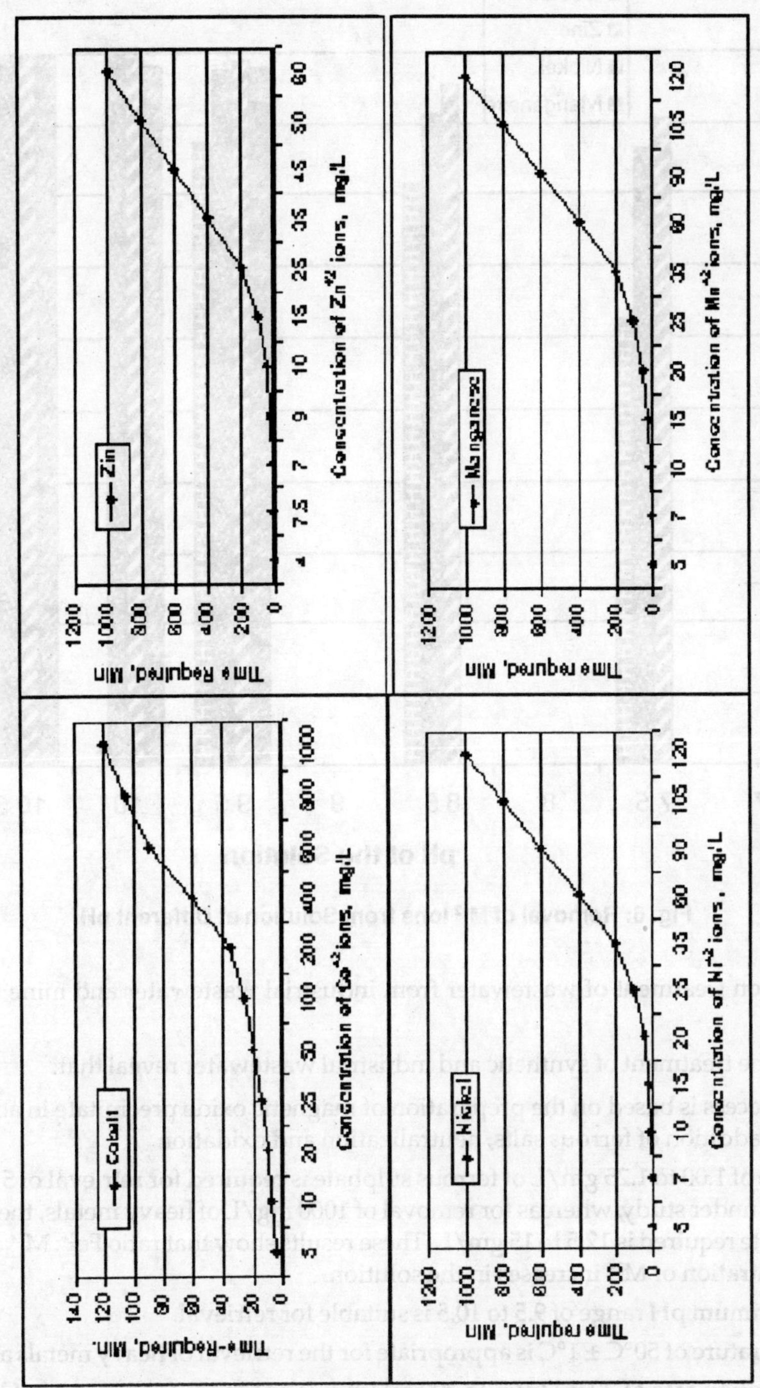

Fig. 5: Time Taken for Formaton of Ferrite Phase and Filtrate as a Function of M^{+2} Ion Concentration

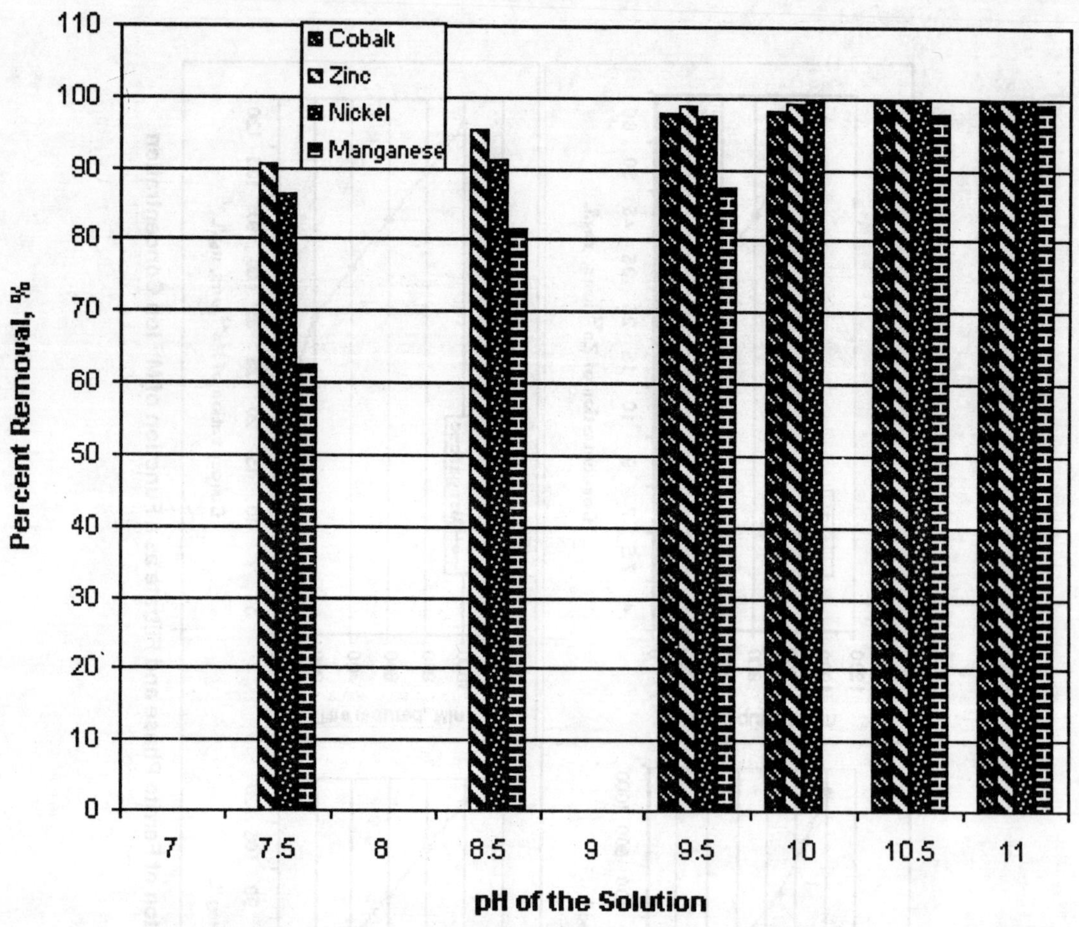

Fig. 6: Removal of M^{-2} Ions from Solution at Different pH

The results on treatment of wastewater from industrial wastewater and mine wastewater are given in Table 3.

Results on the treatment of synthetic and industrial wastewater reveal that:

1. The process is based on the preparation of magnetic oxide precipitate in aqueous solution by the addition of ferrous salts, neutralization and oxidation.
2. A dose of 1.00 to 1.25 gm/L of ferrous sulphate is required for retrieval of 5 mg/L of heavy metals under study, whereas for removal of 1000 mg/L of heavy metals, the dose of ferrous sulphate required is 12.5 to 15 gm/L. These results show that ratio Fe^{2+}: M^{2+} decreases as the concentration of M^{2+} increases in the solution.
3. An optimum pH range of 9.5 to 10.5 is suitable for retrieval.
4. Temperature of 50°C ± 1°C is appropriate for the retrieval of heavy metals as ferrite.
5. An aeration rate of 20 ml/min to 200 ml/min is optimum for retrieval of heavy metals in aqueous solutions.

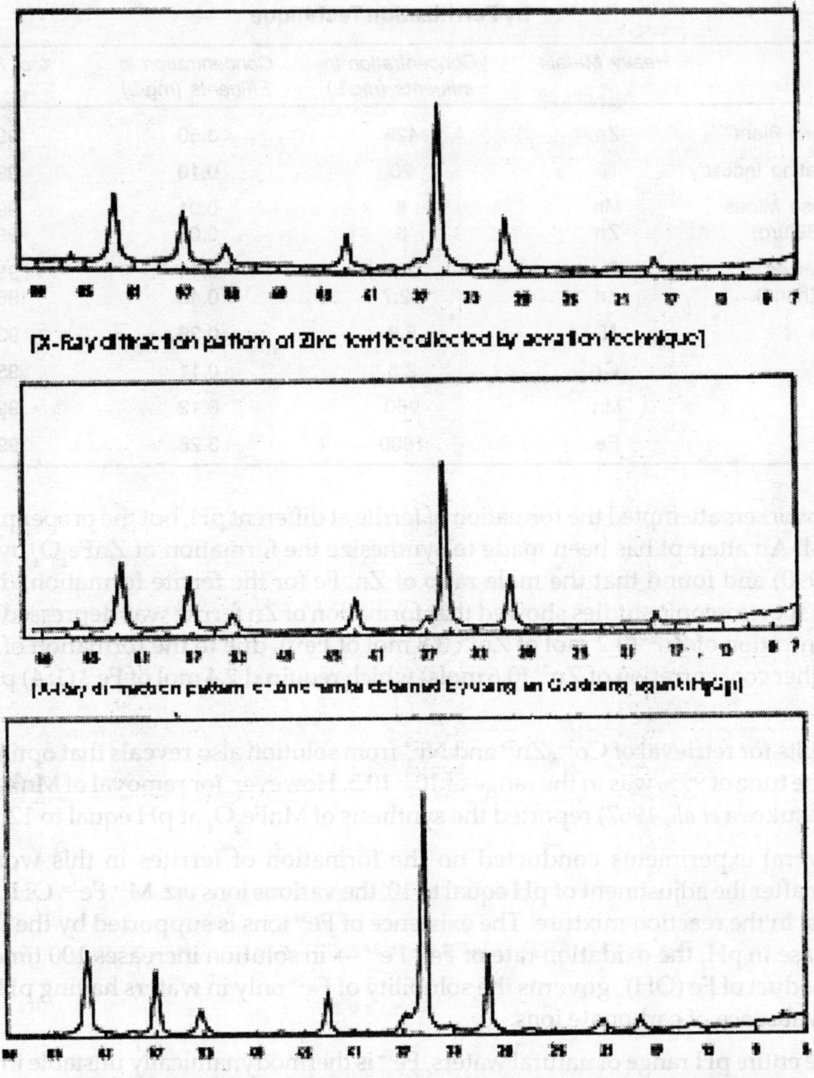

Fig. 7: X-ray Pattern of $ZnFe_2O$, Collected in Solution by Various Ways of Oxidation

6. The formation of ferrite of the type MFe_2O_4 depends on the ratio of Fe^{2+}: M^{2+} during treatment or else mixed ferrite of the type $MxFe_{3-x}O_4$ is formed.
7. The process has also been applied for treatment of wastewater from electroplating industry galvanizing industry, manganese mines and copper mines. The treatment is effective.
8. The efficiency of this process in removal of heavy metal ions from industrial effluents can be visualized from the results given in Table 3.

Table 3: Retrieval of Heavy Metals from Effluents Obtained from Industries and Mines by Ferritization Technique

Industry	Heavy Metals	Concentration in Influents (mg/L)	Concentration in Effluents (mg/L)	% of Removal
Galvanising Plant	Zn	425	3.50	99.18
Electroplating Industry	Ni	20	0.10	99.95
Manganese Mines (Dongri-Bujurg)	Mn	8	0.01	99.88
	Zn	6	0.03	99.50
Copper Mines (Malanj Khand)	Cu	211	0.14	99.93
	Zn	12.7	0.45	96.45
	Ni	5.8	0.36	93.83
	Cr	2.5	0.11	95.49
	Mn	960	0.12	99.98
	Fe	1800	3.28	99.91

Several workers attempted the formation of ferrite at different pH, but the proper pH range has not been defined. An attempt has been made to synthesize the formation of $ZnFe_2O_4$ by this technique (Kanzaki, 1990) and found that the mole ratio of Zn: Fe for the ferrite formation, depends on pH. Mössabauer spectroscopic studies showed that formation of Zn ferrite was depressed at pH >= 10 for lower concentration of Zn^{2+} (0.2 mol of Zn^{2+}: 3.8 mol of Fe^{2+}), due to the formation of α – Fe O (OH). Whereas higher concentration of Zn^{2+} (0.6 mols) which required 2.4 mol of Fe^{2+} (1: 4) preferred higher pH.

The results for retrieval of Co^{2+}, Zn^{2+} and Ni^{2+} from solution also reveals that optimum pH for the removal to the tune of 99% was in the range of 10 – 10.5. However, for removal of Mn^{2+} optimum pH is 10.5 to 11 (Yasukova *et al.*, 1967) reported the synthesis of $MnFe_2O_4$ at pH equal to 12.

The several experiments conducted on the formation of ferrites in this work reveals that immediately after the adjustment of pH equal to 10, the various ions *viz.* M^{2+} Fe^{2+}, OH^-, H^{+1} along with Fe^{3+} ions exist in the reaction mixture. The existence of Fe^{3+} ions is supported by the fact that for one unit of increase in pH, the oxidation rate of Fe^{2+} $Fe^{3+} \rightarrow$ in solution increases 100 times. At high pH, solubility product of Fe $(OH)_2$ governs the solubility of Fe^{2+} only in waters having pH in the range of 9 – 11 and in absence of carbonate ions.

Over the entire pH range of natural waters, Fe^{2+} is thermodynamically unstable in the presence of dissolved oxygen.

$$4Fe^{2+} + O_2 + 10H_2O \rightarrow 4Fe\,(OH)_3 + 8\,H^+$$

Thus, when the oxygen is supplied to the system by aeration firstly, $FeFe_2O_4$ is formed when the ratio of M^{2+}: Fe^{2+} is controlled and oxidation conditions are critical, the respective ferrite M Fe_2 O_4 is formed *i.e.* this the critical stage, at which the heavy metal ion to be retrieved from wastewater gets locked into the cubic lattice of $FeFe_2O_4$ forming M Fe_2 O_4. This accommodation of the heavy metal ion in to the cubic lattice of $FeFe_2O_4$ depends on the ionic radius of the heavy metal.

$$\frac{d[Fe^{2+}]}{dt} = K.Fe^{2+}.Po_2(OH^{2-})$$

Advantages of the Process are:

1. Various heavy metals can be removed simultaneously in one stage reaction.
2. Heavy metals in wastewaters can be reduced to an extremely low concentration, regardless of their original contents.
3. The separated heavy metals ferrite is environmentally stable and is not redissolved in water under normal conditions (Mandavgane, *et al.*, 1989, Dharmadhikari *et al.*, 2001).
4. Suspended solids (Hayashi, 1975) in the wastewaters can be removed simultaneously.
5. The existence of high concentration salts has no influence.
6. The process is easy to operate and is cost effective.
7. The process can be used economically and efficiently for any scale of operations in treatment of heavy metal laden industrial wastes.
8. In large scale treatment, "Pickle liquor" (Okuda *et al.*, 1975) which is available as another waste in industrial metal plating operations, can be used as a source of ferrous ions Fe^{2+}, thereby reducing the treatment cost further.
9. In case of small scale industries, which do not normally possess technical expertise or adequate financial resource, the room temperature ferritization process (ageing process) offers a boon.
10. What all should be done is only to mix the heavy metal-laden wastewater with adequate quantity of solution containing Fe^{2+}, ions ($FeSO_4.7H_2O$, or pickling liquor (waste)) and allow to stand for few days after adjusting the pH 9.5. The heavy metal ions present in wastewaters are converted to their respective metal ferrites, which are environmentally stable.
11. The final precipitate, ferrite, can be removed magnetically. It has many potential applications, for instance as microwave absorbing materials, in electronic appliances in antenna rods, radios, televisions, computers because of their magnetic property and as a scavenger material, as a catalyst for oxidation reactions etc. (Johnson *et al.*, 1990).
12. The "Magnetite" and other spinel ferrites formed in the process absorbs-algae, bacteria, viruses and can scavenge them out effectively (Mitchall *et al.*, 1974).
13. This technique, which is investigated in the present studies, not only removes toxic heavy metal ions from industrial wastewaters to the discharge limits stipulated by the Environmental Protection Agencies, but also recovers them in the form of commercially useful metal ferrites.
14. The sludge produced by the present process is immediately usable ferrite ferromagnetic material. This is quite an important fact, which enables conversion of waste products having very low cost to such useful materials as Ferro fluid and microwave absorption sheet.

Acknowledgement

Dr. D.M. Dharmadhikari is thankful to Director, NEERI for giving permission to publish this work.

REFERENCES

1. Brooks C.S., 1991, Metal Recovery from Industrial Waste, Lewis publishers, Chelsea; Mich, USA.
2. Dharmadhikari D.M., 1978, Structural and Electrical Properties of Franklinite ($ZnFe_2O_4$) and Related Oxides, Ph.D Thesis, Nagpur University, Nagpur (India).

3. Dharmadhikari D.M., Dhopte S.M., Mandaokar S.S. and Dara S.S., 2001, Recycling and Reuse of Heavy Metals in Wastewater in the form of Environmentally Stable Ferrites, *J. IAEM*, 28, 106-108
4. Dharmadhikari D.M., Kulkarni D.K. and Deshpande V.V., 1986, Structural and Electrical properties of scandium substitute zinc ferrites, *Ind. Journal of Chemistry*, 25A (2) 175-177
5. Gorter, E.W., 1954, *Philips Res. Rept*. 9, 295.
6. Hayashi Saburo, 1975, Heavy metal treatment by ferritization, *Chemical Economy and Engineering Review*, 7 (5), 32-7, 43.
7. Johnson R., Kumar R. and Dharmadhikari D.M., 1990, Catalytic Oxidation of Carbon Monoxide and Hydrocarbons on Substituted Zinc Ferrites, *J. of Institute of Public Health Engineers (India)*, 90 (4), 1 – 7.
8. Kazanki Tadao (Iwaki Meisei Univ, Iwaki Japan 970), 1990, Mossbauer spectra at 77° K of products formed during transformation of Iron $^{(II)}$ hydroxide containing Zn $^{(II)}$ ion to Zn-bearing ferrite in aqueous suspension by air oxidation. *J. Electron, Spectros, Relat. Phenomenon*, 50 (4), 309-21 Eng.
9. Krishnan E.R.1993, Recovery of Metals from sludges and wastewaters, *Tech. Report*, Noyes Data Corp., Park Ridge N.J., USA.
10. Mandaokar S.S., Dharmadhikari D.M. and Dara S.S., 1994, Retrieval of Heavy Metal Ions from Solution via Ferritization, *Environmental Pollution (U.K.)* 83 (4), 277-282.
11. Mandavgane S.K., Dharmadhikari D.M. and Dara S.S., 1989, *J. IAEM*, 16, 203 – 204.
12. Mitchall R, Bitton G, De Latour and Maxwelle, 1974, *Proceedings 7th Int1 Conf.* on Water Pollution Research.
13. National report, Hazardous Waste Generation and Treatment, 1991, U.S. EPA, Report no. EP A530 SW 4160 Washington D.C.
14. Okuda T, Sugano I, Tsuji T., 1975, Removal of heavy metals from wastewater by ferrite Co-precipitation, Filtr-sep, 12 (5), 472-478, Eng.
15. Okuda T., Sugano, I., Tsuji, T., 1975, Removal of Heavy Metals from Wastewaters by Ferrite Co-precipitation. *Report*, Central Research Lab. 57, Nippon Electric Co. Ltd., Japan.
16. Peters R.W., Ku. Y. and Bhattacharyya D., 1985, Separation of Heavy Metals and other Trace Contaminants, AICHE symposium series 243, 81, New York, 165-203.
17. Sharma V.N., 1965, *J. Appl Phys*. 36, 108-114.
18. Smit J. and Win, 1945, Advance Electron Phy. *J. HPJ.*, 6, 83-87.
19. Technical report, 1991, Resource Recovery Project Technology Demonstrations, series no. 11, REP # 48255, MSE. Inc. Butte, USA.
20. Yasuoka H. and Hirai A., 1967, NMR Determination of metal ion distribution in Manganese ferrite prepared from Aqueous solution. *J. Physical Soc. of Jpn.*, 22 (1), 174-180.

49

Recovery of Quality Water from Tannery Wastewater Adopting Photochemical Oxidation and Membrane System

S. RAJAMANI, E. RAVINDRANATH, R. SUTHANTHARARAJAN, B. UMAMAHESWARI & K. CHITRA

Department of Environmental Technology, Central Leather Research Institute, Adyar, Chennai – 600 020

ABSTRACT

Conventional effluent treatment system with physio-chemical and biological components could remove the organic and inorganic pollutants such as Suspended Solids (SS), Biological Oxygen Demand (BOD), Chemical Oxygen Demand (COD), heavy metals such as chromium etc., but not the Total Dissolved Solids (TDS) and chloride. Due to the inherent nature, the effluent from tanneries contains high TDS and other non-degradable matter. The ions that contribute towards TDS are sodium, chloride, and sulphate etc. Due to inadequate water source and poor quality of the available ground water, many tanneries in South India transport water through tankers at a high cost. The treated effluent from Effluent Treatment Plant (ETP) is not suitable for reuse in the process due to the presence of high TDS. Therefore to recover water from treated effluent from conventional treatment plants and reuse it in tanning process a pilot integrated tertiary treatment system including photochemical oxidation, two stage Reverse

Osmosis (RO) system etc. has been designed and installed for a capacity of 20,000 litres/day, in one of the tanneries in Melvisharam, Ranipet, India. The quality of recovered water meets the drinking water standard and is used in wet finishing operations in the tannery. This integrated plant, which is first of its kind in the world, is implemented with technical and financial support of United Nations Industrial Development Organisation (UNIDO), Chennai.

Keywords: Photochemical Oxidation, Reverse Osmosis, Water Recovery

Introduction

Wastewater from the tanneries is treated by physio-chemical and biological treatment methods for removal of organic and inorganic pollutants. The treated wastewater from effluent treatment plant contain high amount of TDS in the range of 5000 – 12000 mg/L. The ions that contribute towards TDS are Sodium, calcium, chloride and sulfate. The treated effluent from conventional physio-chemical and biological treatment water is not suitable for reuse in the leather finishing process due to the presence of high TDS with residual COD and other non-degradable constituents. Therefore to recover water from treated tannery wastewater and reuse it in tanning process a pilot scale RO system integrated with tertiary treatment units such as photo chemical oxidation, micro filter etc. was installed, operated and studied for a period of nearly two years in one of the commercial tanneries in Melvisharam, Ranipet, India.

Objective of the Study

The objective of the study is to investigate the suitability of applying Reverse Osmosis (RO) system with other tertiary treatment units such as photochemical oxidation, micro filter etc. required prior to RO unit for the removal of residual suspended organic and inorganic matter, for recovery and reuse of quality water in tanning process from biologically treated tannery effluent.

Unit Operations

The important units of the selected system are collection tank (CT), pressure sand filter (PSF), photo-chemical oxidizer (PCO), activated carbon filter (ACF), softener, microfilter, dosing units, Intermediate storage tanks and two stage RO units. The schematic process flow diagram is shown in Fig. 1. The integrated system with treatment capacity of 20,000 litres per day was designed to operate with the influent characteristics is given in Table 1.

Table 1: Basic Design Parameters of Integrated Membrane System

Sl. No.	*Parameters*	*Value*
1.	pH	7.0 – 9.0
2.	COD	< 400 ppm
3.	Suspended Solids	< 150 ppm
4.	Total Dissolved Solids	5600 – 8000 ppm
5.	Silica	< 20 ppm
6.	Permeate recovery	80-85 %
7.	Total Dissolved Solids in permeate	< 700

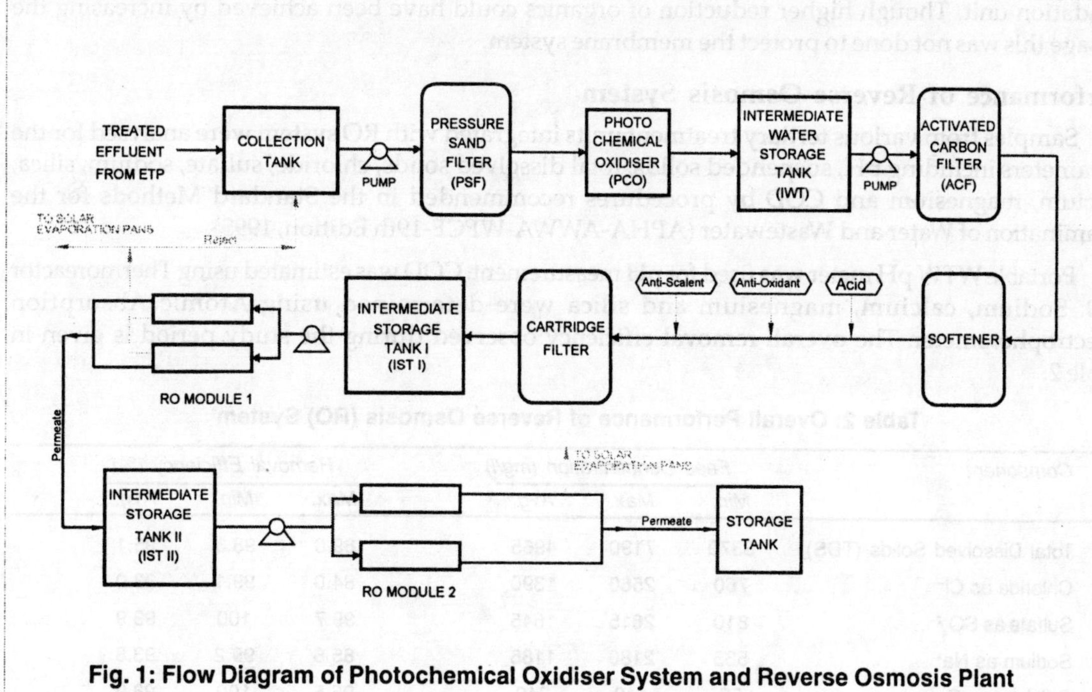

Fig. 1: Flow Diagram of Photochemical Oxidiser System and Reverse Osmosis Plant

Operation of the System

Treated effluent from the conventional effluent treatment plant was taken up for the study. Wastewater from the secondary settling tank is pumped to the collection tank of the RO system. Wastewater from the collection tank is pumped to Intermediate Water Tank (IWT) through Pressure Sand Filtration and Photo Chemical Oxidation. From IWT wastewater is pumped into Intermediate Storage Tank (IST-I) through Activated Carbon Filtration, softener and cartridge filter. Anti-scalent, anti-oxidant and acid were added to softener using online dosing pumps before cartridge filter.

All the units were operated at ambient temperature of 28 – 35°C on batch wise upto two shifts in a day. Wastewater from IST-I is pumped using high pressure pump with a hydraulic pressure of 10-14 psi to RO module I. Effluent from IST-II is pumped using high pressure pump with a hydraulic pressure of 12-16 psi to RO module II. Part of the reject-I from RO module I continuously flows to collection tank. 100% Reject-II from RO module II is mixed with wastewater from PCO in IWT.

Performance of Photochemical Oxidation System

Prior to Photo Chemical oxidation system, wastewater was passed through pressure sand filter for removal of suspended solids. Photo Chemical Oxidizer system with UV lamp and hydrogen peroxide dosing was installed for destruction of organics prior to first stage membrane system. The dosing of hydrogen peroxide was monitored closely to avoid excess free oxygen radical getting into the subsequent treatment units, especially the first stage membrane system. From the analytical results, it was observed that COD concentration reduced from 200 mg/L to 100 mg/L after Photo Chemical

Oxidation unit. Though higher reduction of organics could have been achieved by increasing the dosage this was not done to protect the membrane system.

Performance of Reverse Osmosis System

Samples from various tertiary treatment units integrated with RO system were analysed for the parameters including pH, suspended solids, total dissolved solids, chloride, sulfate, sodium, silica, calcium, magnesium and COD by procedures recommended in the Standard Methods for the Examination of Water and Wastewater (APHA-AWWA-WPCF-19th Edition, 1995).

Portable WTW pH meter was used for pH measurement. COD was estimated using Thermoreactor 300. Sodium, calcium, magnesium and silica were determined using Atomic Absorption Spectrophotometer. The overall removal efficiency observed during the study period is given in Table 2.

Table 2: Overall Performance of Reverse Osmosis (RO) System

Component	*Feed Concentration (mg/l)*			*Removal Efficiency (%)*		
	Min.	*Max.*	*Avg.*	*Max.*	*Min.*	*Avg.*
Total Dissolved Solids (TDS)	3370	7190	4965	89.0	98.3	95.1
Chloride as Cl^-	760	2560	1390	84.0	99.1	93.0
Sulfate as SO_4^{2-}	810	2615	1645	99.7	100	99.9
Sodium as Na^+	535	2180	1185	85.6	99.2	93.8
Calcium as Ca^{++}	52	440	240	96.5	100	98.9
Magnesium as Mg^{++}	7	38	23	90.9	100	97.0
Silica as SiO_2	5	25	12	81.2	100	93.4

Part of the reject from RO Module I was discharged into solar evaporation pan. The flow was controlled manually. The characteristics of concentrate discharged into solar pan is given in Table 3.

Table 3: Characteristics of Concentrate

Parameters	*Reject*		
	Min.	*Max.*	*Avg.*
Total Dissolved Solids (TDS)	7960	27830	12830
Chloride as Cl^-	1540	4285	2670
Sulfate as SO_4^{2-}	1855	13640	6060
Sodium as Na^+	1100	5920	2750
Calcium as Ca^{++}	8	950	360
Magnesium as Mg^{++}	6	285	86
Silica as SiO_2	6	100	33

Note: All values are expressed in mg/L.

Increase in concentrate volume was observed due to fouling in the RO module and this resulted in increasing the hydraulic pressure of RO module I. This resulted in slow filling up of permeate in IST II.

Water Recovery

The Total Dissolved Solids concentrations in the treated tannery effluent during the period was in the range of 3300-7200 mg/L. TDS concentration in the permeate was about 88-440 mg/L. The average recovery of permeate was 60%. Characteristics of permeate collected during the study period is given in Table 4.

Table 4: Characteristics of Permeate

Parameters	*Permeate Characteristics*		
	Min.	*Max.*	*Avg.*
Total Dissolved Solids (TDS)	88	440	225
Chloride as Cl^-	17	190	82
Sulfate as SO_4^{2-}	BDL	8.2	2
Sodium as Na^+	15	122	60
Calcium as Ca^{++}	BDL	11	3
Magnesium as Mg^{++}	BDL	1.1	0.5
Silica as SiO_2	BDL	3.0	1.0

BDL: Below Detectable Limit;
All values are expressed in mg/L

Recovered water from RO module II is collected in syntex tank and reused in wet finishing operations in tannery.

Cleaning of Tertiary Treatment Units

At the end of the day's operation, almost all the units in the RO system were washed. PSF and ACF are back washed with the wastewater in IWT tank to remove the suspended solids. Both the RO modules are flushed with permeate water. Softener was regenerated with about 15-20 kg of commercial sodium chloride everyday. Brine liquor from softener after regeneration is discharged into solar evaporation pan.

Cost Analysis

Total capital cost of the system: Rs. 1.2 millions (or say 30,000 USD)

Fixed operating cost per 1000m^3 of water: Rs. 20,000 (or say 450 USD)

Variable operating cost per 1000m^3 of water: Rs. 20,000 (or say 450 USD)

Total recovery cost per 1000m^3 of water: Rs. 40,000 (or say 900 USD)

Conclusions

Ionic removal efficiency of reverse osmosis membranes from treated tannery wastewater is more than 93% and the recovered water met the drinking water standards. Average water recovery limited to 60% because of the dye present in the treated tannery wastewater and membrane fouling. Recovered water was used for wet finishing operations in tannery regularly.

Acknowledgements

Acknowledgement is made to the United Nations Industrial Development Organizations (UNIDO) for the technical and financial support. Contributions of M/s A.T.H Leder Fabrik Tannery, Melvisharam towards implementation and support for the operation of the integrated RO system are greatly appreciated.

REFERENCES

1. Ravindranath, E., R. Suthanthararajan, B. Umamaheswari, K. Chitra and S. Rajamani, 2000, "Recovery and reuse of tannery wastewater adopting commercial scale RO system" in *Third Annual Conference of the Indian Environmental Association, Enviro Vision 2000*, on Industrial Wastewater Recycle and Reuse at Mumbai, May 5-6.

2. Rajamani, S., Jakov Buljan, 1996, UNIDO, Technology Package-System for recovery and reuse of chromium from spent tanning liquor using magnesium oxide and sulphuric acid TECHPACK/ UNIDO/RePO/1.

3. Shweta Singh, Rajamani, S. & Ramasami, T., 2001Sustainability of Leather Tanning Small and Medium Enterprises (SMEs) through Environmental Compliance in Developing Countries-XXVI IULTCS Congress, Cape Town, 7-10 March.

4. Sampling, Testing and Evaluation of the pilot Membrane System (RO) unit for processing treated tannery effluent at A.T.H Leder Fabrik, Melvisharam, March 2000.

5. Indian Journal of Environmental Health, Vol. 41, October 1999.

6. Membrane Filtration, R G Gutman, 1987.

50

Development of Solid Based Activated Carbons: An Approach Towards Waste Management Practices

D.S. RAMTEKE, M. MUKHERJEE, R. MALIK & R. SARIN

National Environmental Engineering Research Institute, Nehru Marg, Nagpur - 20

ABSTRACT

Waste generation is an inevitable by-product of the rapid and enhanced industrialization. These waste cause serious and irreparable damage to the environment and are not amenable to conventional method of waste treatment. Recycling is perhaps the most positively perceived and doable of all the sustainable waste management practices. Moreover, the choice between recycle, recovery and reuse of valuable materials from waste and disposal of waste seems to depend mainly on two basic factors, economics and technology. Economic is the main constraint to the adoption of recycle or recovery techniques for waste by-products. Improving the available technology, however, can probably reduce the high costs of recycle and recovery techniques. Such an attempt has been made to develop different low cost adsorbents from industrial waste materials like paper mill sludge, coke fines, water hyacinth and walnut shell. The total work is focused at the potential of waste materials for the development of valuable product like activated carbon, which will be cost effective and better replacement of the commercially available activated carbon besides its future and expansion as valuable waste elimination method. Different processes have been developed based on raw waste material being used for the preparation of activated carbon. Activation conditions were optimised to increase the pore size structure and the products obtained were physico-chemically characterised and applied in the wastewater treatment process. An attempt has also been made to replace the applied carbonisation technology with the non-conventional microwave technology to prove its better future prospects in the carbon manufacturing process.

Keywords: *Solid waste, carbonisation, activated carbon, adsorption, wastewater treatment*

Introduction

Sustainable economic growth in this century requires development of new industrial technologies in harmony with the preservation of global environment. In this context, an approach was made towards solid waste management in terms of waste minimization, reuse and recycling. Utilization of waste material as an alternate raw material for the production of better product by adopting new technology has an immense importance in the modern era. Keeping in view, the Pyrolysis technique (carbonisation) was adopted for the development of activated carbon from different industrial and solid wastes like paper mill sludge, coke fines, water hyacinth and walnut shell.

Potential of Waste Material Generation

Paper Mill Sludge

Pulp and paper manufacturing is one of the largest industries in the world and has shown rapid and steady growth. In India there are total 34 mills with a capacity of 1,551,600 TPA (Wanumanulu *et al.*, , 1975). The current productions of all the mills are around 850,000 TPA as against the installed capacity of 1080,000 TPA (DGTD, 1078). During the manufacturing processes large amount of primary and secondary sludge is generated (Subramanyam *et al.*, , 1972) accounting 0.5 tonne of dry sludge per tonne of paper (Mohandas Rao *et al.*, 1972). The present practices is dewatering of sludge followed by land disposal or incineration and are not cost effective. The sludge is fibrous and cellulose in nature.

Coke Fines

Coke fines are waste product generated during coal (caking type) carbonisation for making coke, which is used in the steel plant. Quantity of coke fines produce at coal carbonization industry is roughly about 40 MTPD. Since the base material is coal it constitutes a good carbonaceous material containing high percentage of carbon (80-90%). The present disposal method is to use as fuel or dispose off on roadsides. It could be better used as a raw material for the preparation of activated carbon.

Water Hyacinth

Water Hyacinth widespread in the tropical countries is considered as one of the most obnoxious weeds creating long time problem for eradication. It is a tropical plant of exquisite beauty possessing good potential for reduction of pollution load of sewage, industrial effluents and also a source of energy, food and livestock food. In India, it grows profusely in the northwestern region and regarded as a menace in agricultural production. Although the present practice is to use as local manure and biogas production, the majority of these are wasted. It is carbonaceous in nature containing 60-70 % of carbon on dry basis.

Walnut Shell

Walnut shell is a solid waste material produced from dry fruit processing industries creating lots of disposal problems in Gulf countries. The present production rate in India and world wise is 14,000 MTPA and 676,260 MTPA respectively. The present disposal practice is to dump in a wasteland side creating nuisance conditions due to microbial degradation or use as a fuel. It is also a carbonaceous material containing 60-70 % carbon and could be used as an alternate raw material for the production of better product.

Modern trends in the waste management are to salvage and recycle the wastes. If this aspect is to be imbibed in solid waste management practice, then pyrolysis (carbonisation) assures significance. Through this technique a carbonaceous solid waste or industrial waste materials could be converted into valuable product like activated carbons. In developing countries like India, industries can not afford to use conventional wastewater treatment chemicals *viz.* alum, ferric chloride, polymer flocculants and commercial grade activated carbons because they are not cost effective. Inexpensive and more easily available adsorbents would make the removal of pollutants an economically viable alternative. The above solid/industrial waste materials being carbonaceous in nature could be used as alternate raw materials in place of costly raw materials being used for the preparation of activated carbon. In view of this an attempt was made to develop the process for the preparation of activated carbon from above said raw materials and evaluate their efficiency in wastewater treatment processes. A comparison was made with the commercial grade granular activated carbon (CGAC), which enable us to study its use in large scale.

Material and Methods

Process Developed for Activated Carbons

The waste materials selected are carbonaceous in nature. Carbonization of such waste materials under controlled conditions may prove to be an economic method of disposal without creating any environmental problem. Carbonization is a pyrolytic method. It usually refers to chemical decomposition of organic materials heated in an environment having insufficient oxygen for combustion (NTIS Report, 1973). Operating conditions for pyrolysis depends on the ratio of flow and feed material. It is carried out in two different sequences known as Low Temperature Carbonization (LTC) and High Temperature Carbonization (HTC) in the range 250-600°C and 600-1000°C respectively. Low temperature carbonisation produces more char and less gaseous products, while high temperature carbonization produces less residue and large quantity of fuel gases. A few intermediate products (liquid) are also obtained in both the processes.

Pyrolysis Unit

The pyrolysis unit (Fig. 1) consists of a stainless steel cylindrical reactor having 10.2 cm in diameter and 91.5 cm in length. It is wounded accurately with heating material like nicrome wire and insulated by asbestos powder. For temperature measurement, that was connected with two thermo couple Ni-Cr/Ni-Al as a multiple temperature indicator. The reactor was connected at the top through heat resistant tubing to a water condenser. A tar and oil tap was kept beneath the condenser to collect the condensate. Further, it was connected with three glass tubes connected in series and kept in freezing mixture to collect the incondensable products. These are further connected with the water scrubber to wash the generated gases and then measured with the gas meter attached to the system. The complete unit was kept airtight during experiment.

Process

Different raw waste based activated carbons were prepared on laboratory scale. The process details are as follows:

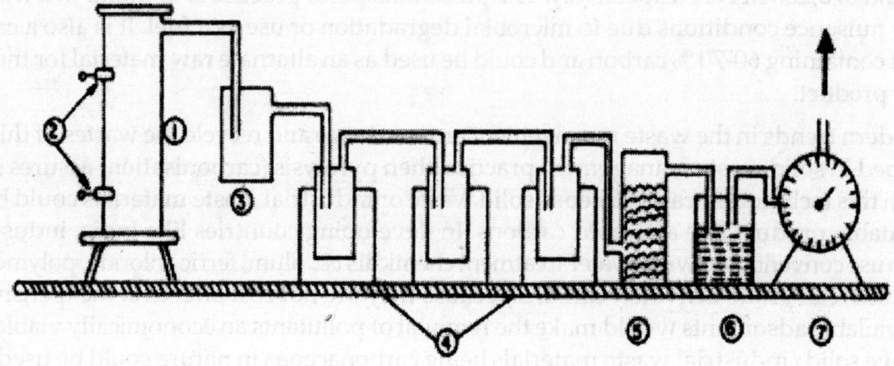

1. Reactor
2. Thermocouple
3. Tar Trop
4. Receives Kept in Freezing Mixture
5. Scrubber
6. Water Trap
7. Gas Meter

Fig. 1: Pyrolysis Unit

Activated Carbon from Paper Mill Sludge

One kg. of the uniformly reduced material was taken for pre-treatment. It was mixed thoroughly with0.1 per cent zinc chloride solution. The impregnation material was kept for soaking for 24 hours with regular mixing and then dried in an hot air oven at 110°C for 24 hours. The dried impregnated material was then fed in to a closed airtight reactor of the pyrolysis unit and heated to 700°C. The reactor temperature was maintained at 700°C for one hour. After one hour, the reactor was allowed to bring at room temperature. The residue was removed from the reactor and weighed. The liquid obtained was thoroughly mixed and subjected to the fractional distillation to obtain different organic fractions. The gases escaping out of the wet test gas meter were collected and analysed for the various constituents.

The product (residue) was further washed with water to remove the excess of zinc chloride used during pre-treatment. The wash water was collected for the recovery of zinc chloride for reuse. It was further washed with hot hydrochloric acid (1+1) for the removal of traces of zinc chloride and the mineral ingredients. After the acid treatment, the material was again washed with hot distilled waster until the acidity and the chloride content was completely removed.

The washed product was then dried in hot air oven at 110°C for 24 hours. This was further sieved through 325-mesh sieve and stored in a plastic bottle for further use. The flow sheet diagram for preparation of activated carbon is presented in Fig. 2.

Activated Carbon from Coke Fines

Coke fines based activated carbon is prepared by adopting single carbonisation process followed

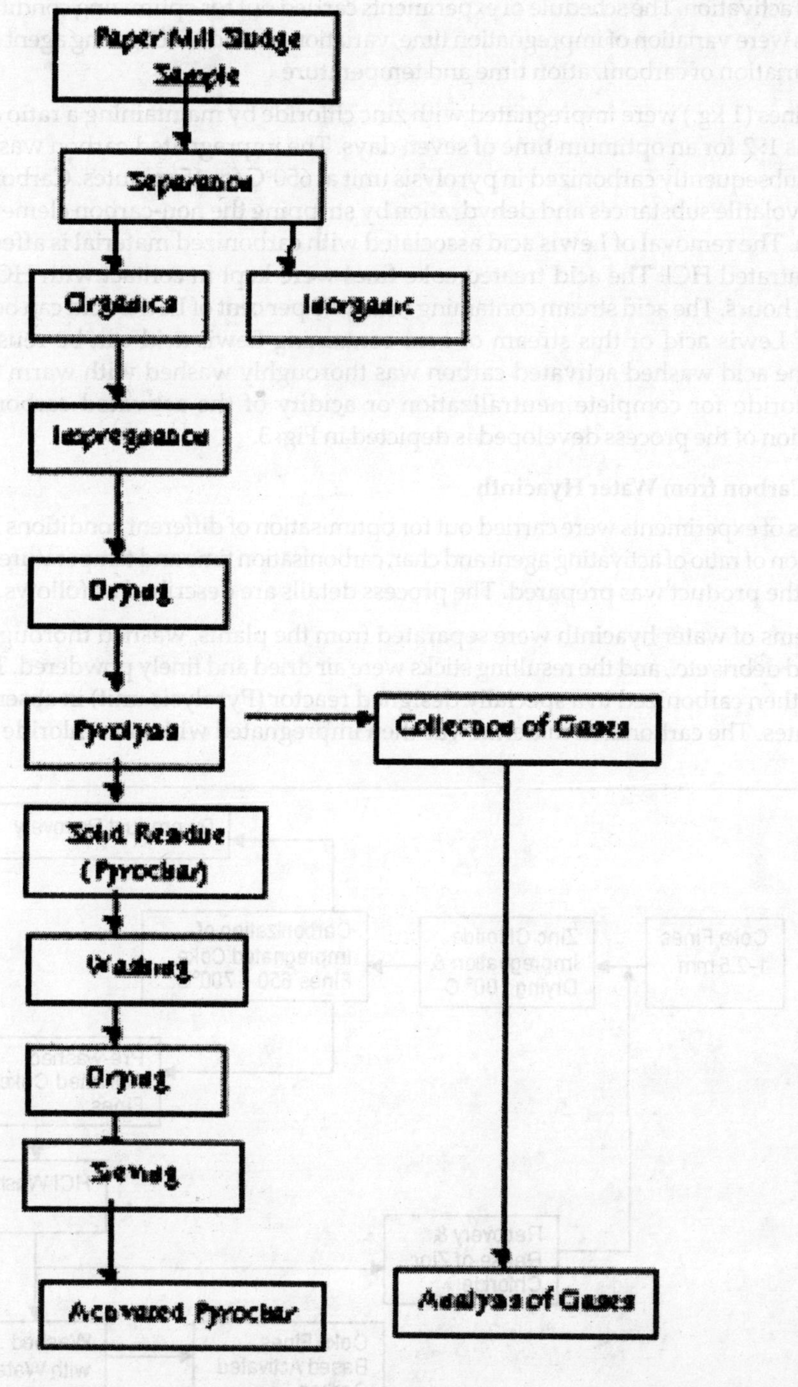

Fig. 2: Flow Sheet Diagram for Preparation of Paper Mill Sludge Based Activated Pyrochar

by chemical activation. The schedule of experiments carried out for optimizing conditions for activation of coke fines were variation of impregnation time, variation of ratio of activating agent and carbonaceous material, variation of carbonization time and temperature.

Coke fines (1 kg.) were impregnated with zinc chloride by maintaining a ratio of Lewis acid and coke fines as 1: 2 for an optimum time of seven days. The impregnated carbon was dried at 100°C to 120°C and subsequently carbonized in pyrolysis unit at 650°C for 15 minutes. Carbonisation results in removal of volatile substances and dehydration by stripping the non-carbon elements like hydrogen and oxygen. The removal of Lewis acid associated with carbonized material is affected by treating it with concentrated HCl. The acid treated coke fines were kept in contact with HCl for a minimum period of 24 hours. The acid stream containing nearly 90 per cent of Lewis acid can be concentrated for recovery of Lewis acid or this stream of acid containing Lewis acid can be reused for activation purpose. The acid washed activated carbon was thoroughly washed with warm distilled water to remove chloride for complete neutralization or acidity of the activated carbon. The schematic representation of the process developed is depicted in Fig. 3.

Activated Carbon from Water Hyacinth

A series of experiments were carried out for optimisation of different conditions like impregnation time, variation of ratio of activating agent and char, carbonisation time and temperature. Under optimised conditions the product was prepared. The process details are described as follows.

The stems of water hyacinth were separated from the plants, washed thoroughly with water to remove mud debris etc., and the resulting sticks were air dried and finely powdered. The dried material (1 kg.) was then carbonized in a specially designed reactor (Pyrolysis unit) in absence of air at 400°C for 30 minutes. The carbonized material was then impregnated with zinc chloride with a optimised

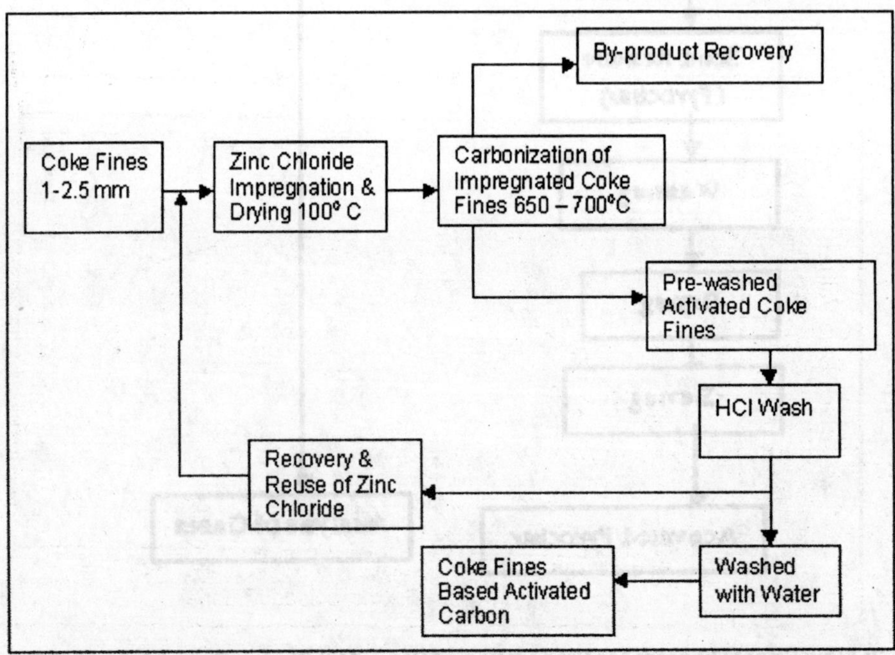

Fig. 3: Schematic Representation for Preparation of Activated Coke Fines

ratio of 1.33 for zinc chloride to carbonised char. The impregnated material was dried at 100°C and subsequently carbonized at 700°C for 30 minutes. The product was treated with hydrochloric acid (1+1) for the removal of excess of zinc chloride. The acid stream containing nearly 92 per cent of zinc chloride can be concentrated for recovery of zinc chloride or this stream of acid containing $ZnCl_2$ can be reused for activation purpose. The acid washed product was thoroughly washed with hot distilled water to remove acidity and chlorides.

Walnut Shell based Activated Carbon

The process developed for the preparation of activated carbon is based on double carbonisation technique. The shredded and sized walnut shell (1 kg.) was carbonised at 400°C for 30 minutes in a Pyrolysis unit to prepare char. Experiments were conducted in the activation process for optimising different conditions for the preparation of better product.

Optimisation of Different Conditions in Activation

Mole Ratio of $ZnCl_2$/carbonised Char (Impregnation)

As a pre-requisite for activation, the ratio of $ZnCl_2$ to char was varied and carbonised at temperature 700°C for half an hour. For each variable ratio the yield of the activated product was noted and the adsorption efficiency in terms of Methylene blue number was determined.

The results of the variation of the ratio of $ZnCl_2$/carbonised char with respect to Methylene blue value of the active carbon produced is graphically represented in Fig. 4. It was observed that the adsorptive power of the carbon produced increased with the increase in $ZnCl_2$ concentrations and reached a maximum at $ZnCl_2$/carbonised char ratio of 1.33. The yield of activated carbon decreased with temperature of activation. It was further observed that Methylene blue values increased with increase in temperature and reached to a maximum 278 at 700°C.

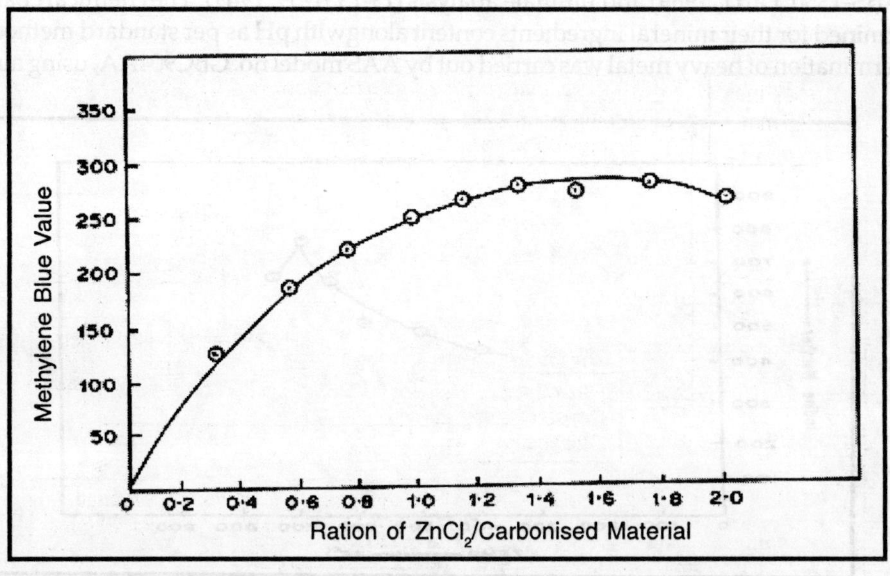

Fig. 4: Effect of Variation of Ration of $ZnCl_2$/Carbonised Material

Activation Temperature

At the mole ratio of 1.33 ($ZnCl_2$/carbonised char) and duration of heating (30 minutes) samples of active carbons were produced at various temperatures (400-750°C) and evaluated for their Iodine numbers and graphically represented in Fig. 5. It was observed that the adsorptive activity of the produced carbon increased gradually upto 650°C and reached a maximum at 700°C. The yield of activated carbon produced decreased with increase in temperature of activation in the range of 500-750°C.

Carbonisation Duration

Keeping the ratio of $ZnCl_2$/carbonised char at 1.33 and temperature at 700°C, samples of activated carbon were produced at different time intervals and results are presented in Fig. 6. It was observed that with increasing time of activation, the iodine number increased sharply from 7-25 minutes and then slowly upto 30 minutes. The yield of the carbon produced decreased considerably over the whole range of activation period.

From the experiments conducted, following conditions were optimised.

1. Ratio of $ZnCl_2$/carbonised char: 1.33
2. Temperature of activation: 700°C
3. Time of activation: 0.5 h

The schematic representation of the process developed for water hyacinth and walnut shell based activated carbons is shown in Figure 7.

Characterisation of the Product

The physical characteristics of the activated carbons were determined by estimating their particle size (USEPA, 1973), bulk density (Hassler, 1974) and surface area (Giles *et al.*, 1969) followed by proximate (IS-1350, Part1, 1969) and ultimate analysis (IS-1351-59, 1969). The chemical characteristics were determined for their mineral ingredients content alongwith pH as per standard methods (APHA, 1989). Determination of heavy metal was carried out by AAS model no. GBC904AA, using air-acetylene flame.

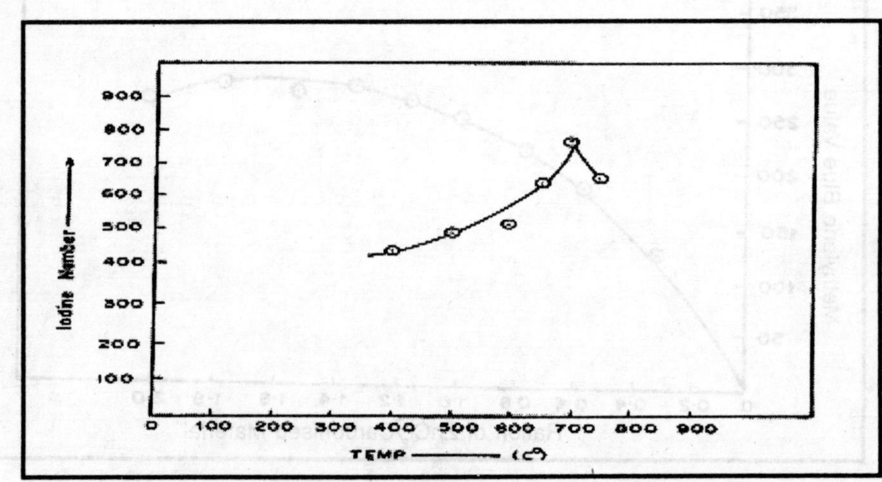

Fig. 5: Effect of Variation of Temperature

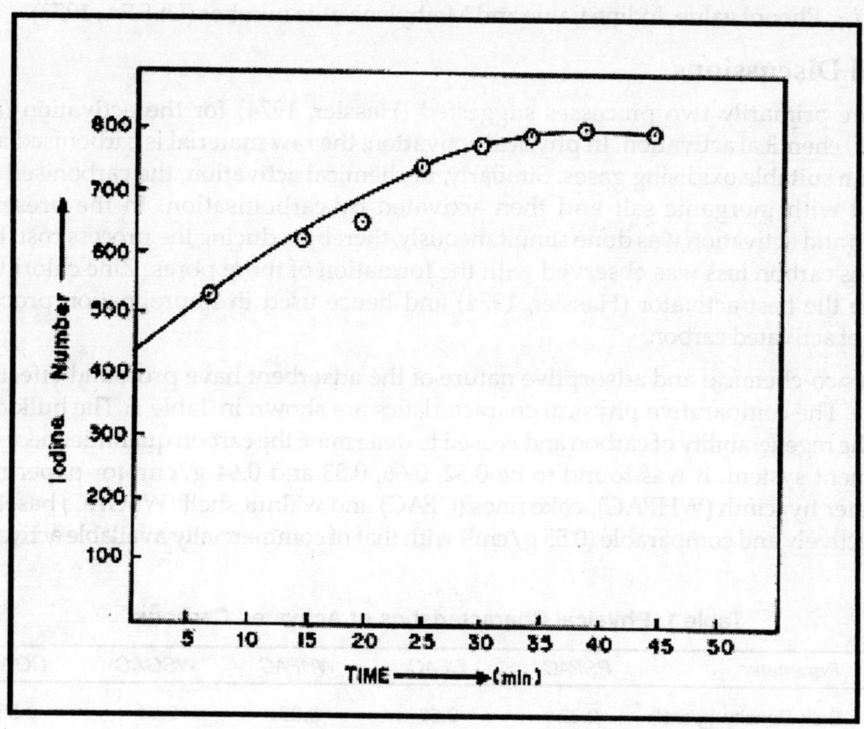

Fig. 6: Effect of Variation of Time on Activation

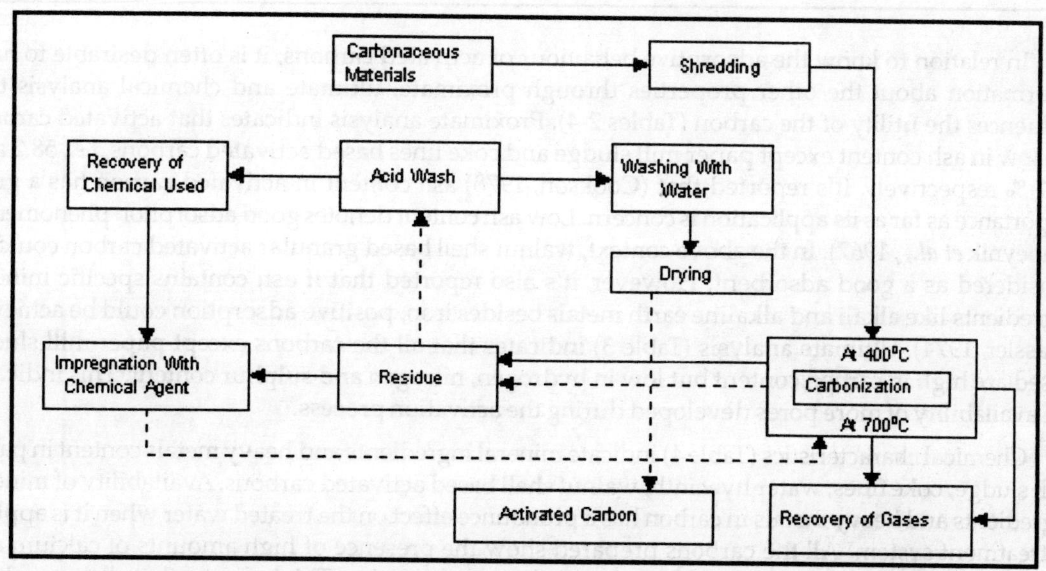

Fig. 7: Schematic Representation for Preparation of Activated Carbon from Water Hyacinth and Walnut Shell

All the products were characterised for their adsorptive characteristics in terms of quality control parameters, *i.e.*, Phenol value, Iodine value and Methylene blue number (USEPA, 1973).

Result and Discussions

There are primarily two processes suggested (Hassler, 1974) for the activation *(i)* physical activation, *(ii)* chemical activation. In physical activation, the raw material is carbonised and further activated with suitable oxidising gases. Similarly, in chemical activation, the carbonised material is impregnated with inorganic salt and then activated by carbonisation. In the present process carbonisation and activation was done simultaneously, thereby reducing the process cost. In chemical activation less carbon loss was observed with the formation of more pores. Zinc chloride has been proved to be the best activator (Hassler, 1974) and hence used in impregnation process for the preparation of activated carbon.

The physico-chemical and adsorptive nature of the adsorbent have profound effect on overall adsorbability. The comparative physical characteristics are shown in Table 1. The bulk density is a measure of the regenerability of carbon and is used to determine the carbon quantities necessary to use in the treatment system. It was found to be 0.32, 0.66, 0.33 and 0.64 g/cm³ for paper mill sludge (PSPAC), water hyacinth (WHPAC), coke fines (CFAC) and walnut shell (WSGAC) based activated carbon respectively and comparable (0.55 g/cm³) with that of commercially available activated carbon (CGAC).

Table 1: Physical Characteristics of Activated Carbons

Sl. No.	*Parameter*	*PSPAC*	*CFAC*	*WHPAC*	*WSGAC*	*CGAC*
1.	Bulk Density (g/cm³)	0.32	0.66	0.33	0.64	0.55
2.	Particle size (mm)	0.044	1-2	0.044	0.80	0.60
3.	Surface area (m²/g)	478	526	1247	750	900

In relation to know the adsorptive behaviour of activated carbons, it is often desirable to have information about the other properties through proximate, ultimate and chemical analysis that influences the utility of the carbon (Tables 2-4). Proximate analysis indicates that activated carbons are low in ash content except paper mill sludge and coke fines based activated carbons, *i.e.*, 58.2 and 29.7 % respectively. It's reported that (Cookson, 1978) ash content in activated carbon has a great importance as far as its application is concern. Low ash content denotes good adsorption phenomenon (Snoeynik *et al.*, , 1967). In the above context, walnut shell based granular activated carbon could be considered as a good adsorbent. However, it's also reported that if ash contains specific mineral ingredients like alkali and alkaline earth metals besides iron, positive adsorption could be achieved (Hassler, 1974). Ultimate analysis (Table 3) indicates that all the carbons except paper mill sludge based are high in carbon content but low in hydrogen, nitrogen and sulphur content. This indicates the availability of more pores developed during the activation process.

Chemical characteristics (Table 4) indicate mineral ingredients and heavy metals content in paper mill sludge, coke fines, water hyacinth, walnut shell based activated carbons. Availability of mineral ingredients and heavy metals in carbon has a pronounce effect on the treated water when it is applied in treatment system. All the carbons prepared show the presence of high amounts of calcium and magnesium, however heavy metals are present in low concentration. This indicates that all the products can be successfully applied to the wastewater treatment, as there are negligible chances to leach out any heavy metal from the product to the treated water.

Table 2: Proximate Analysis of Waste Based Activated Carbons

Sl. No.	*Parameter*	*Value (%)*				
		PSPAC	*CFAC*	*WHPAC*	*WSGAC*	*CGAC*
1.	Moisture	1.0	8.4	2.6	8.0	1.8
2.	Ash	58.2	29.7	8.8	5.5	4.3
3.	Volatile Matter	9.3	4.3	2.1	15.9	7.2
4.	Fixed Carbon	31.3	57.6	86.5	70.6	86.7

Table 3: Ultimate Analysis of Waste Based Activated Carbons

Sl. No.	*Parameter*	*Value (%)*				
		PSPAC	*CFAC*	*WHPAC*	*WSGAC*	*CGAC*
1.	Carbon	29.92	54.80	79.80	79.20	92.30
2.	Hydrogen	1.15	0.86	2.80	4.20	1.12
3.	Nitrogen	Traces	0.98	3.60	0.16	Traces
4.	Sulphur	Traces	0.23	2.40	0.54	Traces

Table 4: Characteristics of Water and Acid Soluble of Waste Based Activated Carbons

Sr. No.	*Parameter*	*Water extract*					*Acid extract*				
		PSPAC	*CFAC*	*WHPAC*	*WSGAC*	*CGAC*	*PSPAC*	*CFAC*	*WHPAC*	*WSGAC*	*CGAC*
1	pH	7.8	7.2	7 3	6.82	5.80	-	-	-	-	-
2	Sodium	12.0	8.0	3.00	14.00	12.00	12	9.20	0.04	18.00	14.00
3	Potassium	4.0	1.00	ND	21.00	2.00	4.3	0.8	Traces	35.00	2.00
4	Sulphate	ND	80.15	131.00	2.40	34.00	ND	95.00	3.13	2.64	38.00
5	Phosphate	3.80	2.40	0.30	4.82	0.32	3.92	2.48	0.008	25.76	0.36
6	Nitrate	0.23	0.32	0.68	0.40	0.42	-	-	-	-	-
7	Calcium	58.3	10.60	28.00	52.30	48.12	72.12	18.2	0.96	54.80	62.44
8	Magnesium	41.7	22.10	26.8	50.90	16.34	46.16	24.62	0.24	59.84	23.62
9	Heavy Metals (%)										
	Cadmium	ND	ND	ND	0.001	ND	0.002	0.001	ND	0.002	0.001
	Chromium	ND	ND	ND	0.024	ND	ND	ND	ND	0.246	ND
	Copper	0.002	ND	ND	0.007	ND	0.004	0.001	0.002	0.047	0.004
	Lead	ND	ND	0.001	ND	ND	ND	ND	0.002	ND	0.001
	Iron	0.072	0.084	0.074	0.051	0.032	0.088	0.092	0.076	1.820	0.046
	Manganese	0.002	0.006	0.003	0.046	0.002	0.005	0.008	0.004	0.784	0.062
	Zinc	0.272	0.342	0.81	0.160	0.072	0.358	0.436	0.2121	1.120	0.326

All the values except pH and Heavy Metals are in mg/l

The adsorptive properties of a particular carbon is expressed in terms of its phenol, Iodine and methylene blue values, collectively known as adsorption quality control parameters and presented in Table 5. Phenol values are found to be 55, 17, 25 and 48 respectively in case of PSPAC, WHPAC, CFAC and WSGAC as compared to commercial grade GAC (35). On the other hand, the iodine values were observed as 658, 785, 620 and 664 respectively, which are also comparable to commercial grade GAC (750). Methylene blue numbers of all the products are comparable. This indicates the better adsorption efficiency for removing taste and odour, colour and low molecular organic substances when applied in the water and wastewater treatment system.

Table 5: Adsorption Characteristics of Activated Carbons

Sl. No.	*Parameter*	*PSPAC*	*CFAC*	*WHPAC*	*WSGAC*	*CGAC*
1.	Iodine Number	658	620	785	664	750
2.	Phenol Number	55	25	17	48	35
3.	Methylene Blue Number	242	248	278	285	280

Product Efficiency in Wastewater Treatment

Activated carbons were further applied in the wastewater treatment for the removal of certain pollutants like colour (Dyes), Dissolved organics and heavy metals. The results are presented in Table 6. It was observed that, all the carbons are having better adsorption efficiency as compared to commercial grade activated carbon and could be used successfully in the wastewater treatment system. This is evident from the comparison made for the parameters evaluated from the adsorption model.

Table 6: A Comparative Removal Efficiency Of Different Solid Waste Based Activated Carbons

Sl.No.	*Activated Carbon*	*Dyes*			*Dissolved Organic (In terms of COD)*	*Heavy Metals*	
		Basic	*Neutral*	*Acidic*		*Chromium*	*Cadmium*
1.	PSPAC	84	76	72	84	90.2	92
2.	CFAC	92	82	78	92	88	86
3.	WHPAC	94	86	74	82	96	94
4.	WSGAC	97	99	89	80	100	60
5.	CGAC	100	95	93	76	100	80

All the above data are in optimized conditions and are in per cent removal.

Conclusion

Study indicates the potential utilisation of different solid waste materials (Paper mill sludge, Coke fines, Water hyacinth and walnut shell) as an appropriate and alternate raw material for the preparation of activated carbons. Different processes for the preparation of activated carbon were developed by optimising carbonisation conditions in a specially designed pyrolysis unit. The products were characterised for their physico-chemical and quality control parameters and successfully applied in the adsorption system for the removal of different pollutants like colour, dissolved organics and heavy metals.

The overall work signifies a major contribution towards integrated waste management by utilising waste materials for the development of better product *i.e.* activated carbon. The total potentials for the development of activated carbon from solid waste by both conventional and modern techniques were explored.

Activated Carbon Production-Modernisation of Technology: A Future Prospect

Any approach towards solid waste management should be economic and technology based. The products *i.e.* activated carbons prepared so far are cost effective interms of its nil raw material costs. The activated carbons prepared were based on the commercial method of carbonisation, however, an approach has been made towards the most cost effectiveness by improving the available technology using a non-conventional energy source, *i.e.*, the microwave energy under the concept of *Green Chemistry*. The application of microwave technology, in the carbon production has several advantages interms of economics, energy and efficiency over the conventional method of carbonisation.

In future, the conventional method of carbonisation could be replaced by the new technique that would not only help to improve the overall quality of the adsorbent, but also in the overall production market. In this context, research work is being initiated to develop the activated carbon from groundnut shell by applying both conventional and non-conventional (Microwave) carbonisation techniques and to establish its applicability in water and wastewater treatment system.

Acknowledgement

The authors are thankful to Director, NEERI, Nagpur for providing facilities to carryout this research work and kind permission for this publication.

REFERENCES

1. Cookson, J.T., 1978, "Carbon Adsorption Handbook"; Cheremisinoff P.M. and Ellerbusch, F. Eds.; Ann Arbor Science Publishers; INC; 241-280;

2. Directorate General of Technical Development (DGTD), 1978, Pulp and Paper Directorate, Govt. of India, New Delhi.

3. Giles, C.H. and Trivedi, A.S., 1969, "Specific Surface Area by Methylene Blue Method"; Chem. Ind.; 40; 1426.

4. Hassler, J.W., 1974, "Purification with Activated Carbon"; 3rd.Ed.; Publishing by Chemical Publishing Company, New York,

5. Mohandas Rao, K.P., Chhabaria, N.D., 1972, "Pollution Control In Pulp And Paper Mill "; IPPTA; 4 (11),

6. NTIS Report, (1973, Cincinnati, Ohio.

7. "Process Design Manual for Carbon Adsorption", 1973, U.S. Environmental Protection Agency, Technology Transfer EPA; 625; 1-71-0029.

8. "Proximate Analysis, IS-1350 (Part 1)", 1969, Publications of Indian Standard Institute, New Delhi;

9. Snoeynik, V.L. and Weber, W.J. (Jr.), 1967, "The Surface Chemistry of Active Carbon-A Discussion of Structure and Surface Functional Groups"; Environ. Science Technol. 1,228.

10. "Standard Methods for the examination of waste and wastewater"; APHA, AWWA and WPCF; 1989, Published by American Public Health Association; New York; 17th Ed.
11. Subramanyam, P.V.R.; Parekh, R.C, and Mohanrao, G.H., 1972, "Low Cost Methods For Treating Pulp and Paper Mill Effluents"; Indian Pulp and Paper; 9; 6.
12. "Ultimate Analysis of Coal and Cokes"; IS-1351-59, 1969, Publications of Indian standard institute, New Delhi.
13. Wanumanulu, V. and Subramanyam, P.V.R., 1975; "Planning Of Water Pollution Control Measures for an Integrated Kraft Pulp and Paper Mill", IPPTA Souvenir 138-152.

51

Alum Recovery by Liquid Ion Exchange Technique: Bench Scale Pilot Studies

S.S. DHAGE, S. PANDEY, R. RAVINDAR RAO, R. PARAMASIVAM

Scientist, National Environmental Engineering Research Institute (NEERI)

ABSTRACT

After establishing the technical feasibility of Liquid Ion Exchange (LIE) process, studies were undertaken to delineate process criteria and optimum test conditions for recovery of alum from water works waste with continuous bench scale operation. Pretreatment of water works waste by gravity thickening in a laboratory model has resulted in significant concentration of suspended solids. Acidification of this concentrate by sulphuric acid was studied for dissolution of alum.

The impact of significant factors governing the efficiency of LIE process have been studied in details. With synthetic alum solution, the optimum test conditions were derived to achieve more than 80 per cent recovery.The suitable modifications in the stripping stage were incorporated to reduce unreacted acid in the recovered alum which otherwise would adversely affect coagulation process.

The developed optimum process criteria when applied to real life aluminium bearing waste showed marginal redu ction in the efficiency of alum recovery.

Keywords: Recovery and reuse, waterworks waste, alum sludge treatment, Liquid Ion Exchange.

Introduction

Alum is the most widely used coagulant in the purification of surface waters. The resultant alum sludge is commonly disposed into the nearby water bodies without giving any treatment. Such disposal practices may lead to environmental degradation and hence not acceptable. Moreover this mode of sludge disposal will also loose the recoverable material. Therefore alternative methods (1,2,3) should be adopted for conserving natural resources. Recently evidence has been presented to support a relationship between aluminium and both alzheimer's diseases and dialysis encephalopathy in human (4,5,6). There is also a felt need for recovery and reuse of the coagulant.

The technical feasibility of the Liquid Ion Exchange (LIE) process for recovery of alum from water works waste under laboratory conditions has been established (7,8,9). Further studies with continuous bench scale operation have been undertaken with the objective of developing process criteria and optimum operating conditions for alum recovery by LIE technique. This paper presents the findings of the experiments conducted for coagulant recovery by LIE.

Materials and Methods

The study comprised of three major components *viz.* Pretreatment of waterworks sludge, bench scale continuous operation of the process with synthetic alum and application of derived optimum conditions to actual waste.

Pretreatment of Waterworks Waste

Thickening

The waterworks waste generated during water purification is usually thin and needs to be concentrated. Based on the settling characteristics of the waste, a laboratory thickener has been designed and fabricated adopting Coe and Clevanger method (10).

The selected flow rate was 5 litres per hour with two hours detention time and Surface Overflow Rate (SOR) of 8 m^3/day/m^2. The height of the fabricated circular unit was 66 cm with 14 cm diameter. Thickening studies were conducted with actual waterworks sludge having suspended solid concentration in the range of 500 to 3800 mg/L. The experimental setup is depicted in Figure 1. The results of the continuous operation are presented in Table 1.

Table 1: Quality of Raw and Thickened Sludge

	Raw Waste	*Effluent*	*Thickened Sludge*
Turbidity (NTU)	320 – 2600	70 – 240	—
Total Solids (mg/L)	500 – 3800	100 – 632	26000 – 125000
Aluminium (Al-mg/L)	19 – 63	3 – 5	650-2200

Acidification

In order to bring the alum in liquid form, acidification of the thickened sludge was studied with process variables like pH, type of mineral acid and impact of mixing. Three mineral acids *viz.* Sulphuric, hydrochloric and nitric of 6 N strength were tried. The findings are presented in Table 2. The studies were conducted with different quality of sludges to find out optimum pH for acidification and the results are presented in Table 3. The experimental setup for acidification is shown in Figure 2.

Fig. 1: Experimental Setup for Thickening

Fig. 2: Pretreatment of Watermarks Waste: Thickening and Acidification

Table 2: Results of Acidification of Sludge Using Different Acids
Initial Aluminium Concentration – 650 mg/L

Test pH	*% Aluminium Dissolution with*		
	H_2SO_4	*HCl*	*HNO_3*
1	85	85	87
2	70	63	65
3	45	38	38

Table 3: Effect of pH on Aluminium Dissolution
Initial Aluminium Concentration in Sludge 650-2240 mg/L

Test pH	*% Aluminium Dissolution*
1	85 – 95
2	70 – 83
3	45 – 61

Bench Scale Operation of LIE Process with Synthetic Alum Solution

The process of LIE consists of two main circuits, extraction circuit in which the aluminium ions from aqueous phase are selectively transferred to organic phase and stripping circuit in which the aluminium ions are concentrated into aqueous phase leaving the organic phase for re-circulation.

Bench scale mixer and settlers were designed and fabricated to study the impact of the variables on the efficiency of extraction and stripping processes. The details of the experimental setup are presented in Table 4 and Figure 3.

Table 4: Set Up for Extraction and Stripping

Design Criteria	*Mixer*	*Settler*
Flow Rate	40 ml/min	40 ml/min
Detention Time	20 min	20 min
Diameter	10.5 cm	5.5 cm
Height	16.0 cm	35.0 cm
L/D Ratio	0.87*	
Location of Outlet	10 cm from bottom	30.0 cm from the bottom
Speed of Mixer		
a) Extraction	1200 rpm	—
b) Stripping	1500 rpm	—
Agitators Type	Propeller with 3 blades located at interphase	
Diameter of Agitator	5 cm	

* The standard range is 0.5 – 1.5.

Fig. 3: Bench Scale Assembly for Continuous Extraction and Stripping Operation

The following variables have been considered while optimizing the process:

Parameters	*Range*
Initial aluminium concentration	500 mg/L –3000 mg/L
Organic to aqueous phase ratio in stripping	3: 1, 6: 1, 10: 1
Number of stages required for extraction	1-stage
Stripping	2-stages
Strength of extractant stripping media	0.1–0.3M Di 2 Ethyl Hexyl Phosphoric Acid (Di 2 EHPA) 3 to 6 N Sulphuric Acid (H_2SO_4)
Effect of other metals on extraction	Ca, Mg, Fe, Mn, Zn in the concentration range of 20-50 mg/l for each metal
Extractant recycling	1-10 cycles

Bench scale studies were carried out using synthetic solution of aluminium sulphate at pH 2.0. The extractant used was 0.15 molar solution of Di-2-Ethyl Hexyl phosphoric acid prepared in hexane. The observations of above study are summarized in Table 5. Schematic flow sheet of the experimental setup is depicted in Figure 4.

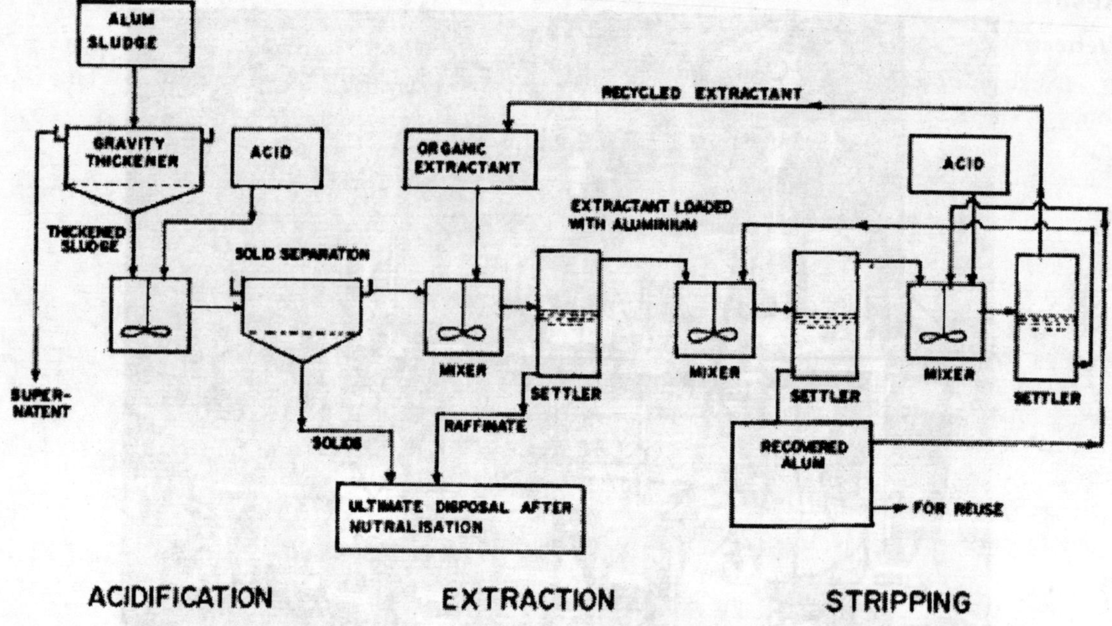

Fig. 4: Alum Recovery from Water Works Slude by Liquid-ion Exchange Technique (LIE)

Table 5: Derived Optimum Test Conditions

Variables	*Extraction Circuit*	*Stripping Circuit*
Initial aluminium feed concentration	1000 ppm	—
Organic to aqueous phase ratio	1: 1	10: 1
No. of stage required	1	2
Concentration of extractant media	0.15 M D2EHPA	5 N H_2SO_4
Effect of other metals on extraction	Calcium, Magnesium, Manganese, Zinc not extracted. Iron partly extracted	—
Extractant recycling	Upto 10 cycles comparable efficiency obtained to get 80-85% transfer	—

Application of the Optimized Process Conditions to Waterworks Sludge

The derived optimum test conditions were applied to the actual thickened and acidified sludge with aluminium concentration in the range of 650 – 900 mg/l.

The quality of recovered alum was tested for its coagulation efficiency by performing Jar tests. An attempt has also been made to work out economic feasibility of the process.

Results and Discussions

Pretreatment

During thickening process, waterworks waste with less than 0.5 per cent suspended solids was concentrated to yield a sludge with suspended solids concentration upto 12.5 percent. The corresponding aluminium content in the sludge was increased from 19-63 mg/L to 650-2200 mg/l. (Table 1). The suspended solids concentrations in the effluent has been reduced by about 80% thereby reducing the pollutional impact on receiving waterbody.

Acidification of thickened of sludge by 6 N sulphuric acid was found to be most effective and economical for dissolution of aluminium at pH value of 2 and 3 (Table 2 and 3). Maximum dissolution had occurred at pH 1. However, for 12 to 15 per cent aluminium dissolution, the acid requirement had increased by a factor of three or even more when compared with the quantity required at pH 2. It was observed during batch experiments (9) that extraction efficiency in LIE process was 67 per cent at pH 1 as against 83 per cent at pH 2. Moreover, lowering pH to 1 required higher quantity of lime for neutralisation of residual sludge. Hence, pH 2 was considered optimum for acidification.

Extraction and Stripping Circuit for LIE Process

The following observations emerged from the study on the effect of variables on the extraction and stripping efficiency:

Initial Aluminium (Al) Concentration During Extraction

It has been observed that under the selected conditions (Table 4), about 80 per cent transfer of aluminium was achieved from the solution with 1000 mg/L Al concentration with single stage operation. Phase ratio of 1: 1 was maintained. With higher initial concentration of Al, the extraction efficiency had reduced thereby indicating the need for multiple stage operation. Therefore 1000 mg/L concentration of aluminium was considered optimum for single stage process.

Organic to Aqueous Phase Ratio

Extraction

The experiments[3] had indicated that a phase ratio of 1: 1 for initial aluminium concentration of 1000 ppm resulted in about 80 per cent transfer. The difficulties like third phase formation, phase separation, etc. were not observed at this phase ratio.

Stripping

Experiments with phase ratios (organic: aqueous) ranging from 3: 1 to 6: 1 resulted in about 84 per cent efficiency. However, it was observed that a considerable amount of aluminium still remained in the organic phase reducing the efficiency of organic phase during recycling, in subsequent extraction circuit. Hence, higher phase ratios were tried. A phase ratio of 10 : 1 was considered optimum. However, the single stage stripping was inadequate to obtain an organic extractant suitable for recycle.

Number of Stages During Extraction and Stripping

With initial Aluminium concentration of 1000 ppm and phase ratio of 1: 1, single stage extraction could achieve about 80 per cent transfer. However, during stripping for a phase ratio of 10: 1 double stage stripping was required to achieve maximum transfer efficiency for aluminium, to get organic extractant acceptable for recycling and to obtain alum concentration of required strength.

Strength of Extractant and Stripping Media

At high concentration, molecules of extractant – alkyl phosphoric acid – have a tendency to polymerize thereby reducing the active sites available for aluminium chelation. Also the degree of polymerization is a function of Aluminium ion to alkyl phosphoric acid concentration. 0.15 M solution of extractant was found suitable with hexane as the diluent which avoided the problem of polymerization.

It was observed that with 5 N sulphuric acid as stripping agent, about 90 per cent efficiency was achieved and hence it was considered as the optimum strength.

Effect of Other Metals on Extraction

The extractant Di 2 EHPA is selective for Aluminium. However, other metals like calcium, magnesium, iron, manganese, zinc are also likely to be extracted at the selected optimum conditions. Therefore, the effect of these metals in concentration of 20 to 50 mg/L with 950 mg/L of aluminium was studied. It was observed that calcium, magnesium, manganese, zinc are not transferred to organic phase. Iron get extracted to the extent of 34 per cent. However, as ferric alum is also used as a coagulant and presence of iron in the recovered alum will not adversely affect the quality of the product.

Extractant Recycling

The efficiency of organic extractant after stripping was tested by recycling it several times. It was observed that upto 10 cylces comparable efficiency was achieved with 80-85 per cent transfer of aluminium. During the experiments on recycling, a loss of about 0.5 to 1.0 per cent of organic phase was observed, due to a number of reasons like evaporation, loss of organic phase in the extraction and stripping.

The overall recovery of alum obtained from synthetic solution was about 80 per cent. But the same conditions when applied to actual waste gave marginally less recovery of about 73 per cent.

The optimized process is quite complex and requires regular and accurate operation and monitoring. The chemists at the pilot plant has to supervise the process very critically so that recovered alum of required strength and purity is obtained.

Coagulation Ability of Recovered Alum

The quality of recovered alum when tested for its coagulant ability was found comparable to that of commercial alum. However, considerable free acid concentration was observed in the recovered alum resulting in lowering the pH of treated water. A modified approach was developed to recirculate recovered alum. This had resulted in increased in alum concentration and considerable reduction in unreacted acid.

The economic feasibility of the process based on the operational cost was assessed. The cost of recovered alum was higher than commercial product. However, economics of this process should be evaluated more critically in terms of pollution abatement benefits in addition to recovery aspects.

Summary and Conclusions

1. During pre-treatment waterworks waste with less than 0.5% suspended solids was concentrated upto 12.5% suspended solids with aluminium concentrations upto 2000 mg/L.
2. Acidification of thickened sludge with 6N sulphuric acid has achieved 70 – 83 % dissolution at pH 2.0.

3. Detailed investigations on extraction and stripping circuits in LIE process has resulted in formulation of optimum test condition for continuous operation of the process.
4. About 80% alum was recovered from synthetic alum solution with the developed process package. The efficiency of recovery has marginally reduced while processing actual waste.
5. Suitable modifications were done to overcome the problem of unreacted acid in recovered alum.
6. Coagulant ability of recovered alum was tested and found satisfactory.
7. Careful monitoring and accurate control is essential for proper functioning of the process.

Acknowledgements

The authors are grateful to Director, NEERI for providing facilities to carry out the research work and permission to publish the paper.

REFERENCES

1. Mrs. S.S.Dhage, R. Ravindar Rao, S.P. Andey and R. Paramasivam, 1984, "Waterworks Wastes – A State of the Art Review" JIWWA, Vol. XVI, No. 2, pp 183-188.
2. Mrs. S.S.Dhage, R. Paramasivam, S.P.Andey and R. Ravindar Rao, 1985, "Pollution Potential of Water Works Waste – A Case Study" JIWWA, Vol. XVI, No. 1, pp 17-22.
3. Mrs. S.S.Dhage, R. Paramasivam, S.P.Andey and R. Ravindar Rao, 1985, "Water Works Waste – A Pollution of Receiving Water Body" Current Pollution Researches in India Eds R.K. Trivedy and P.K.Goel Environmental Publications, Karad.
4. Crapper, D.R., Krishnan, S.S., & Dalton, A.J., 1973, "Brain Aluminium in Alzheimer's Disease and Experimental Neurofibrillary Degeneration, Sci., 180: 511.
5. Aluminium Neurotoxicity (L. Liss, editor), Pathotox Publ., Inc., Park Forest South, III, 1980.
6. Davison, A.M. *et al.*, 1982, "Water Supply Aluminium Concentration Dialysis Dementia, and Effect of Reverse Osmosis Water Treatment", Lancet, 785.
7. Cornwell, D.A. & Lemunyon, R.M., 1979, "Feasibility Studies of Liquid Ion Exchange for Alum Recovery from Water Treatment Plant Sludge" Jour. AWWA, 72: 12: 741.
8. David A Cornwell, Garry C. Cline, John M. Przybyla and David Tippin, 1981, Demonstration Testing of Alum Recovery by Liquid Ion Exchange, JAWWA.
9. Mrs. S.S.Dhage, R. Paramasivam, R. Ravindar Rao and S.P.Andey, 1985, "Recovery of Alum from Water Treatment Plant Sludge by Liquid Ion Exchange (LIE) Technique" JIWWA, Vol. XVII, No. 2, pp 193-199.
10. Walter L. Badger and Julius T. Banchera, "Introduction to Chemical Engineering"

52

Metals Recovery from Metal-Laden Insoluble Agro Based Xanthate

M.L. AGRAWAL & VINOD TARE

Department of Civil Engineering, Indian Institute of Technology, Kharagpur - 721 302 (W.B.)

ABSTRACT

Heavy metals contamination due to industrial waste can be better controlled at source rather than after they are dispersed. Thus success in minimizing the adverse impact depends largely on the ability to design treatment process that isolate, remove and recover the heavy metals. Various treatment processes have been used in practice and many are coming up. One of the most promising amongst several emerging/innovative technologies for heavy metal removal and recovery from industrial effluents utilizes substances with xanthate as functional group.

The present paper describes the studies carried out to investigate the release of heavy metals from metal-laden insoluble agro based xanthate (IAX). The investigation was conducted by extending the postulation made for metal-insoluble starch xanthate interaction to metal-insoluble agro based xanthate. It was hypothesized that metal release from metal-laden IAX could be affected by providing aqueous acidic condition for decomposition of free xanthate. It was envisaged that metal ions as well as the acid used for providing acidic environment could play a significant role.

Laboratory scale experiments using non-flow agitated system were conducted to load the metal ions on IAX as well as to evaluate the extent of metal release from metal-laden IAX. Four typical heavy metals, namely Hg (II), Cu (II), Cd (II) and Ni (II) were used. The base material used for IAX preparation was wood shavings. Sulphuric, hydrochloric and nitric acids in varying strengths were used to maitain acidic environments. The metal recovery process was studied on single metal-laden IAX and also multi metal-laden IAX The metal release phenomenon from metal-laden IAX is found to be similar to that from metal-insoluble starch xanthate. Metals' recovery was found to be proportionate with the oxidizing power of acids and in the order Ni > Cd > Cu > Hg. The interaction between metal-metal in multi metal-laden IAX is found to have influence on recovery process.

Keywords: Heavy metals, Metal recovery, Xanthate decomposition, Insoluble Agro based Xanthate, Wood Xanthate.

Introduction

One of the typical properties of an industrialized and affluent society is its relatively high consumption of various metals. Metals with connotation of toxicity are referred as heavy metals. Heavy metals have drawn increasing human attention because of certain incidents, such as Minamata disaster, Itai-Itai disease, etc. With the introduction of environmental regulations, various treatment technologies have been developed. Any treatment method must enable: achievement of set discharge limits; the recovery of precious metal ions; and be economically viable to the industry. None of the various available treatment technologies can fulfil these three requirements simultaneously.

Of the several emerging/innovative technologies, xanthate process has focused attention as one of the most promising one. Xanthates are polymers with specific functional group to bind metal ions. They are recognized as a class of sulfur compound having starch and cellulose base (Swanson *et al.*, 1973). Xanthates, soluble in water before reacting with metal ions have been termed as soluble xanthates (SX), while insoluble xanthates (IX) are the ones, which can be easily separated by simple settling. The insoluble xanthates synthesized by using naturally available low cost agro-based materials containing large amount of cellulose are termed as insoluble agro based xanthate (IAX). At present very limited information is available on recovery of metals from metal-laden IAX (Agrawal, 1997). The present paper describes the role of various parameters in recovery of metals from metal-laden IAX.

Materials and Methods

The laboratory experiments were carried out in four parts. The first part was mainly aimed at synthesis of IAX. The second part experiments were carried out to determine the Iodine value of IAX's. The third part experiments were carried out to analyze the metal loading on IAX's. The metal loading was done for individual metal (mercury, copper, cadmium and nickel) and also for multi-metal mixture (mercury + copper + cadmium + nickel). The fourth part of experiments were conducted to analyze the recovery of different metals from IAX-metal sludge using different recovery agents (RA).

Synthesis of IAX

Synthesis of wood xanthate (WX) was done according to the procedure given in literature (Chaudhari, 1993; Wing *et al*, 1978). 30g crushed/grinded base material (wood savings) was soaked in 500ml sodium hydroxide (160 g/l) and stirred (Remi strrer, 1/8 HP Remi Udyog Ltd., Bombay). To alkaline base material 25 ml of carbon disulfide was added and stirred for four hours. The contents were then allowed to settle for two hours and supernatant was decanted. The insoluble solids were washed with distilled water to remove excess sodium hydroxide and then subsequently washed with 250 ml acetone. The prepared xanthate was then stored in tightly stoppered bottles at low temperature (6-8°c).

Estimation of Iodine Value

Iodine value of IAX is determined by estimating the reducing sulfur group by oxidation-reduction titration using iodine. A mixture of 10% (v/v) 10 ml acetic acid and 10 ml of 0.1N iodine solution were mixed in distilled water. The mixture was taken in 100 ml Polyethylene bottles and 2.5 g/l of WX was added. The bottles were kept on an end-over-end shaker to permit mixing for 240 minutes. After mixing, the solution mixture was titrated with 0.1 N sodium thiosulfate using starch indicator to determine residual iodine. A blank (without IAX) solution mixture of same composition and amount was kept to estimate available iodine. The difference in iodine amount in blank and with IAX expressed as meq of iodine consumed per g of IAX is referred as iodine value.

Loading of Metal on IAX

This was done in two parts. In the first part of loading experiments different metals (mercury, copper, cadmium, nickel) were loaded individually on the IAX. In the second part of loading experiments multi-metal mixture (mercury, copper, cadmium, nickel) was loaded on IAX. The process adopted was as follows:

Loading of Individual Metal on IAX

Loading of individual metal was done by taking solution of 120-150 mg/l initial concentration, maintaining pH at 5, ionic strength at 0.01 with a mixing time of 240 minutes using end-over-end shaker. The WX dose was taken 5 g/l.

Loading of Multimetal Mixture on IAX

Loading of multimetal mixture was done in such a way so that the resultant loading of each metal in mM per g of IAX was nearly equal. For this all parameters, reported previously for single metal loading, were same except the initial concentration of different metals, which were Hg = 53 mg/l, Cu = 23 mg/l, Cd = 95 mg/l, Ni = 102 mg/l. The WX dose was taken 20 g/l.

Recovery of Metals from Metal-IAX Sludge

This part of experiments was conducted separately for individual metal loaded IAX and multi-metal loaded IAX. The method adopted was as follows:

Recovery of Metal From Metal-IAX Sludge (Single Metal-laden)

Recovery of metal was done using different recovery agents (RA), namely sulfuric acid, hydrochloric acid and nitric acid each of four different concentrations (0.5 M, 2 M, 4 M, 6 M). 50 ml of recovery agent and 0.5 g of metal loaded xanthate were taken in a polyethylene bottle and was allowed to soak for 24 h. The samples were taken at different times and were subjected to metal analysis using atomic absorption spectrophotometer.

Recovery of Metals from Metal-IAX Sludge (Multimetal-laden)

The recovery of metals from metal-IAX sludge was done using different recovery agents (RA) of only that concentration, which was found to be most effective and of lowest strength in case of recovery from individually loaded metal on IAX. And rest of the procedure was similar to that is reported in section (IV) A.

Analytical Techniques Used

Parameter	Instrument/Analytical Techniques	Reference
pH	Digital pH meter, Model No. 335 Systronics, Ahmedabad, India	APHA-AWWA-WPCF, (1989)
Total Metal Analysis {Cu, Cd, Ni)	Atomic Absorption Spectrophotometer Varian Model Spectra AA 20 BQ	APHA-AWWA-WPCF, (1989)
Total Metal Analysis {Hg)	Vapor Generation Assembly VGA-76	APHA-AWWA-WPCF, (1989)
Iodine Value	Iodometric Titration for equivalent of sulfur group	Chakraborty, (1996)

Results and Discussions

To determine the optimum minimum concentration of recovery agents for maximum percentage recovery four concentrations of each recovery agents (0.5 M, 2 M, 4 M, and 6 M) were used for kinetic study with all four metal-laden IAX individually and it was found that 0.5 M sulfuric acid, 2 M hydrochloric acid and 2 M nitric acid were minimum concentration for maximum possible recovery. These optimum strengths of different recovery agents were used for all further study.

Comparison of Different Recovery Agents (RA) for Metal Recovery

Figure 1 presents the comparison of kinetics of metal recovery by different recovery agents (0.5 M sulfuric acid, 2 M hydrochloric acid, 2 M nitric acid), for recovery of different metals (mercury, copper, cadmium and nickel) from metal-laden WX. 2 M nitric acid was found to be most effective for the recovery of copper, cadmium and nickel but for recovery of mercury hydrochloric acid was found to be most effective. This may be due to the fact that chloride complexes of mercury are strong.

Comparison of Different Metals Recovered Using Different RA's

Figure 2 presents the comparison of recovery of all four metals (mercury, copper, cadmium and nickel) using different recovery agents from individual metal-laden WX as a function of contact period. The percentage recovery of metals with all three RA's is found to be in order Ni > Cd > Cu > Hg. The contact period of 3 h is found to be appropriate for copper, cadmium and nickel but for mercury a longer contact period is required. The results also indicate that the percentage recovery of copper, cadmium and nickel is appreciable but the recovery of mercury is least of all four metals.

Comparison of Metal Recovery from Single and Multimetal-laden WX

Figure 3 presents comparison of percentage recovery of metals from single metal-laden WX and multi metal-laden WX using different recovery agents. The results indicate that the percentage recovery from single metal-laden WX is more as compared to the recovery from multi-metal laden WX for all four metals and with all three RA's keeping all other conditions same. The order of percentage recovery of metals from single metal-laden WX is Ni > Cd > Cu > Hg whereas the order from multimetal-laden WX is Ni > Cu > Cd > Hg. This difference in percentage recovery and the order of recovery may be due to, the interaction between metal-metal on WX; and the relative affinity of different metals with xanthate and availability of less free xanthate groups thereby reducing the decomposition of xanthates responsible for release of metals.

Conclusions

Based on the results of the present study and the synthesis of the results of the information available in the literature, following conclusions may be drawn:

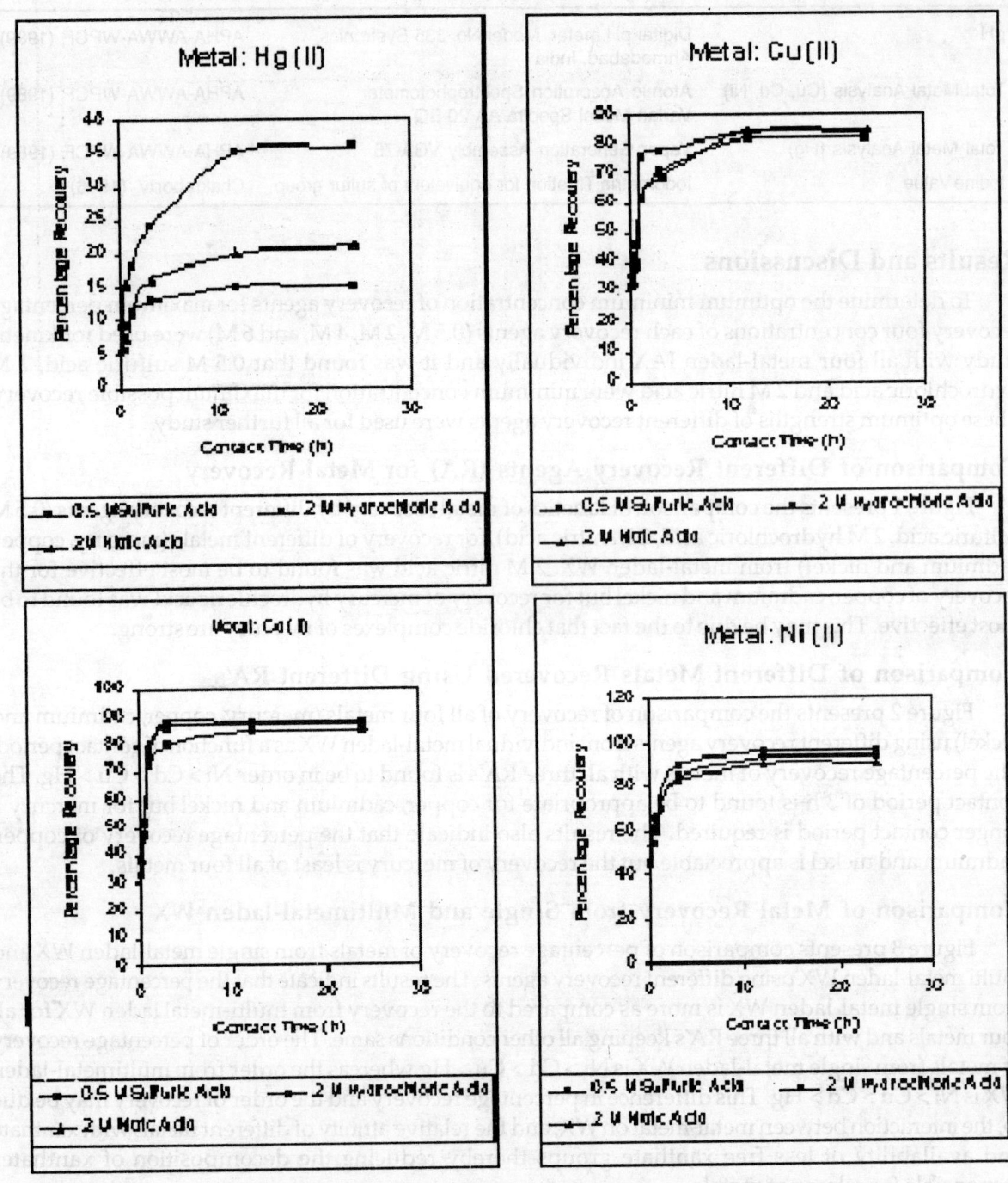

Fig. 1: Effects of Metal Recovery from Metal-laden WX Using Recovery Agents

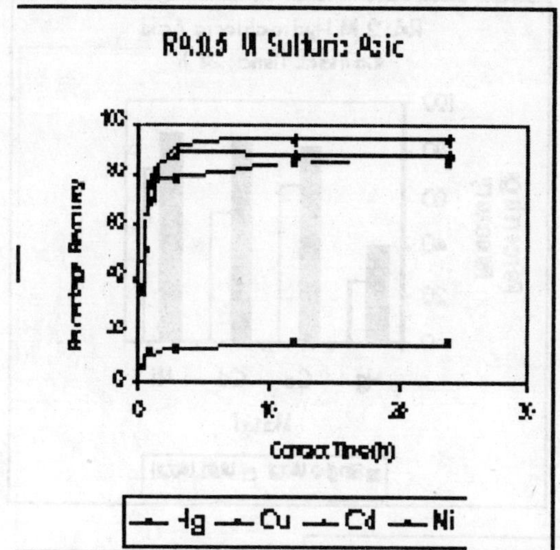

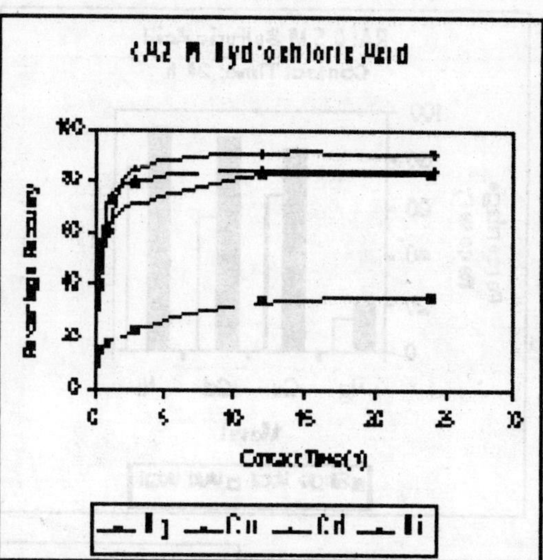

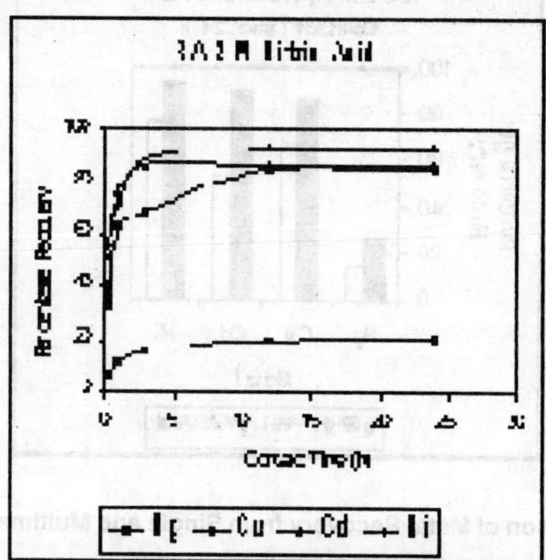

Fig. 2: Comparison of Different Metal Recovered Using Different Recovery Agents

1. The release of metal ions from metal-laden IAX depends on the possible degree of decomposition/destruction of free xanthates by recovery agents.
2. Among strong acids, metal recovery increased with increase in oxidizing power of the acids as well as the ability of associated anion to form soluble metal complexes.
3. The kinetics of metals release from metal-laden IAX appears to be similar for the three metals, namely Cu (II), Cd (II), and Ni (II). For Hg (II) it found to be slower as compared to other three.

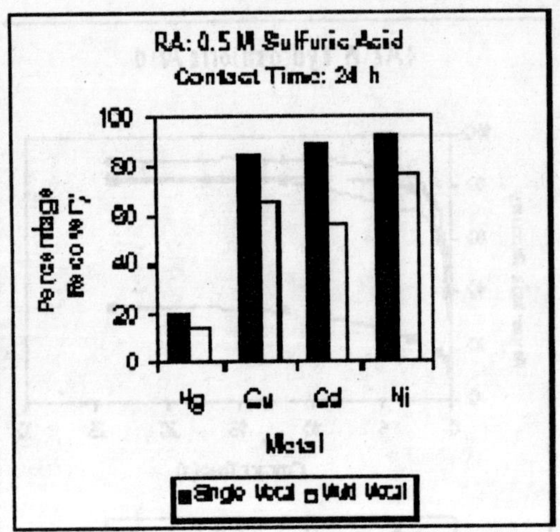

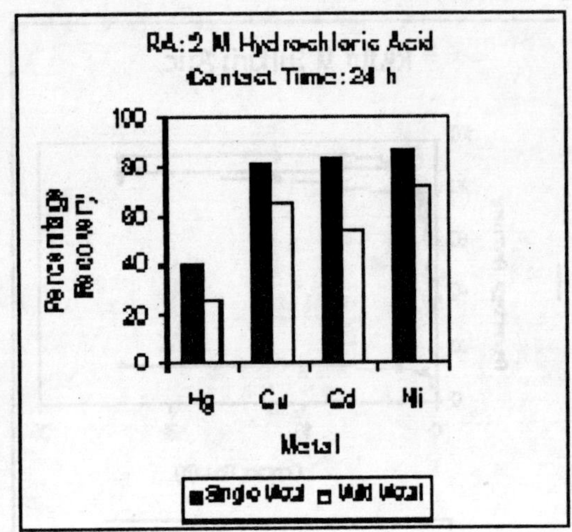

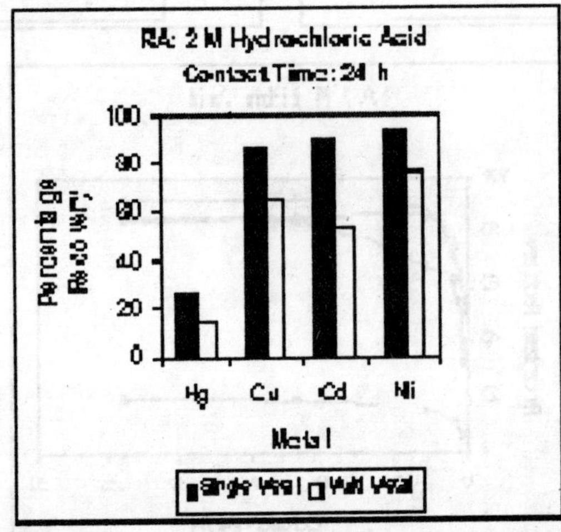

Fig. 3: Comparison of Metal Recovery from Single and Multimetal-laden WX

4. In general, the metal-xanthate interaction in metal release mode in IAX appears to be similar to that hypothesized for metal-insoluble starch xanthate (Zang *et al.*, 1988).

REFERENCES

1. Agrawal, M. L., 1997, *Metals Recovery from Metal-Laden Insoluble Agrobased Xanthates*, M. Tech Thesis, Dept of Civil Engineering, IIT Kanpur.
2. APHA-AWWA-WPCF, 1989, *Standard Methods for Examination of Water and Wastewater*, 17th Ed., Washington, D.C.
3. Chakroborty, S., 1996, *Role of Various Parameters in the Synthesis of Insoluble Xanthates for Heavy Metals Removal from Waters*, M. Tech Thesis, Dept of Civil Engineering, IIT Kanpur, India.

4. Chaudhari, S., 1993, *Analysis, Evaluation and Application of Xanthate Process for Heavy Metal Removal and Recovery from Wastewaters,* Doctoral thesis, Dept of Civil Engineering, IIT Kanpur, India.

5. Swanson, C.L., Wing, R.E., Donae, W.M. and Russell, C.P., 1973, "Mercury Removal from Wastewater with a Starch Xanthate-Cationic Polymer Complex", *Environ Sci and Tech,* Vol. 7, No. 7, pp. 614-617.

6. Wing, R.E., Jasberg, B.K. and Navickis, L.L., 1978, "Insoluble Starch Xanthate: Preparation, Stabilization, Scale-up and Use", *Staerke,* Vol. 30, No. 5, pp. 163-170.

7. Zhang, L., Zhang, L., Zhang, B. and Holl, W., 1988, "Properties of Improved Insoluble Starch Xanthate (IISX) ", *Huanjing Huaxue,* Vol. 7, No. 1, pp. 57-62 (cited from Chem. Abs. No. 109: 43008v).

53

Microbial Phytase: A Tool for Enhanced Biological Phosphorus Removal

K. MADHAVAN NAMPOOTHIRI, KAMAL BEGUM, A. SABU, GEORGE SZAKACS & ASHOK PANDEY

Biotechnology Division, Regional Research Laboratory (CSIR), Trivandrum – 695 019

ABSTRACT

High public concern for enhanced phosphorus level in animal excreta and the subsequent eutrophication of the water bodies near to the poultries and piggeries has driven the biotechnological development of phytase and its application in animal nutrition. The paper summarizes the results of vital parameter optimization for the production of fungal phytase from mesophilic and thermophilic cultures. Studies were performed in solid-state as well as in submerged fermentation. Utilization of packed column fermenter for phytase production is also mentioned. Supplementation of the initial wheat bran based medium (67 % initial moisture) with 1% W/W yeast extract resulted in a maximum enzyme activity of 29 U/gds in SSF using a mesophilic *A.ficuum*.Heat stable phytase is produced more efficiently in submerged fermentation using *T.aurantiacus.* Effects of surfactants, inorganic phosphorus supplementation on phytase production were evaluated. In the case of thermophilic fungal phytase more than 75 % of the enzyme activity was remained even at 70 ºC.

Keywords: *Phosphorus pollution, Phytic acid, Phytase, Solid state fermentation, Packed coloumn fermenter.*

Introduction

Phytate is the major phosphorus contatining constituent (20-70 %) of many cereal seeds, legume seeds, oil seed meals and tubers (1, 2). Its role is a P store of plant primarly for germination. Phytic acid from feedstuffs of plant origin may form insoluble complexes with minerals as well as with dietary protein, amino acid and starch (3). Apart from that it also inhibits the activity of proteolytic enzymes such as pepsin and trypsin, as well amylase (3-5). During the last decade many experimental evidences are documented regarding the implications of phytates in binding dietery macro-microelemets (Ca, Mg, Fe, Zn, Cu, Mn, Mo and Co) thus reducing their solubility and bioavailabilty (6). Majority of the monogastric animals such as pigs and poultry lack or nearly lack the enzyme mechanism for effective digestion of phytic acid and most of the undigested phytic acid is excreted in their manure. In order to compensate the free available phosphate level and also the antinutritional effect of phytate, the farmers used to supplement inoraganic phosphorus in the feed stuff. As a result, together with undigested phytate the excess P will also pass through the intestinal tract of these non-ruminants and this excreta which contains an alarming level of phosphate causes many environmental problems such as algal blooms in areas of intensive livestock production.

One possible remedy to tackle this situation of environmental hazard due to enhanced accumulation of phosphorus is the exploitation of the phytase enzyme. It provides a clean ecofriendly technology for efficient removal of biological phosphorus. Phytase (E.C.3.1.3.8) are classified under the family of histidine acid phosphatases (7) and are found abundantly in microrganisms and plants. It catalyse the hydrolysis of phosphate from phytic acid to inorganic phosphate and myoinositol phosphate derivatives (8). Supplementation of phytase with the feed can thus release free phosphorus from phytic acid and it reduces the requirement of inorganic phosphorus supplimentation, which is the main reason for additional cost and environmental problems. The enzymatic degradation of phytic acid will not produce mutagenic and highly toxic by products and thus the industrial sector is more and more get attracted towards the production and processing of phyase as a feed enzyme.

Phytase can be produced from microorganisms including bacteria, yeast and fungi (9). Submerged fermentation is employed mainly as the production technology. However, in recent years Solid state fermentation (SSF) has gained much interest for the production of microbial metabolites. (10).

The paper deals with the optimization of process parameters for fungal (mesophilic as well as thermophilic) phytase production in Solid state fermentation as well as in Submerged fermentation.

Materials and Methods

Microorganism and Maintenance

Two fungal strains namely *Aspergillus ficuum* (NRRL 3135) and *Thermoascus auruantiacus* (TUBF 43) were used for the present study. The former is a mesophilc culture and the later is thermophilic. The Cultures were maintained on Potato Dextrose Agar (PDA) medium. Fully sporulated slants were obtained after 7 days of incubation at 30°C for mesophilic culture and at 45°C for thermophilic. Strains were subcultured at regular interval of time and stored at 4°C.

Preparation of Inoculum

To a fully sporulated PDA slant, 10 ml of sterile distilled water containing 0.1 % tween 80 was added and the spores were carefully dislodged using an inoculation loop in aseptic conditions. The spore suspension was appropriately diluted for required density of spores and was used as the inoculum. Spore count was determined by serial dilution followed by plate count.

Phytase Production in Solid State Fermentation

Erlenmeyer flasks (250 ml) containing 5 g wheat bran moistened with salt solution (g/L, NH_4NO_3 5; $MgSo_4.7H_2O$ 1 and Nacl 1) and water was autoclaved at 121 °C for 20 min, cooled and inoculated with 1 ml spore suspension containing 7X10^7 spores of *A.ficuum*.The contents of each flask were mixed thoroughly for uniform distribution of fungal spores in the medium. The flasks were incubated at 30°C. The samples were removed at regular interval over a period of time and were used for phytase extraction followed by activity determination.

Coloumn Fermenter for Solid State Fermentation

The coloumn bioreactor used for the present study include an aerator, an air saturation unit, an air distributor, vertical cylindrical glass coloumn (37 × 3.5cm) as fermenter and an air exit unit. The column was packed with the preinoculated substrate (40 g, 25 % V/W inoculum). Moist air was passed through the coloumn and fermentation was carried out at 30°C. Separate columns were harvested after 48 and 72 hours of incubation. Three independent sets were carried out for each time interval to finalize the data. The CO_2 level within the column was monitored using a carbon dioxide analyser (RIKEN, Model RI 411 A, Japan).

Phytase Production in Submerged Fermentation

The semi synthetic medium (g/L, Starch 28; Glucose 5; Peptone 1; KCl, 0.5; $MgSo_4.7H_2O$ 1.5,KH_2Po_4, 1,$CaCl_2.H_2O_2$) supplemented with different percentages of wheat bran extract (5-60 % V/V) was used as the production medium for phytase in submerged fermentation.250 ml Erlenmeyer flasks containing 40 ml production medium were autoclaved at 121 °C for 20 minutes, cooled and inoculated with 1 ml of fungal spore suspension (7X10^7/ml spores of *A.ficuum* and 6x10^7/ml spores of *T.aurantiacus*). The flasks were incubated at 30°C for *A.ficuum* where as at 45 °C for *T.aurantiacus* with a constant shaking of 180 rpm for a desired time interval.

Enzyme Extraction

After SSF, 100 ml of distilled water containing 0.1 % Tween 80 was added in to the fermented sample and it was mixed thoroughly. Flasks were kept on a rotary shaker operated at 180 rpm for 30 minutes at room temperature for the extraction of enzyme from fermented Koji. At the end of extraction the whole suspension wad squeezed through a double layer of muslin cloth and it was centrifuged at 8000 rpm for 15 minutes at 4°C. The clear supernatant was used as the crude enzyme. For column bioreactor the enzyme extraction was carried out from three different regions, the top, middle and bottom.

In SmF, after incubation for a desired time interval, the flasks were removed and the whole fermented sample was centrifuged at 8000 rpm for 20 minutes at 4°C.The supernatant was collected and it was used as the crude enzyme.

Phytase Assay

Phytase activity was determined after enzymatic hydrolysis of Sodium phytate under controlled conditions and measurement of the amount of ortho-phosphate released (11). One unit of enzyme activity was defined as the amount of phytase required liberating I ?mol of inorganic phosphate per minute under standard assay conditions. In SSF Enzyme yield was expressed as Units/gram dry substrate/min (U/gds^{-1}) while in smF it is in U/ml.

Fungal Biomass in Solid Culture

Glucosamine method: Direct estimation of biomass in fungal solid state fermentation process is difficult due to the deep penetration of mycelial cells in to the solid substrate particles. Indirect methods monitor specific components of biomass, metabolic activities etc. Chitin is an insoluble linear polymer of ?-1,4 linked N acetyl glucosamine, predominantly present in the fungal cell wall. The assay involves the hydrolysis of the fermented matter followed by the determination of glucosamine liberated (12). The other method adopted is the evolution CO_2 during the incubation period using a carbon dioxide analyser, which indicates the metabolic activity in relation to fungal growth.

Results and Discussions

Phytase has been found in plant materials, in the intestinal mucosa of various animals, where it is able to release free inositol, orthophosphates and intermediary products including the mono, di-tri-tetra and penta phosphate esters of inositol. (Figure 1). Conventionally phytase activity can be determined by measuring the inorganic phosphorus (Pi) liberated by the action of the enzyme.

This paper describes the results obtained during the process optimization of phytase production from fungi.

Solid State Fermentation for Phytase Production

Initially the various process parameters, which influence phytase production under SSF, were optimized in the case of a mesophilic culture, *Aspergillus ficuum*. SSF was carried out using wheat bran as substrate having an initial moisture of 67 % at 30 °C. Unless otherwise mentioned 1ml fungal spores ($7X10^7$/ml spores) was used as the inoculum. Fermentation was carried out as it is described early. Figure 2 shows that the phytase production increased progressively along with fungal growth and a maximum yield (18.25 U/gds) was obtained after two days of incubation (48 h).As it is shown in Fig. 2 itself the fungal biomass was determined as glucosamine concentration. The maximum enzyme activity was obtained during the exponential phase of the fungal growth, which attained a maximum

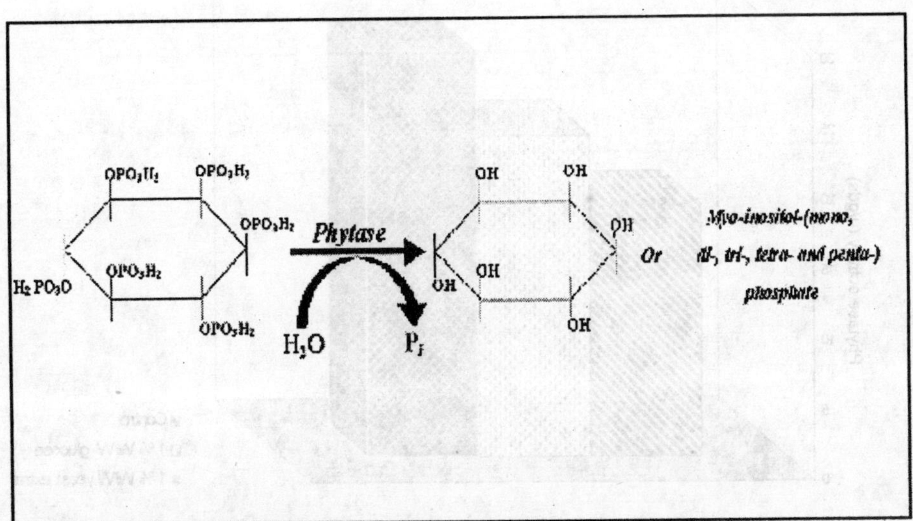

Fig. 1: Hydrolytic Reaction Catalyzed by Phytase

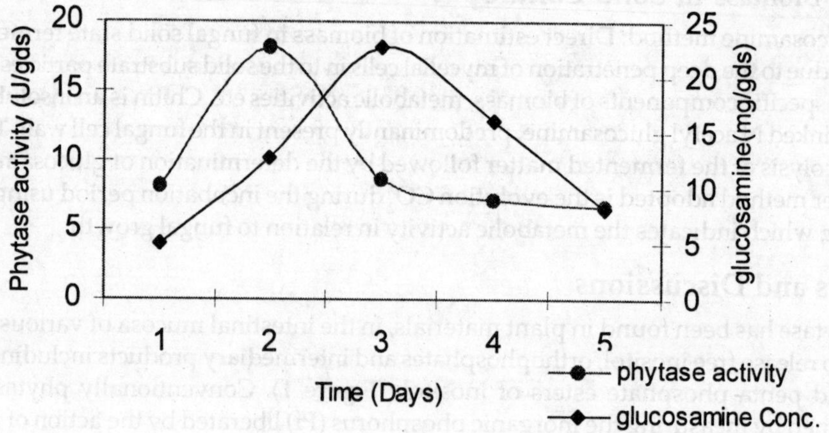

Fig. 2: Fungal Growth and Phytase Production in SSF

(glucosamine level, 22.8 mg/gds) after 3 days of incubation. Glucosamine measurement give a good indication of fungal biomass development and is the most widely used method, as glucosamine content remains constant during fungal growth and does not change with age.

Figure 3 summarizes the best yields obtained from Nutritional parameter studies in SSF. The wheat bran –salt medium was supplemented with various additional carbon and Nitrogen sources (1% W/W). A significant increase in the yield was noticed with the supplementation of either glucose as the additional carbon source (23 U/gds) or yeast extract as the additional nitrogen source (29 U/

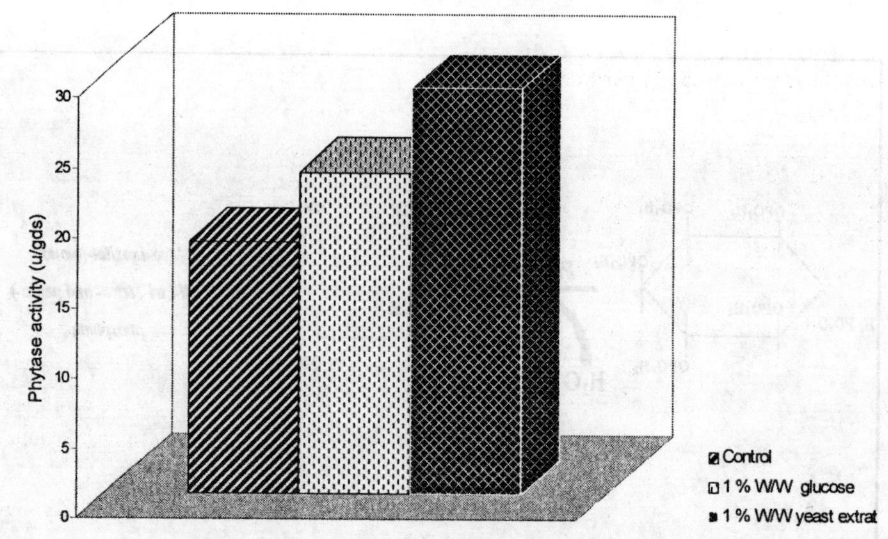

Fig. 3: Effect of Nutrient Supplementation on Phytase Production by *A.ficuum*

gds). On the otherhand, the yield without any kind of additional supplementation was (18U/gds). Supplementation of the wheat bran medium with the readily available glucose might have promoted the growth rate as well as efficient conversion of phytate with the help of more phytase production. Similarly yeast extracts is a cheap nitrogen source of various amino acids, Vitamins, minerals and other growth factors for microorganisms and thus constitute an excellent growth factor for the microorganisms. It is important to mention that beyond 1 % W/W level the supplementation doesn't favoured the enzyme yield in both cases. (data not included).

Several workers for laboratory scale studies have used packed column fermenters in Solid state fermentation (13,14). Figure 4 represents a schematic sketch of the bioreactor designed for the present study. Enzyme extraction was done from three distinguished regions of the column and the yields obtained were shown in Fig. 5. After 48 h a maximum enzyme yield (10.7 U/gds) was obtained from the top region and the lowest was (3.4 U/gds) at the lower position.The yield was reversible after 72 hours. It was more in the lower region (8.5U/gds) compared to top region (3.1U/gds). The variation in yield was comparatively less when the samples were analysed from the middle portion of the column (7.1 U/gds after 48 hours and 9.1 U/gds after 72 hours). The experiment point towards the fact that the variations in moisture level, humidy and temperature might be affecting the enzyme yield. The above parameters might be different at various sectors of the column and logically it would be the top and lower position which will affect more rather than in the middle region where these parameters might be more or less constant throughout the incubation and thus the yield is affected less over a period of time. Figure 6 reveals the level of CO_2 evolved from the bioreactor. The level of CO_2 increased steadily in accordance with the time and it reached the peak level of 7400 ppm after 44 hours of incubation, probably where the metabolic activities and viable fungal growth were at the maximum within the column fermenter.

Submerged Fermentation for Phytase Production

Both mesophilic and thermophilic cultures were used in submerged fermentation for phytase production. SmF was carried out as it was mentioned in materials and method section. Figure 7 shows

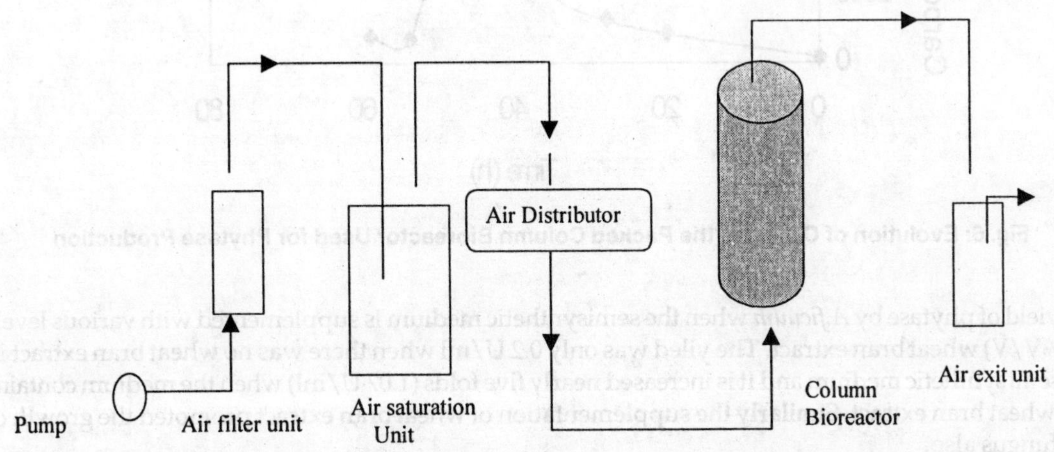

Fig. 4: Schematic Setup of Column Bioreactor for Phytase Production in SSF

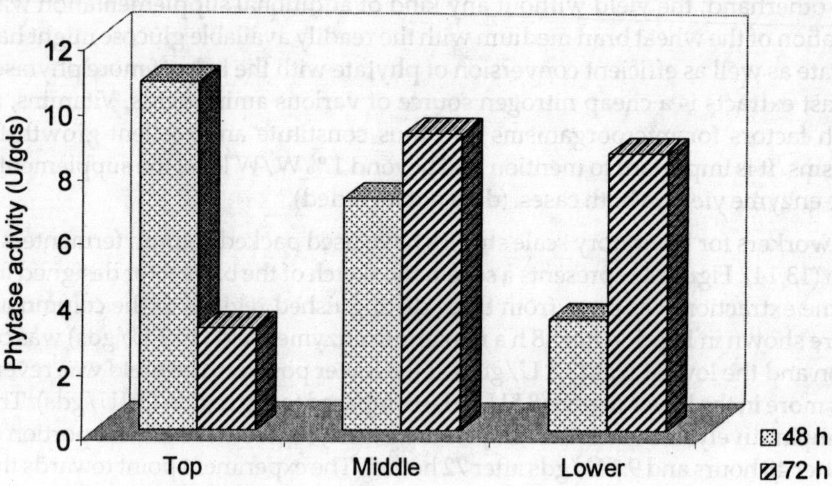

Fig. 5: Phytase Production in Different Zones of the Column Bioreactor

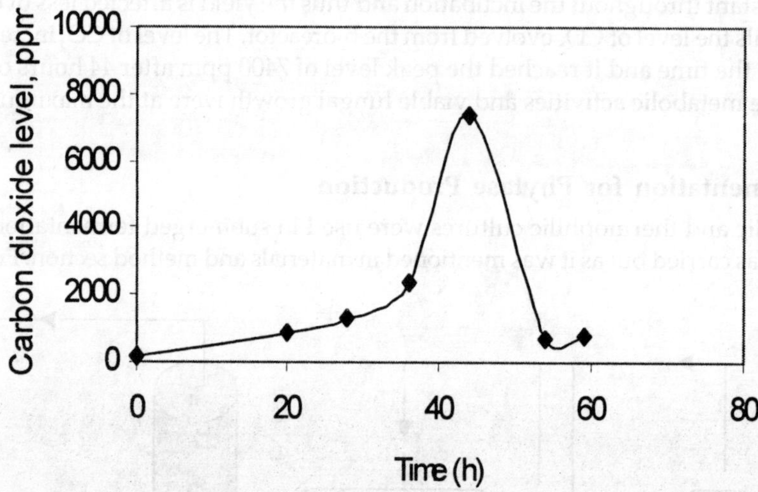

Fig. 6: Evolution of CO_2 from the Packed Column Bioreactor Used for Phytase Production

the yield of phytase by *A.ficuum* when the semisynthetic medium is supplemented with various levels of (%V/V) wheat bran extract. The yiled was only 0.2 U/ml when there was no wheat bran extract in the semi synthetic medium and it is increased nearly five folds (1.07 U/ml) when the medium contains 5 % wheat bran extract. Similarly the supplementation of wheat bran extract promoted the growth of the fungus also.

The growth was maximum when the medium contains more than 10 % V/V of wheat bran extract (Figure 7). But there was a practical difficulty, as the volume of wheatbran extract increases the growth increases and the complete nature fermentation is changed to semi-solid due to high viscosity. We also

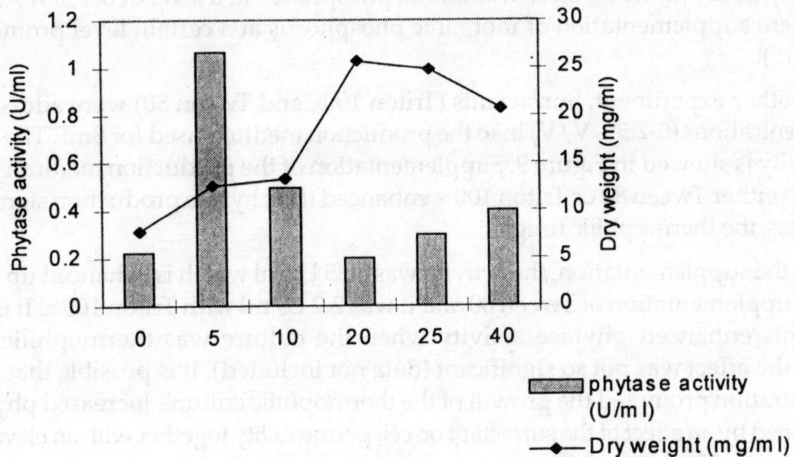

Fig. 7: Effect of Supplementation of Wheat Bran Extract on Fungal Growth and Phytase Production in SmF

noticed that in that case the enzyme yield also starts to decrease and the separation of biomass will also become a difficult task. Hence we found that the optimum level is 5 % V/V of wheat bran extract in the semisynthetic medium which promotes sufficient growth as well as better yield. The experiment shows that, like wheat bran, the extract of wheat bran also contains all the necessary nutrients to promote fungal growth and also the necessary elements to promote phytase production. Hence, this is a kind of method where wheat bran can be utilized in submerged fermentation also rather than using in Solid state fermentation alone.

In an attempt to study the effect of phosphate concentration on enzyme production, the semisynthetic medium was supplemented with various levels of phosphate concentration (0-1 %W/V).The enzyme production increased as the phosphate concentration increased till 0.5 % W/V as it is showed in Figure 8.The effect was more evident in the case of *T.auruntiacus* which is a thermophilic culture. The activity was 0.2 U/ml when there was no additional phoshate in the medium and it

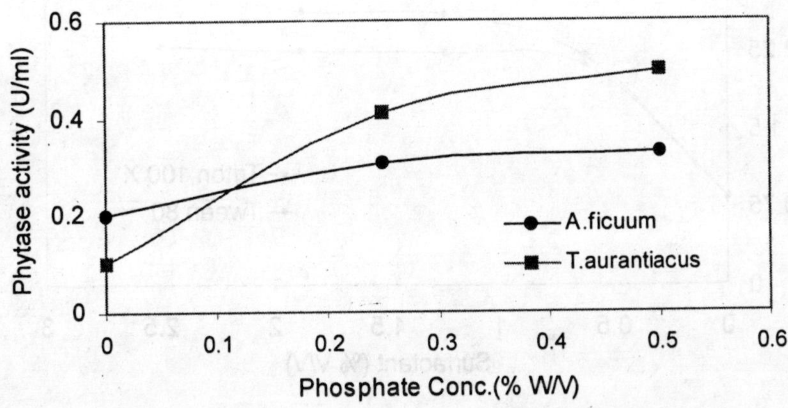

Fig. 8: Effect of Inorganic Phosphate on Phytase Production in SmF

reached nearly 0.5 U/ml when there was added phosphorus at a level of 0.5 % W/V. Similar reports are there where supplementation of inorganic phosphorus at a certain level promoted the phytase production (15).

In an another experiment, Surfactants (Triton 100x and Tween 80) were added individually at various concentrations (0-2.5% V/V) in to the production medium used for SmF. The result in terms of phytase activity is showed in Figure 9. Supplementation of the production medium with 1.25 % V/V of Surfactants either Tween 80 or Triton 100 × enhanced the phytase production significantly the case of *T.aurantiacus*, the thermophilic fungi.

Without the supplementation, the activity was 0.85 U/ml which is enhanced up to 2.6 U/ml with 1.25 % V/V supplementation of Tween 80 and it was 2.2 U/ml with Triton 100 x. It is interesting that the surfactants enhanced phytase activity when the culture was thermophilic. With *A.ficuum* (mesophilic) the effect was not so significant (data not included). It is possible that the surfactant in lower concentration promoted the growth of the thermophilic culture. Increased phytase production may be explained by an effect of the surfactant on cell permeability together with an elevated temperature of 45 °C incubation. There are earlier reports where the use of a surfactant gave higher phytase activity (16).

Figure 10 shows the temperature stability of two different microbial phytases obtained from mesophilic and thermophilic cultures.

For this the enzyme activity was checked at various temperatures such 37, 55 and 70 °C. In both cases the temperature optimum was 55°C (0.84 U/ml for mesophilic culture and 0.88 U/ml for thermophilic culture). It is interesting to note that in the case of mesophilic phytase the activity was drastically decreased when the temperature was above 55°C. On the other hand even at 70°C the phytase from *T.aurantiacus* had an activity of 0.65 units which is more than 75 % of the activity obtained at 55°C. Enzymes from thermophilic organisms are generally more thermostable, better withstand proteolysis and are more resistant to mechanical denaturation than those produced by

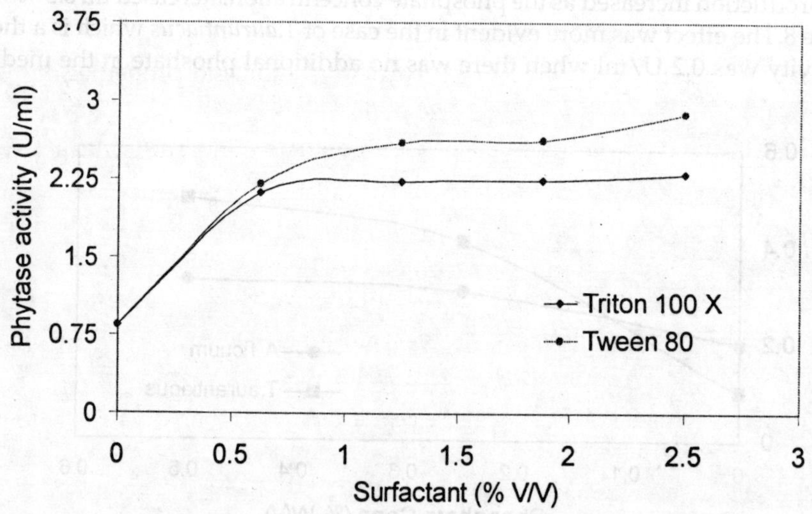

Fig. 9: Effect of Surfactants on Phytase Production in SmF

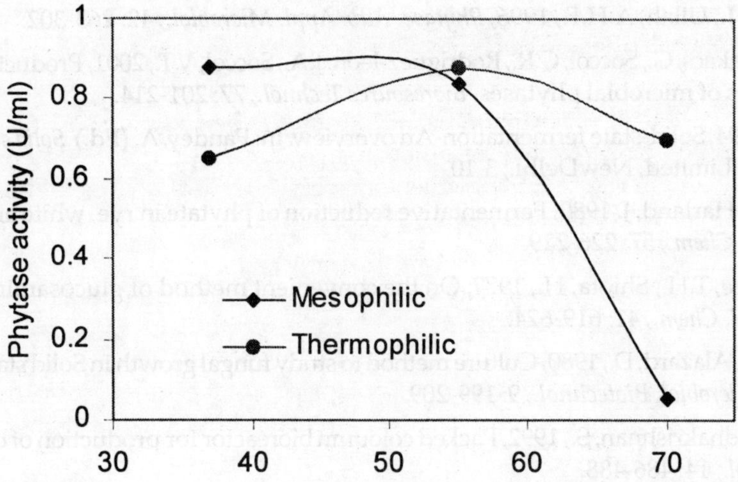

Fig. 10: Temperature Stability of Mesophilic and Thermophilic Phytases

mesophils.Thermostability is considered as an important and useful criterion for industrial application of phytase.

Conclusion

This study demonstrated the impact of various parameter optimizations for effective production of phytase from both mesophilic and thermophilic fungal cultures. Solid state and submerged fermentation were successfully empolyed for the production of extracellular phytase. It also highlights the significance of using cheaply available agro industrial residues such as wheat bran for value added product such as phytase, the feed enzyme, having tremendous potential for cleaning the environment from enhanced phosphorus pollution.

References

1. Bolland, A.R. de, Garner, G.B., O'Dell, B.L., 1975, Identification and properties of 'phytate'in cereal grains and oilseed products. *J. Agric. Food Chem.*, 23: 1186-1189.
2. Common, R.H., 1940, The phytic acid content of some poultry feeding stuffs. *Analyst*, 65: 79-83.
3. Graf, E., 1986, *Phytic acid-chemistry and application*. The Pillsbury Co Pilatus Press, Minneapolis, 42-44
4. Nair, V.C., Duvnjak, Z., 1991, Phytic acid content reductions in canola meal by various microorganisms in solid state fermentation process. *Acta Biotechnol.*, 11: 211-218.
5. Sharma, C.B., Goel, M., Irshad, M., 1978, Myoinositol hexaphosphate as a potential inhibitor of ?-amylase of different origins.*Phytochemistry*, 17: 201-204
6. Erdman, J.W., Poneros-Schneier, A., 1989, Phytic acid interactions with divalent cations in food and in the gastrointestinal tract. *Adv. Exp. Med. Biol.*, 249: 167-171.
7. Mitchell, D.B., Vogel, K., Weimann, B.J., Pasamontes, L., Vanloon, A.P.G.M., 1997, The phytase subfamily of histidine acid phosphatases; isolation of genes for two novel phytases from the fungi *Aspergillus terreus*and *myceliophthora thermophilia*. *Microbiology*, 143; 245-252

8. Wodzinski, R.J., Ullah, A.H.F., 1996, *Phytase. Adv. Appl. Microbiol.*, 42: 263-302.

9. Pandey, A., Szakacs, G., Soccol, C.R., Rodriguez-leon, J.A., Soccol, V.T., 2001, Production, purification and properties of microbial phytases. *Bioresource Technol.*, 77: 201-214.

10. Pandey, A., 1994, Solid-state fermentation-An overview. In: Pandey, A. (Ed.) *Solid state fermentation.* Wiley Eastern Limited, NewDelhi., 3-10.

11. Harland, B.F., Harland, J. 1980, Fermentative reduction of phytate in rye, white and whole wheat breads. *Cereal Chem.*, 57: 226-229.

12. Sakurai, Y., Lee, T.H., Shiota, H., 1977, On the convenient method of glucosamine estimation in kojii. *Agri. Biol. Chem.*, 41: 619-624.

13. Raimbault, M., Alazard, D., 1980, Culture method to study fungal growth in Solid state fermentation. *Eur. J. Appl. Microbiol. Biotechnol.*, 9: 199-209.

14. Pandey, A., Radhakrishnan, S., 1992, Packed coloumn bioreactor for production of enzyme. *Enzyme Microb. Technol.*, 14: 486-488.

15. Al-Asheh, S., Duvnjak, Z., 1995, The effect of phosphate concentration on phytase production and the reduction of phytic acid content in canola meal by *Aspergillus carbonarius* during a solid-state fermentation process. *Appl. Microbiol. Biotechnol.*, 43: 25-30.

16. Ebune, A., Al-Asheh, S., Duvnjak, Z., 1995, Effectsof phosphate, surfactants and glucose on phytase production and hydrolysis of phytic acid in canola meal by *Aspergillus ficuum* during solid-state fermentation. *Bioresource Technol.*, 54: 241-247.

54

Artificial Neural Networks for Design Monitoring and Control of Activated Sludge Process

H. ISHRAF ALI AHAMMAD, ASOK ADAK, G. BANERJEE, MANAS BANDYOPADHYAY

Department of Civil Engineering, Indian Institute of Technology, Kharagpur - 721 302, West Bengal

ABSTRACT

Artificial neural network (ANN) is a well-established tool for a variety of data analysis. There has been a recent application of ANN to problems related to Environmental Engineering. Solution of the problems related to activated sludge process design, operation and control are complex and time-consuming because of the nonlinearities involved in their formulation. Quick solution to these problems can be obtained through parallel processing. Neural networks provide massive parallel processing facilities and may also be used efficiently to model system with nonlinearities. In the present study the use of back propagation neural network for modeling and control of activated sludge process system has been discussed. Computer programs were developed for: design of completely mixed activated sludge process with sequential data; design of completely mixed activated sludge process with random data; control of extended aeration process for calculating recirculation ratio and sludge volume index (SVI) and monitoring the extended aeration process for calculating food to microorganism ratio (F/M) and SVI. The results of neural networks were compared with the kinetic model and field data obtained from the Southern Sewage Treatment Plant of Tata Iron and Steel Company Limited, Jamshedpur, India.

Keywords: *Artificial neural network, kinetic model, activated sludge process operation, back propagation neural network, extended aeration process, recirculation ratio, sludge volume index.*

Introduction

Neural computing is one of the fastest growing areas of artificial intelligence. The reason for this growth is that neural nets hold great promise for solving problems that have proven to be extremely difficult for standard digital computers. The neural networks have the ability to learn *i.e.* one presents them with a series of examples and from these examples the net learns the relationships involved in the training database. Artificial Neural Network (ANN), a well-established tool for a variety of data analysis, which includes fault detection in the process control systems modeling of nonlinear dynamics, solution of nonlinear programming problems. There has been a recent application of ANN to problems related to Environmental Engineering. This paper describes the application of ANN in the field of Activated Sludge Process (ASP). The attempts have been made to develop computer programs for the neural network architectures for design of completely mixed activated sludge process with sequential data, design of completely mixed activated sludge process with random data, control of extended aeration process for calculating recirculation ratio and sludge volume index and monitoring the extended aeration process for calculating food to microorganism ratio and SVI.

Neural network is defined as a computing system made up of a number of simple, highly interconnected processing elements, which processes information by its dynamic state response to external inputs. Neural nets attempts to simulate the thinking and processing procedures of the human brain by modeling the neuron. The basic component of a neuron is the soma, which is attached to the axon. The axon is electrically active and produces the pulse emitted by the neuron. Electrically passive dendrites receive input from neuron by means of specialized contacts called synapses, which act as weights to the input information. The neuron is fired only when the sum of the weighted inputs is above a certain threshold. Information received by the neuron may be processed in parallel or sequentially or as a combination of both.

The different neural network models are: Back Propagation Model (BPM), Recurrent Network Model (RNM), Change Point Detection Model with BPM (BPM_CPD), Change Point Detection Model with RNM (RNM_CPD), Non Linear Auto Regressive Moving Average Model (NARMA), Sequential Network Construction Model (SNCM), Self Organizing Feature Model (SOFM), and Moving Window Model (MWM). In the present study the back propagation model has been used. Back propagation network always consist of at least three hierarchical layers of neurons: an input layer, a middle layer (hidden layer) and an output layer as shown in the Figure 1. There may be any number of hidden layers depending upon the complexity of the problem. The network is constructed in such a way that each layer is fully connected to the next layer. In other words, every neuron in the input layer will send its output to every neuron in the middle layer-1 and every neuron in the middle layer-1 will send its output to every neuron in the middle layer-2 and so on. Every neuron in the last middle layer will send its output to every neuron in the output layer. The number of neurons in the middle layer or middle layers can be varied based on the complexity of the problem and the size of input information. For a given number of inputs, if the middle layer or middle layers are too large, it may not be possible to develop a suitable pattern. On the other hand, a middle layer or middle layers that are too small will drastically extend the number of iteration required to train the network. The input and output layer size is fixed depending upon the problem.

The general algorithm of back propagation method is described below:

Step 1

Set the initial weights and thresholds for all connections using random number between 0 and 1 or –0.5 and 0.5 and set the number of iterations N to 0.

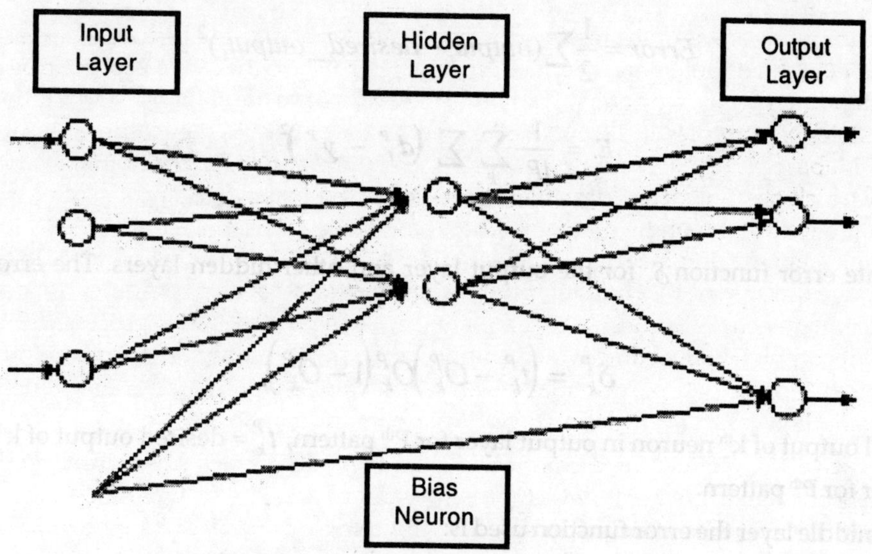

Fig. 1: Schematic of Typical Neural Network

Step 2

Read inputs and desired outputs in appropriate form. The inputs and outputs are taken in normalized form between 0.1 to 0.9 using

$$\bar{X} = \frac{0.8(X - X_{\min})}{(X_{\max} - X_{x\min})} + 0.1$$

X_{max}, X_{min} are the global maximum and minimum respectively in the input set.

Step 3

Find the output of each layer and then final output using the threshold function:

$$O_j^P = f\left(\sum_i W_{ij} X_i^P - X_o\right)$$

f () = Threshold function, P = Pattern, W_{ij} = Weight between node I of input and node j of next layer, X_i = Input to node i, X_o = Threshold set by the auxiliary node connection between the auxiliary node and the node in which we are interested.

Step 4

Find the error level E that is the deviation from the desired output if E < EO (maximum permissible error level) or the number of iteration N > NO (maximum number of iteration permitted) stop processing. Else go to step 5.

$$Error = \frac{1}{2}\sum_i (output_i - desired\_output_i)^2$$

$$E = \frac{1}{iP}\sum_P \sum_i \left(d_i^P - y_i^P\right)^2$$

Step 5

Calculate error function δ for the output layer and other hidden layers. The error function used is:

$$\delta_k^P = \left(t_k^P - O_k^P\right)O_k^P\left(1 - O_k^P\right)$$

O_k^P = actual output of kth neuron in output layer for P[th] pattern, t_k^P = desired output of kth neuron in output layer for P[th] pattern.

In the middle layer the error function used is:

$$\delta_j^P = O_j^P\left(1 - O_j^P\right)\left(\sum_k W_{jk}\delta_k^P\right)$$

O_j^P = output of mode j in middle layer for pattern P, δ_k^P = error function of kth neuron in output layer and for pattern P, W_{jk} = weights between node j in middle layer to node k in the output layer.

Step 6

Modify the weights and thresholds in the network:

$$\Delta W_{ij} = -\eta \frac{\partial E}{\partial W_{ij}}$$

$$\Delta \theta = -\eta \frac{\partial E}{\partial \theta_j}$$

$$\Delta \theta_j = \eta \sum_P \delta_j^P$$

$$W_{ij}(new) = W_{ij}(old) + \eta\left(\sum_P \delta_j^P y_i^P\right) + \alpha\left(W_{ij}(old) - W_{ij}^*\right)$$

Acceleration functions: η = Learning rate, α = Momentum term, W_{ij}^* =Value of weight two steps before current one.

Threshold function: Sigmoid function

$$f(X) = \frac{i}{1 + e^{-\frac{X - X_0}{\theta}}} = \frac{1 + \tanh(X)}{2}$$

X = Variable value, Xo Threshold value, θ = Slope of sigmoid function (assumed as 1), f (X) = 0 as X → α- and X as X → +α, output of auxiliary node (bias neuron) is kept as 1.

These operations are to be performed for every set of input data over which we want to train the neural network. When the training with the first set of inputs is over, we present the new set and desired output for that input set and so on. After all the input sets are entered, the training is repeated with less iteration to improve the performance and fault tolerance of the neural network model. If the domain error occurs re-initialize the weights or save for best weights of earlier training and use those weights for re-initialize the weights.

Kinetic Model of Activated Sludge Process

The activated sludge process is capable of converting most organic wastes to more stable soluble inorganic forms or to cellular mass. A soluble and colloidal organic material is stabilized by a group of microorganisms to CO_2 and H_2O. The stable fraction is converted to cellular mass that can be separated by sedimentation. The assumptions for in the development of ASP kinetic model are as given below:

1. Complete mixing is achieved in the aeration tank.
2. Influent substrate concentration remains constant with time.
3. No microbial solids are contained in the raw wastewater to the aeration tank.
4. No microbial activity occurs in the secondary clarifiers.
5. No sludge accumulates in the secondary clarifier
6. All the biodegradable substrate is in the soluble form.
7. Steady state condition prevails throughout the system.

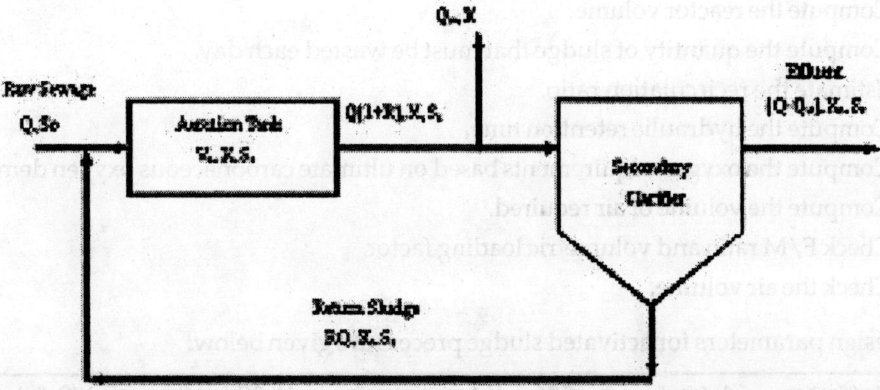

Fig. 2: Schematic of Complete Mixed Activated Sludge Plant

BSRT (biological sludge retention time) is defined as the average time a unit biomass remains in the treatment system.

$$\theta_c = \frac{(X)_T}{\left[\frac{\Delta X}{\Delta t}\right]_T}$$

where,

$(X)_T$ is the total active biomass in treatment system; $\left[\frac{\Delta X}{\Delta t}\right]_T$ is the total quantity of active biomass withdrawn from the system daily.

Material balance equation for biomass around the entire treatment system gives:

$$S_e = \frac{K_s(1 + K_d\theta_c)}{\theta_c(Y_T K - K_d) - 1}$$

Material balance equation for biomass entering and leaving the aeration tank gives:

$$\frac{1}{\theta_c} = \frac{Q}{V_a}\left[1 + R + \frac{RX_r}{X}\right]$$

where,

Q = wastewater flow rate; S_o = substrate concentration in raw wastewater; S_e = steady state substrate concentration after treatment; V_a = volume of aeration tank; X = biomass concentration in aeration tank; RQ = rate of sludge recycled; R = sludge recycle ratio (RQ/Q) ; Q_w = rate of sludge wastage; X_r = biomass concentration in the under flow from secondary clarifier.

The different process design considerations are: loading criteria, selection of the type of reactor, sludge production and process control, oxygen requirements and transfer, nutrient requirement, solid separation and effluent characteristics.

Different types of reactors for ASP are: Tapered aeration, Step aeration, Continuous flow stirred tank, Extended aeration, Pure oxygen process etc.

The design steps of the ASP are as follows:

1. Compute the reactor volume.
2. Compute the quantity of sludge that must be wasted each day.
3. Estimate the recirculation ratio.
4. Compute the hydraulic retention time.
5. Compute the oxygen requirements based on ultimate carbonaceous oxygen demand.
6. Compute the volume of air required.
7. Check F/M ratio and volumetric loading factor.
8. Check the air volume.

The design parameters for activated sludge process are given below:

Process Modification	*θ_c (day)*	*F/M*	*Volumetric loading*	*MLSS (mg/l)*	*V/Q (hr)*	*R*
CFSTR	5-15	0.2-0.6	0.8-2.0	3000-6000	3-5	0.25-1.0
EA	20-30	0.05-0.15	0.1-0.4	3000-6000	18-36	0.75-1.5

Neural Networks in Activated Sludge Process

Planning

To operate the system efficiently and effectively Q, S_o, × has to be forecasted and Y_T, k, k_d has to be evaluated so as to have steady state, dynamic and transient stabilities. Most of these studies are performed off line.

Design

Type of ASP has to be selected (CM, PF, EA, TA, CSB) and aeration volume, secondary clarifier etc. are to be designed.

Control

Closed loop and open loop control have to be used for MLSS control or Q_{wa} or Q_{wr} control or BSRT etc.

Operation

ASP operates with large uncertainties, retrieval of complete knowledge of the relation between the variables to be controlled is very complex because of the size and nonlinear behavior of the system. The decision-making procedure can be thought of as a pattern recognition problem.

Results and Discussion

Architecture-1 Description

In the present work, the network has been used for the design of activated sludge process with the input parameters such as input BOD, output BOD, BSRT, inflow, Y_T, MLSS, K_d, SVI. This network finds volume of aeration tank, waste activated sludge flow, recirculated flow, HRT, desired oxygen, food to microorganism ratio. With this network 10 patterns are tested which are not used in the training phase and which lie within the range (given in standard manuals) of each input variables. The input data for network-1 is developed from random number, which is system independent. The general architecture for network-1 is presented below:

No of input neurons	8
No of middle layers	1
No of neurons in middle layers	10
No of output neurons	6
Value of learning rate (α)	0.1
Value of momentum term (β)	0.9
No of training iterations	20000
No of training patterns used	145
No of testing patterns	49
Learning rule	Delta
Transfer function	Sigmoid
Seed	257

Contd...

Inputs are	Influent BOD (mg/l) : 299.01123 (max), 109.069824 (min)
	Effluent BOD (mg/l) : 39.788818 (max), 10.043335 (min)
	q_c (day) : 14.916992 (max), 5.0720210 (min)
	Influent flow (m^3/d) : 29946.898438 (max), 10366.8212 (min)
	Y_T: 0.799255 (max), 0.402802 (min)
	MLSS in aeration tank (mg/l) : 5965.8812348 (max), 3002.6245120 (min)
	K_d (day^{-1}) : 0.074838 (max), 0.040128 (min)
	SVI: 118.835449 (max), 60.1806640 (min)
Outputs are	Volume of aeration tank (m^3) : 6188.628906 (max), 1341.649048 (min)
	Waste flow (m^3/d) : 952.9558 (max), 97.1132 (min)
	Recirculated flow (m^3/d) : 400.28775 (max), 29.442516 (min)
	θ (HRT) (day) : 0.595958 (max), 0.213189 (min)
	Desired oxygen required (g/d) : 666164.9375 (max), 98661.265625 (min)
	F/M: 0.529054 (max), 0.201163 (min)

Result for this network for the testing pattern-1 is given below:

Input to the Network

S_o = 185.790955, S_e = 24.374374, Q = 23173.390625, Y_T = 0.703674, K_d = 0.070566, X = 3244.907715, SVI = 70.251472

Output from Neural Network

V = 4494.652832, Q_{wa} = 402.143005, Q_{wr} = 205.349243, θ = 0.284843, Desired oxygen required = 370643.093750, F/M = 0.206763

Output from Kinetic Model

V = 4585.917969, Q_{wa} = 399.158264, Q_{wr} = 196.332138, θ= 0.354949, Desired oxygen required = 370643.093750, F/M = 0.212376

Architecture-2 Description

In the present work, the network used for the design of activated sludge process with the input parameters such as input BOD, output BOD, BSRT, inflow, Y_T, MLSS, K_d, SVI. This network finds volume of aeration tank, waste activated sludge flow, recirculated flow, HRT, desired oxygen, food to microorganism ratio. With this network 10 patterns are tested which are not used in the training phase and which lie within the range (given in standard manuals) of each input variables. The input data for network-1 is developed from sequential data. The general architecture for network-2 is presented below:

No of input neurons	8
No of middle layers	1
No of neurons in middle layers	10
No of output neurons	6
Value of learning rate (a)	0.1

Contd..

Value of momentum term (β)	0.9
No of training iterations	20000
No of training patterns used	145
No of testing patterns	49
Learning rule	Delta
Transfer function	Sigmoid
Seed	257
Inputs are	Influent BOD (mg/l) : 300.0 (max), 150.0 (min)
	Effluent BOD (mg/l) : 40.0 (max), 10.0 (min)
	θ_c (days) : 15 (max), 5 (min)
	Influent flow (m³/d) : 30000.0 (max), 10000.0 (min)
	Y_T: 0.8 (max), 0.4 (min)
	MLSS in aeration tank (mg/l) : 6000 (max), 3000 (min)
	K_d (day^{-1}) : 0.075 (max), 0.055 (min)
	SVI: 120.0 (max), 60.0 (min)
Outputs are	Volume of aeration tank (m³) : 6141.176758 (max), 1257.142944 (min)
	Waste flow (m³/d) : 1223.529541 (max), 87.84314 (min)
	Recirculated flow (m³/d) : 550.588257 (max), 19.76706 (min)
	θ (HRT) (day) : 4.912941 (max), 3.017143 (min)
	Desired oxygen required (g/d) : 889623.75 (max), 79505.773438 (min)
	F/M: 0.528846 (max), 0.207386 (min)

Result for this network for the testing pattern-1 is given below:

Input to the Network

$S_o = 300.0$, $S_e = 10.0$, $Q = 10000.0$, $Y_T = 0.8$, $K_d = 0.065$, $X = 6000.0$, SVI 60.0

Output from Neural Network

$V = 2155.249512$, $Q_{wa} = 469.033447$, $Q_{wr} = 195.882004$, $\theta = 4.502432$, Desired oxygen required = 215595.46875, F/M = 0.350251

Output from Kinetic Model

$V = 1874.747314$, $Q_{wa} = 374.948914$, $Q_{wr} = 126.545174$, $\theta = 4.499394$, Desired oxygen required = 185335.546875, F/M = 0.355603

Architecture-3 Description

In the present work, the network has been used for the operation of extended aeration process with the input parameters such as input BOD, output BOD, inflow, MLSS, MLVSS and half an hour settilability. This network finds food to microorganism ratio and SVI. With this network 10 patterns are tested which are not used in the training phase. The input and output data for this network during training and testing is taken from the existing treatment plant of TISCO, Jamshedpur.

No of input neurons	6
No of middle layers	1
No of neurons in middle layers	7
No of output neurons	2
Value of learning rate (a)	0.1
Value of momentum term (b)	0.9
No of training iterations	20000
No of training patterns used	30
No of testing patterns	10
Learning rule	Delta
Transfer function	Sigmoid
Seed	257
Inputs are	Influent BOD (mg/l) : 250.0 (max), 44.0 (min)
	Effluent BOD (mg/l) : 15.0 (max), 1.0 (min)
	Influent flow (m^3/d) : 245000.0 (max), 100000.0 (min)
	MLSS in aeration tank (mg/l) : 8400.0 (max), 1500.0 (min)
	MLVSS in aeration tank (ms/l) : 4210.0 (max), 1180.0 (min)
	Half an hour settilability in 1000 ml (mg/l) : 960.0 (max), 80.0 (min)
Outputs are	F/M: 0.175638 (max), 0.024747 (min)
	SVI: 318.40448 (max), 21.186441 (min)

Result for this network for the testing pattern-1 is given below:

Input to the Network

S_o = 122.0, S_e = 7.0, Q = 110000.0, Half an hour settilability = 920.0, MLSS = 6040.0, MLVSS = 3540.0

Output from Neural Network

F/M = 0.061892, SVI = 152.904129

Output from Kinetic Model

F/M = 0.061892, SVI = 152.904129

Architecture-4 Description

In the present work, the network has been used for the control of extended aeration process with the input parameters such as MLSS, returned sludge suspended solids and half an hour settilability. This network finds recirculation ratio and SVI. With this network 10 patterns are tested which are not used in the training phase. The input and output data for this network during training and testing is taken from the existing treatment plant of TISCO, Jamshedpur.

No of input neurons	3
No of middle layers	1
No of neurons in middle layers	2
No of output neurons	2
Value of learning rate (α)	0.1
Value of momentum term (b)	0.9
No of training iterations	2000
No of training patterns used	47
No of testing patterns	10
Learning rule	Delta
Transfer function	Sigmoid
Seed	257
Inputs are	MLSS in aeration tank (mg/l) : 8504.0 (max), 2800.0 (min)
	Returned sludge suspended solids (mg/l) : 24980.0 (max), 6380.0 (min)
	Half an hour settilability in 1000 ml (mg/l) : 950.0 (max), 90.0 (min)
Outputs are	Recirculation ratio: 1.88 (max), 0.27 (min)
	SVI: 182.0 (max), 0.0 (min)

Result for this network for the testing pattern-1 is given below:

Input to the Network

MLSS = 5399.0, Returned sludge suspended solids = 16880.0, half an hour settilability = 169.0

Output from Neural Network

Recirculation ratio = 0.508182, SVI = 33.202728

Output from Kinetic Model

Recirculation ratio = 0.430, SVI = 31.0

Actual Field Data

Recirculation ratio = 0.470, SVI = 31.0

Figure 3 represents root mean square error Vs number of iteration, from which it is concluded that variation of the curve is large but overall pattern is reducing with number of iteration. Figure 4 represents root mean square error Vs learning rate, from which it is concluded that variation is like parabolic, *i.e.* for small learning rate RMS error is small and for large values of learning rate RMS error achieved is large for the same number of iterations. Figure 5 represents number of iterations to achieve the same RMS error Vs momentum term, from which it is concluded that for small values of momentum term number of iterations required is large and for large value of momentum term number of iteration required to achieve same RMS error is less. This indirectly reflects the time required for executing the program.

The results of network 3 and 4 are compared with kinetic model and field data. It has been found that neural network models are performing better than kinetic models. The results of sequential data

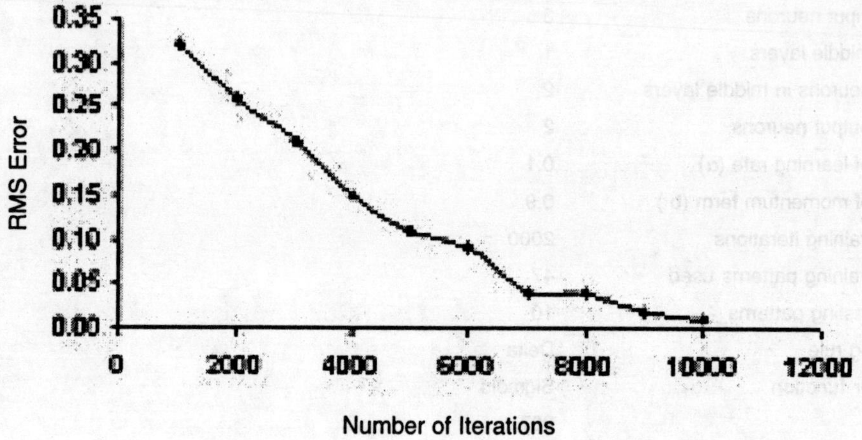

Fig. 3: RMS Error vs. Number of Iterations (Network - 1)

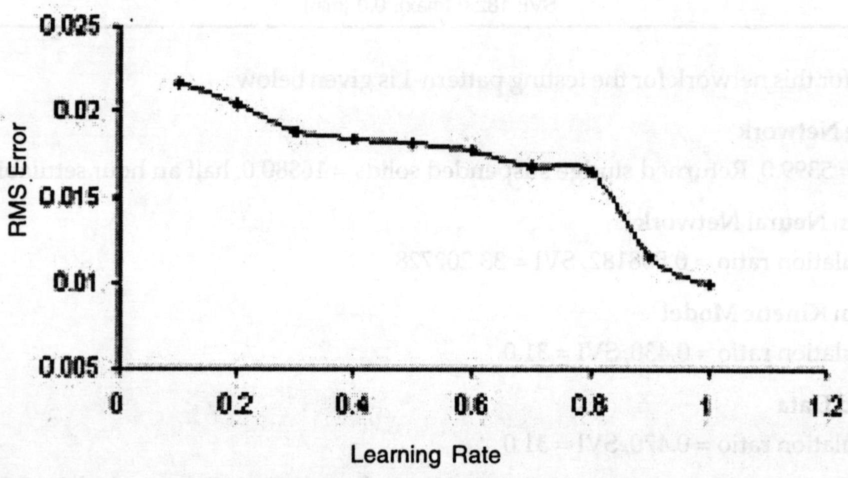

Fig. 4: RMS Error vs. Learning Rate (Network - 1)

and random data shows that if a normalized data used for training should be such that the frequency distribution should be nearly uniform, between any range, if it deviate much then generalized capability is less. Neural network is very effective in handling the problems where the functional relation is not valid, such as for sludge bulking problems in ASP etc., but training should be available of such type extensively for generalization of such system models.

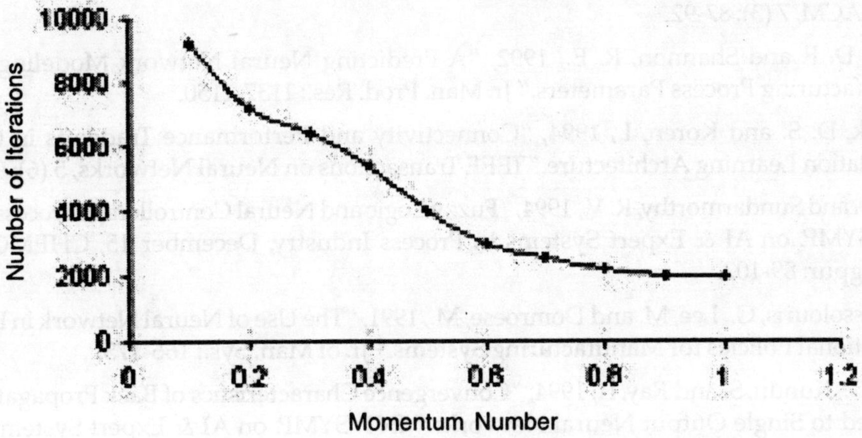

Fig. 5: Number of Iterations vs. Momentum Term (Network - 1)

Conclusions

Activated sludge process is a unique system with variety of process variation, in which modeling of process is complicated when large number of process parameters are considered. Application of ANN simplifies the solution of the differential equations representing such process. As the number of iterations increases the RMS error decreases. For small learning rate the RMS error is small and for large value of learning rate RMS error is large for same number of iterations. For small values of momentum term number of iterations required is large and for large value of momentum term number of iteration required to achieve same RMS error is less. This indirectly reflects the time required for executing the program. Increase of number of nodes in the middle layer will reduce the RMS error with other parameters being fixed to some extent. The method is advantageous because it takes less time for solving complicated nonlinear problems.

REFERENCES

1. Andrew, C. M. and Alonzo, W. L., 1976, "Latest Cost Design of Activated Sludge System." Jr. of WPCF, 48 (5): 889-905.

2. Samanta, A. N. and Roy, S., 1994, "Inferential Control of Refinery Stream Properties using Artificial Neural Network." INT. SYMP. on AI & Expert System in Process Industry, December 15, 1994, CHEMCON 94, IIT Kharagpur: 141-147.

3. Tang, C. C., Brill, D. and Pfeffer, J. T., 1987, "Optimization Techniques for Secondary Wastewater Treatment System." ASCE, Jr. of Env. Engg., 113 (5): 935-951.

4. Giles, C. L. and Chan, H. H., 1995, "Constructive Learning of Recurrent Neural Networks: Limitations if Recurrent Cascade Correlation and a Simple Solution." IEEE Tran. on Neural Networks, 6 (4): 829-836.

5. Leslie, C. P. and Lim, H. C., 1981, "Biological Wastewater Treatment." Marccel Dekker, Inc., New York and Basel.

6. Rumelhart, D. E. and Windrow, B., 1994, "The Basic Ideas in Neural Networks." Communications of the ACM, 7 (3): 87-92.

7. Cook, D. F. and Shannon, R. E., 1992, "A Predicting Neural Network Modeling System for Manufacturing Process Parameters." Jr. Man. Prod. Res.: 1137-1150.

8. Pathak, D. S. and Koren, I., 1994, "Connectivity and Performance Tradeoffs in the Cascade Correlation Learning Architecture." IEEE Transactions on Neural Networks, 5 (6): 930-935.

9. Rao, D. and Sundarmorthy, R. V., 1994, "Fuzzy Logic and Neural Controllers in Processes Control." INT. SYMP. on AI & Expert Systems in Process Industry, December 15, CHEMCON 94, IIT Kharagpur: 89-102.

10. Charyssolouris, G., Lee, M. and Domroese, M., 1991, "The Use of Neural Network in Determining Operational Policies for Manufacturing Systems." Jr. of Man. Sys.: 166-175.

11. Saha, G., Kundu, S. and Ray, P., 1994, "Convergence Characteristics of Back Propagation Method Applied to Single Output Neural Networks." INT. SYMP. on AI & Expert Systems in Process Industry, December 15, CHEMCON 94, IIT Kharagpur: 103-115.

12. Rogers, J. L., 1994, "Simulating Structural Analysis with Neural Network." Jr. of Computation in Civil Engineering, 8 (3): 252-265.

13. Savkovic, J., 1994, "Use of Neural Nets for Process Fault Diagnosis." INT. SYMP. on AI & Expert Systems in Process Industry, December 15, CHEMCON 94, IIT Kharagpur.

14. Dayhof, J. E., 1990, "Neural Network Architecture." Van Nostrand Reinholi, New York.

15. Flood, L. and Kartam, N., 1994, "Neural Networks in Civil Engineering-1: Principles and Understanding." Jr. of Computation in Civil Engineering, 8 (3): 131-148.

16. Flood, L. and Kartam, N., 1994, "Neural Networks in Civil Engineering-2: Principles and Understanding." Jr. of Computation in Civil Engineering, 8 (3): 149-161.

17. Benefield, L. D. and Randall, C. W. "Biological Process Design for Wastewater Treatment." Prentice-hall Inc. Englewood Cliffs N. J.

18. Chao, L. C. and Skibnieuski, M. J., 1994, "Estimation Construction Productivity Neural Network based Approach." Jr. Computations in Civil Engineering, 8 (3): 234-251.

19. Metcalf and Eddy. Inc. "Wastewater Engineering: Treatment Disposal Reuse." Tata McGraw-Hill, New Delhi, 1993.

20. Murtaza, M. B. and Fisher, J., 1994, "Neuromodex-neural Network System for Modular Construction Decision making." Jr. of Computations in Civil Engineering, 8 (3): 231-233.

21. William, N. K. and Bovee, K., 1994, "Neural Networks for River Flow Prediction." Jr. of Computations in Civil Engineering, 8 (2): 201-220.

22. Bhat, N. and Mcavor, T. J., 1990, "Use of Neural Nets for Dynamic Modeling and Control of Chemical Process systems." Comp. Chem. Engg., 14 (4): 573-583.

23. Gray, N. F. "Biology of Wastewater Treatment." Oxford Science Publications, 1989.

24. Gagarin, N. Flood, I. and Albrecht, P., 1994, "Computing Truck Attributes with Artificial Neural Networks." Jr. of Computations in Civil Engineering, 8 (3): 179-199

25. Lisboa, P. G. J. "Neural Networks Current Applications." Chapman and Hall.

26. Bhagat, P., 1990, "An Introduction to Neural Nets." Chem. Engg. Pro., (4): 55-60.

27. Kalara, P. K., Srivastava, A. *et al.*, 1992, "Possible Applications of Neural Nets to Power System Operation and Control." Electrical Power System Resources, 25: 83-90.

28. Hegazy, T. and Moselht, O., 1994, "Analog Based Solution to Markup Estimation Problems." Jr. of Computations in Civil Engg., 8 (2): 72-87.

29. Jalee, V. V. and Sarker, P., 1994, "Application of Artificial Neural Networks in Air Pollution Monitoring." INT. SYMP. on AI & Expert Systems in Process Industry, December 15, 1994, CHEMCON 94, IIT Kharagpur.

30. Lai, W. and Berthonex, P.M. "Testing Expert System for Activated Sludge Process Control." ASCE, Jr. of Env. Engg., 116 (5): 890-909.

31. Carpenter, W. and Jean-Francois, 1994, "Common Misconception about Neural Networks as Approximation." Jr. of Computations in Civil Engg., 8 (3): 345-359.

32. Tarng, Y.S. and Wang, Y.S., 1994, "A Neural Network Controller for Constant Turning Force." Int. Jr. Mach., 34 (4): 453-460.

33. Szewezky, Z. P. and Hajela, P., 1994, "Damage Detection in Structural based on Feature Sensitive Neural Networks." Jr. of Computations in Civil Engg., 8 (3): 163-177.

34. Boger, Z., 1994, "Modeling Process Industry Plants by Artificial Neural Networks." INT. SYMP. On AI & Expert Systems in Process Industry, December 15, 1994, CHEMCON 94, IIT Kharagpur.

55

Biohydrolysis of Urea from Urea-Bearing Wastewater

P.P. PATHE, T. NANDY, S.N. KAUL, C.V. DESHPANDE & L. SZPYRKOWICZ

Scientists, National Environmental Engineering Research Institute, Nagpur - 440 020

ABSTRACT

Biological stabilization of urea is a two staged process; (i) urea hydrolysis and (ii) ammonia stripping/nitrification-denitrification. Hydrolysis of urea is brought out by employing microbiological technique, in which ureolytic bacteria Bacillus Pașteuri convert urea into ammonium carbonate as an intermediate product and then into ammonia and carbon-di-oxide as the end products. Ammonia thus produced is either stripped off by usual methods or after converting into nitrate using chemoautotraphic bacteria. Nitrosomonas sp. which oxidized ammonia into nitrite and further into nitrate by Nitrobacter sp. On dentrification, nitrate is finally converted into nitrogen gas by means of heterotrophic bacteria. Based on microbiological treatment technique urea biohydrolyser was designed and installed for hydrolysis of urea from urea bearing wastewater. Steam under pressure followed by cascade aeration was applied for ammonia stripping. Paddy straw was used as carbon source for microorganisms in urea hydrolyser.

Ammonia free wastewater was routed through series of ponds cultured with algae along with septic tank sewage. The final treated effluent conform standards prescribed by regulatory agencies for discharge of effluent into inland surface water. Details of stabilization of urea from urea bearing wastewater using urea biohydrolyser are presented in this paper.

Keywords: Urea biohydrolyser, urea, ammonia, nitrogenous fertilizer industry wastewater, microbiological treatment, nitrification, denitrification.

Introduction

The production capacity of nitrogenous fertilizer in the country has increased from a modest capacity of 85, 000 tonnes in 1950-51 to 9.480 million tonnes in March, 1997, and the projected production by the end of 2004-05 is 9.940 million tonnes (India 1995 & Monthly Review of Indian Economy, 1997). The consumption of fertilizer per hecter has been steadily rising from mearly 0.55 kg in 1950-51 to 68.00 kg in 1995 to meet high food demand due to high population in the country. Urea and ammonia are the major constituents of wastewater being generated from nitrogenous fertilizer industry which manufacture ammonia and urea as the major products. The volume of wastewater from ammonia production generally varies from 4.9 to 6.7 m^3/tonne of ammonia produced with concentration of ammonia and urea varying from 4.65 to 10.0 and 15.0 to 17.5 kg/tonne of ammonia produced, respectively. Wastewater generation from urea plant varies from 8.0 to 12.0 m^3/tonne of urea produced depending on the quantity of raw material consumption, with ammonia and urea concentrations varying between 200-1500 mg/L and 340-20,000 mg/L, respectively (IS-9841, 1981).

Urea as such is non-toxic for human, animal and aquatic life even at high concentrations. The test fish-*Creek Chub* in the Detroit river were observed to survive even at higher dose of more than 1000 mg/L of urea (Gillette *et al*, 1952). However, urea under favourable conditions hydrolyse to ammonia and carbon-di-oixde, and ammonia concentration of 1.2 to 3.0 mg/L is toxic to fish.

Evans *et al*. (1973) reported that 4 to 6 days are required for urea degradation at 20°C with urea concentration ranged between 1 to 15 mg/L and 14 days or more at a temperature lower than 8°C. This indicates that urea hydrolysis is a very slow process and can takes place only in presence of urase, otherwise the hydrolysis may assumed to be negligible.

Process of Nitrogen Removal

Nitrogen removal through microbiological treatment of wastewater containing urea involves the following unit processes:

Urea Hydrolysis

Urea hydrolysis is brought out by microorganisms mainly ureolytic bacteria *Bacillus pasteurii* as follows:

$$O=C(NH_2)_2 + H_2O \rightarrow (NH_4)_2CO_3 \rightarrow 2NH_3 + CO_2 + H_2O$$

Urea | Ammonium carbonate | Ammonia | Carbon-dioxide

Ammonia and ammonium ions is aqueous solution remain in equilibrium as follows:

$$NH_4^+ \text{ ====== } NH_3 + H^+$$

At pH 7.0 and temperature 0 to 250°C, over 99% of ammonical nitrogen is in ammonium (NH_4^+) form. Autotrophs include the nitrifying bacteria which obtain energy from oxidizing inorganic nitrogen (chemoautotrophs) and algae which obtain energy from sunlight (photoautotrophs).

Nitrification

Ammonia is biologically oxidised to nitrate in two stage process by two groups of chemoautotrophic bacteria which operates in sequence. The first step is oxidising ammonia into nitrite by members of genus *Nitrosomonas sp*. Nitrite is further oxidised to nitrate by members of genus *Nitrobacter sp*. as follows:

$$2NH_4^+ + 3O_2 \xrightarrow{\text{Nitrosomonas Sp.}} 2NO_2 + 4H^+ + 2H_2O + \text{Energy}$$

$$2NO_2 + O_2 \xrightarrow{\text{Nitrobacter Sp.}} 2NO_3 + \text{Energy}$$

Energy released in this oxidation is used in synthesizing cell materials from carbon-di-oxide. It is evident from these equatios that oxygen is required for nitrification process which gives rise to oxygen demand in wastewater. Liberation of hydrogen ion tends acid production during nitrification, which may result in pH drop in poorly buffered wastewater, resulting in loss of process stability.

Denitrification

Denitrification of nitrate is brought out by heterotrophic bacteria first into nitrite and then to nitrogen gas. As the energy yield is lower than that from oxygen respiration, denitrification is effective only at low dissolved oxygen concentration. An external carbon source is necessary to obtain maximum reaction rate. Methanol can be used as one of the carbon source and the reaction can be represented as follows:

$$NO_3 + CH_3OH \rightarrow N_2 + CO_2 + H_2O + OH^-$$

Urea bio-hydrolyser breaks urea molecule into ammonia and carbon-dioxide with the help of ureolytic bacteria *Bacillus pasteurii*. After hydrolysis the ammonia produced is stripped off. The remaining ammonia nitrogen undergoes nitrification and nitrate is formed. Nitrate and ammonia are utilized by algae for growth. Unutilised nitrate is converted into nitrogen gas after denitrification. Thus nitrogen free effluent is produced.

Brief Details of Nitrogenous Fertilizer Industry Under Study

The major products being produced at the coal based nitrogenous fertilizer industry were ammonia, urea, nitric acid, ammonium nitrate, ammonium bicarbonate, coke and tar with rated capacities of 900, 1000, 200, 35, 12, 325 and 10 tonnes/d, respectively. The total water requirement of the fertilizer industry was around 79124 m^3/d which includes water for process, cooling and domestic purposes. Wastewater generation from the various unit processes of the industry was measured. The total wastewater generation, which was in the range of 34608-37944 m^3/d. Table 1 presents the quantity of wastewater, generated from the various units of the industry.

This paper deals with studies on a full-scale effluent treatment plant comprising of urea bio-hydrolyser unit and additional online lagoons in series cultured with algae to treat nitrogenous wastewater from a fertilizer industry.

Material and Methods

Full Scale Urea Bio-hydrolyser Plant

The nitrogenous fertilizer industry had a full-scale effluent treatment plant comprising of a urea bio-hydrolyser unit followed by lagoons in series. Urea bio-hydrolyser plant installed comprised of 16 compartments of different sizes and arranged in 8 horizontal rows. Total volume of all these compartments was 6012 m^3 with detention period of 90 hours. The layout plan of Urea bio-hydroliser is shown in Fig. 1. Pressurised steam was introduced at the bottom of compartment Nos. 11, 12, 13,14,15 and 16 for stripping of ammonia produced during urea hydrolysis. In addition the treated effluent from hydrolyser plant was routed through cascade aeration system to reduce ammonia concentration in the effluent and further treated through 10 laoongs in series. Ammonia stripped effluent was received in lagoon 1 and after routing through other lagoons was being finally discharged into lagoon 9. Lagoon 9 has a holding capacity of 1,00,000 m^3. This lagoon also received combined wastewater from other process units except steam generation and power plant. Septic tank effluent from industry and township was also routed through the ETP and was being discharged into lagoon 4. Total detention time of all these lagoons was 25 days. The schematic of modified Effluent Treatment Plant is depicted in Fig. 2.

Table 1: Quantity of Wastewater Generated from Various Process Units

Sl. No.	*Process Unit*	*Quantity, m³/day*			
		Range			*Average*
1.	Oil Gasification	1656	-	1752	1680
2.	Rectisol	792	-	936	840
3.	DM				
	Acidic	1344	-	1560	1440
	Alkaline	864	-	1056	984
4.	Steam Generation	1584	-	1656	1608
5.	Compressor House	384	-	528	456
6.	Coal and handling house	384	-	576	480
7.	Urea	1800	-	1968	1896
8.	Coke Oven	1152	-	1296	1224
9.	Nitric Acid	360	-	528	432
10.	Ammonium Nitrate	288	-	432	360
11.	Ammonium Bi-carbonate	168	-	264	216
12.	Power Plant	9600	-	10800	10128
13.	Septic Tank Effluent	14322	-	14592	14424
	Combined Effluent	34608	-	37944	36168

Microbial culture produced during pilot plant studies was used for startup of the urea bio-hydrolyser. Chopped paddy straw of 10 to 50 mm sizes was used as the carbon source for microorganisms. As urea-bearing wastewater was phosphate deficient super phosphate was added in requisite doses. Algae *chlorella* was cultured in lagoon 5,6,7 and 8 to utilize ammonia.

The performance of the urea bio-hydrolyser unit and effluent treatment plant were monitored. Wastewater samples were collected from each compartment of the urea hydrolyser plant aι one-hour interval over a period of 8 hours for 7 days. The samples were analysed for pH, TDS, SS, total ammonical nitrogen, free ammonical nitrogen, total kjeldhal nitrogen, nitrite nitrogen, nitrate nitrogen as per Standard Methods, 1989. Urea was analysed as per method described by Watt and Chrisp, 1954. Characteristics of inlet and outlet wastewater from each compartment of urea bio-hydrolyser are presented in Table 2. Hourly samples from each lagoon were also collected over a period of 8 hours for 7 days and analysed for usual parameters. Characteristics of influent and effluent from urea bio-hydrolyser and effluent characteristics from various lagoons are in Table 3. The characteristics of

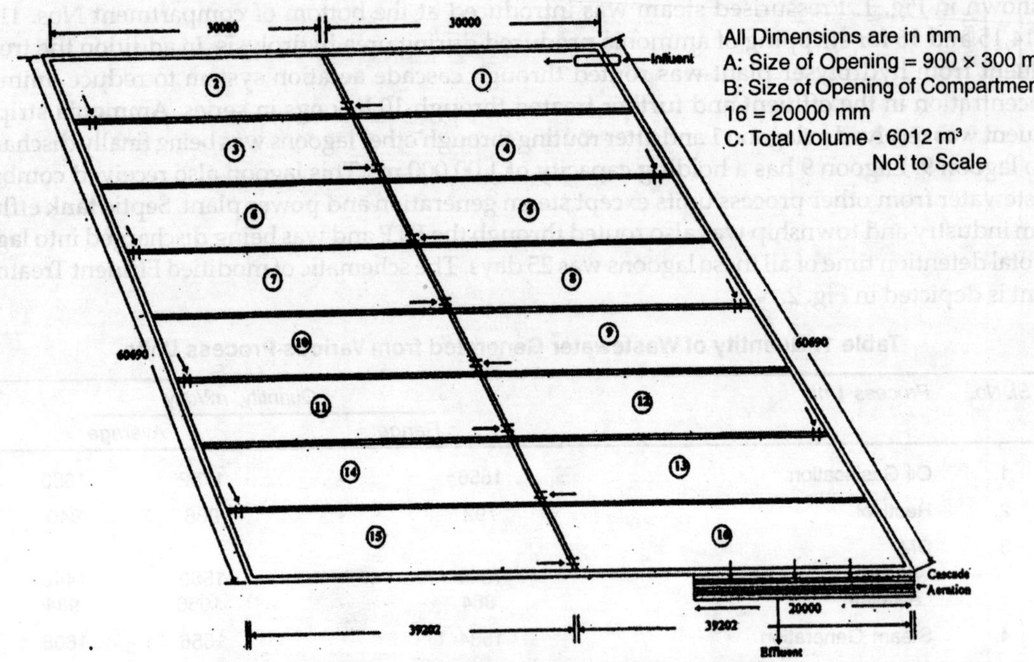

Fig. 1: Layout Plan of Biological Urea Hydrolyser Plant

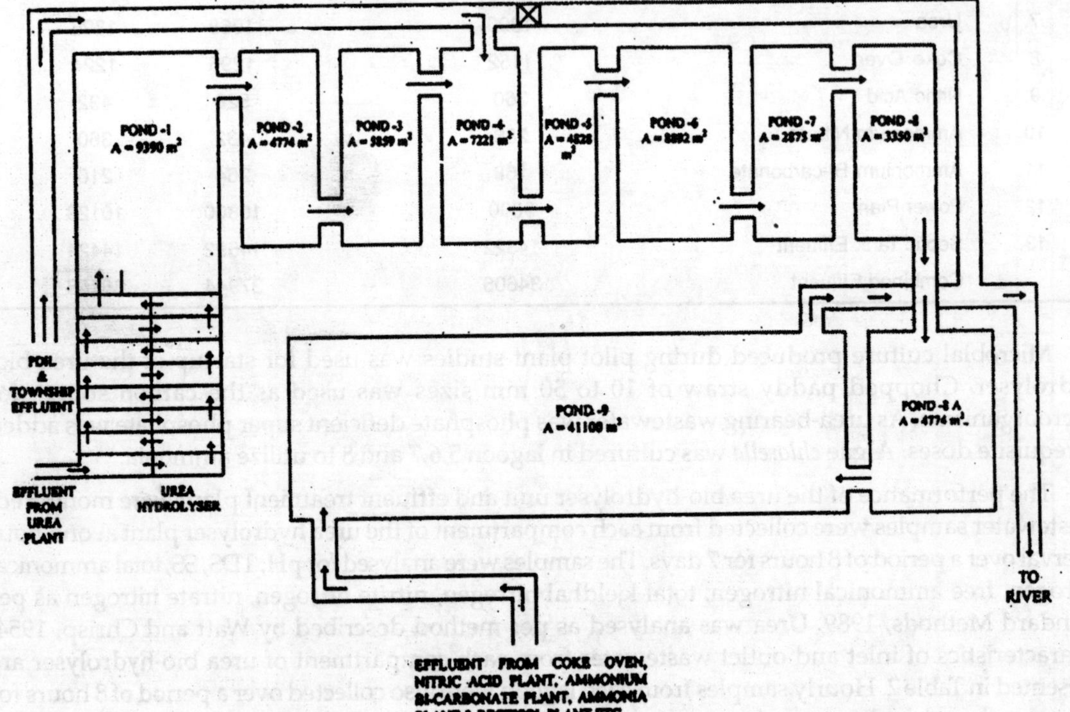

Fig. 2: Schematic Flow Diagram of Modified Effluent Treatment Plant

Table 2: Characteristics of Influent and Effluent from each Compartment of Urea Bio-hydrolyser Plant

Effluent Type	*Parameters*									
	Temp, °C	*pH*	*TDS*	*SS*	*Total Ammonical N*	*Free Ammonical N*	*Total Kjeldhal N*	*Urea*	*Nitrite (NO_2)-N*	*Nitrate (NO_3)-N*
Compartment No. 1 Influent	24 ± 3	10.1 ± 0.3	1726 ± 120	126 ± 18	975 ± 55	794 ± 106	1840 ± 168	1230 ± 182	0.9 ± 0.10	2.0 ± 1.0
Compartment No. 1 Effluent	25 ± 3	10.1 ± 0.3	1694 ± 104	794 ± 76	984 ± 64	804 ± 101	1794 ± 152	1200 ± 176	1.28 ± 0.10	2.1 ± 1.1
Compartment No. 2 Effluent	25 ± 3	10.1 ± 03	1538 ± 102	748 ± 84	996 ± 58	864 ± 94	1678 ± 152	1200 ± 176	1.28 ± 0.11	2.2 ± 1.1
Compartment No. 3 Effluent	25 ± 3	10.1 ± 0.3	1420 ± 96	680 ± 92	1034 ± 68	936 ± 88	1476 ± 136	854 ± 122	1.36 ± 0.11	2.3 ± 1.1
Compartment No. 4 Effluent	25 ± 2	10.2 ± 0.2	1392 ± 90	656 ± 74	1082 ± 78	996 ± 92	1282 ± 140	628 ± 108	1.40 ± 0.14	2.4 ± 1.2
Compartment No. 5 Effluent	25 ± 2	10.2 ± 0.2	1356 ± 92	632 ± 68	1132 ± 80	1074 ± 86	1204 ± 132	440 ± 82	1.46 ± 0.12	2.5 ± 1.2
Compartment No. 6 Effluent	25 ± 2	10.3 ± 0.2	1318 ± 91	600 ± 52	1158 ± 84	1092 ± 90	1196 ± 124	356 ± 64	1.52 ± 0.10	2.6 ± 1.2
Compartment No. 7 Effluent	25 ± 2	10.3 ± 0.2	1250 ± 66	580 ± 48	1004 ± 92	978 ± 84	1042 ± 116	334 ± 58	1.64 ± 0.11	2.7 ± 1.1
Compartment No. 8 Effluent	25 ± 2	10.3 ± 0.2	1222 ± 78	528 ± 50	820 ± 68	798 ± 70	886 ± 110	310 ± 50	1.76 ± 0.13	2.7 ± 1.2
Compartment No. 9 Effluent	25 ± 3	10.2 ± 0.2	1204 ± 62	462 ± 38	806 ± 52	728 ± 46	865 ± 112	294 ± 56	1.88 ± 0.12	2.8 ± 1.2
Compartment No. 10 Effluent	25 ± 3	10.2 ± 0.2	1178 ± 72	370 ± 42	712 ± 46	641 ± 48	830 ± 98	278 ± 52	2.06 ± 0.14	2.9 ± 1.1
Compartment No. 11 Effluent	25 ± 3	10.1 ± 0.2	1144 ± 64	296 ± 38	604 ± 50	558 ± 42	782 ± 84	264 ± 64	2.18 ± 0.18	3.0 ± 1.1
Compartment No. 12 Effluent	25 ± 3	10.1 ± 0.2	1109 ± 70	252 ± 32	534 ± 38	496 ± 44	622 ± 76	238 ± 56	2.26 ± 0.20	3.2 ± 1.1
Compartment No. 13 Effluent	24 ± 3	10.0 ± 0.2	1090 ± 62	224 ± 36	478 ± 32	442 ± 40	572 ± 64	222 ± 60	2.56 ± 0.24	3.5 ± 1.2
Compartment No. 14 Effluent	24 ± 3	10.0 ± 0.2	1058 ± 74	208 ± 32	414 ± 30	394 ± 42	500 ± 58	196 ± 70	2.64 ± 0.26	3.7 ± 1.2
Compartment No. 15 Effluent	24 ± 3	10.0 ± 0.2	1028 ± 68	196 ± 24	382 ± 34	346 ± 36	442 ± 56	178 ± 62	2.66 ± 0.28	3.8 ± 1.2
Compartment No. 16 Effluent	24 ± 3	10.0 ± 0.2	994 ± 60	138 ± 16	346 ± 32	304 ± 32	400 ± 52	160 ± 52	2.70 ± 0.30	4.0 ± 1.2

All values except pH and temperature are in mg/L; Hourly sample collection over a period of 8 hours for 7 days

Table 3: Characteristics of Influent and Effluent from Urea Bio-hydrolyser Plant and Effluent Characteristics of Various Lagoons at ETP

Effluent Type	*Parameters*														
	pH	*TDS*	*SS*	*DO*	*COD*	*BOD*	*Total Ammonical-N*	*Free Ammonical-N*	*Total Kjeldahl-N*	*Urea*	*Nitrite (NO_2) -N*	*Nitrate (NO_3) -N*	*Oil & Grease*	*Total Chromium (Cr)*	*Cyanide (CN)*
Urea Bio-hydrolyser Influent	10.1 ± 0.3	1726 ± 120	126 ± 18	ND	332 ± 56	200 ± 25	975 ± 55	794 ± 106	1840 ± 168	1230 ± 182	0.9 ± 0.10	2 ± 1	45 ± 10	ND	ND
Urea Bio-hydrolyser Effluent	10.0 ± 0.2	994 ± 60	138 ± 16	0.5 ± 0.1	258 ± 48	145 ± 20	346 ± 16	304 ± 32	400 ± 52	160 ± 52	2.70 ± 0.30	4 ± 1	40 ± 8	ND	ND
Lagoon 1 Effluent	9.9 ± 0.2	750 ± 22	100 ± 8	0.6 ± 0.2	216 ± 32	105 ± 16	316 ± 20	282 ± 30	382 ± 48	142 ± 46	4.40 ± 0.8	4 ± 1	34 ± 9	ND	ND
Lagoon 2 Effluent	9.7 ± 0.3	700 ± 24	74 ± 6	1.1 ± 0.2	186 ± 36	85 ± 20	288 ± 22	264 ± 28	367 ± 44	104 ± 32	6.90 ± 0.6	5 ± 1	32 ± 8	ND	ND
Lagoon 3 Effluent	9.6 ± 0.3	644 ± 16	53 ± 7	1.3 ± 0.3	158 ± 24	58 ± 13	256 ± 28	240 ± 24	350 ± 38	56 ± 16	6.80 ± 0.7	5 ± 1	30 ± 7	ND	ND
Lagoon 4 Effluent	9.6 ± 0.4	604 ± 20	35 ± 4	2.0 ± 0.3	102 ± 12	32 ± 6	234 ± 24	212 ± 22	371 ± 34	32 ± 8	8.15 ± 0.85	5 ± 1	25 ± 5	ND	ND
Lagoon 5 Effluent	9.6 ± 0.3	584 ± 16	25 ± 3	2.5 ± 0.4	62 ± 10	28 ± 4	208 ± 22	194 ± 20	320 ± 30	20 ± 6	9.35 ± 0.85	6 ± 1	19 ± 3	ND	ND
Lagoon 6 Effluent	9.5 ± 0.4	542 ± 14	17 ± 3	3.1 ± 0.3	50 ± 7	20 ± 2	184 ± 20	160 ± 16	250 ± 26	12 ± 5	11.50 ± 0.5	7 ± 2	15 ± 2	ND	ND
Lagoon 7 Effluent	9.5 ± 0.3	504 ± 10	10 ± 2	3.6 ± 0.3	38 ± 5	15 ± 2	162 ± 22	124 ± 16	198 ± 25	8 ± 4	12.20 ± 0.9	9 ± 2	11 ± 2	ND	ND
Lagoon 8 Effluent	9.4 ± 0.3	480 ± 12	7 ± 2	4.0 ± 0.3	34 ± 6	11 ± 3	138 ± 20	82 ± 18	132 ± 24	5 ± 2	14.00 ± 1.0	10 ± 3	7 ± 2	ND	ND
Lagoon 8A Effluent	9.0 ± 0.3	462 ± 18	5 ± 2	4.9 ± 0.2	30 ± 5	7 ± 2	112 ± 20	60 ± 16	108 ± 20	2 ± 1	16.75 ± 0.75	12 ± 3	5 ± 1	ND	ND
Factory combined Effluent to Lagoon 9	7.6 ± 0.7	952 ± 96	140 ± 8	1.9 ± 0.2	284 ± 54	130 ± 18	85 ± 16	42 ± 12	88 ± 16	1 ± 1	8.95 ± 0.95	45 ± 8	13 ± 2	0.07 ± 0.02	0.10 ± 0.03
Lagoon 9 Effluent	8.1 ± 0.2	946 ± 84	36 ± 6	4.7 ± 0.3	42 ± 8	14 ± 4	76 ± 6	7 ± 2	52 ± 12	ND	13.10 ± 2	15 ± 1	7 ±0.01 0.01	ND	0.03

ND-Not Detectable; All values except pH are in mg/L

combined final treated effluent from the fertilizer industry being discharged into river are presented in Table 4.

Table 4: Characteristics of Combined Treated Effluent Discharged into River

Sl. No.	Parameter	ETP Combined Treated Effluent	Standards for Treated Effluent (Inland Surface Water)	
			MEF*	SPCB#
1.	pH	7.5-8.0	5.5-9.0	6.5-8.0
2.	Total Dissolved Solids	750 ± 72	-	-
3.	Suspended Solids	46 ± 14	100	100
4.	DO	4.8-5.2	-	-
5.	COD	54 ± 18	250	250
6.	BOD_5d,20°C	16 ± 5	30	30
7.	Total Ammonical-N	40 ± 5	50	75
8.	Free Ammonical-N	2.3 ± 0.3	5	4
9.	Total Kjelhdal-N	52 ± 12	100	100
10.	Nitrite (NO_2)-N	7.2 ± 0.4	-	-
11.	Nitrate (NO_3)-N	9.0 ± 0.8	10	10
12.	Urea	ND	-	-
13.	Total Phosphate	0.08 ± 0.3	5	-
14.	Total Phenolics	0.3 ± 0.1	1	-
15.	Cyanides	0.02 ± 0.01	0.2	-
16.	Fluorids	ND	2.0	-
17.	Oil & grease	4.6 ± 2.1	10	10
18.	Heavy Metals			
	-Iron (Fe)	0.52 ± 0.08	3	3
	-Lead (Pb)	0.002 ± 0.001	1	1
	-Zinc (Zn)	0.11 ± 0.04	5	5
	-Nickel (Ni)	ND	3	3
	-Chromium (Cr)	ND	2	2

ND: Not Detectable
All values except pH are in mg/L
* Ministry of Env & Forests, Govt. of India; # State Pollution Control Board

Results and Discussions

The wastewater from urea and ammonia sections having urea concentration ranging between 1048 to 1410 mg/L was reduced to 108 to 212 mg/L in the urea bio-hydrolyser unit of the ETP. Around 85 to 90 per cent urea hydrolysis was achived. The reaction of urea was brisk upto compartment No. 6 reducing around 70 to 73 percent urea in first 30 hours. The remaining 15 to 17 percent urea was reduced in the remaining ten compartments of the hydrolyser plant. The treated effluent from urea bio-hydrolyser content ammonia nitrogen ranging between 272 to 336 mg/L, removing around 61.3 to 62.9 percent of free ammonia.

The urea bio-hydrolyser unit was designed to accept initial urea concentration up to 6000 mg/L and to hydrolyse 90 to 95 percent urea from urea bearing wastewater. The flowsheet of urea bio-hydrolysis and removal of nitrogen from the wastewater is as follows:

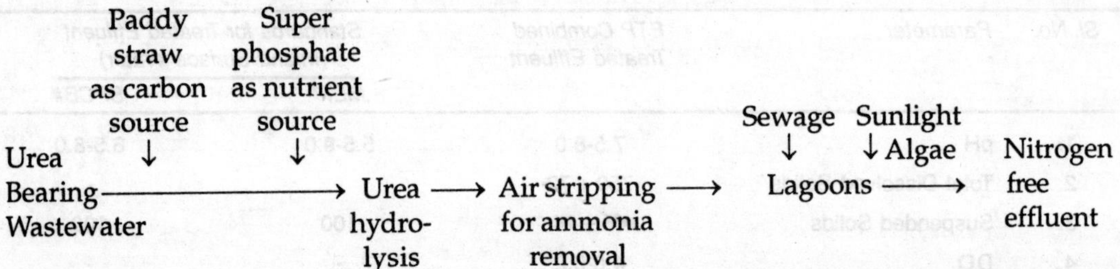

Paddy straw served as carbon source to uerolytic bacteria *Bacillus Pasteurii* and algae utilized ammonia nitrogen for their growth. Urea is being hydrolysed into ammonia, which stripped off by steam and also utilized by algae in lagoons. Upto compartment No. 6 constants increase in ammonia concentration and constant decrease in urea concentration indicating brisk reaction of urea hydrolysis. However, the concentration of ammonia in these compartments showed lower values as compared to stricometric values, due to ammonia stripping at high wastewater pH. At pH above 8.6, ammonia is released from ammonia solution into the atmosphere. Thus, by maintaing high pH, ammonia stripping was achieved. Urea bio-hydrolser removed about 61-63 per cent ammonia nitrogen due to ammonia stripping. The reduction in dissolved solids in the effluents from different compartments indicates the release of ammonia nitrogen into atmosphere. Increase in concentration of suspended solids in the effluent was mostly due to addition of chopped rice straw, which floats on water surface and escape through effluent. The insignificant increase in nitrite and nitrate concentrations in effluent from various compartments of urea bio-hydrolyser indicate that nitrification of ammonia was not takeing place. The slow build-up of nitrite and nitrate concentration in the lagoons indicated nitrification of ammonia during wastewater stabilization. The urea concentration in lagoon effluent was reduced due to the presence of ureolytic bacteria present in septic tank effluent from the industry and township being discharged into lagoon 4. Reduction of ammonia nitrogen was also obtained due to utilization of ammonia by algae *chlorella* which was present by virtue of addition of algal culture in these lagoons. Algae utilized ammonia and other nitrogen species for cell synthesis and growth. The concentration of COD and BOD also reduced significantly. The characteristics of final treated effluent being discharged into river conformed to the discharge standards prescribed by regulatory agencies into inland surface water.

Conclusions

Based on the studies on full scale urea bio-hydrolyser plant to treat urea bearing wastewater from a fertilizer industry it was concluded that ureolytic bacteria *Bacillus pasteurii* could hydrolyse 85 to 90 per cent urea from wastewater. Urea hydrolysis process was brisk at initial period of 30 hours, hydrolysing around 70 to 73 per cent urea in the first six compartments of the urea hydrolyser plant. Ammonia nitrogen produced during urea bio-hydrolysis could be stripped off at pH 10.2±0.2 without applying mechanical means. Injecting pressurised steam at the bottom of latter compartments facilitate 44 to 48 percent ammonia removal by stripping. Paddy straw, an agricultural residue can profitably be used as an carbon source for microorganisms as against costly chemicals like methanol and other organics. Ammonia nitrogen trapped in the form of inorganics further nitrified with help of microorganisms into nitrate via nitrite in lagoons. Algae *chlorella* present in lagoons reduce ammonia nitrogen and nitrate nitrogen from the wastewater. Nitrate nitrogen was denitrified by microorganisms into nitrogen gas to obtain nitrogen free wastewater. BOD and COD was also reduced in the lagoons.

Urea bio-hydrolyser cum stripper reduce significant concentration of urea and ammonia in the wastewater from nitrogenous fertilizer industry to conform standards prescribed by regulatory agencies for discharge of treated effluent into inland surface water.

REFERENCES

1. India, 1995, "A reference annual", Published by Ministry of Information and Broadcasting, Government of India (1995), pp. 545-546.
2. Monthly Review of Indian Economy, Economy Intelligence Service series, Published by Centre for Monitoring Indian Economy (CMIE) Pvt. Ltd., February 1997, p. 58.
3. IS: 9841 "Guide for Treatment and Disposal of Fertilizer Industry", Indian Standards Institution, New Delhi, 1981.
4. Gillette, L.A., Miller, D.L. and Redman, H.E., 1952, "Appraisal of Chemical Waste Problem by Fish toxicity tests", Sewage & Industrial Waste, 24: 1397.
5. Evans, W.H., David, E.J. and Patterson, S.J., 1973, "Biodegradation of urea in river water under controlled laboratory conditions", Water Res., 7: 975.
6. "Standard Methods for the examination of water and wastewater", 17th Edition, APHA, AWWA, WPCF, Washington, DC, 1989.
7. Watt, G.W. and Chrisp, J.D., 1954, "Spectrophotometric method for determination of urea", Anal. Chem. 26: 452.

56

Preliminary Studies on Bacterial Decolorization of H-Acid Based Azo Dye-Reactive Black 5

SAGARIKA MOHANTY, N.N. RAO & S.N. KAUL

National Environmental Engineering Research Institute, Nehru Marg, Nagpur - 440 020

ABSTRACT

A co-culture acclimatized to H-acid was used to degrade Reactive Black 5 (RB 5), a bis azo dye having central H-acid function. The effect of substrate concentration, pH and medium composition on the decolorization has been investigated. Decolorization was found independent of pH. Luria-Bertani broth favored decolorization over Yeast Extract; however further decolorization experiments have been conducted using yeast extract. The Michaelis-Menten model is found to describe the dependence of specific decolorization rate on the RB 5 dye concentration.

Keywords: *Reactive Black 5, Decolorization, Bacterial culture.*

Introduction

Color, the most noticeable contaminant in wastewater is mostly due to the presence of dyes and dyestuffs discharged from textile, food, paper-making and cosmetic industries. The presence of even small amounts of dyes in water is highly visible and affects the aesthetic merit, water transparency and gas solubility in lakes, rivers and water bodies (Banat *et al.*, 1996). Hence, color removal has become the focus of contemporary research interests. Researchers over the past two decades have developed several physico-chemical processes like adsorption (Ramakrishna, K *et al.*, 1997), Flocculation, Chemical Coagulation, Precipitation (Vandeviere, 1998), Electrochemical oxidation (Lin, S.H, 1994), Photocatalytic oxidation (Davis *et al.*, 1994; Zielinska *et al.*, 2001), Electro-Fenton oxidation (Bose, 2002; Huang, 2001) etc. Apart from cost of these technologies, some of them even generate secondary pollution such as sludge. Therefore, developing biological methods to remove color becomes important in view of cost-effectiveness as well as technological feasibility.

In the past, biological decolorization methods have been investigated as a method to transform, degrade or mineralize azo dyes (Banat *et al.*, 1996). Dyes are mainly difficult to treat because of their synthetic origin and complex aromatic molecular structures (Banat *et al.*, 1996). H-acid based azo dyes are a group of synthetic dyes characterized by the presence of H-acid moiety at the center and one or two azo bonds. These chemical structures render the dyes stable to biological as well as chemical degradation to some extent. Thus, their treatment would require specific bacterial cultures.

Bacterial decolorization of reactive azo dyes may occur due to sorption (Hu, 1992), aerobic reduction (Coughlin, 1997) or anaerobic reduction (Haug, 1991). Azo dye degradation normally involves a two step process including an initial azo-reductase catalyzed anaerobic reduction of the azo bond (-N=N-) followed by aerobic or anaerobic reduction of the resulting amines (Seshadri, 1994). Knowledge of bacterial decolorization is still inadequate and hence it is required that endeavors are made to develop bacterial strains which would effectively remove color from the industrial wastewater. In this study we have investigated the effect of different operational parameters (pH, dye concentration and medium composition) on the kinetics of decolorization of H-acid based azo dye-Reactive Black 5 (RB 5) using a co-culture acclimatized to H-acid.

Materials and Methods

H-acid and Reactive Black 5 were procured from a dye manufacturing unit in Gujarat. It is water-soluble with a molecular formula presented in Fig. 1. Different concentrations of RB 5 were scanned in the visible range (400 –700 nm) to obtain the absorption spectra. The wavelength at which maximum absorption (λ_{max} = 597) occurred was used for quantitative analysis for decolorization in the study. All the nutrient media components were procured from Hi – Media Pvt. Ltd, Mumbai).

Microorganism and Cultivation

Two bacterial strains acclimatized to H-acid were developed. A system consisting of soil contaminated with H-acid wastewater procured from M/s Rohan Dyes Ltd., Ahmedabad (10 g/L), di-ammonium phosphate (1 g/L), H-acid (10 mg/L) with a supporting aeration system was incubated at 30°C for one month. Regular transfer of supernatant (5 ml) to a fresh medium in every two days was

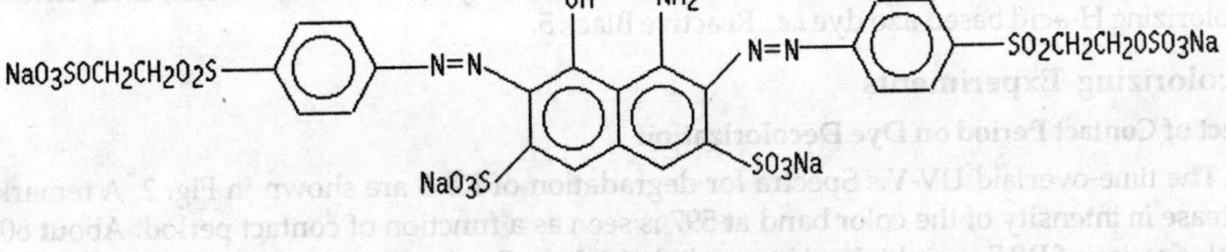

Fig. 1: Structure of Reactive Black 5

done. After one month the supernatant was further transferred to another medium consisting of H-acid (50 ppm), Phosphates (K_2HPO_4, 0.615 g/L; KH_2PO_4, 0.385 g/L), yeast extract (10 mg/l) and the process was continued for 10 days. Finally, two bacterial strains, DRB1 & DRB2, were isolated by spreading 0.1 ml of supernatant onto nutrient agar plates. The two bacterial isolates were subjected to Gram stain and further tests towards identification of these strains are in progress and will be reported in a separate paper.

Measurement of Cell Biomass

The cell biomass of the bacterial co-culture was obtained by method of mixed liquor suspended solids (MLSS). Bacterial co-culture (25 ml) was filtered through a pre-weighed membrane filter paper (0.45 Mm). The filter paper was further dried at 103°C for one hour. The weight of the filter paper was taken after drying and the cell mass was obtained by deducting the initial weight of filter paper from the final weight.

Decolorization Experiments

Two types of media were used for the purpose: Growth medium (Nutrient broth) for mass culture and decolorization medium (Yeast Extract, 5g/L+ Dye) for decolorization experiments. Aerobic cultivation of co-culture was done in growth medium by inoculation of 10 ml each of stock cultures of individual strains for 24 h. Fresh cells were then harvested by centrifugation (5000 rpm, 20 min, 4°C), and used as inoculum for decolorization experiments. The decolorization experiments were carried out in 250 ml Erlenmeyer flasks (reaction volume = 50 ml) in batch mode in static incubation conditions at 30°C. Experiments were carried out for a period of 4 days at different initial dye concentrations (25-125 ppm) and different pH values (5.0-9.0). Decolorization of RB 5 was also performed using Luria-Bertani (LB) broth as the decolorization medium. Samples collected at different time intervals were centrifuged (5000 rpm, 20 min, 4°C) and the supernatants were analyzed for the change in concentration of the dye.

Measurement of Dye Concentration

The concentration of the dye was determined before and after incubation periods by measuring the absorbance of the supernatant of the samples at 597 nm on a UV-Vis spectrophotometer (Systronics, Model-119).

Results and Discussions

Bacterial Isolates

Two bacterial strains (DRB1 & DRB2) were isolated which were found to be tolerant to lower concentrations of H-acid. Gram reaction suggests DRB1 is Gram positive cocci whose colony appears to be off-white and DRB2 to be Gram negative short rod and the colony is colourless and transparent. The weight of the cell biomass of the co-culture taken for each decolorization experiment was 0.236 ± 0.005 g/L. Since these strains were tolerant to H-acid, they were chosen to test, their efficacy in decolorizing H-acid based azo dye *i.e.*, Reactive Black 5.

Decolorizing Experiments

Effect of Contact Period on Dye Decolorization

The time-overlaid UV-Vis Spectra for degradation of RB 5 are shown in Fig. 2. A remarkable decrease in intensity of the color band at 597 is seen as a function of contact period. About 80% of decolorization of RB 5 was obtained in a period of 4 days. Further, this band exhibited blue shift. This

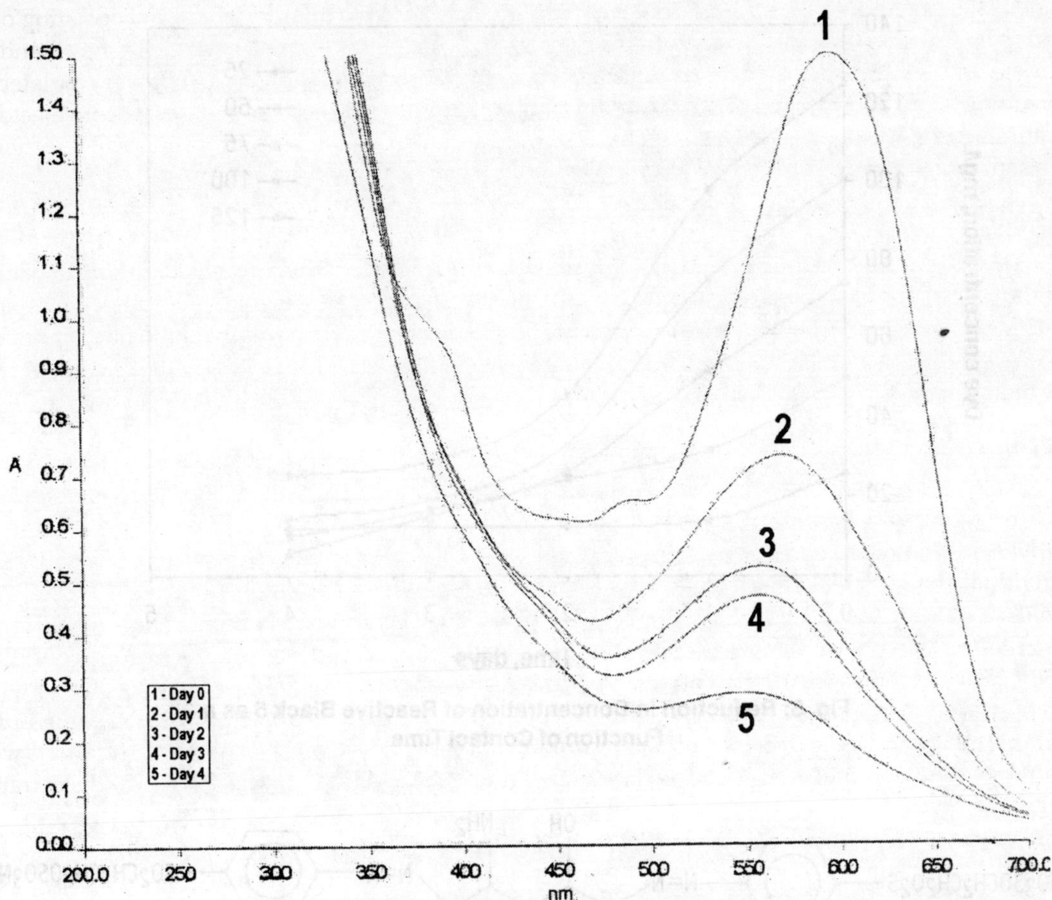

Fig. 2: UV-Visible Spectra for Decolorization of RB 5

shift in band position towards higher energy is indicative of a chemical transformation of the chromophore responsible for colour of the Reactive Black 5, namely azo groups. A characteristic band of H-acid at λ_{max} = 358 nm was not observed. A change in the energy status of the molecules formed could be due to localization of electrons loosely bound in the azo bond. Breakage of the azo bonds could lead to the formation of amines and H-acid. In the present study, based on the spectral information pertaining to changes in RB 5 molecule, it can be suggested that the azo groups of RB 5 undergo partial reduction but the azo groups are not dissociated from the parent dye molecule, as could be understood from the absence of absorption band of H-acid molecule. A tentative mechanism for decolorization of RB 5 is given in Chart 1.

Effect of Dye Concentration on Decolorization

The decrease in dye concentration in a period of 4 days in case of different initial dye concentrations is shown in Fig. 3. A rapid initial decrease in dye concentration in the first 2-3 days followed by slow rate of decolorization in the next 1-2 days can be seen. The dependence of the specific decolorization rate (r_{dye}) on the substrate (dye) concentration is shown in Fig. 4. The r_{dye} increased as the concentration

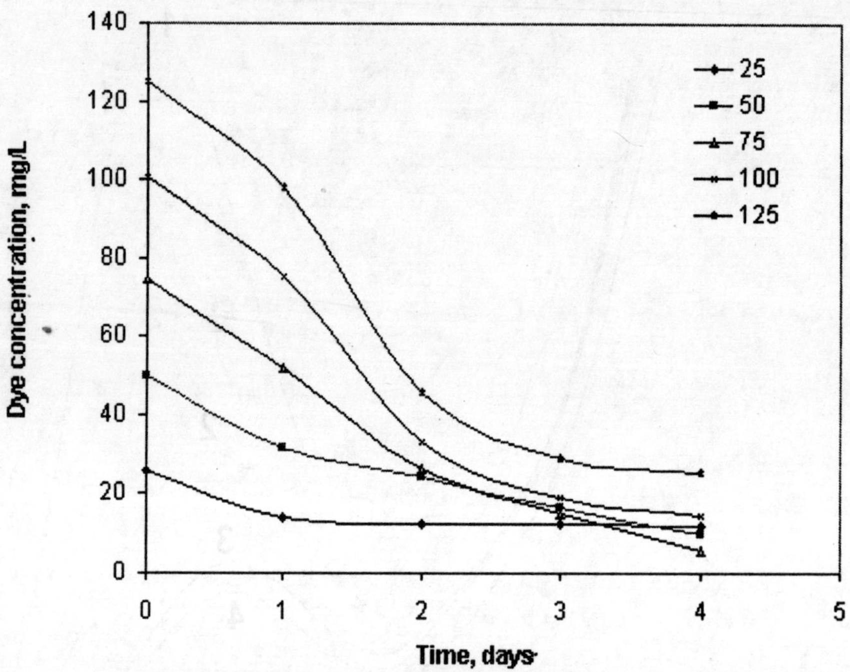

Fig. 3: Reduction in Concentration of Reactive Black 5 as a Function of Contact Time

OH NH2

NaO3SOCH2CH2O2S — N=N — N=N — SO2CH2CH2OSO3Na

NaO3S SO3Na

DRB1

Reduction

DRB2

OH NH2

NaO3SOCH2CH2O2S — NH—NH — NH—NH — SO2CH2CH2OSO3Na

NaO3S SO3Na

Chart 1: Proposed Mechanism for the Azo Bond Reduction of Reactive Black 5

of RB 5 was increased from 0 to 125 mg/L. The co-relation between specific decolorization rate and dye concentration can be described by Michaelis-Menten kinetics relation:

$$r_{dye} = \frac{r_{dye\ max}\,[Dye]}{K_m + [Dye]}$$

where

$r_{dye\ max}$ denotes maximum specific decolorization rate (mg dye d^{-1}g $cell^{-1}$), K_m is Michaelis constant (mg/L). The kinetic constants estimated from the experimental data (Fig. 4) are 16.89 mg dye d^{-1}g $cell^{-1}$ for $r_{dye\ max}$ and 1.21 mg/L for K_m.

Considering the typical enzyme – substrate interaction and product formation as given below, the constant K_m can be represented as $k_1/(k_2+k_3)$. The low K_m value found in the present case suggests that the rate of formation of Enzyme–Substrate (ES) complex and the sum of rates of reaction pathways leading to enzyme and product formation are comparable. Thus, there is always surplus concentration of ES.

$$E + S \underset{k_2}{\overset{k_1}{\rightleftharpoons}} ES \xrightarrow{k_3} E + P$$

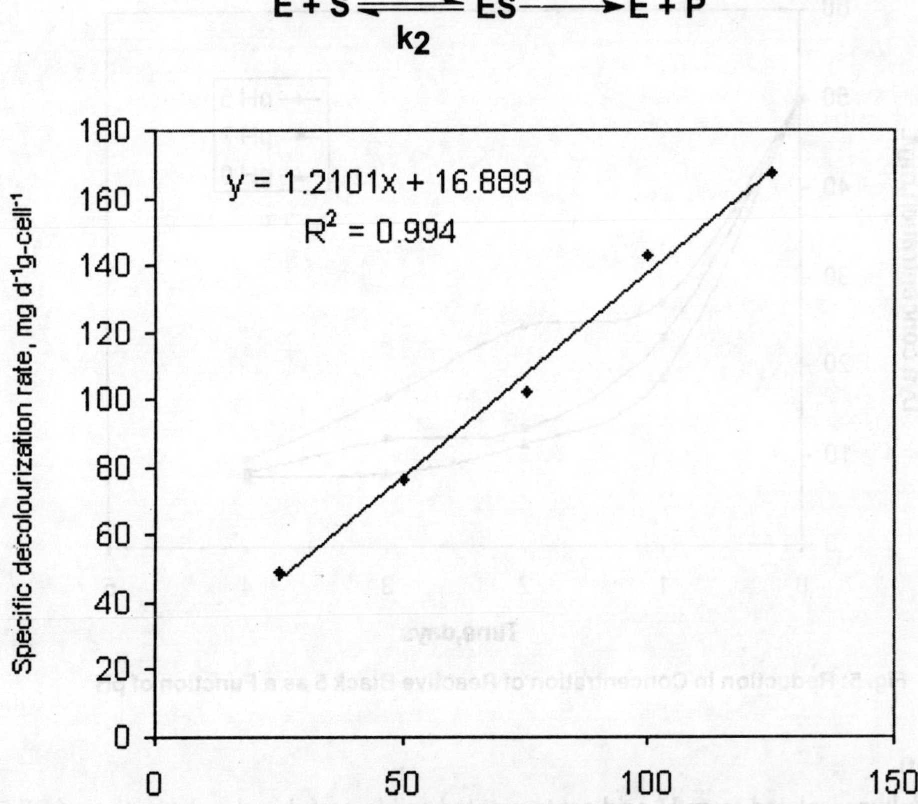

Fig. 4: Dependence of Specific Rate on the Dye Concentration

Effect of pH on Decolorization

In the experiments with different pH, the initial dye concentration was fixed at 50 mg/L. The percent decolorization was around 80% during 4 days irrespective of the pH change. The dependence of specific decolorization rate on pH can be shown in Fig. 5. However, the r_{dye} value increased with the increase in pH at 8.876 mg dye d^{-1}g $cell^{-1}$ per unit change in pH value. The final pH at the end of four days in each case was found 9.0. This Increase in pH could be due to formation of CO_2.

Effect of Medium Composition on Decolorization

Fig. 6 indicates the efficiency of dye removals in different media *i.e.* yeast extract (YE) and LB broth. It has been found that about 85% of dye removal occurred in 3 days in case of LB broth whereas only 65% of dye removal in case of yeast extract. The specific decolorization rate was found higher in LB broth over yeast extract medium. Thus, the specific decolorization rate with LB is 135.7 mg dye d^{-1} g $cell^{-1}$ whereas it is 96.2 mg dye d^{-1} g $cell^{-1}$ using YE. LB broth favoring decolorization could be due to presence of both tryptone (nitrogen source) and yeast extract (growth factor), which are essential to activate the necessary co-enzyme producing metabolic pathway for azoreductase activity, the key to decolorization activity. In this study yeast extract was selected as the decolorization medium for kinetic studies as it is much cheaper than tryptone.

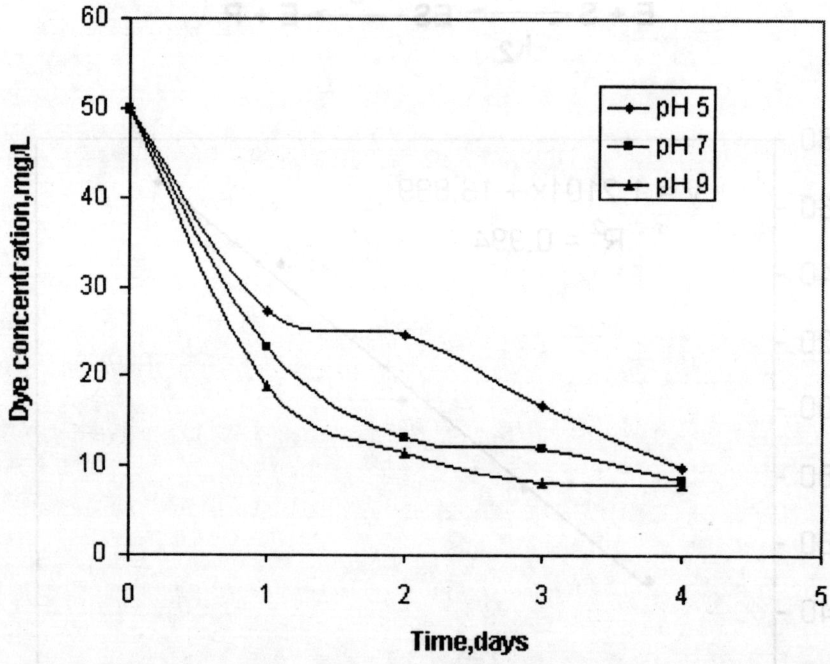

Fig. 5: Reduction in Concentration of Reactive Black 5 as a Function of pH

Conclusion

The co-culture isolated from H-acid contaminated soil is useful in decolorization of RB 5. The decolorization of RB 5 was found independent of pH & greatly dependent on nutrient composition.The dependence of specific decolorization on the substrate (dye) concentration can be described by Michaelis-

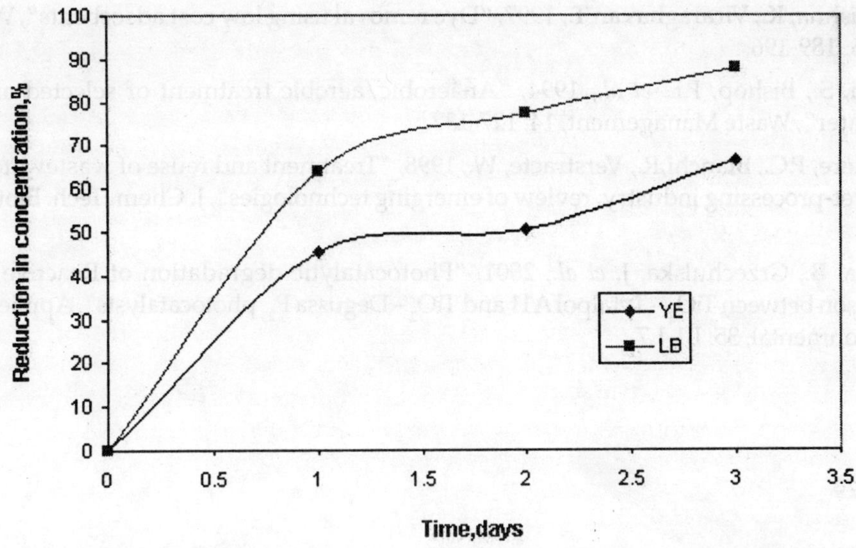

Fig. 5: Decolorization of Reactive Black 5 in Different Nutrient Medium

Menten kinetics. The decolorization of RB 5 appears to proceed through partial reduction of azo groups without scission of azo linkages.

REFERENCES

1. Banat, I.M.,Nigam, P., *et al.*, 1996, "Microbial decolorization of textile –dye containing effluents: a review", Biores. Tech., 58: 217-227.

2. Bose, G., 2002, "Comparison of Fenton and Electrofenton methods for oxidation of H-acid and Reactive Black 5", B.Tech. Dissertation report, Inst. of Tech., BHU.

3. Coughlin, M.F., Kinkle, B.K. *et al.*, 1997, "Charaterization of aerobic azo dye-degrading bacteria & their activity in biofilms",Water Sci. & Tech., 36(1): 215-220.

4. Davis, R.J., Gainer, J.L., *et al.*, 1994, "Photocatalytic decolorization of wastewater dyes", Water Environ. Res., 66: 50-53.

5. Haug, W., Schmidt, A., Nortemann, B. *et al.*, 1991, "Mineralization of the sulphonated azo dye mordant yellow 3 by a 6-aminonaphthalene-2-sulphonate-degrading bacterium consortium", Appl. Environ. Microbiol., 57: 3144-3149.

6. Hu, T.L., 1992, "Sorption of reactive dyes by *Aeromonas* biomass", Water Sci. & Tech., 26(1-2): 357-366.

7. Huang, Y.H. *et al.*, 2001, "Comparison of novel Electrofenton Method with Fenton's Reagent in treating a highly contaminated wastewater.", Water Sci. & Tech., 43: 17-24.

8. Lin, S.H. and Peng, F.C., 1994, "Treatment of textile wastewater by Electrochemical methods", Water Res., 28: 277-282.

9. Ramakrishna, K., Viraraghavan T., 1997, "Dye removal using low cost adsorbents", Water Sci. & Tech., 36: 189-196.

10. Seshadri, S., Bishop, P.L. *et al.*, 1994, "Anaerobic/aerobic treatment of selected azo dyes in wastewater", Waste Management, 14, 127-137.

11. Vandeviere, P.C., Bianchi,R., Verstracte, W., 1998, "Treatment and reuse of wastewater from the textile wet-processing industry: review of emerging technologies", J. Chem. Tech. Biotechnol, 72: 389-302.

12. Zielinska, B., Grzechulska, J. *et al.*, 2001, "Photocatalytic degradation of Reactive Black 5 A comparison between TiO_2 – Tytalpol A11 and TiO_2 – Degussa P_{25} photocatalysts", Applied Catalysis B: Environmental, 35: L1-L7.

57

Modelling the Performance of Phenol Utilizing Bioreactor in Presence of Secondary Toxic Substrates

A. SARGAONKAR, H.J. PUROHIT & R.N. SINGH

National Environmental Engineering Research Institute, Nehru Marg, Nagpur - 440 020

ABSTRACT

Biological activity in common effluent treatment plants depends upon the overall understanding of the different waste streams and their relative toxicity to developing biomass. In this context, an attempt is made using mathematical modeling to evaluate the toxic level of two substrates *viz.* 2,4-Dinitrophenol and o-chlorophenol during their co-metabolism. The selected substrates are toxic even at low levels and are known inhibitors of aerobic metabolism in microorganisms. Batch model simulation study for these two structurally similar toxic compounds indicate that substrate utilization in the mixture is either simultaneous or preferential and hence biomass growth depends upon the interaction amongst the two substrates. Therefore, mathematical model of interacting substrates has been applied in the study with the selected toxic substrates. Sensitivity of the system of interacting substrates has been worked out for a model system of activated sludge pilot plant treating phenol-laden wastewaters. This gives a concentration range for primary as well as secondary substrate over which the system may be operated for a given initial biomass condition to avoid washout (total failure) of the system, and hence provides basis to design operational guidelines for the treatment plant. Study shows that the dilution rate significantly drops from 0.083/h to 0.003/h when both the toxic substrates were included in the waste stream at concentration of 100ppm.

Keywords: Wastewater treatment, substituted phenols, Inhibitiòn constants, co-metabolism, competitive inhibition.

Introduction

Recent trend in developing countries shows installation of common effluent treatment plants. These facilities cater the need of various small industries which otherwise find difficult to support their own effluent management options. However, the efficiency of such treatment facility depends upon the overall understanding of the different waste streams and their relative toxicity. In this context phenols and phenolic wastes constitute a major portion of industrial wastewaters of which nitro and chloro substituted phenolics are highly toxic to the aquatic community as well as to the consumers of water. Therefore, before these are discharged into surface waters *i.e.* either treated completely and discharged or given pre-treatment and discharged into the municipal sewer, it is expected that the treatment provided reduces toxicity of the waste to acceptable level.

In biological treatment processes, the microorganisms use toxic waste as a major substrate or sole source of carbon, and reduce their toxicity to acceptable level (Rozich *et al.*, 1983, Gaudar *et al.*, 2000) *e.g.* degradation of chlorophenol leading to the production of innocuous end products (*e.g.* CO_2) is possible (Shieh *et al.*, 1990). The microorganisms degrading CP under aerobic conditions have been reported as Pseudomonas sp., Nocardier sp. Arthrobactor sp. Mycobacterium Coeliacum. Pathway of PCP degradation by Chlorophenolicus has been established. Anaerobic degradation of CP has also been observed to occur in municipal sludges and laboratory experiments using mixed bacterial culture.

We are working on developing a culture collection from the environmental niches contaminated with phenolic and substituted phenolic waste; and having cultures for treatment of PNP (Qureshi, *et al.*, 2001), chlorophenols (Atuanya *et al.*, 2000), aminophenols (Kutty *et al.*, 2000) and are developing strategies to track phenols in the environment (Kapley and Purohit, 2000). We are also demonstrated that depending upon the availability of the pollutants a consortium with specific ration of microorganisms establishes. We have shown that a bacterial consortium comprising of Pseudomonas spp. Strain S2 could convert p-nitrophenol to 4-nitrocatechol (4NC) after pre-exposure to phenol, when PNP was used as the only carbon source in the medium. *Pseudomonas* spp. strain S2, when mixed with strain S1 in the ratio of 1: 5 respectively decolorized PNP completely (Qureshi and Purohit, 2002).

It is also reported that many toxicants can be metabolized after prolonged acclimation, but these substances remain inhibitory. Therefore kinetic equations of non-toxic substances can not be utilized to describe the kinetic behavior of the microbial population on toxic substrates *i.e.* Monod equation giving rectangular hyperbolic curve for growth can not be utilized for toxic substances (Rozich and Gaudy, 1985; Gaudy *et al.*, 1988). Fig. 1 shows the conceptual diagram of microbial growth on non-toxic and toxic substrates. Since the composition of waste responsible for environmental pollution contains mixture of toxic and non-toxic, easily biodegradable compounds; study of cell growth on multiple substrates is becoming popular. Eventhough the degradation kinetics of individual compound follows a typical pathway, how the interaction among these compounds affects the metabolism in a complex mixture is of practical interest. Therefore, interacting substrate models for multisubstrate system describing the kinetics of biodegradation of individual compounds in a mixture have been

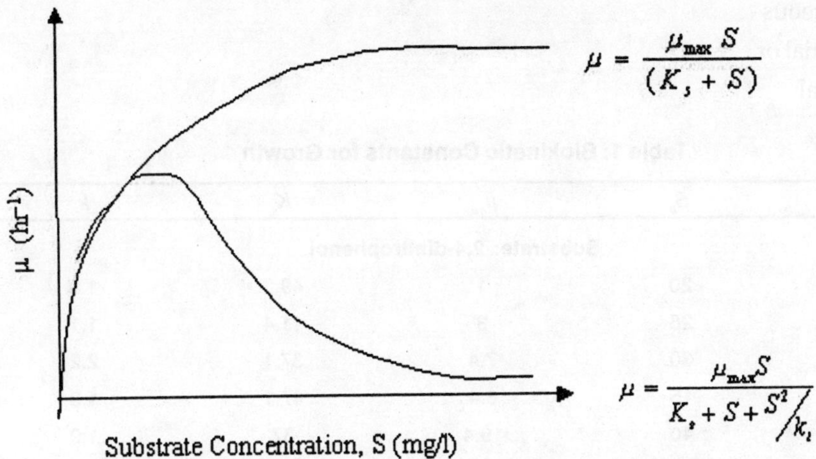

Fig. 1: Relationship Between Substrate Concentration and Specific Growth Rate

reported for mixture of toxic-toxic; toxic-non-toxic; multiple toxic with non-toxic compounds (Klecka and Maier, 1988; Arcangeli and Arvin 1995; Wang and Loh, 2000).

The objective of present study was to investigate and quantify the kinetics of biodegradation of mixture of two structurally similar toxic compounds 2,4-Dinitrophenol (2,4NP) and o-chlorophenol (*o*-CP). The study shows mathematically how these two compounds could be degraded simultaneously, and predicts their toxicity level. This helps in selection of mathematical formulation for interacting substrates. Further, the simulation studies were also done using interacting substrate model to show if a phenol utilizing biomass receives one of these pollutants as the co-substrate, then how it affects the overall performance of the treatment.

Nitrophenol and Chlorophenol as Substrates

Number of studies reported regarding the kinetics of phenol degradation indicate that phenol and phenolic compounds are inhibitory in nature and microbial growth on inhibitory substrates can be described by Andrew's (Haldane) equation. Another important aspect in predictive modeling is the use of biokinetic constants. As model predictions are sensitive to input conditions and other operating parameters, they are also sensitive to the variation in bio-kinetic constants. Therefore, batch studies are often performed over a wide range of substrate concentrations to estimate kinetic constants. Table 1 gives the reported parameters for biokinetic constants maximum growth rate μ_{max}, half saturation constant K_s and inhibition coefficient k_i estimated from batch studies for 2-4 DNP and *o*-CP (Gaudy *et al.*, 1988). This data has been utilized in the present study for simulation of batch model as well as mixed substrate model.

Batch Model Simulation

Occurrence of multi-substrate utilization phenomenon is common in biological wastewater treatment system. What is important from modeling point of view is that whether the substrate utilization is:

1. simultaneous
2. preferential or
3. sequential

Table 1: Biokinetic Constants for Growth

Test No	S_o	μ_{max}	K_s	k_i
		Substrate: 2,4-dinitrophenol		
1	20	1	49.3	1.3
2	26	3	11.4	1.1
3a	30	7.4	37.1	2.2
3b	15	8.4	47.1	1.3
4a	40	19.4	37	1.9
4b	25	13.1	24	1.9
Average Value	8.7	34.3	1.6	
		Substrate: Orthochlorophenol		
1a	30	7.0	12.6	1.1
1b	13	2.9	12.1	1.2
2	40	18.6	26.6	1.7
Average Value	9.5	17.1	1.33	

The diauxic growth observed by Monod in *E. coli* suggested that the presence of a particular substrate in wastewater might prevent in organism from acclimatizing to another substrate until the first one has been completely metabolized (Deshpande *et al.*, 1987). The blockage of metabolism of one compound by another may lead to preferential or sequential substrate removal from a multisubstrate environment. Therefore, in order to work out the toxic level of one substrate on another *i.e.* the type of relationship amongst the nitrophenol and chlorophenol during their biodegradation in a mixture, model simulations were obtained for batch processes using the kinetic coefficients from Table 1 in the following equations for 2,4-DNP (S_1) and *o*-CP (S_2) as:

$$\frac{dS_1}{dt} = \frac{\mu_1}{y} X \qquad (1)$$

$$\frac{dS_2}{dt} = \frac{\mu_2}{y} X \qquad (2)$$

$$\frac{dX}{dt} = \mu X \quad k_d X \qquad (3)$$

where,

$$\mu_1 = \frac{\mu_{max1} S_1}{K_{s1} + S_1 + S_1^2 / k_{i1}} \qquad (4)$$

$$\mu_2 = \frac{\mu_{max2} S_2}{K_{s2} + S_2 + S_2^2 / k_{i2}} \quad (5)$$

$$\mu = \mu_1 + \mu_2 \quad (6)$$

In the present study, it is hypothesized that both the toxic compound 2,4-NP and *o*-CP irrespective of their concentration levels are utilized as substrates (sources of carbon) by the microbial population. Hence the kinetics of utilization for the two substrates is applied independently and an additive growth rate has been used. In order to simulate the effect of *o*-CP on degradation of 2,4-NP in batch reactor, 2,4-NP concentration was kept fixed to 30 mg/l and *o*-CP concentration increased in steps of 5 mg/l at a time. Accordingly, kinetic constants of Test 3a in Table 1 were used for 2,4-NP degradation and kinetic coefficients for *o*-CP were selected from Table 2 as per the initial concentration of Cp (*i.e.* constants of Test 1b when 0≤o-CP≤ 20 and Test 1a when 20 ≤o-CP ≤30). Even though a value k_d=0.002 hr^{-1} has been suggested for heterogeneous population (Klecka and Maier 1988), decay rate coefficient is considered as 0 due to short time period of the experiment in batch study.

Table 2: Batch Simulation Results

Initial Condition	*Simulation Results at time=15hrs*		
Chlorophenol (mg/l)	*Nitrophenol (mg/l)*	*o-Chlorophenol (mg/l)*	*Biomass (mg/l)*
Simulation 1: 2,4-Dinitrophenol concentration = 30mg/l			
0	0.0034	0	90.09
5	0.0019	0.0	93.44
10	0.0012	0.0	96.79
15	0.001	0.0003	100.14
20	0.0012	0.2501	103.33
25	0.0016	10.89	99.54
30	0.0006	0.0001	110.20
35	0.0009	6.40	109.25
Simulation 2: Chlorophenol concentration = 30mg/l			
Nitrophenol (mg/l)			
0	0.000	0.1256	90.01
5	0.0	0.0214	93.43
10	0.0	0.0038	96.79
15	0.0001	0.0011	100.14
20	15.78	0.0597	92.88
25	12.88	0.0177	98.20
30	27.01	0.0737	92.04
35	27.44	0.0338	95.14

Simulation Experiment 1

Equation (1) and (2) were solved using the ode solver in MATLAB and kinetic constants were varied with respect to initial concentration of o-CP in successive simulations. Results of simulation are shown in Table 2. It is observed that initially both the substrate 2,4-NP and *o*-CP are utilized simultaneously by the microorganisms and increase in *o*-CP concentration increases 2,4-NP utilization. However, *o*-CP concentration of 25 mg/l is a toxic level and decreases NP utilization. Also, it becomes self inhibitory at this point. Table 2 shows that *o*-CP utilization reduces from approximately 98% to 56% and prefers NP utilization. Thus, at NP=30 and *o*-CP > 20 mg/l, system changes from simultaneous to preferential utilization. Thus, model simulations for *o*-CP < 25 mg/l indicate the zone of interaction of 2,4-NP and *o*-CP, where, *o*-CP stimulates the utilization of NP and there is no inhibition of any compound, whereas, at *o*-CP=25mg/l, there is sudden decrease in biomass indicating toxic effect of *o*-CP. Since, 2,4-NP concentration was same in all the simulations, change in biomass represents the effect of *o*-CP concentration. This is shown in Fig. 2. Since such a situation is quite likely to occur in wastewater treatment plant due to complex structure of/mixture of number of toxic compounds and heterogeneous microbial population, formulation of an interacting substrate model is more realistic for simulation of degradation of mixture of aromatic compounds like *o*-CP and 2,-NP.

Simulation Experiment 2

Similar exercise was done for initial *o*-CP concentration as 30 mg/l and NP concentration increased in steps of 5mg/l. Interaction amongst the two substrates is reflected in the increase and decrease in the biomass, since the utilization varies depending upon the kinetic coefficients used with reference to initial concentration of the two substrates. Thus, considering the growth kinetics in the mixture same as that on the individual substrates and additive growth model for biomass increase, the increase and subsequent decrease in the biomass seems to be the result of interaction amongst the two substrates. This is shown in Fig. 3. The change in biomass in the simulation experiments is considered as indicative of interaction amongst the two substrates as either stimulating or inhibiting,

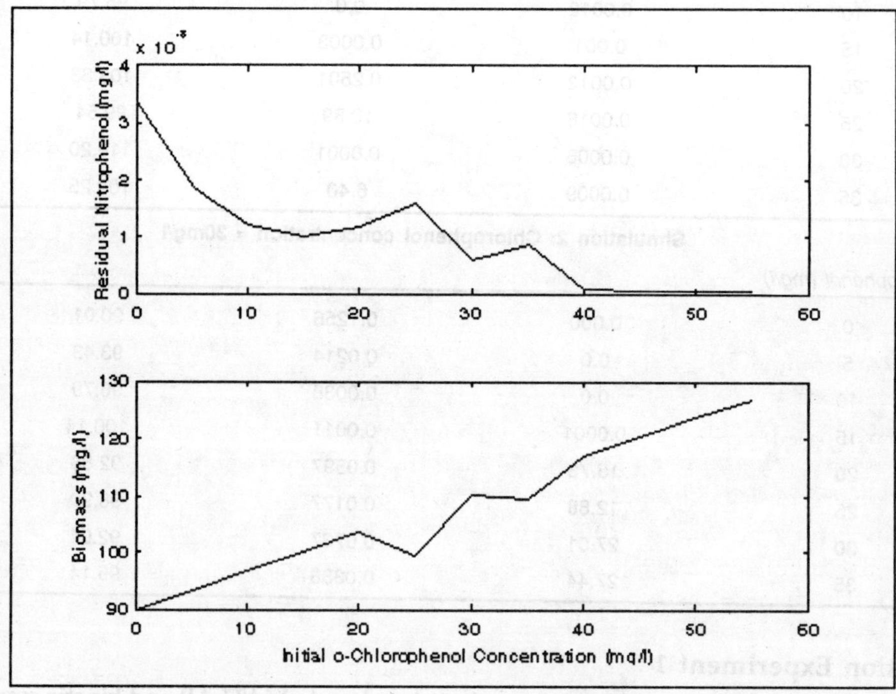

Fig. 2: Effect of o-CP Concentration on 2,4-NP Utilization- Batch Model

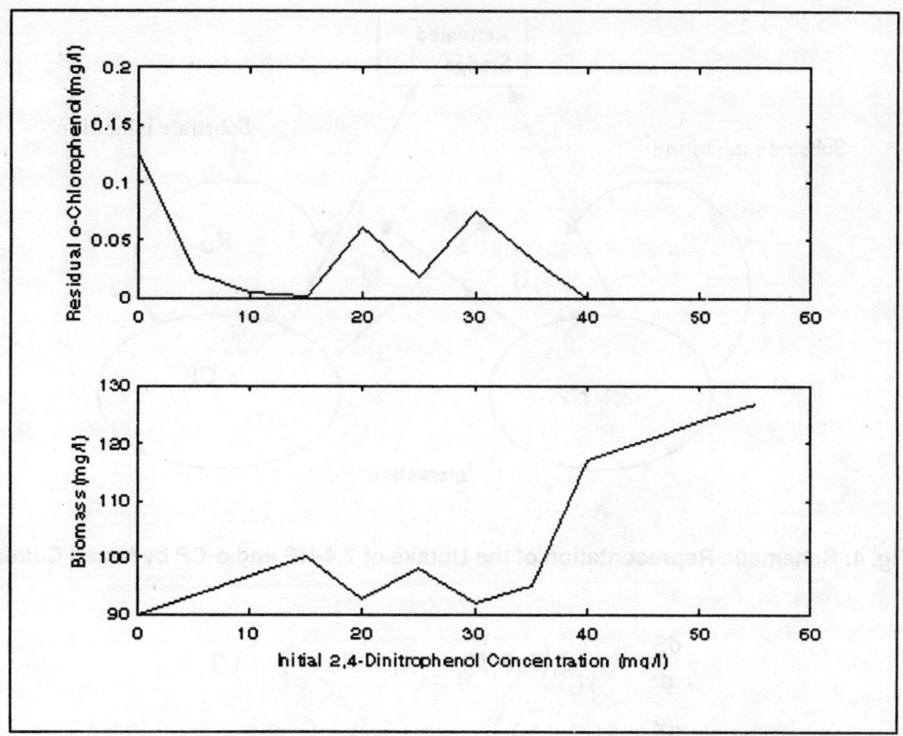

Fig. 3: Effect of 2,4-NP Concentration on o-CP Utilization- Batch Model

and hence it is concluded that microbial growth on a mixture of toxic compounds can be represented by interacting substrate model than by two independent growth models *i.e.* μ_1 as f (S_1) and μ_2 as f (S_2).

Interacting Substrate Model

Substrate interaction studies as discussed earlier have been the topic of research in number of studies. These studies cover structurally similar as well as dissimilar substrates or toxic and non-toxic substrates. Biodegradation study of toxic substance in the presence of non-toxic, easily biodegradable organic compound is interesting because one can understand if the mixture of two structurally different substrates (toxic compound coming from industry and conventional coming from agriculture or urban area) can simultaneously be biodegraded or not; and secondly if it is possible then can biodegradation of toxic pollutants be enhanced by addition of simple substrates in the waste (Wang *et al.*, 1996). However, biodegradation studies of structurally similar compounds such as phenol & chlorophenol, phenol & 4-chlorophenol, pentachlorophenol & chlorinated aromatic compounds have also been reported (Beltrame *et al.*, 1988; Wang and Loh, 2000; Klecka and Maier, 1988).

In general simultaneous utilization of both the substrates results in enhanced removal efficiencies by production of more biomass than in case of individual substrate (Klecka and Maier, 1988). Similar results were obtained in the present work from simulation experiments 1 and 2, over a certain range of concentration for the co-substrate. Therefore, kinetics of multiple substrate utilization has been modeled by additive growth rate equation with substrate interaction (Klecka and Maier, 1988; Wang *et al.*, 1996) as:

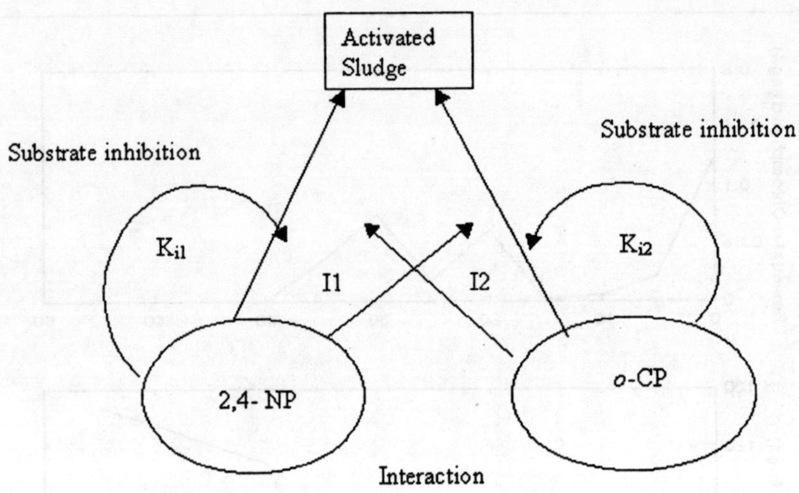

Fig. 4: Schematic Representation of the Uptake of 2,4-NP and o-CP by Mixed Culture

$$\frac{dS_i}{dt} = \frac{1}{y_i}\mu_i\left(S_i,S_j\right)X, \qquad i,j = 1,2 \quad i \neq j \qquad (7)$$

$$\frac{dX}{dt} = \sum_{i=1}^{2}\mu_i\left(S_i,S_j\right)X \qquad i,j = 1,2 \quad i \neq j \qquad (8)$$

where

$$\mu_i\left(S_i,S_j\right) = \frac{\mu_{imax}S_i}{S_i + K_{si} + \frac{S_i^2}{K_{Ii}} + K'_{3i}S_j + K_{3i}S_iS_j} \qquad i,j = 1,2 \quad i \neq j \qquad (9)$$

Eq. (9) states that in the absence of second substrate, culture grows on single substrate which is represented by Andrews Haldane equation (for finite K_{Ii}) or Monod equation (for infinite (K_{Ii}). Based on enzyme kinetics, following interaction patterns are reported for various combinations of coefficients K_{3i}' and K_{3i} as:

1. Competitive cross-inhibition
2. Uncompetitive cross-inhibition
3. Competitive partially inhibitory
4. Uncompetitive partially inhibitory

Thus, an important consideration in growth model representation is the choice of inhibition kinetics. It has been found from earlier studies that in structurally similar substrates, competitive inhibition is observed in cometabolism process, whereas structurally dissimilar substrates (substitutable resources) exhibit uncompetitive interactions (Wang and Loh, 2000; Wang *et al.*, 1996). The model equations for substrate interaction kinetics are also reported by Scott and Ollis (1996), Ollis (2001) and many others.

In the present study, since 2,4-NP and *o*-CP are structurally similar and both the substrates have inhibition effect on utilization of one another, probably they share same key enzyme and hence competitive cross-inhibition inhibition is used to describe their co-metabolism. Index of cross-inhibition between the two substrates is taken as f, where f=f (S_2) in μ_1 and f=f (S_1) in μ_2, representing:

$$\mu_1 = \frac{\mu_{max1} S_1}{K_{s1} + S_1 + S_1^2 / k_{i1} + f(S_2)} \quad (10)$$

$$\mu_2 = \frac{\mu_{max2} S_2}{K_c + S_2 + S_2^2 / k_{i2} + f(S_1)} \quad (11)$$

where

$$f(S_2) = K_{s1} S_2 / K_{i2}$$
$$f(S_1) = K_{s2} S_1 / K_{i1}$$

Predictive Modelling for Interacting Substrates

Mathematical models are being developed and verified with the objective of obtaining design and operational guidelines for the treatment of toxic wastes by microbial population. The activated sludge reported by Rozich *et al.* (1983) ; Rozich and Gaudy (1985) was developed for shock-load study of phenol using an initial inoculum from primary settling tank effluent at the municipal wastewater treatment plant. The cells were initially fed glucose and gradually acclimated to phenol as a sole source of carbon and energy.

The experimental study with this activated sludge by Rozich and Gaudy (1985) in an activated sludge pilot plant (ASPP) considered a reactor of volume V where, S is the substrate concentration and × is the biomass. Flow F into the reactor brings in the phenol concentration S_i. The recycle flow ratio α and recycle cell concentration X_R were used as selectable engineering control parameters for any dilution rate *D*. Cell recycle was considered because it is more comparable to field situation. The mathematical representation of this system has been reported as follows

$$\frac{dX}{dt} = \alpha DX_R + \mu X \quad k_d X \quad (1+\alpha)DX \quad (12)$$

$$\frac{dS}{dt} = DS_i \quad D(1+\alpha)S \quad \frac{\mu X}{y} \quad (13)$$

The activated sludge pilot plant system reported in their study with system parameters *viz.* dilution rate, recycle ratio and recycle sludge concentration as D=0.083 hr^{-1}, α =0.25, and X_R =6280 mg/l has been considered as a model system in the present work for treatment of mixture of substrates. Simulation study for interacting substrates considers mixture of phenol and 2,4-NPl; phenol and *o*-CP and that of phenol,2,4-NP and *o*-CP. Since the system under consideration took the shock-load of phenol from 500-2000mg/l, this concentration level was considered for phenol in the present study for simulation of mixture of substrates. The corresponding kinetic coefficients for phenol were used as reported (μ_{max} = 0.181 hr^{-1}, k_d =0.02 hr^{-1}, y=1.02, K_s=62 mg/l, k_i=175 mg/l).

Mathematical formulation for interacting substrates considered Eq. (13) to describe the kinetics of phenol (S_1) and 2,4-NP (S_2) or Phenol (S_1) and *o*-CP (S_2), where, μ_1 and μ_2 are defined by Eq. (10) and (11), respectively. Biomass growth is given by Eq. (12) where, μ is given by additive growth model. Average values of kinetic coefficients reported for 2,4-NP and *o*-CP were used in all the following simulations in ASPP study. Two types of simulation studies were conducted. First, under the given condition of biodegradation of phenol, whether the system can sustain the shock load of second toxic substrate or not; and second, what initial conditions are required to be maintained in order to sustain a particular waste load.

Results and Discussion

In order to obtain the model simulations for the system of interacting substrates in an activated sludge pilot plant, phenol concentration was considered as 500 mg/l. The concentration of co-substrate either nitrophenol or chlorophenol was varied from 100 mg/l to 500 mg/l. Whether the system will attain the steady-state under these shock-load conditions for initial biomass of 2000 mg/l in the system has been worked out. Function *fsolve* in MATLAB has been used by considering the rate of change zero at steady state. Model simulations for the steady states attained by the system under these shock-load conditions are given in Table 3. It can be seen that as the concentration of co-substrate either 2,4-NP or *o*-CP increases, degradation of phenol reduces. In the absence of co-substrate complete degradation of phenol was possible, whereas in presence of 2,4-NP or *o*-CP phenol utilization reduces to 97% when co-substrate concentration is 100 mg/l and to 95% when co-substrate concentration is 500 mg/l. However, system attains the steady states in all the conditions for dilution rate less than 0.083 hr^{-1}.

Table 3: Steady-states for ASPP System Treating Mixture of Inhibitory Substrates

Interacting Substrates: Phenol and 2,4-Dinitrophenol

Phenol Concentration = 500 mg/l

Initial Condition: 2,4-Di Nitrophenol Concentration	*Steady State Condition*		
	Phenol	*2,4-Di Nitrophenol*	*Biomass*
100	14.8	0.4	1427
200	17.1	0.8	1469
300	19.2	1.2	1512
400	21.3	1.6	1555
500	23.3	2.1	1598

Interacting Substrates: Phenol and Orthochlorophenol

Phenol Concentration = 500 mg/l

Initial Condition: Orthochlorophenol Concentration	*Steady State Condition*		
	Phenol	*Orthochlorophenol*	*Biomass*
100	15.2	0.4	1425
200	17.9	0.8	1469
300	20.4	1.2	1511
400	22.7	1.6	1597
500	25.0	2.0	1597

The conditions under which the system will exhibit washout have been simulated and the results are shown in phase-plane diagrams (Bequette, 1998). It is observed from Fig. 5 that in case of interacting substrates, the selected system can sustain the shock-load of 2,4-NP < 700 mg/l. However, for phenol

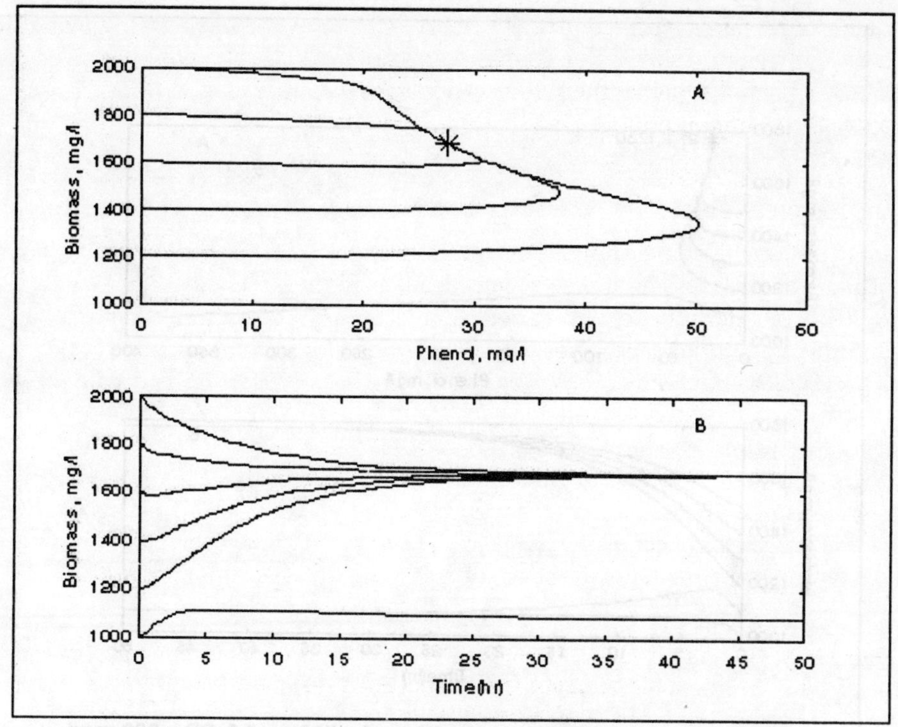

Fig. 5: Phase-plane Diagram for Phenol = 500 mg/l & NP = 700 mg/l

concentration of 500m/l and 2,4-NP 700 mg/l, initial biomass in the system must be maintained more than 1000 mg/l, otherwise system shows washout. Similarly, in case of phenol and *o*-CP, system can sustain the shock load of *o*-CP <900mg/l. For *o*-CP concentration of 900 mg/l, initial biomass in the system should be more than 1000mg/l. This is shown in Fig. 6. These results are indicative of the toxicity level of co-substrate or secondary substrate upon the degradation of primary compound in the mixture.

On the similar lines simulations were obtained from interacting substrate model for phenol concentration of 2000 mg/l and concentration of the co-substrate varied in steps. Results of simulation indicate that system could attain the steady-state for 2,4-NP concentration < 400 mg/l. However, for 2,4-NP ≥ 400mg/l, initial biomass in the system must be more than 2000 mg/l (Fig. 7). Similarly, in case of *o*-CP as co-substrate, initial biomass must be more than 2000mg/l to sustain the shock load of 800 mg/l (Fig.8).

The same system, however, does not respond to the mixture of phenol, nitrophenol and chlorophenol concentration of 500, 100 and 100 mg/l respectively with initial biomass 2000 mg/l, for the settings under consideration. For very low dilution rates D In the range 0.001 – 0.003 hr^{-1}, and only certain shock-loads in terms of 2,4-NP and *o*-Cp as 100mg/l, system can attain the steady-state (0.2, 0, 0, 1636) for initial phenol 500mg/l.

Conclusion

Batch model simulations performed for the two toxic substrate 2,4-NP and *o*-CP confirmed the

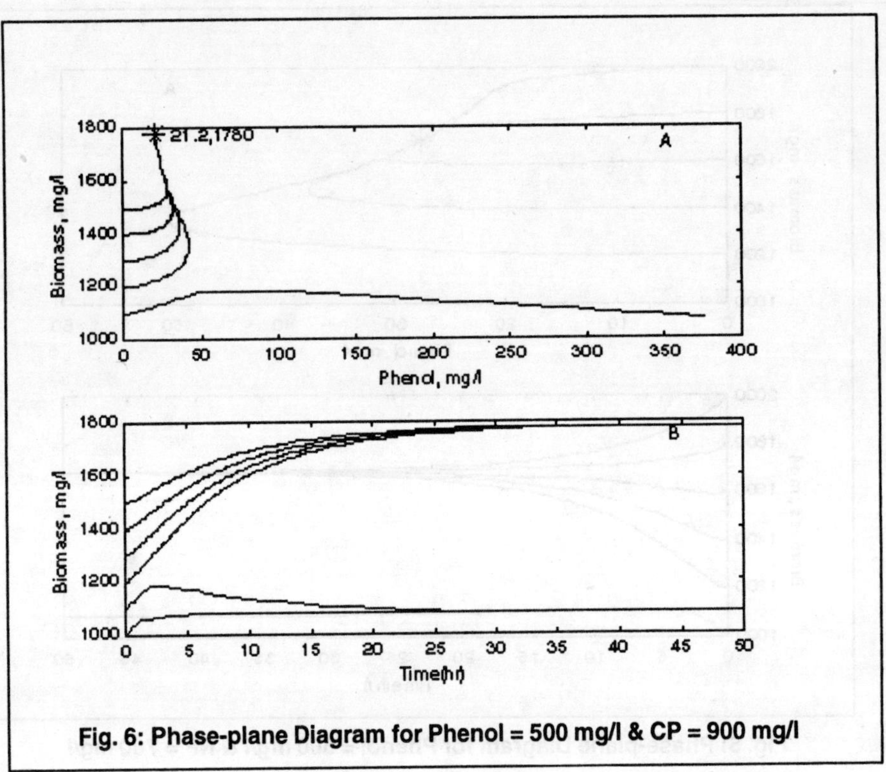

Fig. 6: Phase-plane Diagram for Phenol = 500 mg/l & CP = 900 mg/l

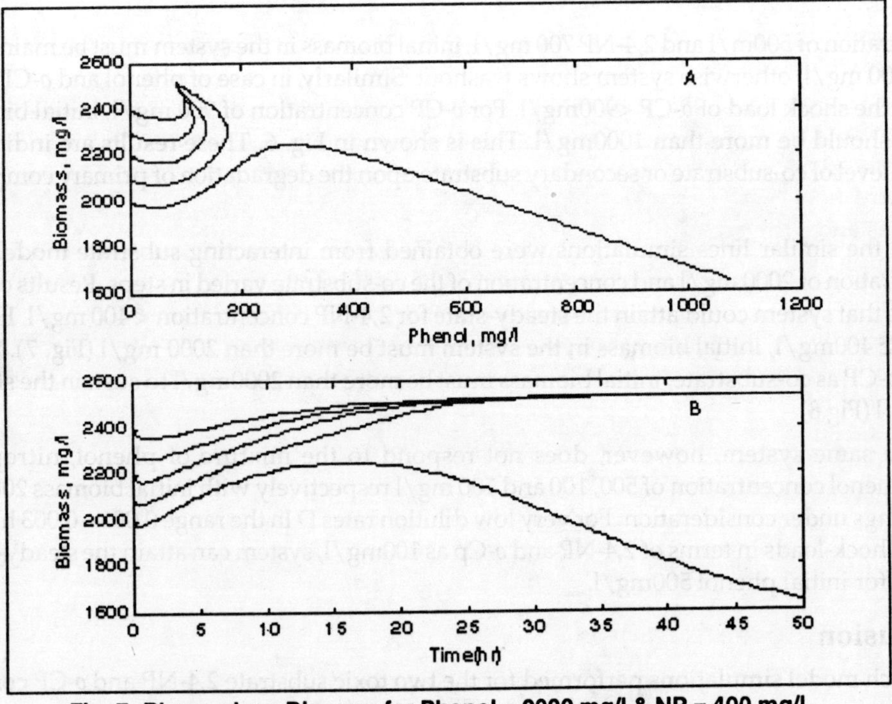

Fig. 7: Phase-plane Diagram for Phenol = 2000 mg/l & NP = 400 mg/l

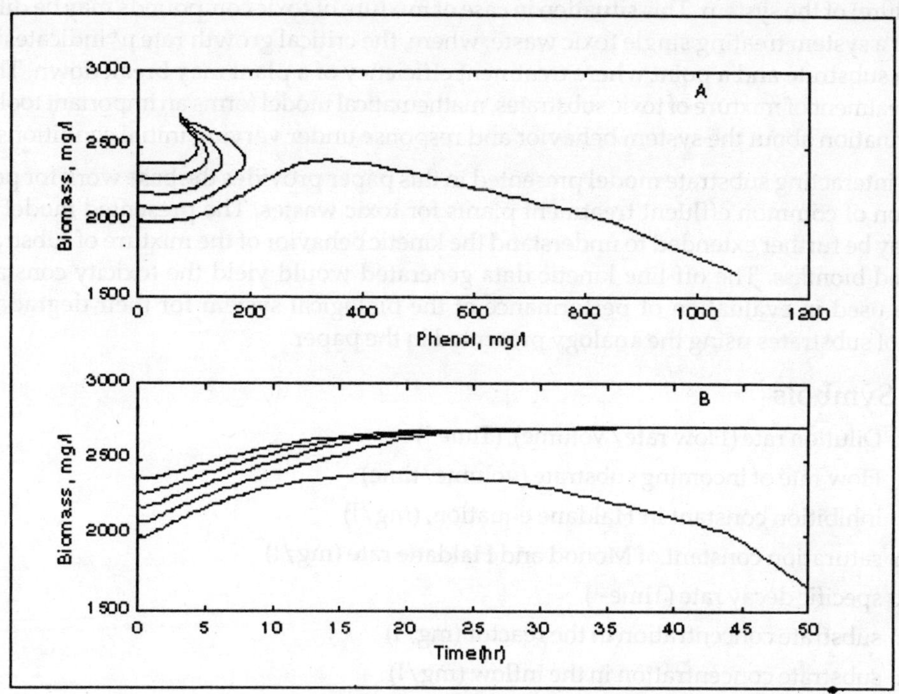

Fig. 8: Phase-plane Diagram for Phenol = 2000 mg/l & CP = 800 mg/l

interaction amongst these compounds, which affects their utilization as simultaneous or preferential. This necessitates estimation of index of cross-inhibition for the interacting substrates to account for the stimulation or inhibition by the co-substrate during their co-metabolism. In this context, mathematical formulation for competitive cross-inhibition which represents the substrate interactions during co-metabolism and accounts for the toxic level of interacting substrates has been used in the present study. Since the presence of multiple substrates in different waste streams coming to common effluent treatment plant is quite likely situation, evaluation of type of interaction amongst different substrates and proper choice of mathematical formulation is necessary and more realistic way of representing biodegradation in wastewaters.

Another important consideration in operation of treatment plant is the effect of changes that may take place in the numerical values of inflow rate, waste strength and hence lead to either failure of the system or low performance of the system. In this context, model simulations obtained for different waste strength in terms of co-substrate concentration *viz.* 2,4-NP or *o*-CP, for a fixed concentration of primary substrate *i.e.* phenol for which the system or treatment plant was designed indicate the reduced performance of the system for phenol removal. Mathematical modelling of multi-substrate system therefore forms an important tool to work out the effect of co-substrate concentration on the degradation of primary substrate.

Sensitivity of the system for treatment of mixture of substrates is also evaluated from the mathematical model. This is shown in terms of one of the important parameter as the initial biomass in the system. The concentration range for primary as well as secondary substrate over which the system may be operated for a given initial biomass condition has been estimated to avoid the washout (total failure) of the system. This situation in case of mixture of toxic compounds may be different than in case of a system treating single toxic waste, where, the critical growth rate μ^* indicates of the toxic level of a substrate and a point where treatment efficiency of a plant may break down. Therefore, in case of treatment of mixture of toxic substrates, mathematical model forms an important tool in deriving the information about the system behavior and response under varying initial conditions.

The interacting substrate model presented in this paper provides the base work for performance evaluation of common effluent treatment plants for toxic wastes. The presented model simulation study may be further extended to understand the kinetic behavior of the mixture of substrates for any developed biomass. The off-line kinetic data generated would yield the toxicity constants, which could be used in evaluation of performance of the biological system for their degradation in the mixture of substrates using the analogy presented in the paper.

List of Symbols

- D: Dilution rate (Flow rate/Volume), (Time^{-1})
- F: Flow rate of incoming substrate (volume/time)
- k_i: inhibition constant in Haldane equation, (mg/l)
- K_s: saturation constant of Monod and Haldane rate (mg/l)
- k_d: specific decay rate (Time-^{1})
- S: substrate concentration in the reactor (mg/l)
- S_i: substrate concentration in the inflow (mg/l)
- V: volume of the reactor (liters)
- X: biological solids concentration in the reactor (mg/l)
- X_R: biological solids concentration in the recycle flow to the reactor (mg/l)
- Y: cell yield coefficient (mg/mg)
- M_{max}: the maximum specific growth rate (Time^{-1})
- μ: specific growth rate (Time^{-1})

REFERENCES

1. Arcangeli J. and Arvin E., 1995, Biodegradation rates of aromatic contaminants in biofilm reactors. Wat. Sec. Tech., Vol. 31, No.1,pp. 117-128.
2. Atuanya E.I., Purohit H.J. Chakrabarti T., 2000, Anaerobic and aerobic biodegradation of chlorophenols using USAB and ASG bioreactors. World J. Microbio. & biotechnol., Vol. 16, pp. 95-98.
3. Beltrame P., Beltrame P.L, Carniti P., Guardione D. and Lanzetta C., 1988, Inhibition action of chlorophenols on biodegradation of phenol and its correlation with structural properties of inhibitors. Biotechnol and Bioeng., Vol 31., pp. 821-828.
4. Bequette B.W., 1998, Phase Plane Analysis In Process Dynamics: Modelling, Analysis, and Simulation, Prentice Hall International Series in the Physical and Chemical Engineering Sciences, pp. 303-330.
5. Deshpande S.D. Chakrabarti T, and Subrahmanyam P.V.R., 1987, Mixed substrate utilization by acclimated activated sludge in batch and continuous flow stirred tank reactors. Environ.Sci. technol., 1987, Vol 21, No.10, 1003-1008.

6. Gaudar CT, Ganji SH, Pujar BG and Strevett KA, 2000, Substrate inhibition kinetics of phenol biodegradation. Wat. Environ. Res., Vol 72, No. 1, pp. 50-55.

7. Gaudy A.F. Jr., Lowe W., Rozich A. and Colvin R., 1988, Practical methodology for predicting critical operating range of biological systems treating inhibitory substrates. J. WPCF, Vol. 60,No.1 pp. 77-85.

8. Kapley A. and Purohit H.J., 2000, Tracking of phenol degrading genotype. Environ. Sci. & Pollt. Res., Vol. 16, pp. 89-90.

9. Klecka G.M. and Maier W.J., 1988, Kinetics of microbial growth on mixtures of pentachlorophenols and chlorinated aromatic compounds, Biotechnol. And Bioengg., Vol. 31, pp. 328-335.

10. Kutty R. Purohit H.J. and Khanna P., 2000, Isolation and characterization of Pseudomonas specie strain PH1 for utilizating meta-aminophenol. Can. J. Microbio., Vol. 46, pp.211-217.

11. Ollis D.F., 2001, On the need of engineering models of integrated chemical and biological oxidation of wastewaters. Wat. Sci and Technol., Vol. 44, No. 5, pp. 117-123.

12. Qureshi A. Prabhu S.K. and Purohit H.J., 2001, Isolation and characterization of Pseudomonas strain for utilization of 4-nitrophenol. Microbes and Environments, Vol 16, 49-52.

13. Qureshi A. and Purohit H.J., 2002, Isolation of bacterial consortium for degradation of p-nitrophenol from agricultural soil. Annals of Applied Biology, Vol 140, 159-162.

14. Rozich AF and Gaudy AF Jr., 1985, Response of phenol-acclimated activated sludge process to quantitative shock loading. J. Wat Poll Control Fed., Vol. 57, No. 7, pp. 795-802.

15. Rozich AF, Gaudy AF Jr, and Adamo PD., 1983, Predictive model for treatment of phenolic wastes by activated sludge. Wat Res., Vol 17, pp. 1453-1466.

16. Scott J.P.and Ollis D.F., 1996, Engineering models of combined chemical and biological processes. J. Env. Engg., pp. 1110-1113.

17. Shieh W.K. Puhakka J.A. Melin E. and Tuhkanen T., 1990, Immobilized cell degradation of chlorophenols. J. Env. Engg., Vol 116, No.4, pp. 683-697.

18. Wang K.W., Baltzis B.C. and Lewandowski G.A., 1996, Kinetics of phenol biodegradation in the presence of glucose.Biotechnol.and Bioeng., Vol 51, pp. 87-94.

19. Wang Si-jing and Loh Kai-chee, 2000, Growth kinetics of Pseudomonas putida in cometabolism of phenol and 4-chlorophenol in the presence of a conventional carbon source. Biotechnol. And Bioeng. 2000, Vol. 68, No.4, 437-447.

58

Application of Vermitechnology for Management of Wastewater

R.V. BHOYAR & A.V. SHEKDAR

Solid Waste Management Division, National Environmental Engineering Research Institute, Nagpur - 440 020

ABSTRACT

Wastes (solid/liquid state) containing biodegradable solids require treatment/processing from environmental as well as sanitation point of view. Bioprocessing/treatment is one of the approach for treatment of waste. Existing bioprocessing approaches require skilled personnel/ operators, are energy oriented and involves large expenditure. Vermitechnology is an emergent discipline which is eco-friendly and significantly less energy input is required for treatment. It is cost-effective and skilled personnel and operators are not essential. Vermiconversion is influenced by temperature, moisture, stability of biodegradable solids, aerobic environment, toxicity, etc.

In India, very few cities and towns are sewered where sewage treatment is practiced. Normally, community wastewaters are let drained in water courses. Under such circumstances vermitechnology application is appropriate for treatment of wastewater. To arrive at feasibility of vermitreatment of wastewater preliminary experiments were conducted using filtered and Perionyx excavatus worm species. The results points that treatment of solid phase of wastewater is feasible by vermitreatment.

Introduction

Everyday human activity in urban areas generate enormous quantities of waste material which could be broadly categorized as solid waste and wastewater. As per their state *i.e.* solid or liquid, these wastes are collected and managed separately. In case of solid wastes these are discretely collected and normally without any treatment openly dumped in the low lying areas in an around Indian urban areas. Recently for the degradable organic matter, which is main constituent of municipal solid waste, biological treatment has been proposed to convert it into compost. Further, if the source of these solid waste is like vegetable markets, large kitchen etc., producing relatively homogeneous degradable organic matter in sizeable quantities, vermicomposting has been proved to be a better approach for bioconversion of these materials into organic manure.

In some of large Indian cities underground sewerage system are provided and wastewater treatment plants have been constructed to treat the wastewater before its discharge in to the environment. In most of these treatment plants invariably sizable quantum of sludge is produced which needs appropriate treatment for its use or disposal. In the present paper an attempt has been made to explore the possibilities of use of vermiconversion technology to produce manure from the wastewater (solid phase) directly in order to develop it as a low cost treatment alternative, especially where the quantity produced is comparatively less.

Vermitechnology for Bioconversion

Vermitechnology is an emerging discipline for conversion of biodegradable organics into end products-the castings. Degradation of organics occur due to the combined activity of microbes and earthworms. This conversion is influenced by temperature, aerobic/anaerobic environment, toxicity of heavy metals as well as waste decomposition products, etc. It is a complex system which involves chemical, biochemical, physiological processes etc. and interaction with microbial consortia (Doube, 1998) within the controlled environment. Vermiconversion efficiency also depends on degree of stabilization of the biodegradable organics, microbial pathogenicity to worms as well as survival of pathogens content in the waste during vermiconversion.

Earthworms are highly adopted for burrowing and feeding activities. As feed substrates are sucked in the gut, particle size and brittleness of the waste influences feeding activity. Digestion begins as the food enter the mouth. Particles that are swallowed are subjected to grinding in the organ of gut called gizzard. In gizzard, ingested organic feed particles get further reduced in size due to the combined action of ingested mineral particles and muscular movement of the gizzard. The feed substrates are metabolised through the activity of enzymes secreted in the intestine of the earthworms. The digested residue is thrown out from the anus as castings.

Information on integumentary adsorption of digestible components is very scanty. Adsorption of soluble amino acid has been reported by Sylvia and Arme (1982). Earthworms though are aerobic, they can survive and are active in air with low partial pressure (P_2O). It has been reported that per cent of the total oxygen transported by earthworm haemoglobin (for *lumbricus worm* sp.) is highest at 38 mm partial pressure (Kerkul 1963). Earthworms are very sensitive to stability of waste. Mortality of earthworm has been reported for materials such a cellulose, peptone, yeast extract, fly ash, anaerobically digested residue etc. whereas they survive in soil, compost, manure, tree bark etc. (Shanthi, *et al.*, 1993). Most of the practices related with vermiconversion are simply natural conversion involving earthworm activity. Temperature exceeding 30°C is lethal to worms. Moisture range for worm survival is related with stabilization level of feed substrates. There is often considerable debate about defining the differences between casting and vermicompost. Some uses the term interchangeably where some refers vermicomposts to casting. Thus vermiconversion is a complex system influence by environmental variables, climatic conditions, feed composition and its palatability, pathogenicity to earthworms, survival of pathogens during vermiconversion of organics etc.

Sludge formation is normally associated with conventional treatment of wastewaters which requires further treatment from environmental as well as sanitation point of view. Vermiconversion leads to high level of stabilisation of biodegradable solids. As in other biological treatment processes vermitechnology has many advantages and limitations also.

Merits

1. Sludge formation is minimum.
2. Vermitechnology does not require any complex physical treatment. Moreover, operating expenses are limited.
3. Skilled personnel and professionals are not essential.
4. Vermitreatment is without odour and insect nuisance.
5. Financial input is much less compared to other treatment processes.
6. Energy requirement is less.

Limitations

1. Controlling aerobic condition in the system may pose operational constraints.
2. System is highly sensitive to the variation in climatic conditions.
3. Techno-economic viability has always been situation specific.
4. Toxic effect of heavy metals and decomposition products may be inhibitive to the system.
5. Activity zone of earthworm is very limited.

Feasibility of Sludge for Vermitechnology

Soil is the natural habitat for earthworms dwelling. In the favourable environmental conditions like temperature, moisture, etc. earthworms are active in soil. Inherent tendency of worm for feeding and converting biodegradable organics into castings is utilised for stabilisation and vermiconversion of solid biodegradable organics. Soil environment is relatively static as compared to other waste where physico-chemical conditions are in dynamic state. Characteristics of soil, municipal solid waste and sewage, relevant for vermiconversion are indicated in Table 1:

Table 1: Comparison of Different Media for Vermitreatment

Parameter	*Soil*	*Municipal Solid Waste*	*Sewage*
Volatile solids (% of total solids)	1-4	30-50	40-60
Particle	Small	Large	Small
Moisture (%)	5-50	30-50	96-98
Stabilization level of biodegradable organics	Partially stabilised	Not stabilised	Partially stabilised
Decomposition product toxicity	Less	More	More
Heavy metal toxicity	Less	More	More
Metabolic heat	Negligible	More	More

The above comparison reveals that the solid phase of wastewater can be vermiconverted controlling excess moisture, toxicity and providing aerobic environment. Thus, to explore the use of this technology, an exhaustive literature was reviewed.

Literature Survey

Detailed literature has been review to evaluate current practices of vermitechnology. Prior to 1980 trend of research was limited to physiology, population, dynamics, biology of worms etc. and mostly confined to soil media (Krishnamoorthy 1990, Ochiai, 1982, Senapati *et al.*, 1993). Most of the researches regarding application of earthworm activity for conversion of biodegradable occurs in post 1980. Edward and Bohlen (1996) suggested vermiculture ecosystem for utilization of organic wastewater. In this system combine activity of earthworms and microbes along with filtration plays a role. Studies relate to vermiconversion of sewage sludge using earthworm (Loehr *et al.* 1988), and activated sludge with emphasis on accumulation of heavy metals (Hartenstein *et al.* 1980). Hartenstein and Hartenstein (1981) observed the physico-chemical changes in activated sludge during vermiconversion. The researcher observed 1-3 fold mineralization of nutrients and 2% assimilation by earthworm. Bissesi (1990) conducted studies for management of biological sludges under dynamic conditions of temperature and seasonal changes. Treatment of liquid sludge from fish farms has been studied utilizing filter bed (Koster *et al.*, 1988).

In most of these attempts the vermiconversion has adopted as an add-on treatment to existing wastewater treatment plant for the treatment of sludge alone. This does not make any alteration/ improvement in the prevailing wastewater treatment system. Hence, an attempt to develop a vermiconversion plant where wastewater is directly fed to a vermiconversion reactor with the following objectives.

Objectives

1. Monitoring survival of earthworms in wastewater.
2. Feasibility of combined dewatering and stabilization.
3. To find out the feasibility of filter bed approach.

Materials and Methods

Earthworms survival in sewage columns of varying depth were initially studied and observed that the worms survived for a period studied upto 24 hrs. Then sewage treatment using vermibed filter was adopted. A Plexiglass reactor (28.5 cm dia and 60 cm high) with working volume of 25 litres was used. Filter bed consists of 4" layer of pebbles and coarse sand followed by 12" of fine sand (Figure 1). Sewage input was done from the top and effluent was taken out from the bottom. The studies conducted in three runs as a batch process with 24 hrs detention time. Approximately 30 earthworms (*perionyx excavates*) collected from the peripheral soil of oxidation ponds were added to the filter bed. Control experiment (with earthworm) was simultaneously conducted. Chemical analysis of various parameters for raw sewage and effluents from filler bed are indicated in Table 2. Casting were observed in the solid phase retained on the surface of filter bed.

Results and Discussion

The study showed that the earthworms survive for detention period of 24 hours maintained during the experimentation. The experiments further reveals that combined dewatering and stabilisation occurs in the single step in the vermifilter bed. The results of analysis from the probing studies (Table 2) reveals that reduction in COD takes place in vermifilter bed effluent. High concentration of nitrate and water soluble phosphours are observed in the effluent. Similar observation for increased concentration of soluble phosphate and nitrate have been observed during vermiconversion of biodegradable solid organics (NEERI report unpublished data). Detention time for filter bed approach

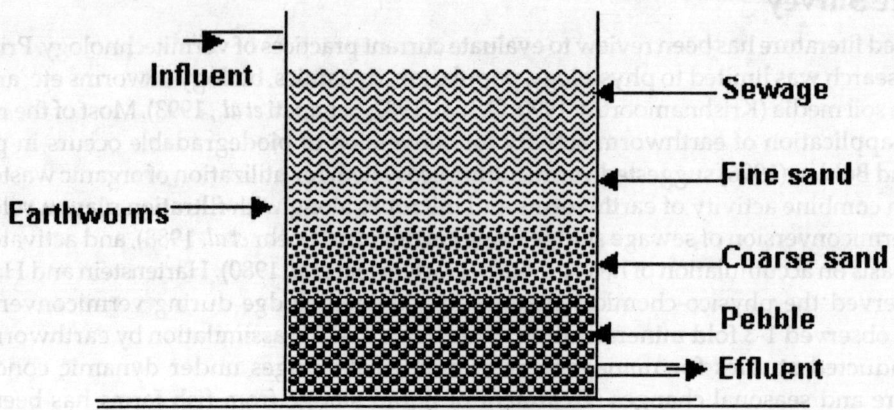

Fig. 1: Schematic Presentation of Vermifilter Bed

has important role in respect of earthworms survival and conversion of solids. The findings of the experimental runs points that vermitreatment of sewage is feasible. However, detailed investigation is essential for developing and establishing vermitreatment approach.

Table 2: Physico-chemical Analysis of Wastewater

Parameter	*Treatment*		
	Raw Sewage	*Control*	*Filter Bed Effluent*
pH	7.22	7.38	7.28
	7.23	7.60	7.68
	7.23	7.21	7.20
Conductivity (μs/cm at 25°C)	68.20	69.0	69.0
	58.70	63.6	61.8
	67.00	64.00	63.6
COD (mg/l)	119.04	46.00	46.09
	358.8	100	50.00
	620.00	207	92.00
	960	80	30.00
Dissolved phosphorus (mg/l)	3.52	2.05	3.11
	3.01	1.50	2.20
	4.35	1.40	2.25
Nitrate (mg/l)	0.167	0.18	0.37
	0.21	0.17	0.36
	0.61	0.33	0.49
TS (g/l)	1.06	0.67	0.53
	0.54	0.44	0.46
	0.72	0.48	0.50

Concluding Remarks and Recommendations for Future Study

Vermiconversion is an established process for conversion of degradable matter into manure. This method could be applied to wastewater treatment in order to develop a cost-effective treatment system for low capacity situation. The comparison of process parameters of the vermiconversion process for soil, solid waste and wastewater indicates the process feasibility. Further globally, a number of attempts indicates the feasibility of use of this process to sludge. However, in order to replace the prevailing wastewater technology an attempt to validate the direct application of wastewater to a vermireactor for detailed investigation is necessary. The results of these experiment are quite encouraging as the system reduces the COD load considerably. However, further attempts are necessary to decide loading rate versus conversion into casting, earthworm population dynamics in the system, detention time, bed thickness, etc.

References

1. Hartenstein R. & Hartenstein F., 1981, Physiochemical changes effected in activated sludge, by the Earthworm *E. Foetida*, J. Environmental Quality 10 (3), 377-381
2. Edward C.A. & Prohlen P.J., 1996, Biology & ecology of earthworms, Chapman & Hall London
3. Loehr R.C., Martin J.H., Neuhauser, E.F., 1988, Stabilization of liquid municipal sludge using earthworms, Earthworms in Environmental & Waste Management, Edwards C.A. & Neuhauser, E.F. (eds), SPB Academic Publishing, The Netherlands, 9-20
4. Shanthi, N.R. *et al.*, 1998, Vermicomposting of vegetable waste: Compost Science and Utilisation, 1 (4), 27-30
5. G.A. Kerkul (ed), 1963, Problem in biology, Vol.1, Pergamon press
6. Sylvia R.K. and Arme, C., 1982, Integumentary uptake of dissolved organic matter by earthworms lumbricidac, Pedobiologia, 23 (5), 358-366
7. Krishnamoorthy R.V., 1990, Mineralization of phosphorous by faecal phosphates of some earthworms of Indian tropics, Proceeding Indian Academic Science, (Anim. Sci.) 99 (6), 509-518
8. Hartenstein R., Neuhauser, E.F.& Collier, J., 1980, Accumulation of heavy metals in the earthworms, Eisenia Foetida J. Environmental Quality, 9 (1), 23-26
9. Takehiko Ochiai and Tohru Yamakuru, 1982, Oxygen equilibrium characteristics of Eisenia Foetida Hemoglotin I, Bohr effect. Personal Comunication
10. Senapati B.K & Sahu S.K., 1993, Population, Biomass & Secondary production in earthworms, Earthworms Resource & Vermiculture, 57-78
11. Doube B.M.& Brown G.G., 1998, Life in complex community functional interactions between earthworms, organic matter, microorganism and plants, Earthworm Ecology, C.A.Edward (ed), St. Lucile Press, New York, 179-212
12. Bisesi M., 1990, Vermial & microbial management of biological sludges under dynamic condition of temperature and seasonal changes, Indiana State Uni. Terre Haunted, In USA; Biological Waste, 32 (2), 99-109
13. Koster, T.W., Vroon, D.A., Dorendeur, 1988, Earthworm filters for treatment of liquid sludge from fish farm, Proceeding of the fifth international solid waste conference, Vol. 2, (Lizzi Anderson *et al.*, eds.

59

Review of Performance Data of Few Pilot and Full Scale Aquatic Weed Based Waste Treatment Plants from India

R.K. TRIVEDY

1, Sheila Apartment, Karad - 415 110

ABSTRACT

Performance data of 3 water hyacinth based and four emergent plant species based wastewater treatment plants in India is reviewed in the present paper. Water hyacinth based systems were employed for treatment of sewage and industrial wastes. All the three systems are full-scale plants. The performance of hyacinth based systems was good despite poor designing and vegetation management especially with regard to reduction in BOD, COD and total nitrogen. Reduction in solids was not good in all the systems. Considerable variations in performance was noticed which is attributed to different hyacinth coverage maintained and also the seasonal influences besides the level of housekeeping.The emergent plants based system (out of 4 one was a full scale plant and remaining three experimental level pilot systems) were found to be very efficient in BOD, COD and phosphorus removal but poor in nitrogen and solids reduction. The performance of these systems, however, gives very encouraging signals and more research is urgently needed on these systems especially with regard to system design and choice of plant species.

Introduction

In the last four decades, there has been considerable emphasis on use of aquatic plants in pollution control. Large amounts of data have been generated in this period and thousands of aquatic weeds based treatment plants have been established all over the World. Boyd (1970), Cornwell *et al.* (1977), Sheffield (1967) and Yount (1964) were among the first scientists to demonstrate the pollution abatement potential of aquatic plants subsequently Reddy and Smith (1987) ; Gopal and Sharma (1981) and Trivedy and Joag (1989) have listed a number of studies.

Aquatic plants can be tremendously useful in pollution control in developing countries especially in tropical areas due to following reasons:

1. Lesser cost in erection.
2. Monitoring and consumables requirements are extremely low.
3. Power requirements are negligible.
4. The aquatic plants can be easily harvested and put to some use.
5. Useful byproducts are obtained from aquatic plants.

In India the first ever experimental study on use of aquatic plants in pollution control was of Sinha and Sinha (1969) who reported successful treatment of sugar industry waste by using water hyacinth. Since then a large number of experimental studies have been carried out on use of floating plants in wastewater treatment. The studies are reviewed by Trivedy & Gudekar (1985) ; Trivedy & Joag (1989) ; Trivedy (1998) & Trivedy (2001).

The concept of root zone (constructed wetlands using emergent macrophytes) for wastewater treatment is very new to India. The very first plant was established in 1994.

Table 1 gives list of some root zone installations in India. Very few experimental studies on this technology are published from India.

Table 1: Reference List of Root Zone Biofilters Installations in India

Sl. No.	*Name*	*Co-Group*	*Flow*
1.	Associated Cement Co. Gagal	ACC	100 cu.m/day
2.	Bush Betta Jungle Resort, Bandipur, Near Mysore	An R.C.I.Affiliated resort	50 cu.m/day
3.	Meadows Resort, Aurangabad	Meadows Group	40 cu.m/day
4.	Larson & Tubro	L & T Group	40 cu.m/day
5.	Kimberly Clark Lever Ltd. Sanaswadi, Pune	JV of KC & Lever	30 cu.m/day
6.	INS Shivaji, Lonavala, Pune	Indian navy	20 cu.m/day
7.	CTR Mfg (I) Ltd. Chikalthana, Aurangabad	Crompton Greaves Group	20 cu.m/day
8.	Byke Retreat Matheran (Maharashtra)	Byke Group	20 cu.m/day
9.	Z.F. Steering Geers (I) Ltd. Sanaswadi, Pune	Bajaj Tempo Group	20 cu.m/day
10.	AT & T Phenolix	JV of AT & T/Phenolix	10 cu.m/day
11.	Hindustan Lever Ltd., Dabgram, Darjeeling (W.B.).	Hindustan Lever Group	10 cu.m/day
12.	Lever Johnson Consumer Products Ltd., Dhingrajwadi, Pune	Hindustan Lever Group	5 cu.m/day

(Courtesy: Merit Environmental Consultancy, Pune)

In this paper an attempt has been made to present performance data of few pilot and full-scale wastewater treatment plants employing aquatic weds. To the best of my knowledge this is the first such attempt in India.

The treatment plants have been divided into two parts:

1. Water hyacinth based systems.
2. Constructed Wetlands with emergent macrophytes.

The data published by various authors on these treatment plants including author of this paper are presented here.

Table 2 gives salient features of hyacinth based plants; Table 3 reports performance of their systems. Table 4 presents salient features of plants with emergent vegetation and their performance is recorded in Table 4.

Results and Discussion

Water Hyacinth Based Systems

The three systems compared here all belong to same microclimatic area and are employed for treating sewage. In all the systems depth was about 1 m and several ponds were used in series.

Two plants Sangli Municipal Corporation (SMC) and Krishna Hospital (KH) are employed primary sewage. All the hyacinth systems mentioned in the study are more or less same in structure but a better hyacinth coverage was provided in the first system (SMC).

The data show considerable variations in performance of three systems, which is due to seasonal influences and also poor vegetation management and housekeeping, which was noticed, in the latter 2 systems. The changing composition of incoming waste water must have also contributed to variation in performance of these systems. It seems only one time data are provided for the first system hence systems 2nd and third (where monthly data are available for full one year) are more relevant to discuss. The reduction in solids was usually below 50%. Considerable reduction in BOD and COD was however reported in all the systems usually above 70% and maximum up to 93.3%. Total nitrogen reduction is remarkable in all the systems and up to 89.3% reduction was noted in total nitrogen content. Total phosphorous reduction was also appreciable.

The performance of various hyacinth systems was however low with regard to reduction in hardness, chlorides and other mineral ions like sodium, potassium and calcium.

Microbiological data are available for SMC system only. The standard plate count got reduced by 91.3%. There must have been reduction in coliforms too but could not be recorded due to methodological problems.

Emergent Macrophytes Based Systems

This technology originally developed by Kickuth in Germany has become very popular and thousands of systems have been established all over the world. It came to India around 1992 and has rapidly spread since with many environmental consultancy companies actively taking interest in it. Around 25 plants have been established all over India. However, relatively much less published studies are available on these systems. Venkataramani (1992) reports one such application from Shembaganur, Tamil Nadu. The system treats 6000 litre wastewater from a community of 60 individuals and is composed of a 90 sq.m brick and mortar tank with depth of 1m. It has been loosely filled with

Table 2: Performance of Some Full-scale Water Hyacinth Based Treatment Plants in Maharashtra, India

Details	*Sangli Municipal Corporation (Jogalekar & Sonar 1986)*			*S.P.Engineering Karad (Trivedy 2001)*			*Krishna Hospital Karad (Trivedy 2001)*		
Year of Commencement	1986			1994			1994		
Total Wastewater Quantity Treated	33.6 m3d^{-1} in each pond			200 m3d^{-1}			10 m3d^{-1}		
Type of wastewater Treated	Sewage			Sewage + Hospital waste			Sewage + Light engineering industry waste		
System details	Ten ponds each measuring 7.3 × 7.3 × 0.9 m constructed by using RCC panels. Flow rate kept was 33.6 m3d^{-1} in each pond 100 % area coverage by water hyacinth.			Ten ponds (earthen) 10.37 × 6.1 × 91 m each in series. Average cover of water hyacinth 75%			Three ponds (RCC) of 8.0 × 2.5 × 1.0 m (Free Board 30 cm) in series. Average cover of water water hyacinth 41%		
Performance of System	*Inlet*	*Outlet*	*% Reduction*	*Inlet (Range)*	*Outlet (Range)*	*% Reduction (Range)*	*Inlet (Range)*	*Outlet (Range)*	*% Reduction (Range)*
pH	7.5	7.1	-	5.98-7.64	6.71-7.56	-	6.5-8.6	7.2-8.2	-
Total Solids	1470.0	1350.0	8.2	332-2492.0	431.0-1806.0	9.36-49.88	540.0-1055	426-775	6.50-46.1
Total Dissolved Solids	820.00	670.00	18.3	296-1348.0	334.0-1278.0	9.29-38.82	455-880	268-656	0.6-50.37
Total Suspended Solids	190.00	220.00	-	36-1940.0	30-582.0	50.8-90.55	22-448	20-232	17.6-75.6
COD	280.00	80.00	71.4	88-266.6	25.5-368.0	1.06-86.67	126-266	45-180	22.2-67.6
BOD	187.5	20.00	89.3	43-98	16-160	33.3-78.67	72-160	10.5-82	9.1-93.3
TKN	22.5	2.5	88.9	17.6-30.8	2.8-16.1	14.8-89.19	25.6-81.9	15.3-50.5	27.2-64.3
Total P	5.0	2.5	50.0	8.67-32.09	7.5-16.51	9.87-54.18	4.65-18.39	2.1-9.95	38.8-82.81
Total Org. P	36.0	15.6	56.7	3.49-15.99	2.97-13.98	9.69-77.48	2.21-14.2	1.26-6.57	38.9-91.2
Total Hardness	388.0	352.0	9.3	22.0-70.0	19.0-51.0	3.77-64.42	227.0-456.8	225-433.0	0.92-16.8
Chlorides	150.00	124.00	17.3	68.87-150.52	80.2-155.5	6.86-40.57	74.55-150.0	63.19-132.06	4.35-29.64
Sodium	140.00	152.0	—	—	—	—	—	—	—
Potassium	29.0	18.6	35.9	—	—	—	—	—	—
Calcium	33.0	30.0	9.1	2.8-14.8	3.2-6.01	19.9-75.6	76.1-134.5	50.34-95.85	7.51-31.59
Standard Plate Count ml^{-1}	1340 × 108	117 × 108	91.3	—	—	—	—	—	—
MPN of Coliforms 100 ml^{-1}	72400	72400	—	—	—	—	—	—	—

All the values are in mg/L, otherwise stated.

Table 3: Important Highlights of Various Studies on Wastewater Treatment by Constructed Wetlands with Emergent Macrophytes in India

Radha Madhav et al. (2002)	*Meshram (2000)*	*Diwan et al. (2002)*	*Trivedy (2000)*
In all 30 Indian Wetland species were screened for their efficiency in wastewater treatment in constructed wetlands out of which *Phragmites Karka, Ausndo donax* and *Brachiana mutica* were found to be most efficient. There was no reduction in TDS during first and second phases, which might be due to higher evaporative losses during the period of peak summer during which the experiments were done. Red Soil was used as the bed material	*Typha Latifolia* was used in sand and soil beds. Only reduction in total nitrogen and potassium content of soil was studied.	The study reports performance of a full scalel wastewater treatment plant operating in M.P., India treating wastewater from a surfactant manufacturing industry, one of the first few plants established in India. The plant has one equalization tank before reed bed. Reed species used *Phragmites australis* and *Phragmites communis*. The plant was monitored for five months during Jan 1999 to May 1999. The performance of the plant was at par with any conventional plant and all the limits of concerned Pollution Control Board are met. Additional benefits are removal of nutrients like nitrogen and phosphorous and surfactants (90% removal).	Laboratory scale study using a bed of sand and gravel. *Typha latifolia* was used at a density of 160 plants $/m_2$. The system was capable of accepting COD up to 11513 mg/L. Appreciable reduction was noted in BOD, COD suspended solids, total nitrogen and total phosphorous.

soil on which reeds have been planted at the rate of 4 clumps per sq.m. Inlet and outlets have been provided. The author claims excellent performance of the system but data on chemical characteristics and type of reed species planted are not available.

I have reviewed here the performance of rootzone systems on which published data are available out of which Diwan *et al.* (2002) report performance of a full scale plant while rest are experimental laboratory scale studies. According to Radha Madhav *et al.* (2002), thirty wetland species have been screened by them and *Phragmites karka, Arundo donax* and *Brachiaria mutica* were found to be most efficient. The other two experimental studies including one of the author have used *Typha latifolia* while *Phragmites australis* and *Phragmites karka* have been used in the full scale plant. Soil is used in all the systems while sand and gravel is also used alongwith soil in other systems.

However, the data reported are very scanty. Radha Madhav *et al.* (2002) report a COD reduction of 84% in 10 days with upto 2000 mg/L COD but increasing it to 4000 resulted in only 39.65% reduction. Diwan *et al.* (2002) report about 80% reduction in COD at inlet COD of 534.0 to 910.0 mg/L. Trivedy and Nakate (2000) also report about 80% reduction in COD.

TDS reduction was poor in most of the studies. Suspended solid reduction was also poor but was better with more retention time. Nitrogen reduction (nitrates or total nitrogen) was from 35.0% to 50%. Reduction in total phosphorous reported in only one study was appreciable (about 75.0%). Reduction in surfactants concentration was from 198.0-332.89 mg/L to 19.08-36.66 mg/L as reported by Diwan *et al.* (2002).

Table 4: Performance of Some Root Zone Installations and Pilot Projects in India

Studies	Place	System Details	Plant Species Used	Retention Time	Change in Wastewater Character: Inlet	Outlet & % Reduction in Parenthesis
1. Radha Madhav *et al.* (2002)	Hyderbad, India.	Linning HDPE 250 microns Length - Top110l - Bottom90l Width – Top 50l - Bottom 30l Slope – Side 3 : 1 Inlet to Outlet 100:1 Depth 3 l Working Area 3405 sq. ft Working Volume 10455 cu.ft Inlet Area 288 sq. ft Outlet Area 500 sq.ft Flow rate 21.5 cu.m/d Bed filled with Red Soil	Phragmites Karka Arundo donax Brachiaria mutica	10 days	*COD* Ist phase 2000.0 IInd phase 3000.0 IIIrd phase 4000.0 *TDS* Ist phase 9720.0 IInd phase 12520 IIIrd phase 18900	 80 (84%) 1072 (46.4%) 2414 (39.65 %) 13190 14650 15450 (18.25%)
2. Meshram (2000)	Nagpur, India	High density PVC pipes Dia – 30 cm Bed filled with 60% sand and 40% soil.	Typha latifolia	4 days (Total 50 applications)	*Total Nitrogen* Above 15mg/L Potassium Around 50 mg/L	Below 10 mg/L (except 2 observations) usually below 7 mg/L usually below 20 mg/L (Max. reduction 71.5% (Min. reduction 43.1%)
3. Diwan et al. (2002)	Bhopal, India	Full scale plant Treating effluent of M/S P&G, M.P. (Detergent manufacturing unit) Starting date 1994 Surface 200 m² Wetland type - subsurface flow constructed wetland Flow Horizontal System Porosity factor 0.32 Total bed depth 1 m	Phragmites (2 species grown) P. austrails P. cummins	4 days	*pH* 9.4 – 7.15 *COD* 534.0 – 910.0 *TDS* 439.8 – 616.0 *TSS* 176.0 – 256.0 *NO3 – N* 0.91 – 3.55 *Total P* 3.37 – 30.18	 6.4 – 7.8 77.0 – 169.0 523.0 – 695.0 67.0 – 101.0 0.19 – 1.43 0.76 – 5.78

Contd...

Table 4–Contd...

Studies	Place	System Details	Plant Species Used	Retention Time	Change in Wastewater Character: Inlet	Outlet & % Reduction in Parenthesis
		Total dimension of the bed 20m × 10 m Filling material –Loam soil and Sand Bed incorporates drain pipes at various intervals at the bottom of the bed.			*Surfactant* 198 – 332.89 Average 249.39 *BOD* 145.35 – 62.3	 19.08 – 36.66 Average 26.186 12.0 – 30.0
4. Trivedy & Nakate (2000)	Karad, Maharashtra	System composed of galvanized metal sheet. Shape of the bed rectangular (length 1.5 m, breadth 0.3m and height 0.3 m) The bed was filled with 75% sand and gravel and 25% of soil. Wastewater applied after 4 weeks of planting.	Typha latifolia (@ 160 plants/m²)	.10 days	*pH* 5.2 *BOD* 1456 *COD* 11513 *TDS* 2326 *TSS* 988 *Total Nitrogen* 25.91 *Total P* 17.31	 7.2 760 2443 1723 990 21.12 9.8
5. Trivedy & Nakate (2000)	Karad, Maharashtra	System composed of galvanized metal sheet. Shape of the bed rectangular (length 1.5 m, breadth 0.3m and height 0.3 m) The bed was filled with 75% sand and gravel and 25% of soil. Wastewater applied after 4 weeks of planting.	Typha latifolia (@ 160 plants/m²)	15 days (Different parallel set)	*pH* 5.5 *BOD* 685 *COD* 14908.8 *TDS* 2692 *TSS* 928 *Total Nitrogen* 46.6 *Total P* 15.5	 7.5 465 1830.3 1314 454 15.42.00 3.89

Conclusion

The studies reported from India show a remarkable capacity of aquatic weeds based wastewater treatment plants from India. Some of the conclusions, which can be drawn from the present study, are:

1. Water hyacinth based systems are extremely effective in treatment of sewage with regard to reduction in organic matter and nitrogen as well as microbial characteristics. The nitrogen reduction capability of these systems is very high but the systems appear to be poor in reduction of various nutrients including Phosphorus.
2. The emergent plants based systems are extremely efficient in reducing organic matter but poor in reduction of solids. Phosphorous reduction was, however, appreciable.
3. There is a considerable variation in performance of different systems and in different seasons in the same system.

The aquatic weeds based system perform extremely well in India, however, more research is needed on design aspects, selection of different plants species and various other ecological factors.

References

1. Boyd, C.E., 1970, Vascular aquatic plants for mineral nutrient removal from wastewater. *Econ. Bot.*22, 95-103.
2. Cornwell D.A., Zollek, J. Jr., Partrinely, C.D., Furam, T.S. and Kim, J.I. , 1977, Nutrient removal by Water hyacinth. *J.Wat. Poll. Cont. Fed.* 49, 57-65.
3. Diwan,V.,Vyas,V.and Shrivastava, P., 2002, Performance evaluation of a constructed wetland (Rootzone system) for its efficiency to treat industrial wastewater *Asian Journal of Microbiology.Biotech. and Env. Sc.*, 4: 155-160
4. Gopal B.and Sharma, K.P., 1981, *Water hyacinth the most troublesome weed of the world.* Hindasia Publishers, New Delhi.
5. Jogalekar,V.R. and Sonar, A.G., 1986, Application of water hyacinth for treatment of domestic wastewater, generation of biogas and organic manure In *Aquatic plants for water treatment and Resource Recovery* (edited by Hooper, P.F. and Boon, A.G.) Mangolia Publishing Co., U.S.A.
6. Meshram, J.S., 2000, Use of artificial wetlands for the removal of nitrogen and potassium. *Asian Jr. of Microbiology, Biotechnology* and Environmental Sciences 2, 239-241
7. Radha Madhav, R., 2002, Efficiency of constructed wetland to treat domestic sewage and industrial wastewater. *Asian Journal of Microbiology, Biotechnology and Env. Sc.* 4,162-165.
8. Reddy, K.R. and Smith, W.H., 1987, *Aquatic plants for water treatment and Resource recovery.* Mangolia publishing incorporated, Florida.
9. Shefield, C.W., 1967, Water hycainth for nutrient removal. *Hycainth Control Journal* 6, 7-30.
10. Sinha, S.N. and Sinha, L.P., 1969, Studies on use of water hycainth cultures in oxidation ponds treating digested sugar wastes and effluents of septic tanks *Jr. of Environmental Health* 11, 197-207
11. Trivedy R.K. and Gudekar, V.R., 1985, Wastewater treatement using water hycainth: a review of the progress in *Current Pollution Researches in India* (edited by R. K. Trivedy and P.K. Goel), Environmental Publications, Karad, India.

12. Trivedy, R.K. and Joag G.A., 1989, Treatement of Industrial wastewater by hyacinth application in *Pollution management in Industries* (edited by R.K. Trivedy) Environmental Publications, Karad.

13. Trivedy, R.K., 1998, Water hyacinth based systems for wastewater treatment in *Advances in Wastewater Treatment Technologies.* (edited by R.K. Trivedy) Global Science Publications, Aligarh, India.

14. Trivedy, R.K. and Nakate, S.S., 2000, Treatment of diluted distillery waste by using constructed wetlands. IJEP 20, 749-753.

15. Trivedy, R.K., 2001, Use of Aquatic plants in wastewater treatement in *Low Cost wastewater treatement technologies* (edited by R.K.Trivedy and S.N.Kaul). ABD Publishers, Jaipur.

16. Venkatramani, G., 1992, Simple methods to treat wastewater. *The Hindu* (Oct.30).

17. Yount J.L., 1964, Aquatic nutrient reduction potential and possible methods. Rep. 35th ann. meet. Anti Mosquito Assoc. pp. 83-85.

60

Voltammetric Methods for Pollution Measurements

SALVATORE DANIELE, M. ANTONIETTA BALDO AND CARLO BRAGATO

Department of Physical Chemistry, University of Venice, Calle Larga S. Marta, 2137, 30123 Venice, Italy

Introduction

For long time a variety of electroanalytical techniques have been applied for environmental investigation (1). Instrumental analytical techniques based on electrochemical methods while compare well with other instrumental analytical techniques in terms of sensitivity, selectivity and detection limits, offer several advantages as, for instance, simple and compact low cost apparatuses, reliable automatic measurements, high sensitivity, speciation capabilities (2). Moreover, with the advent of microelectrodes (3) (electrodes with sizes over the range 10 nm-50 μm), the analytical applications can be extended to resistive media, not accessible to conventional voltammetry with macroelectrodes (electrodes with sizes > 100 μm). It must be emphasised that the use of microelectrodes reduces the requirement of a well defined cell geometry, thus leading to the simplification of the electroc'.emical apparatus. A two-terminal system is needed, in place of the classical three electrode cell, while the rapid achievement of steady-state conditions provides a means to avoid forced convection when preconcentration steps are needed to increase the sensitivity for trace analysis measurements.

In recent years, there has been considerable interest in the development of probes and analytical methods for *in situ* analysis (1). In situ analysis is attractive because it allows, for instance, elimination of many artefacts due to sample handling, possibility of real time analysis allowing the rapid detection of pollutant inputs in environmental samples, reduction of analysis time.

In general, microsensors and microanalytical systems are key steps in the development of *in situ* determination, and although there is not an ideal probe to this purpose, electroanalytical techniques in conjunction with microelectrodes are probably the most suitable tools.

In this paper we will present the results concerning *in situ* investigations by using microelectrode voltammetry in environmental samples. Note that here *in situ* is synonymous with measurements taken directly in the sample with no or minimum pretreatment. The samples will be considered are either homogeneous or colloidal solutions and "semisolid samples", as, for instance, soils and sediments. The microelectrodes investigated are either naked solid electrodes or covered by a film of mercury. It will be demonstrated that by using these probes it is possible to detect or monitor reliably and directly *in situ* electroactive species from mM down to sub μM levels.

Microelectrodes Under Steady-state Regime

Most of the measurements performed in the investigations were taken under steady-state conditions. Under these conditions, the limiting current, I_d, is a linear function of both geometric characteristics of the microelectrode and analyte characteristics:

$$I_d = k\,n\,F\,D\,C\,a \tag{1}$$

where

k is a constant which depends on the electrode geometry, *a* is the substrate electrode radius, *D* and *C* are the diffusion coefficient and the concentration of the electroactive species, *n* is the number of electrons and *F* is the Faraday constant.

The diffusion limiting current can be measured at the plateau of a wave recorded by linear scan or staircase voltammetry at very low scan rate or frequency (see Fig. 1a) as well as by potential step experiments at long time (see Fig. 1b).

Other experiments, as in stripping analysis, which may not occur under a steady-state regime, exploit the latter conditions during the preconcentration step. This allows to achieve a larger and constant flux at the electrode surface with high precision.

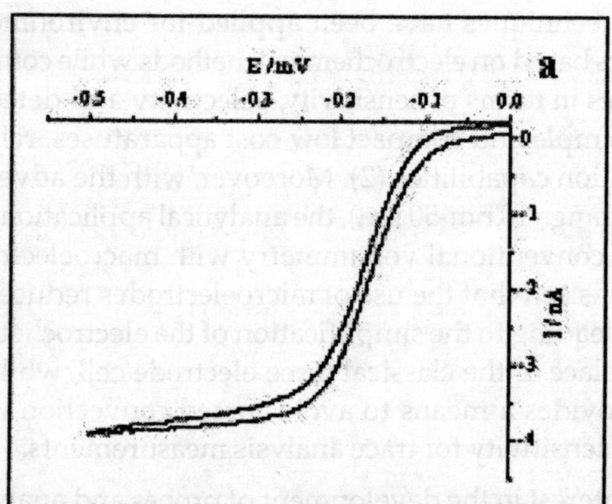

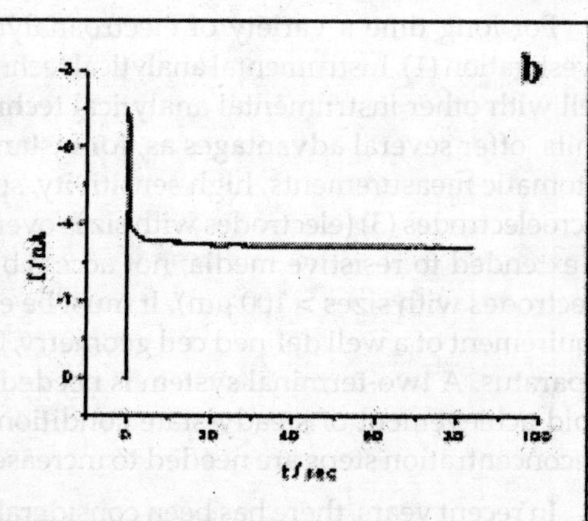

Fig. 1: (a) Cyclic Voltammogram and (b) Cronoamperometry Recorded in 1 μM Ru $(NH_3)_6Cl_3$ + 0.1 M KCl

Results and Discussion

Use of Solid Microelectrodes for Heavy Metal Analysis by Anodic Stripping Voltammetry (ASV)

Stripping analysis has been widely used for measuring trace metals in numerous matrices (4). Mercury-based electrodes consisting of a mercury-drop or-film of conventional dimensions have been traditionally employed. The toxicity of mercury and the growing concerns about on-site trace metal monitoring in environmental samples have prompted also the development of mercury-free sensors for stripping analysis. Solid electrodes represent a different option as substrate for metal preconcentration. However, the interaction among metals deposited on an electrode is a relevant problem in ASV (4). For instance, copper and lead deposited simultaneously on to a solid surface interact to give intermetallic compounds. In analytical practice, real systems containing both lead and copper are often encountered. In this case the use of a naked solid electrode for multielements analysis may bring about several interference. The use of solid microelectrodes for ASV analysis must therefore be properly designed.

The performance of carbon microdisk electrodes for the simultaneous trace analysis of lead and copper by ASV was investigated in detail. The evaluation of the two metal ions in a single step, that is, by employing a plating potential at which both cations are reduced simultaneously gave unpredictable results due to the formation of Pb-Cu intermetallic compounds. In fact, as shown in Fig. 2, multiple and irreproducible stripping peaks for both lead and copper were observed upon spiking the solution with each element.

Moreover, the calibration plots for lead were not linear, while the current stripping peak of copper was always broad and no linearly dependent on concentration. These problems for lead were in part avoided by using a two steps deposition procedure. It is based on the preelectrolysis of copper, performed at a potential at which lead ions are not reduced, followed by the simultaneous reduction of the two cations at large enough negative potentials. In this way lead is deposited onto a layer of copper to form a solid solution (5). Under these conditions linear calibration graphs for lead were obtained. Copper was determined without interference by plating the metal at a negative potential

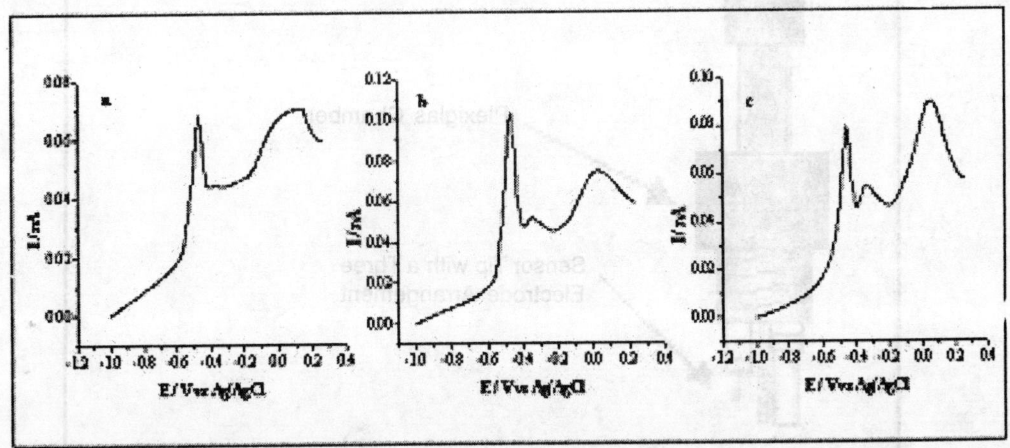

Fig. 2: ASV Recorded with a Carbon Microdisk 4 μm Radius in a Solution Containing: (a) 0.060 μM Pb^{2+} + 0.038 μM Cu^{2+} (b) 0.060 μM Pb^{2+} + 0.075 μM Cu^{2+} (c) 0.060 μM Pb^{2+} + 0.012 μM Cu^{2+}

where lead was not reduced. The reliability of such a procedure was verified in synthetic solutions and applied for the detection of lead and copper directly in rain samples.

Gold microelectrodes can be also employed for trace elements in real samples, particularly for the simultaneous detection of copper and mercury.

Mercury is a very toxic element even at very low levels. Its toxicity actually depends on the chemical form; methyl mercury is more toxic than inorganic mercury. However, inorganic mercury forms extremely toxic methylmercury via an environmental methylation process. Mercury determination in aquatic systems is relevant, as it can enter the food chain through aquatic plants and fish.

Submersible voltammetric probes, for in situ monitoring in aquatic systems of the so-called labile fraction of the heavy metals, which often greatly affects their bioavailability, have been developed recently (6). They rely on a 'mercury-free' gold-fiber stripping electrode as that shown in Fig. 3.

It consists of a three-electrode assembly, which contains a 50 µm-diameter, 1 cm long gold wire working electrode, and a 250 µm-diameter, 1 cm long silver/silver chloride and platinum wires as reference and counter electrodes, respectively. These electrodes are connected through a three-pin environmentally sealed rubber connector to a long shielded cable.

Such in situ monitoring of the labile fraction of the metals by stripping analysis can be complicated by the presence in the samples of oxygen and organic matter. Square wave voltammetry (SWASV) and potentiometric stripping analysis (PSA) can be employed to overcome the influence of oxygen and to address the problem of organic matter (7).

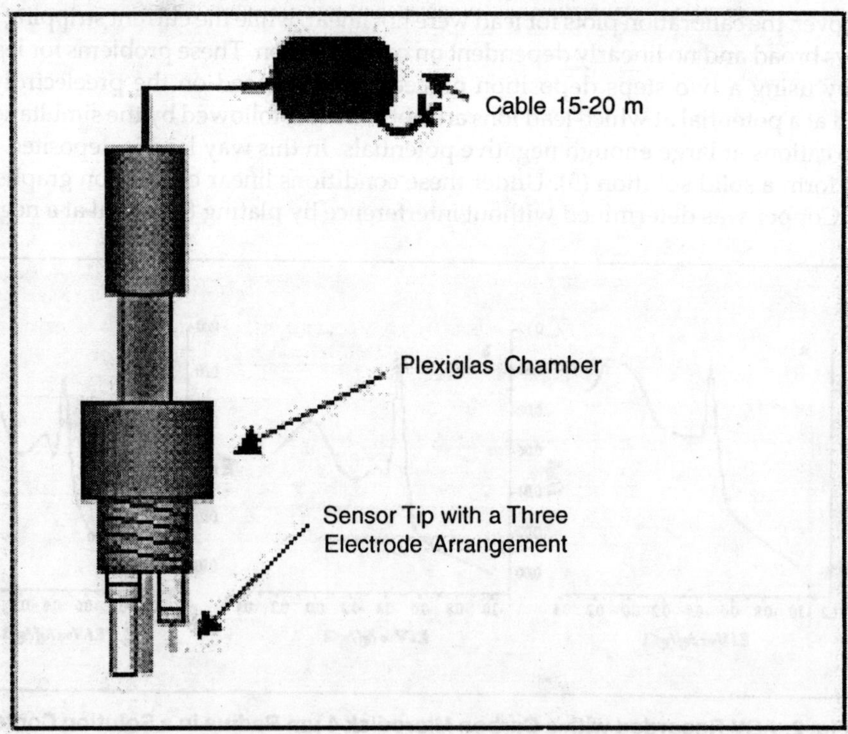

Fig. 3: Scheme of a Submersible Voltammetric Probe

The performance of a 'mercury-free' remote sensor in conjunction with both SWASV and PSA techniques for monitoring the labile (in situ) and total amount of copper and mercury in high-loaded organic-matter sites as those characterising the Lagoon of Venice was investigated.

Fig. 4 shows typical SWASV and PSA responses obtained in synthetic 0.5 M NaCl aqueous solutions containing Hg^{2+} and Cu^{2+} ions.

Preliminary measurements indicated that no interference arose in the simultaneous detection of both analytes. In fact, calibration plots of peak area against metal concentration, over the concentration range 1 – 50 ppb, were linear (the average correlation coefficient was 0.999).

The submersible stripping sensor was then used for the in situ detection of the labile fractions of copper and mercury in different sites of the Lagoon of Venice. These measurements were performed from a boat, while the probe was immersed down to a about 1m depth. Typical responses obtained under these conditions with PSA are displayed in Fig. 5.

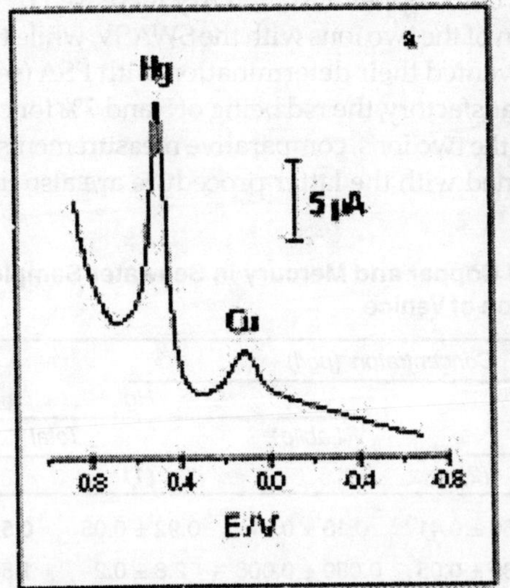

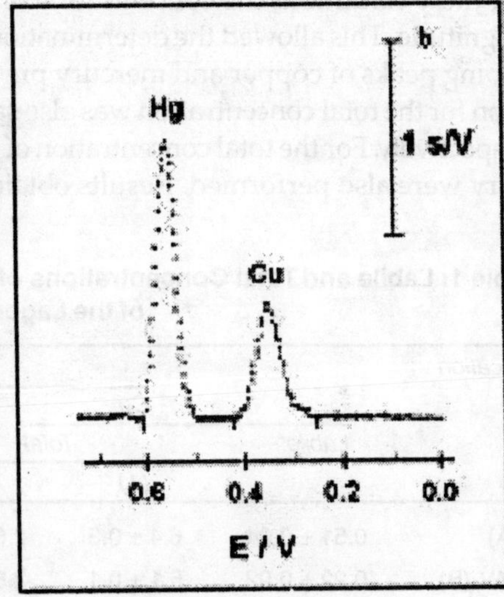

Fig. 4: (a) SWASV in Solution Containing Cu^{2+} 10 ppb and Hg^{2+} 18 ppb, (b) PSA in Solution Containing Cu^{2+} 5 ppb and Hg^{2+} 5 ppb

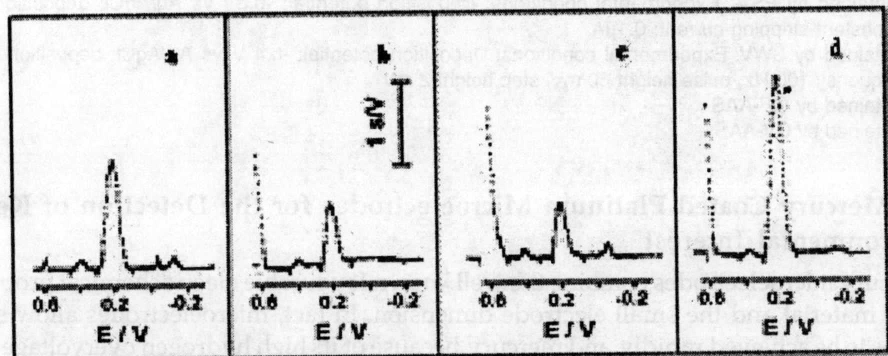

Fig. 5: PSA Recorded in Different Sites of the Lagoon of Venice

This Figure shows that in all cases well-defined responses are recorded. The SWASV obtained under the same conditions (not shown), did not reveal stripping peaks statistically different from the background. It must be noted that the stripping peaks of both mercury and copper recorded in the Lagoon were shifted cathodically with respect to those recorded in the synthetic solutions. This was attributed to the effect of organic matter present in the sea water, which may form complexes with the two metal ions formed at the electrode surface during the stripping step.

All data obtained in the different sites are summarised in Table 1.These results show that using PSA, in situ detection of the labile fraction of both metals can be done down to 0.22 and 0.04 ppb for copper and mercury, respectively. Under these conditions the relative standard deviation (rsd) is within 8 and 10 % for copper and mercury, respectively. Instead, no reliable in situ data could be obtained with SWASV. This was attributed to the lack of a suitable software for baseline treatment with the latter technique. Laboratory measurements were employed for total concentration determination of the metal ions. The total concentration was obtained after acidification of the sample to pH=1 with HCl. Under these conditions the concentration of both copper and mercury increased by about one order of magnitude. This allowed the determination of the two ions with the SWASV, while the overlap of the stripping peaks of copper and mercury prevented their determination with PSA (see Table 1). The precision for the total concentration was also satisfactory, the rsd being of 5 and 7% for copper and mercury respectively. For the total concentration of the two ions, comparative measurements by atomic spectrometry were also performed. Results obtained with the latter procedure are also included in Table 1.

Table 1: Labile and Total Concentrations of Copper and Mercury in Seawater Samples of the Lagoon of Venice

Sample Location	*Concentraton (µg/l)*					
	Cu			*Hg*		
	*Labile**	*Total*		*Labile**	*Total*	
		(1)	*(2)*		*(1)*	*(3)*
Fusina (A)	0.51 ± 0.04	6.4 ± 0.3	2.60 ± 0.41	0.13 ± 0.01	0.92 ± 0.05	0.51 ± 0.04
Arco AMAV (B)	0.22 ± 0.02	5.1 ± 0.1	2.89 ± 0.05	0.080 ± 0.006	2.8 ± 0.2	1.58 ± 0.03
Mentedison (C)	0.35 ± 0.02	6.2 ± 0.2	3.49 ± 0.32	0.070 ± 0.006	0.82 ± 0.04	0.69 ± 0.02
S. Macro Bay (D)	0.50 ± 0.03	7.0 ± 0.03	4.15 ± 0.19	0.040 ± 0.002	0.46 ± 0.02	0.80 ± 0.03

* Obtained by PSA. Experimental conditions: Deposition potential: -0.3V vs Ag/AgCl; deposition time: 5 min; Constant stripping current: 0.1µA

[1] Obtained by SWV. Experimental conditions: Deposition potential: -0.4 V vs Ag/AgCl; deposition time: 10 min;. frequency 100 Hz, pulse height 60 mV, step height 2 mV

[2] Obtained by GF-AAS

[3] Obtained by CV-AAS

Use of Mercury Coated Platinum Microelectrodes for the Detection of Key Species of Environmental Interest

Mercury microelectrodes combine the well known favourable electrochemical properties of the electrode material and the small electrode dimension. In fact, microelectrodes allow steady-state conditions to be achieved rapidly, and mercury, because of its high hydrogen overvoltage, may be the material of choice for the investigation of processes occurring over the negative potential region in aqueous solutions (8).

The development of mercury microelectrodes has been the subject of numerous papers over the past 15 years (8). Stationary mercury microelectrodes are usually prepared by electrodeposition of mercury on metal substrates (8). Considering an inlaid microdisk metal substrate, the mercury deposit is usually in the shape of a sphere cap with the basal plane coincident with the original inlaid disk (see Fig. 6). This is a consequence of the large surface energy of the solution/mercury interface. The sphere cap height (h) can be determined from the total charge spent during the electrodeposition of mercury (8).

Sphere cap microelectrodes with h over a wide range have been prepared, on platinum microdisks. These microelectrodes have shown good stability at room and higher temperature and have found numerous analytical applications. Carbon microdisks can be also used, but because of the lower stability of the mercury deposit, only *in situ* deposition procedures can be employed.

In this section, it is described the use of mercury microelectrodes, prepared by using *in situ* or *ex-situ* procedures, for the determination of key species of environmental interest directly in natural waters, soils and pore water of sediments.

Determination of Heavy Metals in Natural Waters

With the *in-situ* procedure, mercury microelectrodes are formed by codeposition of mercury and the analytes during the deposition step according to the following scheme:

$$Hg^{2+} + 2e \rightleftharpoons Hg(0) \quad (2)$$

$$M^{n+} + ne\ M \rightleftharpoons (Hg) \quad (3)$$

To this purpose, mercury (II) ions are added to the investigated solutions, and the preconcentration

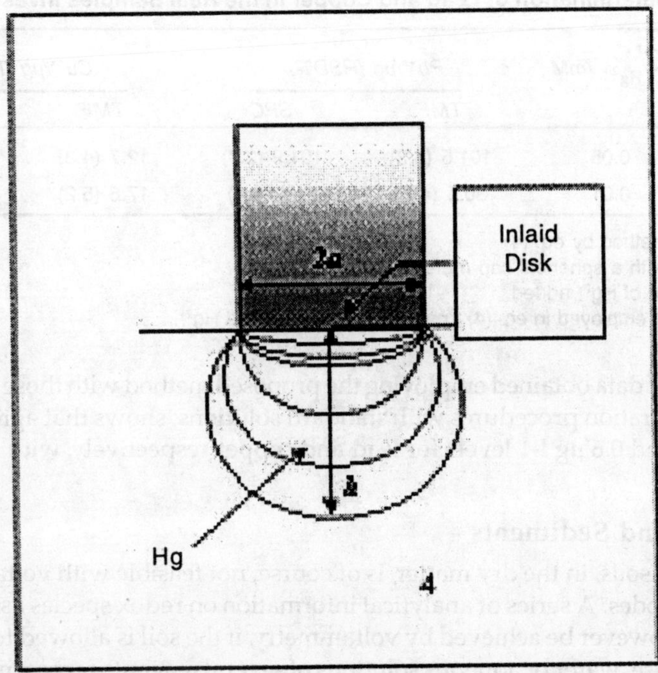

Fig. 6: Schematic View of a Family of Mercury Microelectrodes

step is accomplished at such a negative potential that both Hg^{2+} and M^{n+} ions are reduced simultaneously.

A method, not demanding calibration with standard solutions of the analysed ions, was proposed for the determination of lead and copper in synthetic and real samples, by employing thin mercury film electrodes formed onto carbon disk microelectrodes. The method is based on an equation derived by exploiting the characteristics of the *in-situ* thin mercury film formation and steady state currents achievable at microelectrodes.

$$C^{b}_{M^{n+}} = \left(n_{sHg} / n_{sM}\right)\left(D_{Hg^{2+}} / D_{M^{n+}}\right)\left(Q_{sM} / Q_{sHg}\right)C^{b}_{Hg^{2+}} \quad (4)$$

Well water and rain were investigated and the concentrations of lead and copper ions were determined by eq. 4. Tables 2 and 3 summarize experimental parameters and data obtained in the real samples.

Table 2: Characteristics of the Real Samples Investigated

Sample	*pH*	$C^{*}_{Cl^-}$ *(mM)*	*Conductivity (µS cm⁻¹)*
Well water	7.8	0.45	189
Rain	4.1	0.62	52.3

*Concentration of Cl^-

Table 3: Determination of Lead and Copper in the Real Samples Investigated

Sample	$C^{*}_{Hg^{2+}}$ */mM*	*Pb**/µg (RSD%)*		*Cu**/µg (RSD%)*	
		TMF	*SPC*	*TMF*	*SPC*
Well water	0.05	101.5 (3.2)	101.9 (2.7)	12.7 (4.3)	12.1 (5.1)
Rain	0.01	30.5 (4.7)	31.1 (3.4)	17.6 (5.2)	18.3 (5.4)

TMF, determined method by eq. (4)
SPC, determined with a spherical cap mercury microelectrode
* Bulk concentration of Hg^{2+} added
** other parameters employed in eq. (4) : ns = 2 for Pb^{2+}, $Cu^{2+}$and Hg^{2+}.

The comparison of data obtained employing the proposed method with those taken, or evaluated by using classical calibration procedures with standard solutions, shows that agreement within 11% is found for 2 µg l-1 and 0.6 µg l-1 levels for lead and copper respectively, with a relative standard deviation of about 6%.

Analysis in Soils and Sediments

In situ analysis in soils, in the dry matter, is of course, not feasible with voltammetric methods, even with microelectrodes. A series of analytical information on redox species associated within the solid materials may however be achieved by voltammetry, if the soil is allowed to equilibrate with a solvent. In the latter case, water or aqueous solutions of several reagents represent good examples of systems which close simulate experimental conditions of the environment. The strength of the solvent will determine the degree of dissolution of either natural components or pollutants from the solid particles. In this way information on the chemical equilibria which can be established between solid and solution phases can be obtained. Here electroactive species such as oxygen, reduced sulphur and heavy metals are considered.

The cell apparatus employed for most measurements in soils and sediment samples was an air-tight vessel made of glass, which contained a small Teflon cup 1.5 ml volume, into which the solid sample was placed. A micromanipulator was used to hold the microelectrode in place. The reference electrode was either an Ag/AgCl saturated with KCl or a pseudo reference silver wire of 0.5 mm radius.

Dry soils, as sand, were previously treated with water solutions containing only an inert electrolyte such as NaCl and in addition acids and other reagents. Typically, 10 g of sand were mixed with 3.5 mL of aqueous solutions; the mixture was allowed to equilibrate at least for 24 h.

Fig. 7 shows typical voltammograms obtained on a sand sample prepared with aerated and deoxygenated aqueous solutions. The curve recorded in unpurged solution displays two waves that are due to the irreversible reduction of oxygen to hydrogen peroxide and water.

The involvement of oxygen in the reduction process is confirmed by the lack of waves in mud samples prepared with deoxygenated aqueous solutions. From the height of the first wave of curve *b* and equation 1 (using n=2, D = 2.12x 10-5 cm2 s-1), the concentration of oxygen directly in the sample can be obtained. The oxygen content in the water phase of a soil is important parameter telling on the extent to which oxidation processes may occur at the water-solid interface. For instance, the major changes that occur in redox conditions between oxic waters and anoxic sediments can have profound influences on the speciation and bioavailability of many trace metals.

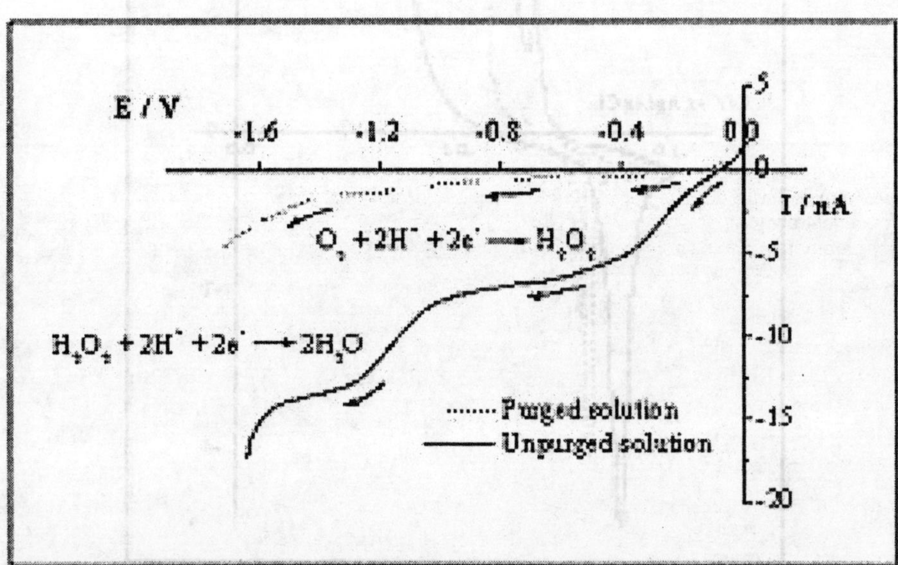

Fig. 7: Steady State Voltammograms Obtained in Aqueous Solution

Inorganic sulphur compounds have been the subject of great laboratory and field interest in the environmental literature. Characterisation and monitoring of such species are in fact critical for understanding of sulphur and metal cycling in suboxic sedimentary environments. For the detection of sulphide ions in aqueous solution, the following electrode process occurring at a mercury electrode is usually exploited:

$$HS^- + Hg^\circ = HgS + H^+ + 2e \qquad (5)$$

Voltammograms that reveal the presence of sulphur species are shown in Fig. 8.

The anodic waves correspond to the formation of various layers of HgS, while the cathodic peaks correspond to the reduction of HgS formed onto the electrode surface, according to the inverse of reaction (2). Both position and height of these waves depended on the sulphide concentration in the solution as well as on the amount of HgS accumulated onto the electrode surface. In particular, it was found that the peak potential shifted towards the less negative values by 30 mV for a tenfold decrease of the sulphide concentration, while the peak shifted towards the more negative potential as the amount of HgS accumulated on the mercury surface increased.

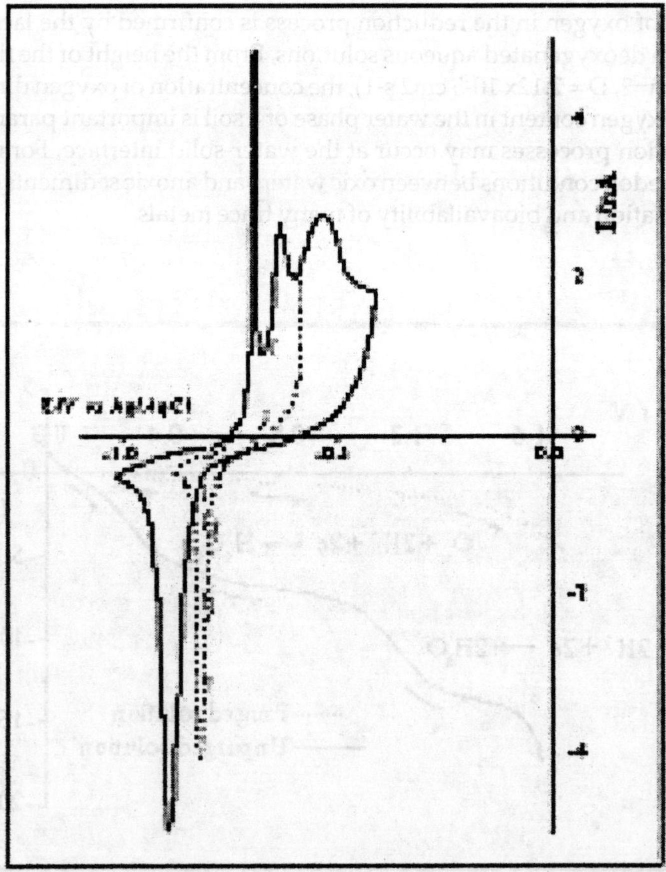

Fig. 8: Cyclic Voltammograms Recorded at 50 mV/sec with a Mercury Microelectrodes in 1.5 mM Na_2S + 0.1 M $NaClO_4$ Aqueous Solution

Well developed anodic peaks were observed only when the concentration of sulphide in the solution was relatively large (> 0.2 mM). Relatively more dilute solutions yielded small broad anodic waves as those shown in Figure 9, that refer to sand samples mixed with 0.01 M Na_2S solutions

Cathodic peaks were in any case observed, even in very dilute aqueous solutions, provided that a HgS film was allowed to form on the electrode surface at a potential ≥-0.4 V.

In spite of the fact that the starting concentration of $HS^-$of the aqueous solution employed to prepare the mud-sand sample was relatively high, the reduction peak due to HgS is rather small. This was the results of chemical equialibria involving the numerous metal cations of the sand to give insoluble sulphide compounds according to the following reaction:

$$2\,M^{n+} + n\,HS^- = M_2Sn + nH^+ \qquad (6)$$

The height of the cathodic peak can therefore be employed to monitor the amount of sulphide in the wet soils.

By using a similar approach, general equilibria involving metal ions between water solution and sand particles can be studied directly in soils. In particular trace metals can be determined by ASV measurements. Heavy metals are naturally present in sand, and the final solution pH of the water phase will determine the amount metal ions released from the soil particles. Fig. 10 shows the effect of mixing sand samples with aqueous solutions of different strenght. No stripping peak is evident even if a preconcentration time as long as 30 min was employed, when neutral NaCl solution was employed. Instead, well defined ASV peaks were obtained when aqueous solutions with 1M HCl were employed to equilibrate the sand samples.

In these types of experiments, the quantification of analytes by the classical methods employed for real sample, i.e, the standard of multiple standard addition method, are not easily applicable.

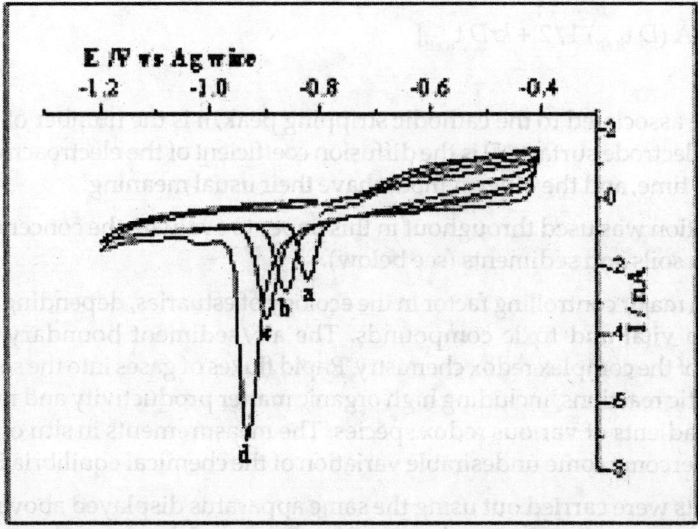

Fig. 9: Cyclic Voltammograms Recorded with a Mercury Microeletrode in a Sand Sample Mixed with 0.01 M Na_2S Aqueous Solution at Different Preconcentration Times: (a) 0, (b) 5, (c) 15 and (d) 30 s

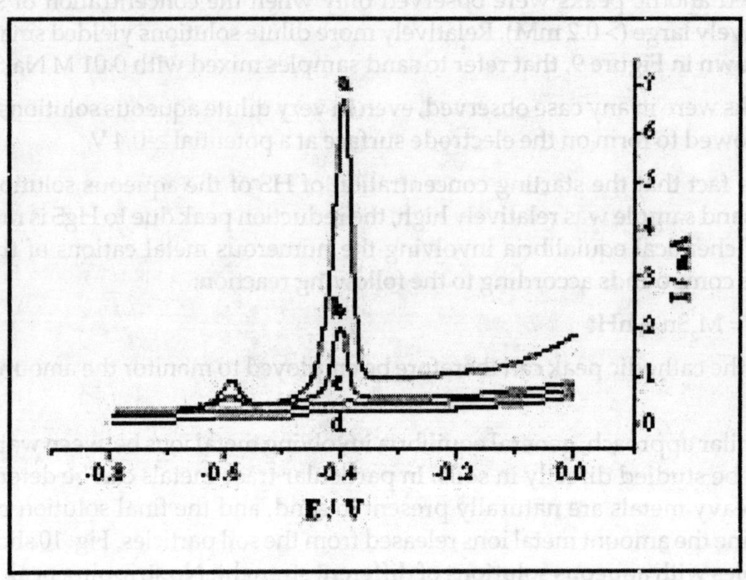

Fig. 10: ASV Recorded on a Sand Sample Equilibrated with (a) 0.5 M NaCl + HCl, pH = 3, t = 5 min, (b) 0.5 M NaCl + HCl, pH = 4.5, t = 5 min, (c) 0.5 M NaCl + HCl, pH = 7, t = 30 min, (d) 0.5 M NaCl, pH = 8, t = 30 min

Thus, for the quantification of the electroactive species in soils, calibrationless procedures were employed. The calibrationless approach is based on the following equation (9):

$$Q_c = nFC_b\ [1.13A\ (D\ t_{d,eff})\ 1/2 + krD\ t_{d,eff}] \quad (7)$$

where

Q_c is the charge associated to the cathodic stripping peak, n is the number of electron involved, $A = \pi\ (a^2 + h^2)$ is the electrode surface, D is the diffusion coefficient of the electroactive specie, $t_{d,eff}$ is the effective deposition time, and the other symbols have their usual meaning.

The latter equation was used throughout in this paper to evaluate the concentration of sulphide and heavy metals in soils and sediments (see below).

Sediments are a major controlling factor in the ecology of estuaries, depending on conditions they can release, or trap vital and toxic compounds. The air/sediment boundary of the salt marsh environment is one of the complex redox chemistry. Rapid fluxes of gases into the sediments combined with biotic and abiotic reactions, including high organic matter productivity and mineral dissolution, give rise to steep gradients of various redox species. The measurements in situ of these species may provide a way to overcome some undesirable variation of the chemical equilibria involved.

The experiments were carried out using the same apparatus displayed above for sand samples, but also directly in core sampling designed on purpose. In general, however, the sampling of the cores were performed in the lagoon of Venice and taken with Polyethylen core tubes approximately 20 cm long and 6 cm wide. Stoppers were placed in the bottoms of the tubes and then taped immediately after the samples were taken, to minimise the sample exposure to air. Samples were then brought to the laboratory. Some of them were analysed within 2 h of sampling, the others were kept at low temperature to avoid degradation.

Fig. 11 shows a series of cyclic voltammograms obtained in sediments.

The sediments investigated soon after they were brought in the lab displayed only cathodic processes due to the reduction of HgS formed at the electrode surface. The effect of keeping the sediment in contact with air is also shown in Fig.11A where the amount of sulphur is monitored over about 7 hours. Interesting is the fate of both sulphur and heavy metals.

To obtain information on both these types of analytes, ASV measurements were performed starting from negative potential which allowed to reduce HgS and, at the same time, to accumulate the metal into the mercury film. A typical voltammogram is shown in Fig. 11B.

It shows the decrease of the cathodic stripping peak of HgS, is accompanied by an increase of the stripping peak of lead.

If measurements were carried out in acidified samples, the metal ion released from the sediment particles can be investigated. A typical ASV obtained under the latter condition is shown in Fig. 12.

Data obtained in the sediments strongly depended on the depth at which they were sampled. Depth profiles of sulphur, oxygen and, eventually, heavy metals were evaluated by using the same mercury microelectrode. Fig. 13 shows depth profiles for reduced sulphur and oxygen obtained either by performing measurements in slides of sediments taken at different depth, or by using properly designed core sampling.

These profiles are similar to others reported in the literature in similar environments. In fact the concentration of sulphide in sediments increases with depth, as a result of sulphate reduction during the decomposition of organic matter; oxygen, instead, is depleted quickly below the sediment-water interface. Small amounts of heavy metal ions could be determined only at the water/sediment boundary after the concentration of sulphide had allowed to decrease to a very low level, by its oxidation in the presence of oxygen from air.

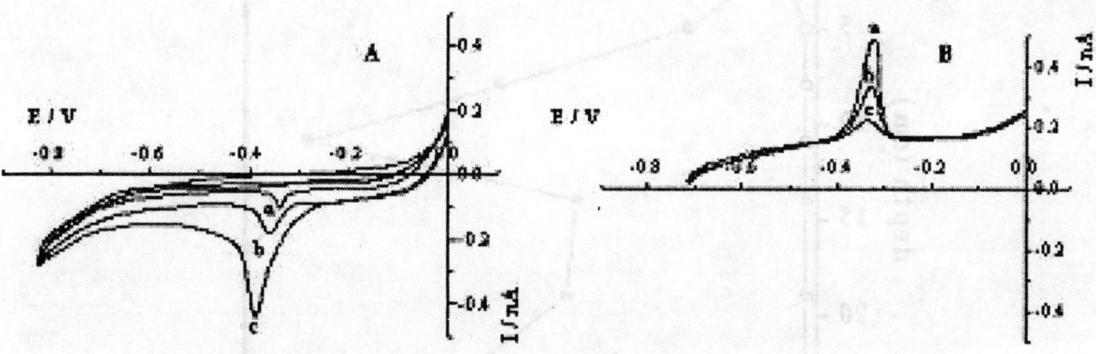

Fig. 11: (A) Cathodic Stripping Voltammetry Obtained on a Sediment Sample at Different Time (B) ASV Obtained on a Sediment Sample at Different Time: (a) 7, (b) 2, (c) 1 h

Conclusions

In this paper it has been shown that microelectrodes are very useful for the determination of analytes of environmental interest. Moreover, investigation on equilibria involved among different

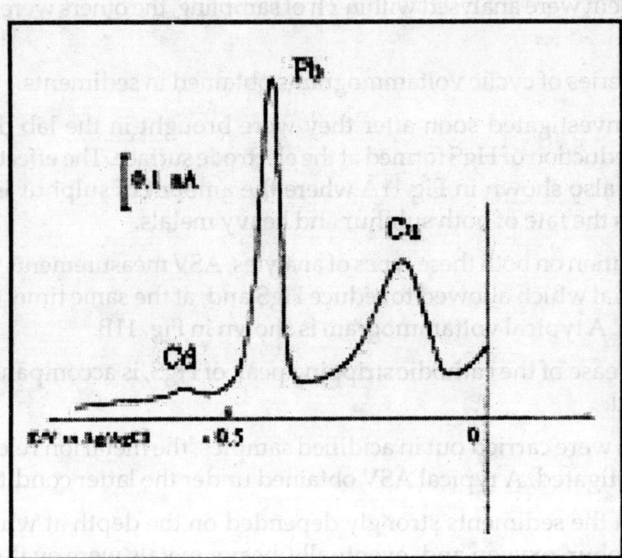

Fig. 12: ASV Obtained on a Sediment Sample at pH = 2

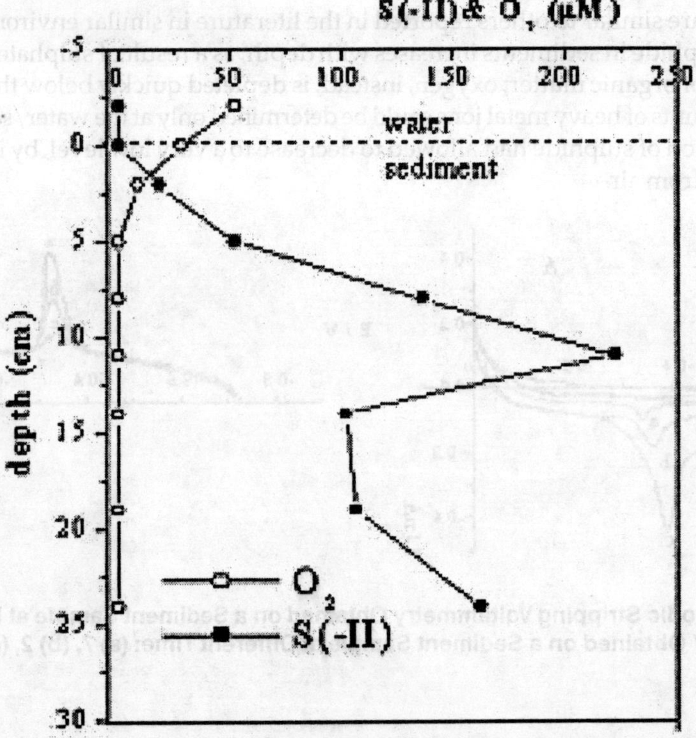

Fig. 13: Depth Profiles for Oxygen and Sulphure on a Sediment Core Sample

phases can be performed. In all cases, the measurements can be carried out directly in the sample even if they are heterogeneous or resistive soils. Finally, thanks to the microelectrode properties, calibrantionless procedures for the quantification of the electroactive species can be applied.

Acknowledgements

Financial support by MURST, Rome (Piano "Servizi al cittadino ed al territorio", Cluster C22, Progetto 20) and CNR is gratefully acknowledged.

REFERENCES

1. P.M. Bersier, J. Howell and C. Bruntlett, 1994, Analyst, 119: 219.
2. M.L. Tercier and J Buffle, 1993, Electeroanalysis, 5: 187.
3. M.I Montenegro, M. Arlete Queiros and J. L. Dashbach (Eds), 1991, *Microelectrodes: Theory and Applications*, NATO ASI Series, Dordrect.
4. J. Wang, 1985, *Stripping Analysis,: Principles, Instrumentation and Applications*, VCH, Deefield Beach, FL.
5. M. A. Baldo, S. Daniele and G. A. Mazzocchin, 1998, Electroanalysis, 10: 410.
6. J. Wang, 1997, *Trends in Anal. Chem.*, 16: 84
7. S. Daniele, C. Bragato, M.A. Baldo, J. Wang and J. Lu, 2000, Analyst, 125: 731.
8. S. Daniele, C. Bragato, M.A. Baldo and G. A. Mazzocchin, 2002, Ann. Chim. (Rome), 92: 203.
9. S. Daniele, C. Bragato, M.A. Baldo and G.A. Mazzocchin, Anal. Chim. Acta, in press.

61

Application of Acetylcholinesterase Biosensor for Detection of Organophosphorus Pesticides

S.R. WATE & R. SARIN

National Environmental Engineering Research Institute, Nehru Marg, Nagpur - 440 020

ABSTRACT

A simple biosensor design for the determination of organophosphorus pesticides, based on the principle of enzyme inhibition, is presented. Membranes prepared by cross-linking AChE with BSA and glutaraldehyde were attached to the bulb of glass pH electrode. The decrease in enzyme activity when immersed in a pesticide solution was detected potentiometrically using acetylthiocholine iodide as substrate. The percentage inhibition of enzyme activity is correlated to pesticide concentration. The procedure for estimation was optimized and the detection limits of pesticides were determined. Pesticides could be detected in the range of 10^{-2} to 10^{-9} M. The inhibited enzyme membranes were regenerated with a dilute solution of 2-PAM and reused for analysis. This biosensor can be applied to field conditions for monitoring of ambient water quality with respect to organophosphorus pesticides.

Keywords: Biosensors, Potentiometry, Enzyme Inhibition, Pesticides.

Introduction

Monitoring of non-point pollution in water bodies due to pesticide residues which pose a potential risk to human health and environment has invited increased scientific interest in recent years (Tran-Minh *et al.*, 1990; Bernabei *et al.*, 1993; Sherma, 1993). In the last few decades organochlorine insecticides (*e.g.* DDT, aldrin and lindane) were progressively replaced by organophosphorus insecticides (*e.g.* paraoxon and parathion) and derivatives of carbamic acid (*e.g.* carbaryl and aldicarb) that have low environmental persistence being biodegradable but cause high acute toxicity. Most of these pesticides are inhibitors of cholinesterase enzymes (Taylor, 1991) which are involved in muscle contraction and impulse transmission in the nervous system. These compounds therefore need to be monitored regularly.

The need for reliability in detection and estimation of organophosphorus compound residues at very low concentrations has led to the development of highly sensitive techniques *viz.*, liquid chromatography (LC) and gas chromatography (GC). These methods distinguish different pesticide residues from a mixture and are highly sensitive. However these are laboratory based and time-consuming methods (Corcia *et al.*, 1991). Therefore methods for the rapid and inexpensive on-line monitoring of pesticides in water need to be developed.

Biosensors, consisting of a biological receptor (enzyme, antibody, microorganisms) in intimate contact with a transducer (electrochemical, optical, gravimetric or thermal) have various options for design and wide range of applications (Kramer *et al.*, 1992). Two biological techniques employing sensors *viz.* immunoassay and immobilized enzyme probes, for the detection of organophosphorus pesticides have been reported to be promising (Kramer *et al.*, 1989; Tran-Minh, 1985). Immobilized enzyme probes have been studied more extensively as it is generally a simple and practicable method.

Several analytical techniques have been developed for the detection of organophosphorus pesticides on the basis of cholinesterase inhibition (Kumaran and Tran-Minh, 1992). Tran-Minh *et al.* (1990) have used pH electrodes coated with acetylcholinesterase, acrylamide/methacrylamide polymer membrane. Kumaran and Morita (1995) have measured organophosphorus pesticides in soil extracts using butyrylcholinesterase immobilized on commercially available, preactivated nylon-polymide membranes held in place at the surface of a pH electrode by a nylon mesh. Ion-sensitive field-effect transistors (ISFETs) have also been used to form potentiometric biosensors for organophosphorus compounds (Nyamsi Hendji *et al.*, 1993). Amperometric sensor for estimation of organophosphates using sequential enzyme reactions catalysed by acetylcholinesterase and choline oxidase have also been reported (Marty *et al.*, 1992; Palleschi *et al.*, 1992).

This study involves estimation of organophosphorus pesticides using a simple biosensor employing enzymatic membranes prepared by cross-linking with brovine serum albumin (BSA) and glutaraldehyde attached to the sensitive surface of glass pH electrodes. Acid formed by the hydrolysis of substrate is sensed by the electrode in a weakly buffered system. This method is expected to be sensitive and accurate for acetylcholinesterase assay and estimation of organophosphorus pesticides.

Experimental

Reagents

Acetylcholinesterase (AChE) (true cholinesterase, EC 3.1.1.7) from electric eel, type V-S, 1120U mg^{-1} was purchased from Sigma Chemical (USA) and was used without any pretreatment. Acetylthiocholine iodide, BSA, glutaraldehyde and the regenerating agent pyridine-2-aldoxime methiodide (2-PAM) were obtained from Sigma Chemical Co.

For preparation of standard pesticide solutions, the organophosphorus pesticides *viz.*, chlorpyriphos, dimethoate and monochrotophos from EPA standards kit and paraoxon from Sigma were used.

The composition of working buffer, used for the preparation of substrate solution and to conduct potential measurements, was $5x10^{-3}$ M phosphate buffer containing $0.8gl^{-1}$ $Na2HPO_4.2H_2O$, $0.18gl^{-1}MgCl_2$, $1.42gl^{-1}NaCl$, $0.1gl^{-1}KCl$ and $0.1gl^{-1}NaN_3$. All chemicals were of analytical reagent grade.

Enzyme Immobilization and Biosensor Preparation

A small volume of enzyme-BSA solution was spread on a glass slide. Glutaradehyde solution (2.5% solution in buffer) in micro quantity was added to the center of the slide. The two solutions were then mixed rapidly with a glass rod. The membranes were allowed to form and dried over a period of 3 hours at room temperature and then washed several times in buffer before use.

Biosensor was prepared by attaching the enzyme membrane to the bulb of a pH (glass) electrode. A cellophane membrane was placed over the enzyme membrane which was held in place with a rubber 'O' ring.

Measurement of Enzyme Activity, Inhibition and Reactivation

All measurements were carried out at 25°C. The sensitive part of the sensor was immersed in the stirred buffer solution and the response in terms of pH was noted. This was treated as the reference value. An appropriate amount of 0.1 M acetylthiocholine iodide solution in buffer was added to obtain the required (10 mM) final acetylthiocholine iodide concentration (total volume of the reaction mixture was 12 ml). A change in proton concentration is produced by the enzymatic reaction at the sensitive surface of the electrode and the corresponding change in the differential output voltage is considered to be a measure of the membrane's enzymatic activity. The sensor was washed with buffer solution after each measurement till the reference pH were attained.

Procedure of pesticide analysis involved the following steps: The response of sensor for 10 mM substrate concentration was recorded and used as a reference. The sensor was then washed and dipped in pesticide solution in the buffer for 10-30 min. After a fixed incubation time the substrate was once again added to the contents of the cell and the response noted. The decline in the output difference is related to the decrease in enzyme activity induced by enzyme inhibition.

To reactivate the inhibited enzyme, the sensor was immersed in 20 MM 2-PAM solution prepared in 0.05 M phosphate buffer at pH 7.8 for 30 min. as reported by Kumaran and Tran-Minh (1992).

Results and Discussion

Choice of Working Conditions

The effect of pH and buffer concentrations on enzyme activity were studied. It was observed that responce of the sensor varied substantially by changing the buffer strength and pH (Figs. 1, 2). Based on the optimal sensor response, pH 7.8 and phosphate buffer of 5 mM strength were maintained for all estimations.

Variation of the reaction rate with substrate concentration is shown in Fig.3. A working substrate concentration of 10 mM was chosen based on optimum response and linearity of the curve.

The operational stability of the sensor was tested under the optimum conditions for pH, buffer strength and substrate concentration. The sensor did not show loss in activity upto 7 assays whereas

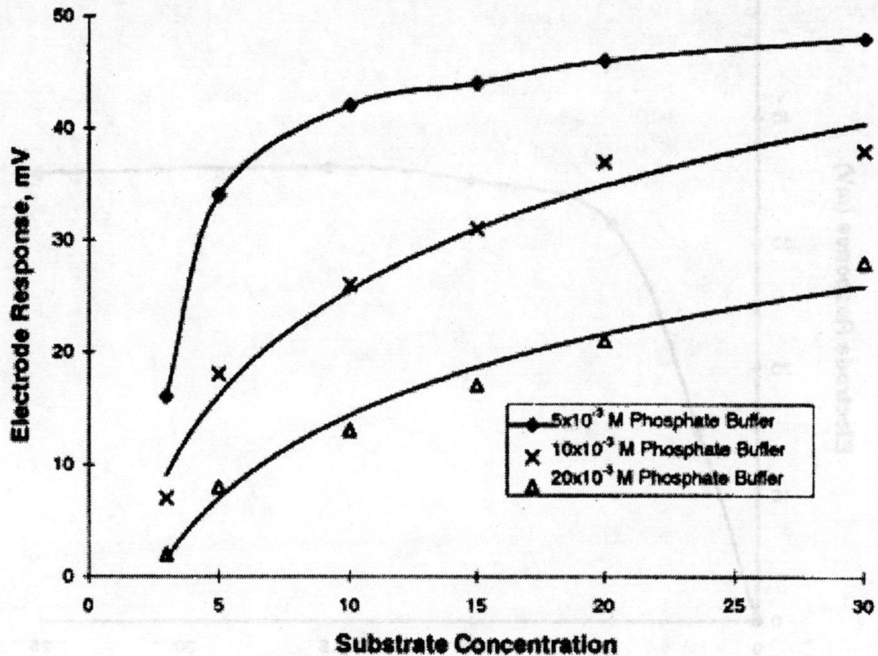

Fig. 1: Effect of Buffer Concentration on Electrode Response (Phosphate Buffer, pH 7.8)

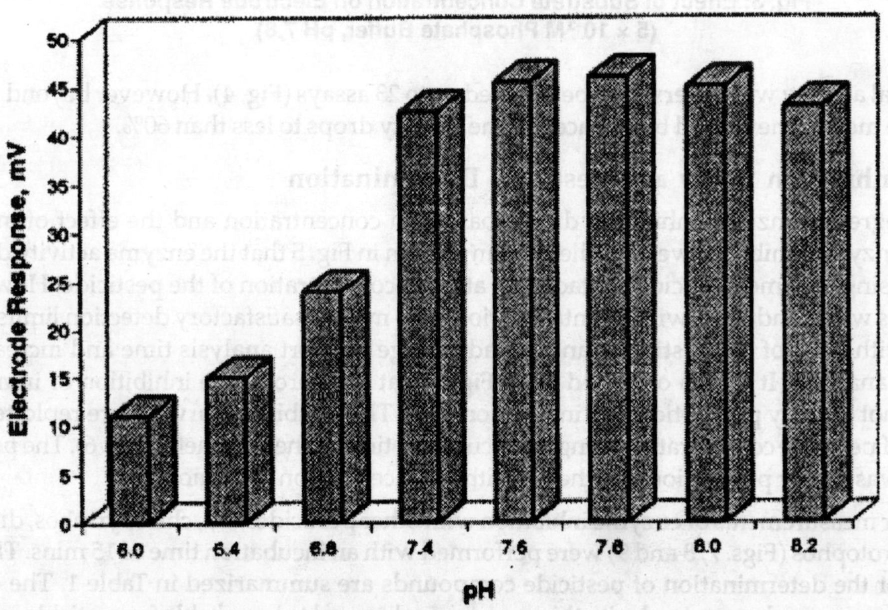

Fig. 2: Effect of pH on the Immobilized AChE on Electrode (5×10^{-3} M Phosphate Buffer)

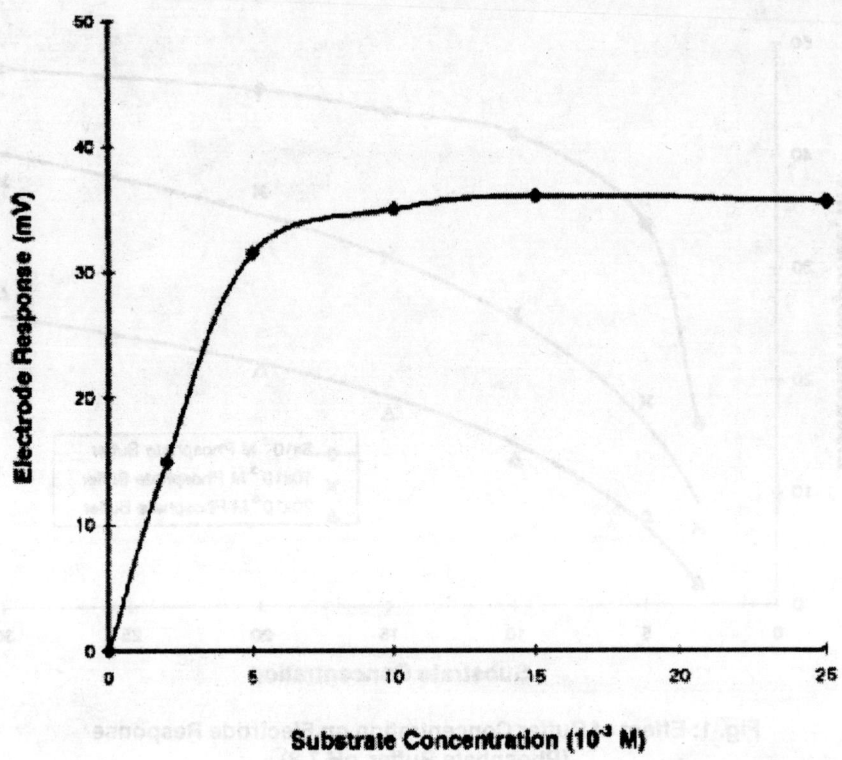

Fig. 3: Effect of Substrate Concentration on Electrode Response (5×10^{-3} M Phosphate Buffer, pH 7.8)

75% of initial activity was observed to be retained upto 23 assays (Fig. 4). However beyond 23 assays the enzyme membrane should be replaced as the activity drops to less than 60%.

Enzyme Inhibition Study and Pesticide Determination

The degree of enzyme inhibition due to paraoxon concentration and the effect of incubation period on enzyme inhibition were studied. It can be seen in Fig. 5 that the enzyme activity decreased with increasing enzyme-pesticide contact time at fixed concentration of the pesticide. However, the experiments were conducted with a contact period of 15 min. as satisfactory detection limits could be achieved with most of the pesticides and the advantage of short analysis time and increase in the number of analyses. It is also observed from Fig. 5 that the percentage inhibition of immobilized enzyme is not directly proportional to incubation time. The inhibition curves were replotted versus logarithm of pesticide concentration using the incubation time as the parameter (Fig. 6). The percentage inhibition was nearly proportional to the logarithm concentration of paraoxon.

Further measurements of enzyme inhibition with other pesticides *viz.*, chlorpyriphos, dimethoate and monocrotophos (Figs. 7, 8 and 9) were performed with an incubation time of 15 mins. The results obtained for the determination of pesticide compounds are summarized in Table 1. The detection limits of the compounds examined with this sensor was observed to be suitable for pesticide monitoring in ambient water samples. The reproducibility and accuracy of the biosensor depends on the characteristics of the enzyme membrane: thickness, enzyme concentration, adhesion, life time etc.

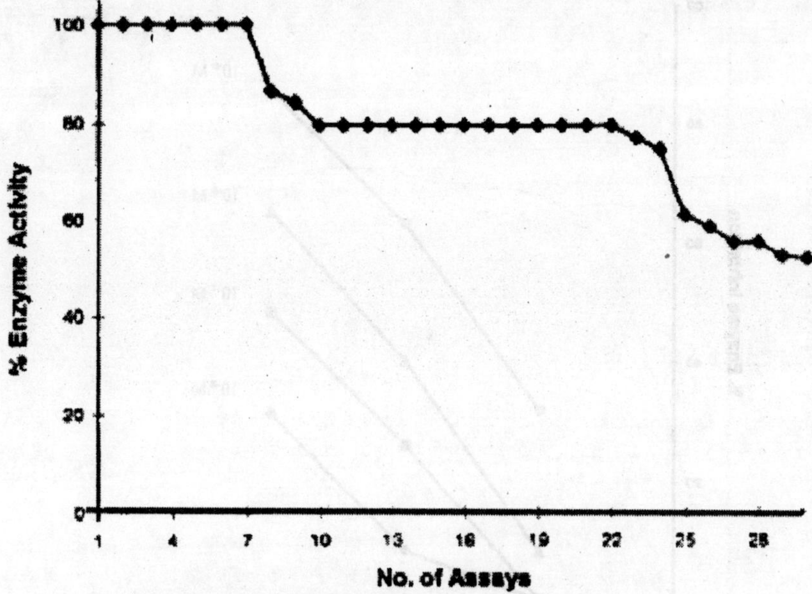

Fig. 4: Operational Stability of AChE Electrode (Enzyme 10 U, Substrate 10 mM, 5 mM Phosphate Buffer, pH 7.8)

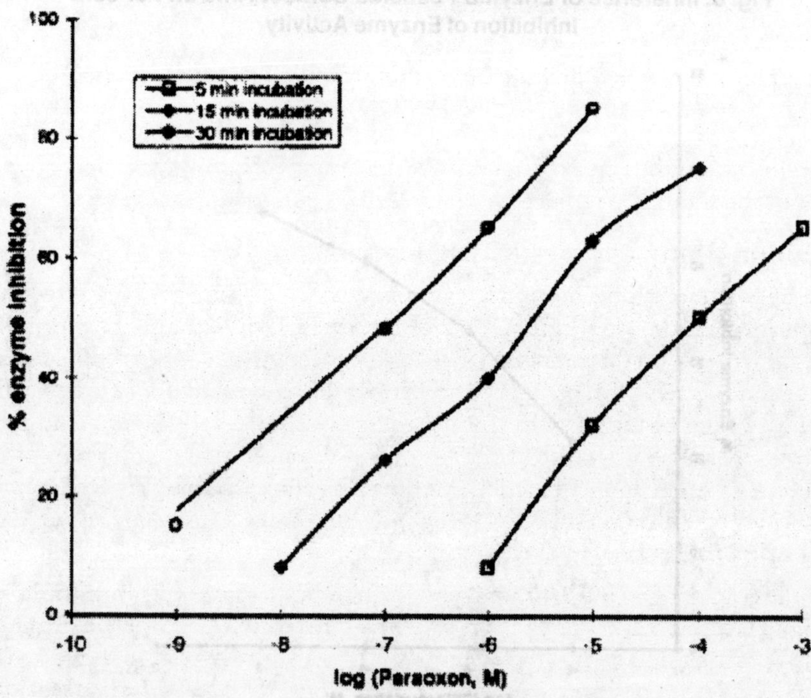

Fig. 5: Per cent Enzyme Inhibition at Various Concentrations of Inhibitor (Paraoxon)

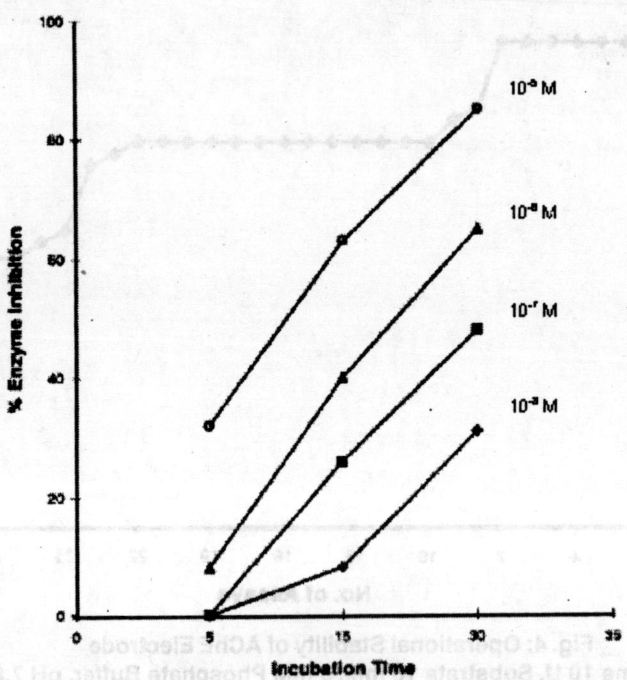

Fig. 6: Inherence of Enzyme-Pesticide Contact Time on Per cent Inhibition of Enzyme Activity

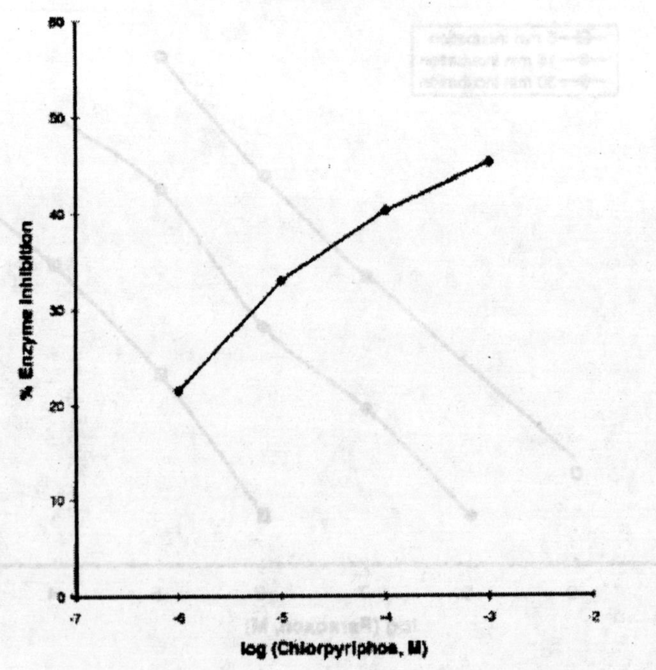

Fig. 7: Response of Sensor to Chlorpyriphos

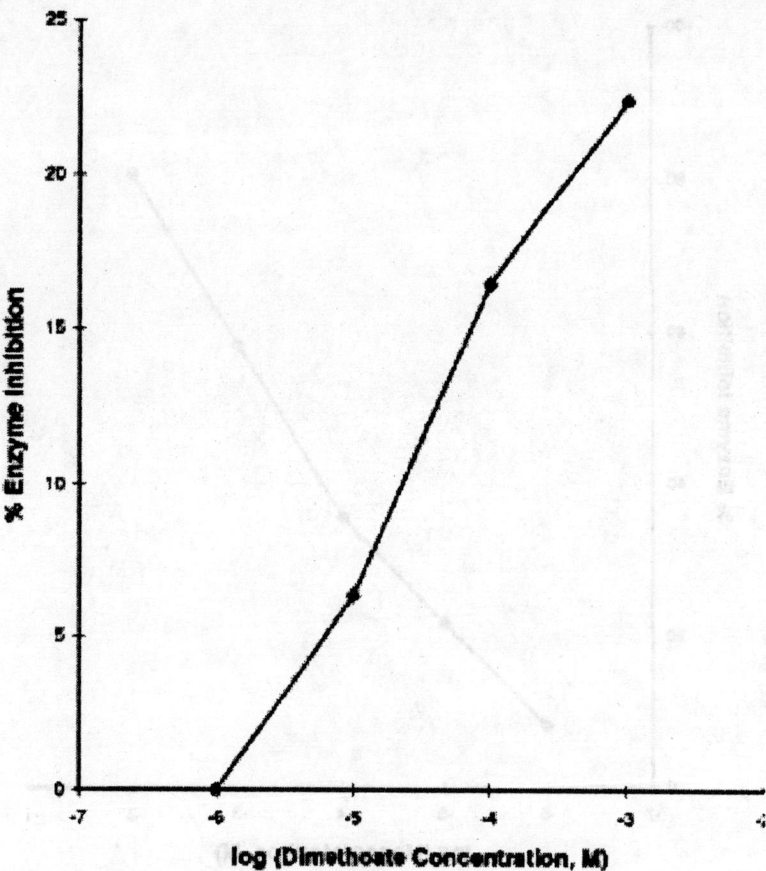

Fig. 8: Response of Sensor to Dimethoate

Table 1: Characteristics of Biosensor for the Detection of Organophosphorus Pesticides

Pesticide	*Detection Limit*		*Dynamic Range*
	µg l⁻¹	*(M)*	*(M)*
Paraoxon	1.0	10^{-9}	10^{-5}–10^{-9}
Chlorpyriphos	350	10^{-6}	10^{-3}–10^{-6}
Dimethoate	2290	19^{-5}	10^{-3}–10^{-5}
Monocrotophos	223	10^{-6}	10^{-2}–10^{-6}

Regeneration of Inhibited Enzyme

The sensor can be regenerated after inhibition with 20 MM 2-PAM in phosphate buffer for 30 min. at room temperature. Only partial enzyme regeneration, upto 60% of the initial enzyme activity, was achieved in the experiments (Table 2).

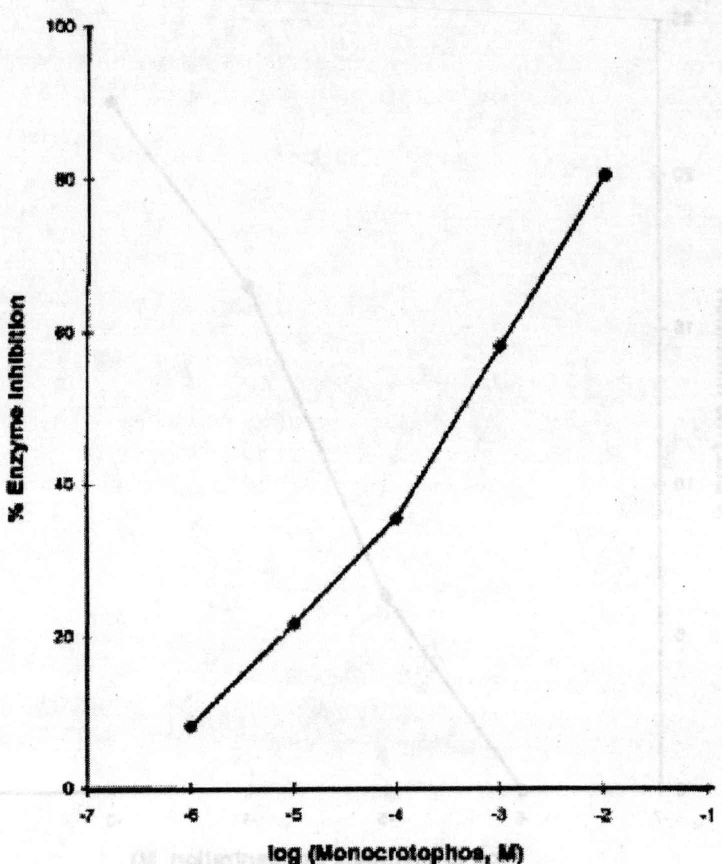

Fig. 9: Response of Sensor to Monocrotophos

Table 2: Reactivation of Enzyme with 20 μM 2-PAM after Inhibition with Paraoxon

Paraoxon	*Inhibition (%)*	*Reactivation (%)*
10^{-8} M	8	100
10^{-7} M	26	100
10^{-6} M	40	82
10^{-5} M	63	76
10^{-4} M	75	60

Interferences

Presence of heavy metal ions, fluorides or organic compounds have been reported to inhibit cholinesterases (Skladal *et al.*, 1994). However, during this study, biosensor was not influenced by heavy metals (Cu, Fe, Cr, Pb, Ni, Hg, As), fluorides and phenols when present within desirable standard limit (Indian Standards for drinking water quality).

Conclusion

AChE biosensor was prepared by cross-linking of the enzyme with BSA and glutaraldehyde on a glass electrode. The operational conditions of the biosensor were:

Substrate concentration: 10 mM

pH: 7.8

Buffer concentration: 5 mM phosphate buffer

Temperature: 25°C

Detection limit: 10^{-2}-10^{-9}M

Enzyme-pesticide contact time: 15 min.

Reactivation time: 30 min. with 20 μM of 2-PAM

The inhibition of enzyme due to organophosphorus pesticides *viz.* paraoxon, chlorpyriphos, dimethoate and monocrotophos was studied and the inhibitory effect was correlated to the pesticide concentration. Detection limits of pesticides and reproducibility of enzyme membrane were also studied. The enzyme membranes showed good operational stability upto 23 assays and no loss in activity was recorded upto 12 weeks when stored dry at 4°C. The enzyme regeneration was found to be 60% of its initial activity. Owing to these factors, the biosensor can be used as an analytical instrument to screen for organopsphorus pesticides.

Acknowledgements

The authors are grateful to Director NEERI for providing the facilities and valuable guidance. The financial assistance rendered by Department of Biotechnology for carrying out this work is also gratefully acknowledged.

REFERENCES

1. Bernabei, M., Chiavarini, S., Cremisini, C. and Palleschi G., 1993, Biosensors and Bioelectronics, 8, 265.
2. Corcia, A.D. and Marchetti, M., 1991, Anal. Chem., 63, 580.
3. Kramer, P., Stocklein, W. and Schmid, R.D., 1989, Wasser, 73, 345.
4. Kramer, P. and Schmid, R.D. in Gopel, W., Hesse, J. and Zemel, J.N. (Eds.), 1992, Sensors, Vol. 3, VCH, Weinheim, 1013.
5. Kumaran, S. and Tran-Minh, C., 1992, Anal. Biochem., 200, 187.
6. Kumaran, S. and Morita, M., 1995, Talanta, 42, 649.
7. Marty, J.L., Sode, K. and Karube, I., 1992, Electroanalysis, 4, 249.
8. Nyamsi Hendji, A.M., Jaffrezic-Renault, N., Martelet, C., Clechet, P., Shul'ga, A.P. and Wlodarsks, W.B., 1993, Anal. Chim. Acta., 281, 3.
9. Palleschi, G., Bernabei, M., Cremisini, C. and Mascini, M., 1992, Sensors and Actuators, B., 7, 513.
10. Sherma, J., 1993, Anal. Chem., 65, 40 R.
11. Skladal, P., Pavlik, M. and Fiala, M., 1994, Anal. Lett., 27 (1), 29.
12. Taylor, P., 1991, J. Biol. Chem., 266, 7.
13. Tran-Minh, C., 1985, Ion-Sel. Electrode Rev., 7, 41.
14. Tran-Minh, C., Pandey, P.C. and Kumaran, S., 1990, Biosensors and Bioelectronics, 5, 4.

62

Flameless Atomic Absorption Spectrophotometric Determination of Toxic Trace Metals in Aqueous Ecosystem

ANIMESH KUMAR, M.Z. HASAN & S.N. KAUL

National Environmental Engineering Research Institute, Nehru Marg, Nagpur – 440 020

ABSTRACT

Flameless atomic absorption spectrophotometry using heated graphite atomizer is the technique of choice for determination of micro and sub-micro quantities of toxic trace metals accumulated in complex forms in the aquatic ecosystem. This technique, however, suffers from interferences by sample matrix components. The author studied metal atomization from water samples and standard metals solutions for making comparison of analyte atomization behaviour in them. Effect of varying concentration of commonly occurring cations and anions in water samples on the atomization of Cd, Cr, Pb, Cu, Mn and Ni were studied. It was found that Ca, Mg, Na, K, Cl, SO_4 and other trace ions interfere in most of the cases. However, these interferences have been eliminated by using analyte/matrix modifications. Appearance temperature of analyte metal in various matrices was experimentally determined. New methods have been developed for interference free determination of Cd, Cr, Cu, Mn, Ni and Pb in natural water samples by using Graphite Furnace tube atomizer AAS.

***Keywords:** AAS, matrix modification, metal determination, matrix interference.*

Introduction

Water is main source of trace metals entering the body systems. Although some of the metals at low concentrations may be valuable micronutrients and essential for life, at higher concentration and in long run the same may become toxic. Pollution of aquatic ecosystems by toxic trace metals has increased many folds in recent years due to increasing anthropogenic activities. Besides industrial discharges, the atmospheric particulates enter water bodies through precipitation as rain or snow. Most of the air pollution abatement processes also add to metal budget in aquatic ecosystem. Lechates from solid waste dumps, agriculture runoff, mining and domestic sewage also contribute metallic pollution. Metals being non-biodegradable, persist in the aquatic ecosystem for long and are capable of affecting all forms of ecological systems. It is obligatory, for the regulating agencies to monitor their levels in water for protecting human health. A critical study of literature published on the determination of toxic trace metals reveals that no sincere efforts have been made to undertake a systematic and coordinated study of the extent of pollution of our aquatic ecosystems by the toxic trace metals.

Atomic Absorption Pectrophotometric Determination

Determination of micro and sub-micro quantities of toxic trace metals accumulated in complex forms in the aquatic ecosystem requires highly sensitive and precise technique. Flameless atomic absorption spectophotometry using heated graphite atomizer is the technique of choice for such studies. This technique, however, suffers from interferences by sample matrix components. Such interferences result from many physico chemical reactions making it impossible to adequately compensate for by purely instrumental means. Chemical modification of analyte or matrix is practiced to isolate metal atomization signal for achieving accurate determination of metal concentration.

Atomization Mechanism

The main step in the flameless AAS is to dissociate the analyte from its matrix and atomize the anlayte in the closed cell geometry of the graphite tube atomizer for making atomic absorption measurements. Various researchers have proposed different atomization mechanisms. Campbell and Ottaway (1974) explained the mechanism of atom formation through reduction of metal oxides by carbon of the graphite tube yielding carbon monoxide and free metal atom. Chakrabarti *et al.* (1978) studied the mechanism of atom formation by using thermodynamic and Kinetic approach. They observed 60% loss of atomic vapour occurs due to the diffusion and condensation of atomic vapour to the cooler extremities of the graphite tube. They also observed that the atomization occurs through 3 major steps, *viz.* thermal dissociation of analyte oxide or halide, carbon reduction of the oxide followed by atomization of free metal. Pinta & Raiandey (1975) found that lead volatilizes in the metallic form from simple media and as oxide from a calcium matrix. Sedykh *et al.* (1979) observed that atomization of lead nitrate takes place through reduction of lead oxide by carbon of the furnace while the atomization of lead chloride occurs mainly due to vaporization of the chloride followed by their dissociation in gas phase. The atomization mechanism of nitrates and chlorides of lead have been studied by Genc *et al.* (1981). L'vov and Ryabchuk (1982) observed that the partial pressure of oxygen in graphite furnace is several order of magnitude higher than explained by thermodynamic equilibrium of the $2O + 2C = 2CO$ reaction. They suggested that the atomization of metal oxide proceeds by their thermal dissociation without direct participation of C in the reduction process. The relation between appearance temperature and atomization energy for various elements have been studied by Ma *et al.* (1983). The effect of heating rate on peak height absorbance has been studied by Zhou *et al.* (1984). Chakrabarti *et al.* (1985) developed models for determining both the gas phase and furnace wall temperature as a function of time.

Matrix/analyte Modification

Although analysis of water samples by direct atomization in graphite atomizer appears attractive, such analysis may give erroneous results. The atomization of analyte present in complex matrix of water sample suffers from sample matrix interferences. Either the matrix may co-volatilize with the analyte atomization or the analyte itself may combine with some other cation or anion in the sample and behave different from standard solution. Hulanicki *et al.* (1985) studied effects of inorganic matrices on determiantion of Cd by ETAAS. The author studied atomization of Pb in presence of various ions generally present in water samples (1986) and reviewed flameless atomic absorption methods for determination of trace metals of environmental significance (1987). Various researchers practiced chemical modifications of matrix/analyte, as tabulated below, to eliminate interference problem.

Element	*Matrix modification Reagents*	*Reference*
Al	Phosphoric acid; Magnesium nitrate	13, 14
As	Nickel; Palladium; Magnesium nitrate; Di-ammonium phosphate; platinum nitrate	19,15,16, 17,18
Cd	Magnesium nitrate; ammonium dihydrogen phosphate; Lanthanum nitrate; nickel nitrate, tri-chloromethane; sodium hydroxide; mixture of ammonium nitrate, Ascorbic acid and Orthophosphoric acid; mixture of ammonium nitrate and Orthophosphoric acid; mixture of ammonium nitrate and di-ammonium hydrogen phosphate; phosphoric acid; ammonium nitrate; Lanthanum chloride	20, 21, 22, 23, 24, 25, 26, 27, 28
Cu	Nickel nitrate; trichloroethane; mixture of ammonium nitrate and tartaric acid	20, 30, 29
Cr	Ascorbic acid; triamine citrate, calcium nitrate; magnesium nitrate; ammonium hepta molybdate	30, 33, 34, 32, 31
Co	Ammonium hexacynocobalt; ferrate; organic acid; lanthanum nitrate	35
Pb	Mixture of Ascorbic acid and oxalic acid; nickel nitrate; ammonium tartarate, magnesium chloride; trichloroethane; ammonium dihydrogen phosphate; Lanthanum; Lanthanum chloride; nickel nitrate; Palladium chloride, magnesium chloride; Palladium nitrate; Magnesium nitrate; Mixture of ammonium nitrate and ammonium dihydrogen phosphate	36, 38, 37, 39, 40, 41, 42, 21, 22, 43, 44, 45, 46, 31
Ni	Ascorbic acid; Lanthanum nitrate; trichloroethane	36
Fe	Nickel nitrate; Palladium chloride; magnesium nitrate; ammonium dihydrogen phosphate; ammonium nitrate	47, 48
Mn	Organic acid; Nickel nitrate, magnesium nitrate; ammonium dihydrogen phosphate; calcium chloride; calcium nitrate; phosphoric acid; Ascorbic acid; mixture of ascorbic acid and ammonium nitrate	49, 17, 50, 52, 51, 53

Experimental Results and Discussion

The author studied metal atomization from water samples and standard metals solutions for making comparison of analyte atomization behaviour in them (Fig. 1).

Effect of varying concentration of commonly occurring cations and anions in water samples on the atomization of Cd, Cr, Pb, Cu, Mn and Ni were studied. It was found that Ca, Mg, Na, K, Cl, SO_4 and other trace ions interfere in most of the cases. However, these interferences have been eliminated by using analyte/matrix modifications. Appearance temperature of analyte metal in various matrices was experimentally determined. The effect of heating rates on anlayte atomization (Fig. 2) has been evaluated. New methods have been developed for interference free determination (Figs. 3 through 8) of Cd, Cr, Cu, Mn, Ni and Pb in natural water samples by using Graphite Furnace tube atomizer AAS. The matrix modification reagents and optimum charring and atomization temperature worked out in this study have been presented in Table 1.

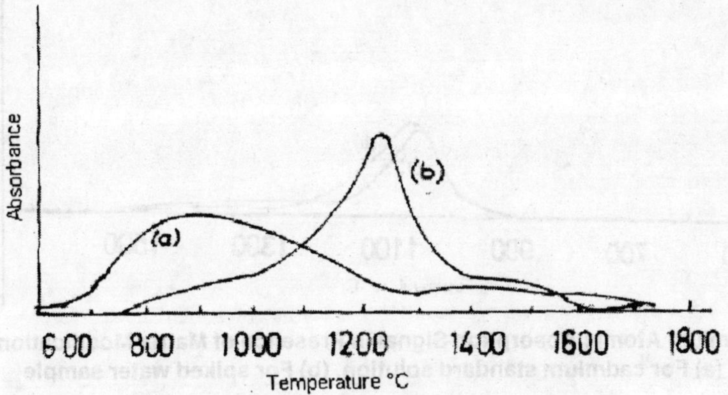

Fig. 1: Cadmium Atomic Absorption Signal
(a) For cadmium standard solution, (b) For spiked water sample

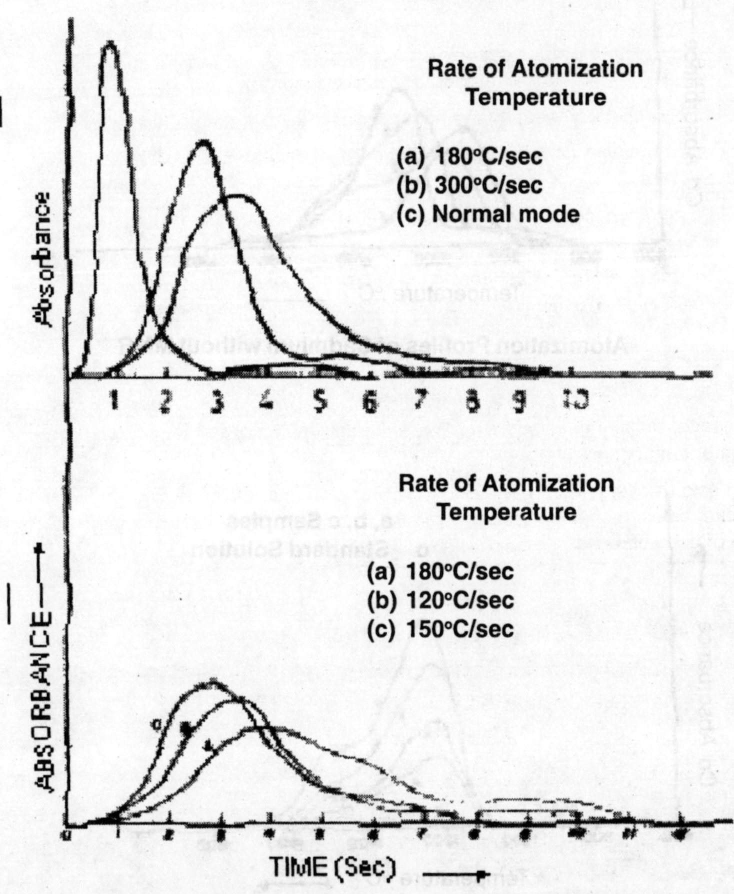

Fig. 2: Effect of Rate of Atomization Temperature on Absorbance Signal of Cadmium

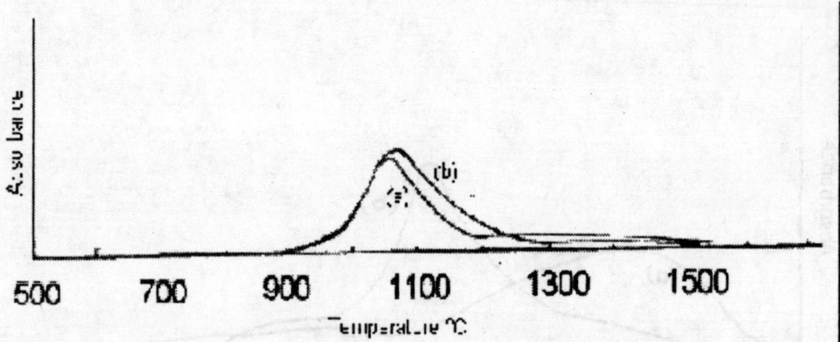

Fig. 3: Cadmium Atomic Absorption Signal in Presence of Matrix Modification Reagent (a) For cadmium standard solution, (b) For spiked water sample

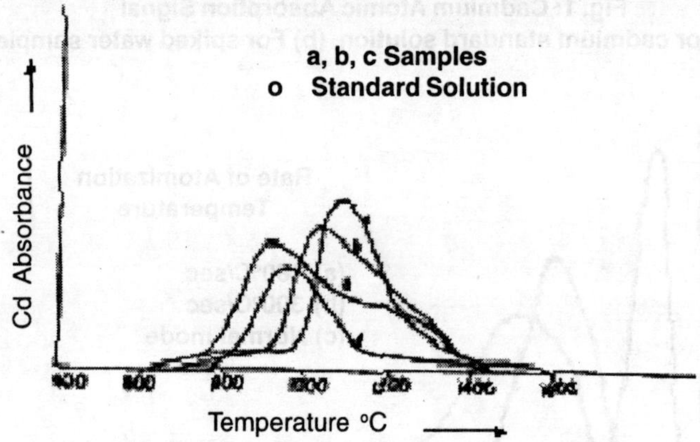

Atomization Profiles of Cadmium without MMR

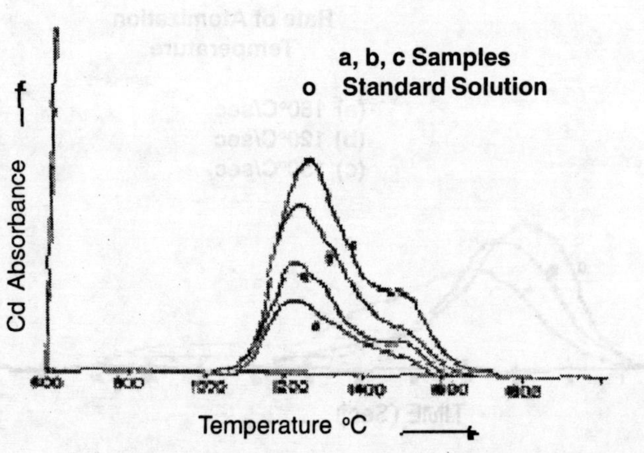

Fig. 4: Atomization Profiles of Cadmium with MMR

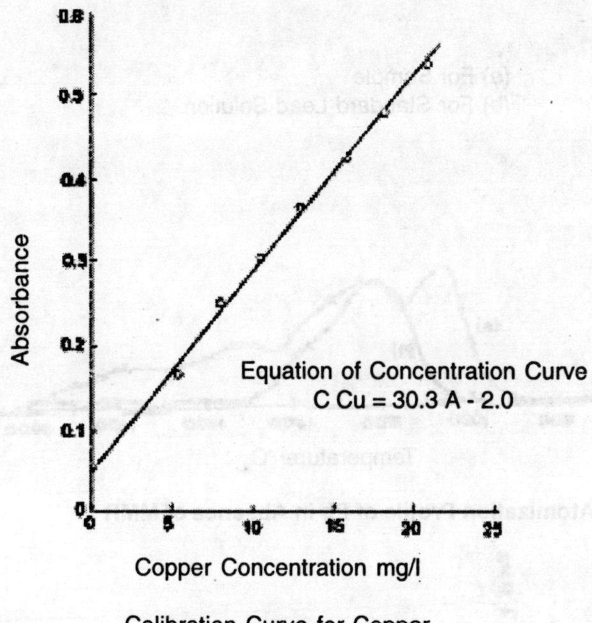

Calibration Curve for Copper

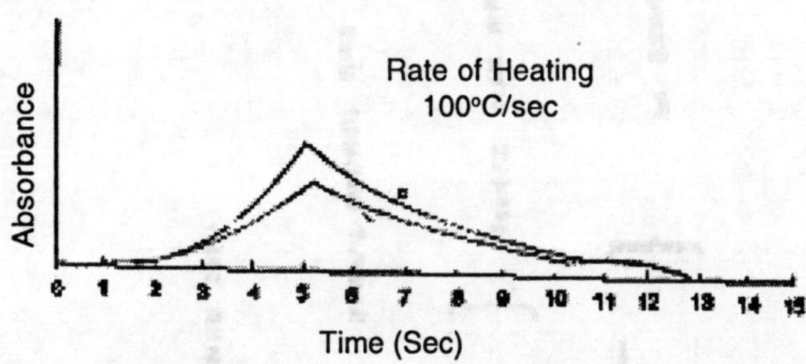

Fig. 5: Atomization Profiles for Copper in Presence of Matrix Modifying Reagent (A) Copper standard solution, (b) Water sample

Table 1: The Matrix Modifying Reagent (MMR) and the Optimum Charring and Atomization Temperature

Metal	*MMR used*	*Charring Temperature (°C)*	*Atomization Temperature (°C)*
Cd	Ammonium nitrates + Di ammonium hydrogen phosphate	650	2300
Cr	Ammonium nitrate + Ammonium hepta molybdate	1300	2500
Cu	Ammonium nitrate + Tartaric Acid	1000	2100
Mn	Ammonium nitrate + Ascorbic acid	1000	2700
Ni	(Use of one blank cycle between sample atomization)	900	2600
Pb	Phosphoric acid + Ammonium di-hydrogen phosphate	750	2200

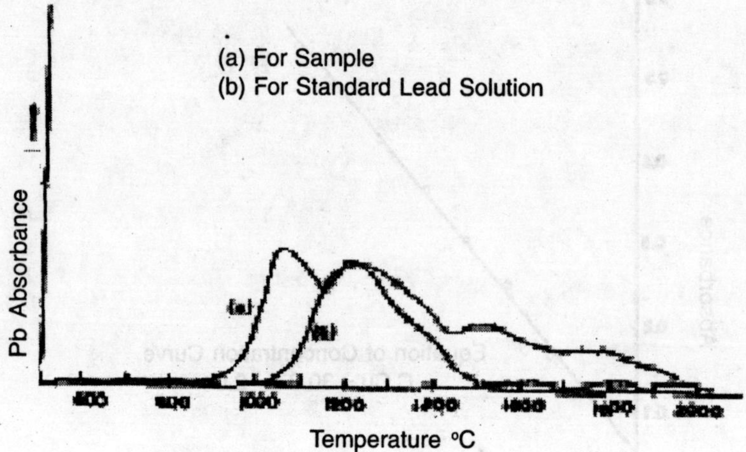

Atomization Profile of Pb in Absence of MMR

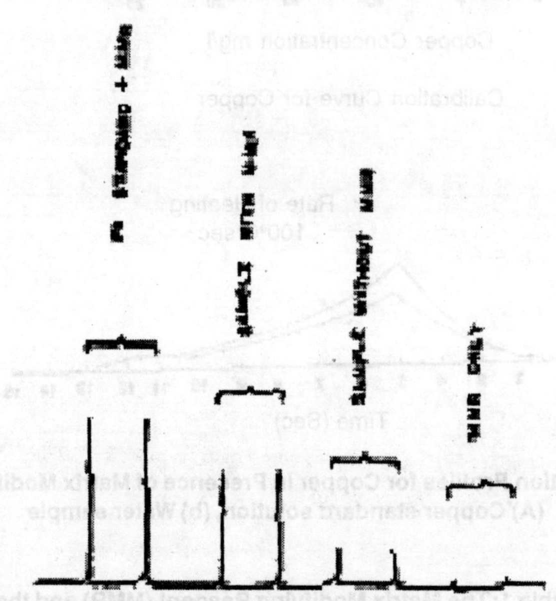

Fig. 6: Effect of MMR on Peak Effect of Lead Absorbance Signal

The methods have been evaluated for accuracy (percentage recovery), precision (RSD, percentage Relative standard deviation), sensitivity (metal concentration for 0.0044 absorbance units), linearity and detection limit (mean value of blank + 3 times standard deviation of blank replicates). The values obtained are shown in Table 2.

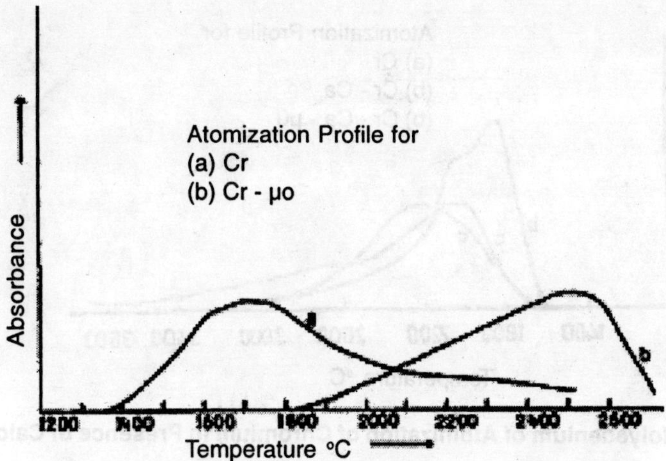

Effect of Molyedenium of Atomization of Chromium

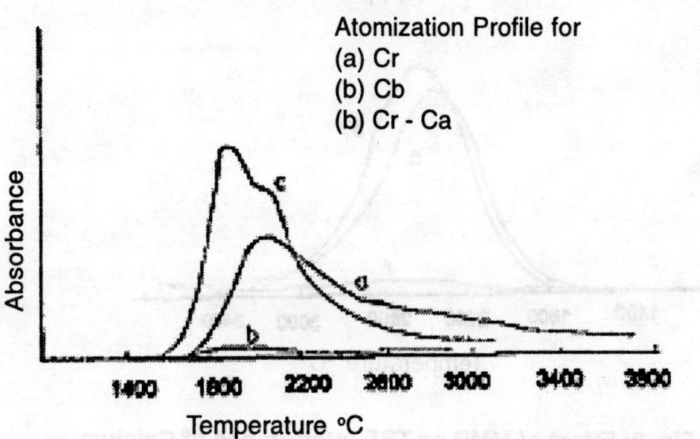

Fig. 7: Effect of Calcium on Chromium Atomization

Table 2: Accuracy, Precision, Sensitivity, Linearity, and Minimum Detection Limit of the Methods

Metal	*Recovery %*	*R S D %*	*Sensitivity µg/L*	*Linearity µg/L*	*Detection Limit µg/L*
Cd	93-103	4.3 - 7.9	0.07	5	0.15
Cr	92-107	2.6 - 5.7	0.30	60	1.30
Cu	94-110	1.6 - 3.8	0.10	20	1.00
Mn	91-107	3.6 - 7.4	0.10	8	0.50
Ni	91-108	4.6 - 8.2	1.10	60	2.00
Pb	95-103	3.4 - 8.4	0.20	200	0.30

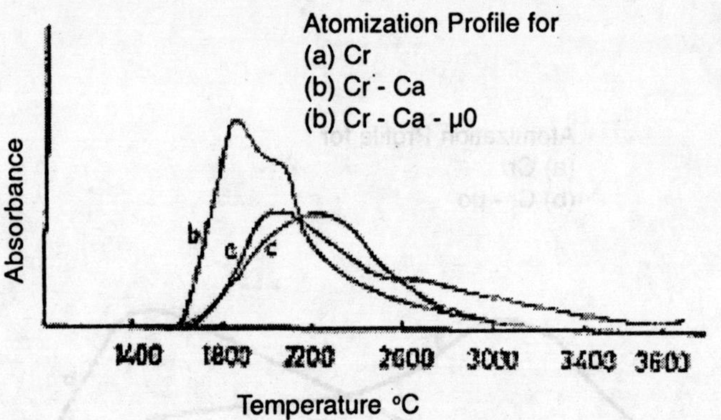

Effect of Molyedenium of Atomization of Chromium in Presence of Calcium

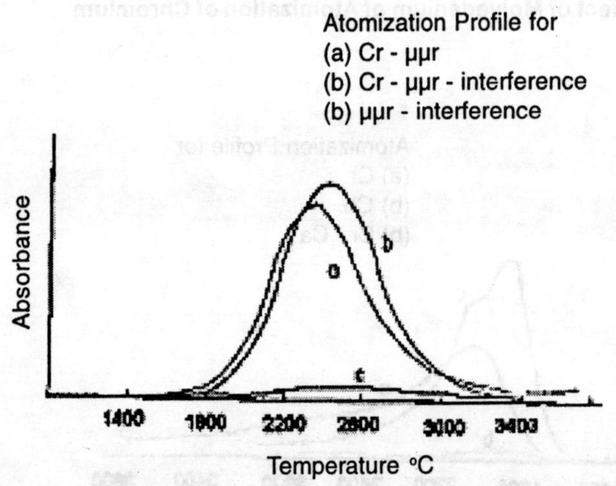

Fig. 8: Effect of MMR on TBE Interference of Calcium in Atomization of Chromium

The above results show that the developed methods have requisite merits for the determination of low concentrations (ppb) of these metals in natural waters.

These developed methods were used in monitoring trace metals in natural water samples. Raw water and treater water samples were collected from water treatment plants at three metropolitan cities namely Delhi, Kanpur and Kolkata. Samples were also collected from various Ghats of Ganga at Kanpur, Ranney wells at Delhi and from various important points on the river Hooghly at Kolkata. Water samples collected from 38 different sources during 1982, 1983 and 1984 were analysed for the determination of Fe and Zn, by flame atomic absorption spectrometry and for Cd, Cr, Cu, Mn, Ni and Pb by graphite tube furnace atomic absorption spectrophotometry using the methods developed above.

The results of analyses of water samples from some of the sources are shown in Table 3. An increasing trend in the levels of these metals with time has been found at all the sources during the three years study. The seasonal studies conducted during the year 1983 indicated that the values obtained in summer season are greater than the values obtained during rainy and winter seasons and the values obtained during rainy seasons are the least. In most of the cases these metals have been found to be present within the permissible limits prescribed for them by WHO and ISI (Table 4).

Table 3: Metal Contents of Some of the Water Sources

Name of the Source	*Year*	*Metal Content*							
		Cd µg/L	*Cr µg/L*	*Cu µg/L*	*Mn mg/L*	*Ni µg/L*	*Pb µg/L*	*Fe mg/L*	*Zn mg/L*
Wazirabad Water Works (Raw Water) Delhi	1982	ND	5.8	24.6	0.09	12.8	5.53	1.2	0.16
	1983	1.20	6.1	25.1	0.10	14.2	6.2	2.8	0.21
	1984	3.40	8.4	30.4	0.12	15.8	8.5	2.9	0.34
Ranney Well Nizamuddin Bridge, Delhi	1982	ND	ND	7.7	0.006	1.0	2.13	ND	0.05
	1983	0.16	ND	8.6	0.006	2.0	3.5	1.0	0.08
	1984	0.34	1.5	10.5	0.006	2.5	6.8	2.0	1.2
Yamuna Water, Shamli-Panipat Road	1982	0.46	ND	0.8	0.05	3.8	2.12	0.18	0.09
	1983	0.71	ND	1.0	0.06	4.6	3.2	0.21	0.10
	1984	1.04	1.1	4.5	0.06	6.4	6.4	0.44	0.25
Parmat Ghat, Kanpur	1982	4.0	8.7	25.9	0.08	22.7	13.5	0.48	0.13
	1983	5.2	10.0	27.1	0.08	23.5	14.2	0.51	0.15
	1984	7.1	12.1	30.1	0.10	25.1	15.1	0.72	0.17
Benezhabar Water Works (Raw Water) Kanpur	1982	2.3	6.6	17.3	0.03	8.6	5.7	0.35	0.13
	1983	4.1	7.5	18.2	0.04	9.5	9.2	0.41	0.15
	1984	3.4	9.5	22.8	0.05	14.6	10.2	0.52	0.18
Dumurdaha, Kolkata	1982	0.1	1.1	5.0	0.06	0.8	0.90	0.09	0.05
	1983	0.2	2.2	6.5	0.08	1.2	0.91	0.10	0.07
	1984	0.3	3.4	6.6	0.08	1.4	0.95	0.12	0.09
Uttarpara Water Works (Raw Water) Kolkata	1982	6.6	3.6	36.0	0.04	1.0	1.0	1.7	0.01
	1983	7.5	5.2	42.1	0.06	2.8	5.8	3.2	0.09
	1984	9.8	7.8	51.6	0.08	4.5	7.2	5.8	0.15

ND = Not Detectable

Table 4: Permissible Limits for Metals

Metal	*WHO Guide Limit 1984 µg/L*	*ISI Limit 1983 µg/L*
Cadmium	5	10
Lead	50	100
Copper	1000	50
Chromium	50	50
Manganese	100	100
Nickel	100	—
Iron	300	300
Zinc	5000	5000

REFERENCES

1. Campbell, W.C. and Ottawal, J. M., 1974, Talanta, 21, 837.
2. Chakrabarthy, C. L., 1978, Can. J. Spectrosc., 23, 134.
3. Pinta, M. and Riandey, C., 1975, Analusis, 3, 86.
4. Sedykh, E. M. and Boluaev, Yu. I., 1979, Zh. Anal, Khim, 34, 1984.
5. Cenc, O. Akman, S., Ozdural, A. R. Ates, S. and Balkis, T., 1981, Spectrochim. Acta, 36B, 163.
6. L'vov, B. V. and Ryabchuk, G. N., 1982, Spectrochim. Acta, 37B, 673.
7. Ma, I. T., Wu, Z., Zhang, Z. and Li, S., 1983, Huaxue Tongbao, 5, 17.
8. Zhou, N. G., French, W. and De, G. L., 1984, Spectrochim. Acta., 39B, 225.
9. Chakrabarti, C. L., WU, S., Karwowska, R. Rogers, J. J. and Dick, R., 1985, Spectrochim. Acta, 40B, 10-12.
10. Hulanicki, Adam. Bulska, Ewa. and Wrobel, Katarzyua, 1985, Analyst, 110, 1141-1146.
11. Animesh Kumar, 1986, Atomic Absorption Spectrophotometric Studies in Environment, Ph. D Thesis, Nagpur University.
12. Animesh Kumar and Hasan, M.Z., 1987, Flameless atomic absorption Spectrophotometric Methods for the Determination of Trace Metals of Environmental significance – A Review, Indian Journal of Pure & applied Physics, 25, 49-57.
13. Craney, C. L., Swatout Kurt, Smith, F. W. and West, C. D., 1986, Atomic absorption analysis of Al determination in water, Anal Chem., 5, 656-660.
14. Quinonero, J., Mongay, C. and Delaguardia, M., 1989, Determination of Al in aqueous solution using magnesium nitrate, Microchem J., 39, 344-350
15. Robert, K. A., Michael Thompson and Elisabeth, C., 1986, Interference effects produced by heavy metals in the determination of As in natural waters, Analyst, 111, 1143-1153.
16. Weigert, P. and Sappl, A., 1983, Determination of As by HGAAS, Fresenius'Z Anal Chem., 316, 306-308.
17. Grobenski, Z, Lehmann, R., Radziuk, B., and Vollkopf, V., 1984, Determination of trace metals in seawater using Zeeman GFAAS. At. Spectrosc., 5, 87-90.
18. Dietz Mark, L., 1985, Determination of As in river water and sediments using matrix modification. At. Spectrosc., 6, 59-62.
19. Matsusaki, K. and Koji, 1982, Mechanism and removal of halide interferences in the determination of metals by ETAAS, Anal Chim Acta, 141, 233-240.
20. Calmano, W., Ahlf, W. and Schilling, T., 1986, Determination of Cd, Cu, Ni using nickel modifier at submicrogram detection in seawater. Fresenius'Z Anal Chem, 323, 865-868.
21. David, T.E., Hunt, D. and Winnard Alan, 1986, Appraisal of selected techniques for the determination of Pb and Cd in waters by GFAAS, Analyst, 111, 785-789.
22. Kumar, Aggrawal, A. L., Hasan, M. Z. and Desmukh, B. T., 1987, Investigation of various modifiers for the determination of Cr, Cd and Pb by ETAAS, Indian J. Pure Appl Phys, 25, 153-160.

23. Lan, C. R., 1993, Sodium hydroxide modification in determination of Cd in seawater. Analyst, 118, 189-192,

24. Oliver, W. K., Reeve, S. Hannrond, K. and Baskuter, 1983, Determination of Cd and Pb using ETAAS with H_3PO_4 matrix modifier. J. Instt Wat Engng Sci, 37, 460-470.

25. Ohta, K., Aoki, W. and Mizuno, J., 1988, Elimination of interferences in ETA AAS of Cd, Talanta, 35, 831-836.

26. Hasan, M. Z., Kumar A, and Pande, S. P., 1982, Determination of cadmium in water by flameless atomic absorption spectrophotometry, Research and Industry, 27, 8-10.

27. Hasan, M. Z. and Kumar A., 1983, Interference due to calcium and magnesium in flameless atomic absorption spectrophotometric determination of cadmium in water; its suppression by matrix modification, Indian J. Environ. Hlth., 25, 161-168.

28. Hasan, M. Z., Kumar A, and Pande, S. P., 1984, Occurrence of cadmium in water: Its determination by flameless atomic absorption spectrophtometry, J. Indian Inst. Sci., 65 (B), 105-110.

29. Deshmukh, B. T., Animesh Kumar and Hasan, M. Z., 1987, Elimination of matrix interferences in flameless atomic absorption spectrophotometric determination of copper in water, Indian Journal of Chemistry, 26 (A), 629-631

30. Apte, S. C., Comber, S. D. W., Gardener, M. J. and Gunn, A. M., 1991, Determination of Cr by GFAAS, J. Anal At Spectrom, 6, 169-172.

31. Kumar A, Aggrawal, A. L. and Hasan, M. Z., 1987, A graphite furnace atomic absorption spectrophotometric determination of lead & chromium a in air & water, Indian Journal of Pure & Applied Physics, 25, 193-196.

32. Miller Donald, B., 1988, Effects of anions interference in the determination of Cr by AA, At. Spectrosc, 9, 43-45.

33. Yang, X., Li, G. Tang, B. and Zhang Z., 1992, Determination of Cr in Seawater samples using ascorbic acid and triamine acetate, Chem Abs, 116, 221141.

34. Gong, B and Lin, Y, 1990, Determination of Cr in seawater, At. Spectrosc, 11, 229-232.

35. Churella D. J. and Copeland, T. R., 1978, The interferences of alkali halides, Anal Chem, 50, 309-311

36. Imai Hoji and Hayashi, Y., 1991, Determination of Pb by GFAAS using ascorbic acid, Anal Chem., 63, 772-775.

37. Byrne J. P. and Chakrabarti, C. L., 1993, Atomic absorption analysis of Pb in acid modifier, Anal Chem, 65, 1267-1269.

38. Bass D. A. and Holocombe J. A., 1988, Determination of Pb in GFAAS, Anal Chem., 60, 578-580.

39. Rettberg T. M. and Holocombe, J. A., 1988, Atomic absorption electrothermal metal analysis. Anal Chem., 60, 600-602.

40. Webster John and Wood Andrew, 1984, Evaluation of an electrthermal atomisation procedure for determination of Pb in potable waters, Analyst, 109, 1255-1258.

41. Apte S. C. and Gunn, A. M., 1987, Rapid determination of Cu, Ni, Pb and Cd in small samples of estuarine and coastal waters by liquid extraction and ETAAS, Anal Chim Acta, 193, 147-156.

42. Bertenshaw, M. P., 1982, Determination of Pb in aqueous samples by AAS with electrothermal atomization and lanthanum treatment. Analyst, 107, 163-165.

43. Bertenshaw, M. P., and Gelsthorpe, D., 1981, Determination of Pb in drinking water by AAS with electrothermal atomization, Analyst, 106, 23-25.

44. Manning, B. E., 1983, Decontamination procedure for GF and AS-40 autosampler after use of nickel nitrate matrix modifier, At Spectrosc, 4, 159-160.

45. Brunbagh, W.G. and Koirtyohan, S. R., 1988, Determination of Pb in $MgCl_2$ matrix, Anal Chem, 60, 1051-1055.

46. Hasan, M. Z., and Animesh Kumar, 1983, Elimination of sodium interference flameless atomic absorption spectrophotometric determination of lead in potable water by matrix modification, J. Indian Inst. Sci., 64 (B), 127-131.

47. Montgomery, J. R. and Peterson, G. N., 1980, Determination of Cu, Fe, Mn in seawater. Anal Chim Acta, 117, 397-400.

48. Standard methods for the examination of water and wastewater, 1992, 18th Edition (Apha, AWWA, WPCF, Washington)..

49. Mc Nally, J. and Holocombe, J. A., 1991, Topology and vapourisation characteristics of Pb, Co, Mn, Al on a graphite surface using ETAAS, Anal Chem, 63, 1918-1921.

50. Hulanicki, A., Bulska, E. and Dittrich, K., 1990, Determination of Mn in aqueous solution using Ca cation and phosphates as modifier, J. Anal. At. Spectrom., 5, 209-213.

51. Kumar S. J. and Gangdharan S. J., 1993, Palladium and ascorbic acid as modifier in the determination of Se., J. Anal At. Spectrom, 8, 615-622.

52. Sinemus, H.W., 1985, Determination of Pb GFAAS, Anal Chem, 57, 2268-2271.

53. Animesh Kumar, and Hasan, M. Z., 1986, Interference-free determination of manganese in water by graphite furnace atomic absorption spectrophotometry, Indian Journal of Pure &Applied Physics, 24, 465-468.

63

Economic Evaluation of Indian Wetlands

J.S. PANDEY, V. JOSEPH & S.N. KAUL

National Environmental Engineering Research Institute, Nagpur - 440 020

ABSTRACT

An exercise of economic evaluation of wetlands in various Indian states has been carried out based on their (wetlands') ecological potential. Although wetlands occupy only about 2% of the world's land area, they are estimated to contain 10%-14% of carbon. Moreover, the aerobic-anaerobic stratification of wetlands' sediments is important far out of proportion to their area because of their role in the global cycling of sulfur, nitrogen, phosphorus and carbon. Conversion of wetlands to other land-use classes releases large quantities of CO_2 to the atmosphere, thus contributing significantly to the greenhouse problem. Therefore, in view of their importance, the economic values of artificial as well as natural wetlands in different Indian states have been estimated. The ratio of artificial to natural wetland values (ANR) has been analyzed in view of its direct bearing on the pollution-control and mitigation activities in the states concerned. As far as economic values of natural wetlands are concerned, Gujarat (5834.5 million US$yr^{-1}), West Bengal (4316.6 million US$ yr^{-1}) and Bihar (3323.4 million US$ yr^{-1}), turn out to be the richest states, whereas Nagaland (3.1 million US$ yr^{-1}) and Madhya Pradesh (7.7million US$ yr^{-1}) have the lowest values. Chandigarh has the lowest value (2.5 million US$ yr^{-1}) amongst union Territories. In terms of ANR-ratio, Madhya Pradesh (564.1) and Karnataka (159.8) happen to be on the top of the list.

Keywords: Ecological potential, natural wetlands, artificial wetlands, Indian states, economic evaluation, ratio of artificial and natural wetland values (ANR-ratio)

Introduction

India (Figure 1) happens to be the second largest country in Asia and the second most populous country in the world, which supports one-seventh of human population on the earth. Moreover, the livelihood and the economy of the majority of this vast population depend very significantly on wetlands and the resources they generate (1).

It (India) is very rich in wetland resources and exhibits significant ecological diversity, primarily because of variability in climatic conditions and changing topography (2). Moreover, wetlands cover approximately 17 million hectares in India. They range from high Himalayan lakes in the north to the salt flats of Kutch, the mangrove swamps of the Bay of Bengal and the innumerable small water reservoirs of the Deccan (3,4). This great diversity of wetlands supports an extremely rich biodiversity of flora and fauna. As national parks and wildlife sanctuaries, many important wetland zones have been well protected. The predominant wetland types in India's geographical zones are (1):

1. Tanks, Reservoirs, and other water bodies of the Deccan peninsula
2. Backwaters and Estuaries of the west coast of the peninsula
3. The Saline expanses of Rajasthan and Gujarat (Rann of Kachch)
4. Freshwater Lakes and Reservoirs from Gujarat eastwards through Rajasthan and Madhya Pradesh

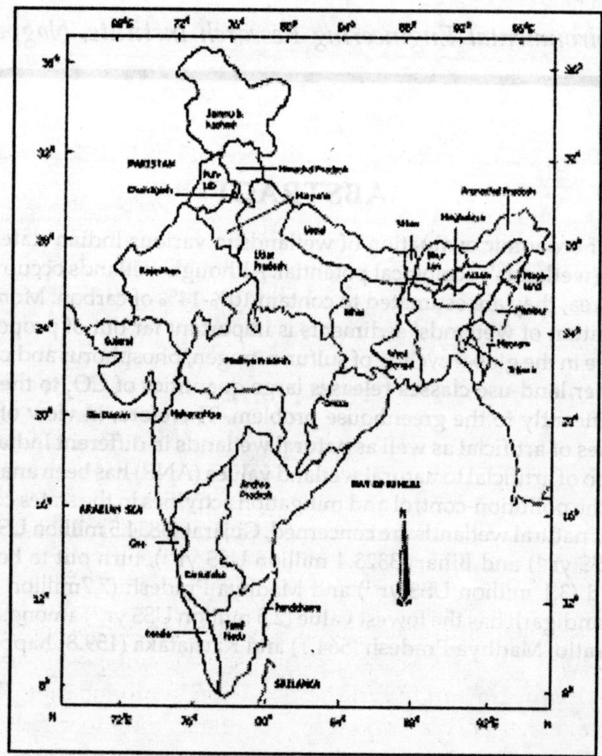

Fig. 1: Different States of India

5. Deltaic Wetlands, Lagoons and Salt Swamps of India's east coast
6. Marshes, Jheels, Terai Swamps, and Chaur lands of the Indo-Gangetic plain
7. Floodplain of the Brahmaputra and the Marshes and Swamps in the hills of northeastern India
8. Lakes and Rivers of Kashmir and Ladakh

Estimates of the types of wetlands in India and the total area under them (1) are given in Table 1. As far as aquaculture (farming of fish and shell fish) production (Table 2) is considered, India happens to be the second largest producer after China (5). In India, intercropped shrimp yields are nearly 200 to 500 kg/ha per crop (5-6).

Table 1: Wetland Types in India and Their Area (in million ha)*

Sl. No.	Wetland Type	Area
1.	Area under paddy cultivation	40.9
2.	Area suitable for fish culture	3.6
	a) fresh water (1.6)	
	b) brackish water (2.0)	
3.	Area under capture fisheries	2.9
4.	Estuaries	3.9
5.	Backwaters	3.5
6.	Man-made Impoundments	3.0
	Total (excluding mangroves)	57.8

* Reference : DIW, [1], p 6

Table 2: Distribution of Global Aquacultural Production*

Srl No.	Country	Global Production (%)
1.	China	57
2.	India	9
3.	Japan	4
4.	Indonesia	4
5.	Thailand	3
6.	United States	2
7.	Phillippines	2
8.	Republic of Korea	2
9.	Other Countries	17

* Reference: Mitsch and Gosselink [5], p 631

In the most intensive aquacultural techniques, wetlands are dredged to form ponds in which water levels are controlled by pumps. "Seed" organisms, the young post larvae, are raised in separate hatcheries. The young organisms are fed in the ponds on synthetic diets, often composed of fish bycatch from commercial fisheries. Water quality is regularly monitored, the ponds are aerated and the wastes are treated. The worldwide boom in aquaculture, fueled by the high demand for fishery products, has led countries such as India to offer large incentives to initiate new fish farms and has drawn large corporations to invest in this industry (5).

In view of the significant role the wetlands play in Indian economy, we have carried out an exercise of economic evaluation of wetlands existing in various Indian states based on their ecological potential. The exercise has mainly taken recourse to the methods currently being used in literature (5, 7-10). Subsequently, the ratio of artificial to natural wetland values (ANR) has been analyzed in view of its direct bearing on the pollution-control and mitigation activities in the states concerned.

Wetlands and their Ecological Role

Wetlands are transitional areas between aquatic and terrestrial ecosystems covered by shallow, fresh water for at least part of the annual cycle (11). They include marshes, swamps, flood-plains, bogs, peatlands, shallow ponds, littoral zones of larger water bodies, tidal marshes etc. (2). Inter alia, they offer excellent economically beneficial possibilities for the control of nutrients and sediments from point and non-point sources (12).

Moreover, although they occupy only about 2% of the world's land area, they are estimated to contain 10%-14% of carbon (11). Their soils contain up to 20% carbon by weight. In addition, the aerobic-anaerobic stratification of wetland sediments is important far out of proportion to their area because of their role in the global cycling of sulfur, nitrogen, phosphorus and carbon. Conversion of wetlands to other land-use classes releases large quantities of CO_2 to the atmosphere, thus contributing significantly to the greenhouse problem (11, 13-14).

During last few decades, many experimental demonstrations have been carried out for exploring the possibilities of effective and purposeful application of wastewater to natural and artificial wetlands. Artificial wetlands have been designed and applied for flood control, water quality enhancement and land stabilization. Moreover, in view of the present crisis attributed to the accumulation of greenhouse gases in the atmosphere, there is an urgent need for controlling these gases either through appropriate measures at the place of their emissions or through enhancing their assimilation by designing and managing natural and artificial ecosystems such as wetlands, which offer enormous potential for the mitigation of greenhouse gases (4).

Economics of Wetlands

In ecological economics, generally two different techniques are used for natural resource accounting and wetland valuation (5, 8). They are: willingness-to-pay (WTP)-valuation and energy-analysis (EA)-valuation. In WTP-valuation, economic values for the benefits are determined by society's willingness to pay for the benefits. WTP, in general, utilizes the concepts of opportunity cost, utilitarian value, option value, existence value and contributory value, while making the evaluation (8).

As far as evaluation based on the Energy Analysis (8-9, 15) is concerned, it looks at the total amount of energy captured by natural ecosystems as an estimate of their potential to do such useful work which is economically beneficial. In this methodology, the total biological productivity of wetland is converted into an equivalent economic value on the basis of the cost, which the society will be required to pay so as to replace this energy source with fossil fuel. In other words, this technique uses the Gross primary Production (GPP) of the ecosystem as an Index of the solar energy captured by the system, and converts this energy value into dollars using energy-dollar conversion factor (8).

Results and Discussion

Environmental decision making cannot do without economic quantification of ecological benefits and costs. However, the estimation of the economic value of natural wetlands is a difficult and complex task. For the economic evaluation of natural wetlands mainly the values given in Costanza and others (7, 8) have been used. As far as costs of artificial (man-made) wetlands are concerned, they were computed on the basis of the model given in Mitsch and Gosselink (5) (*i.e.* Cost = \$ 196336 $A^{-0.511}$, where A = Wetland area in hectares), while for the benefit-valuation for even the artificial wetlands, the values (Table 3) by Costanza and others (7) have only been used. The wetland data for Indian states was taken from the Directory of Wetlands of India (2).

Figure 1 shows different Indian states for which the present wetland-economic-valuation exercise has been carried out. Total economic values of wetlands range from 2.5 million US\$ yr^{-1} for Chandigarh (Northern Zone) to 7896.5 million US\$ yr^{-1} for Karnataka (Southern Zone), which is a very large variation showing clearly a very uneven distribution of wetlands in different zones. Figure 2 shows the values of natural wetlands in different states. While Gujarat, (5834.5 million US\$$yr^{-1}$), West Bengal (4316.6 million US\$ yr^{-1}) and Bihar (3323.4 million US\$ yr^{-1}) happen to be the wealthiest states in terms of natural wetlands, Nagalnad (3.1 million US\$ yr^{-1}) and Madhya Pradesh (7.7 million US\$

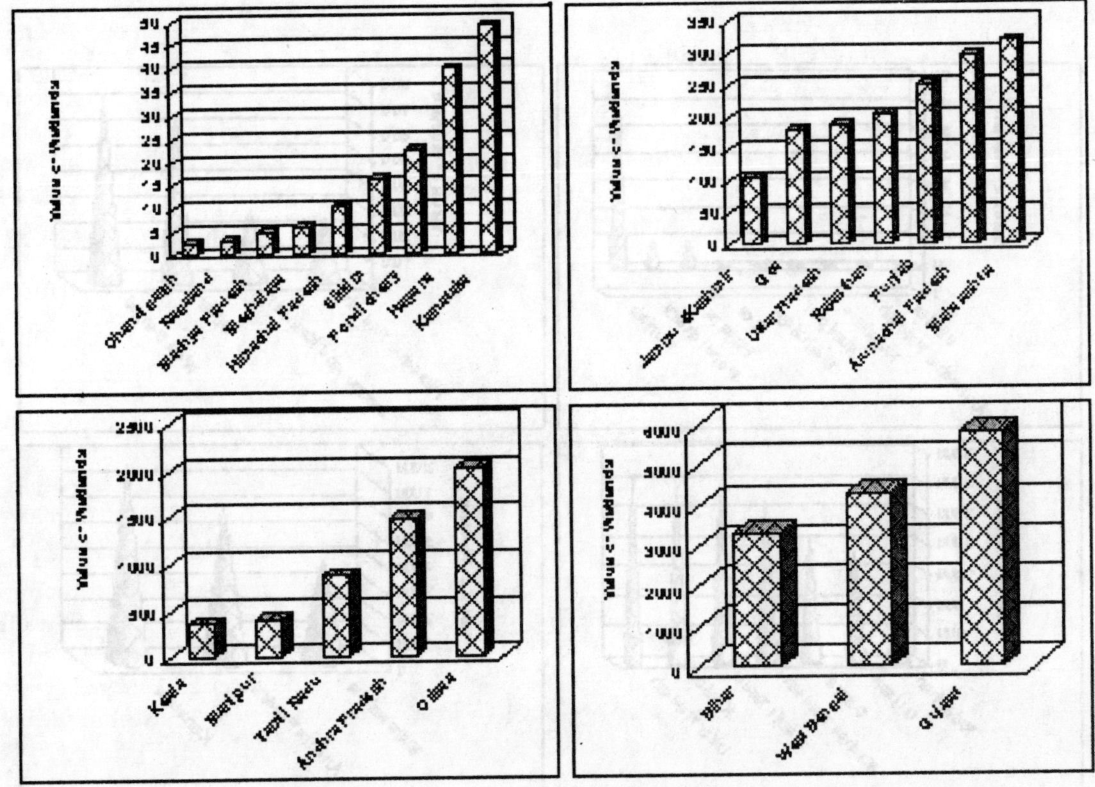

Fig. 2: Economic Values of Natural Wetlands in Different States (million US\$ yr^{-1})

yr^{-1}) have the lowest values. Chandigarh has the lowest value (2.5 million US\$ yr^{-1}) amongst union Territories.

Table 3: Unit Values of Different Ecosystems*

Sl. No.	*Ecosystem*	*Unit Value (US \$ ha^{-1} yr^{-1})*
1.	Estuaries	22832
2.	Wetlands	14785
3.	Lakes/Rivers	8498
4.	Forests	969
5.	Grasslands	232

* Reference : Mitsch and Gosselink, Costanza et al. [5, 7]

Figure 3 shows the values accruing out of artificial wetlands only. In terms of artificial wetlands, Karnataka (7847.4 million US\$ yr^{-1}), Andhra Pradesh (6185.6 million US\$ yr^{-1}) and Maharashtra (4034.9 million US\$ yr^{-1}) happen to be the richest states while Sikkim, Arunachal Pradesh, and Nagaland have the lowest values. In fact, in Arunachal Pradseh and Nagaland there are no artificial wetlands, so the wetland wealth is zero. Moreover, Sikkim has 3.5 ha of artificial wetlands. So the cost (5) of the

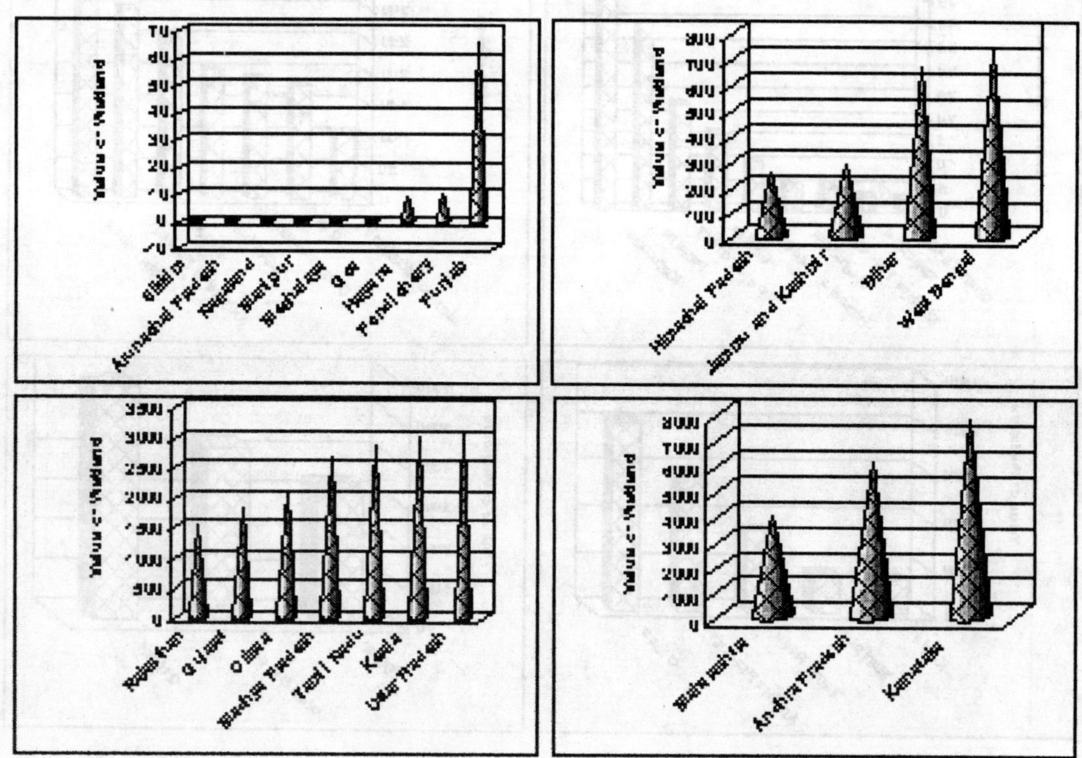

Fig. 3: Economic Values of Artificial Wetlands in Different States (million US\$ yr^{-1})

artificial wetlands works out to be 0.36 million US\$ yr^{-1}, which is more than the benefit (0.05 million US\$ yr^{-1}) due to the wetlands (7).

Figure 4 finally depicts the ratio ANR (economic value of artificial (A) wetlands to the economic value of natural (N) wetlands) and is seen that this ratio is the highest for Madhya Pradesh (564.1) and Karnataka (159.8), while it is lowest for Sikkim (-0.019) and West Bengal (0.17). ANR is closely liked with the states' concerns for environmental issues. And in this sense, Madhya Pradesh and Karnataka can be said to be the most environmentally concerned states.

Ecosystems economic value should include the value of their explicit physical, chemical and biological role in the over-all human system. And, depending upon their importance, it can focus and concentrate on the roles which are relatively more important. However, this process of economic evaluation can carry significant bias depending on the fact whether the valuation is carried out from an individual, societal or regional perspective. This particular aspect calls for the need of utilizing additional mathematical and computational techniques for carrying out sensitivity analysis with respect to various relevant ecological parameters.

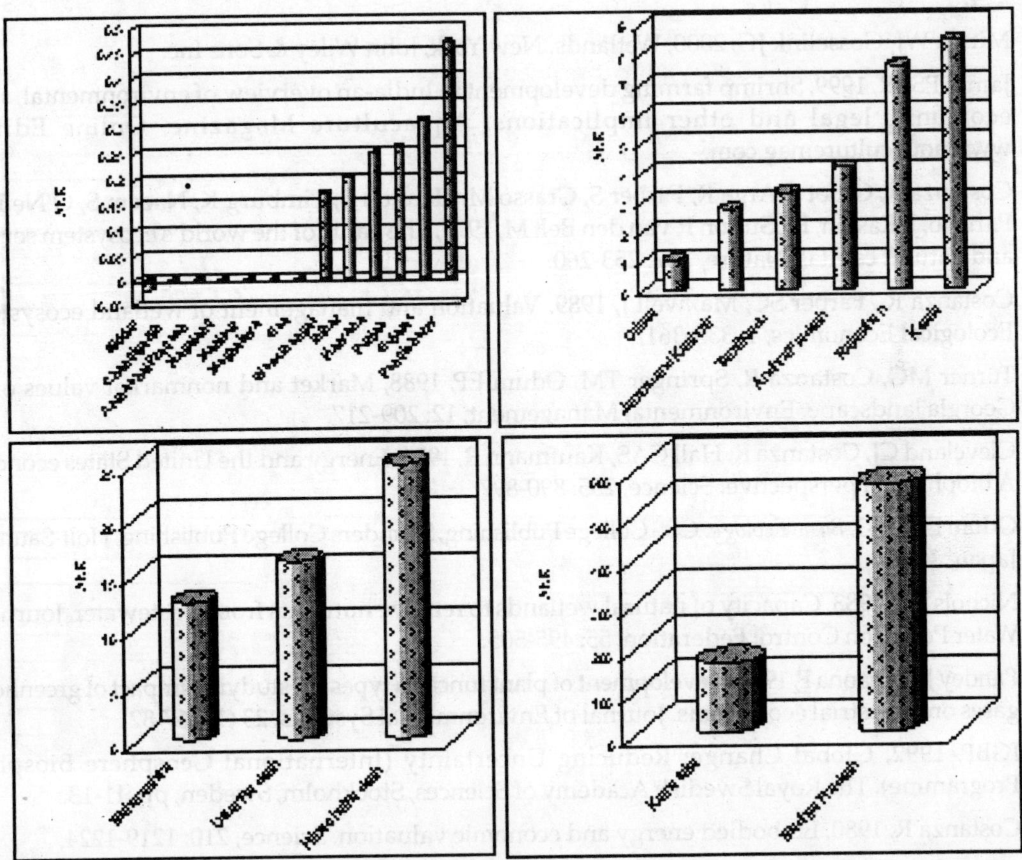

Fig. 4: Ratio (ANR) of Economic Values of Artificial Wetlands to the Economic Value of Natural Wetlands (Dimensionless)

Acknowledgement

The authors are thankful to Dr. R. N. Singh, Director, NEERI for encouragement and infrastructural support.

REFERENCES

1. DIW, 1993, Directory of Indian wetlands. Compiled by the World Wide Fund for Nature (WWF), India in collaboration with the Asian Wetland Bureau.
2. WID, 1990, Wetlands of India-A Directory. Government of India, Ministry of Environment and Forests, New Delhi.
3. AWB, 1989, A Status Overview of Asian Wetlands. Compiled by D.A. Scott and C.M. Poole for the Asian Wetland Bureau (AWB) and World Wide Fund for Nature (WWF). Published by the Asian Wetland Bureau
4. Pandey JS, Deb SC, Khanna P., 1997, Issues related to greenhouse effect, productivity modeling, and nutrient cycling: a case study of Indian wetlands. Environmental Management, 21 (2) : 219-224.
5. Mitsch WJ, Gosselink JG, 2000, Wetlands. New York, John Wiley & Sons Inc.
6. James PSBR, 1999, Shrimp farming development in India-an overview of environmental, socio-economic, legal and other implications. Aquaculture Magazine, Online Edition, www.aquaculturemag.com.
7. Costanza R, Groot d'Arge R, Farber S, Grasso M, Hannon B, Limburg K, Naeem S, O'Neill RV, Paruelo, J, Raskin, RJ, Sutton P, van den Belt M, 1997, The value of the world's ecosystem services and natural capital. Nature, 387: 253-260.
8. Costanza R., Farber SC, Maxwell J, 1989, Valuation and management of wetland ecosystems. Ecological Economics, 1: 335-361.
9. Turner MG, Costanza R, Springer TM, Odum EP, 1988, Market and nonmarket values of the Georgia landscape. Environmental Management, 12: 209-217.
10. Cleveland CJ, Costanza R, Hall CAS, Kaufmann R, 1984, Energy and the United States economy: A biophysical perspective. Science, 255: 890-897.
11. Odum EP, 1983, *Basic Ecology*. CBS College Publishing, Saunders College Publishing, Holt-Saunders Japan, 1983.
12. Nichols DS, 1983, Capacity of natural wetlands to remove nutrients from wastewater. Journal of Water Pollution Control Federation, 55: 495-505.
13. Pandey JS, Khanna P, 1994, Development of plant function types for studying impact of greenhouse gases on terrestrial ecosystems. Journal of Environmental Systems, 23 (1) : 67-82.
14. IGBP, 1992, Global Change: Reducing Uncertainty (International Geosphere Biosphere Programme). The Royal Swedish Academy of Sciences, Stockholm, Sweden, pp. 11-13.
15. Costanza R, 1980, Embodied energy and economic valuation. Science, 210: 1219-1224.

64

Application of Industrial Ecology Concept in Developing SDP Complex with Particular Emphasis on Wastewater Management

S.K. GOYAL

Scientist, National Environmental Engineering Research Institute (NEERI), Nehru Marg, Nagpur - 440 020

ABSTRACT

The concept of industrial ecology has been applied in developing and industrial complex affiliated to sugar industry, where the sugar industry generated two of the major wastes; molasses and bagasse are used as main raw material in distillery and paper industry for manufacturing alcohol and paper products respectively. Further, to meet the energy requirement in the complex, the wastewaters generated from these industries are utilized for biogas production through anaerobic digestion. A systematic approach has been presented in this paper to demonstrate the suitability of industrial integrity of sugar, distillery and paper industry (SDP Complex) with particular emphasis on wastewater management for sustainable industrial development.

Introduction

Industrial ecology is a system framework for managing energy and materials flows for high efficiency and low pollution. It is a system approach that enables executives, managers and designers to balance financial, technical and environmental objectives. Further, application of industrial ecology opens opportunities for cost savings, new revenues, and definition of new product lines.

Essentially, industrial ecology refers to the exchange of materials between different industrial sectors, where the 'waste' output of one industry becomes the 'feedstock' of another. In this approach number of different industries, usually in close proximity to each other, exchange a variety 'waste' outputs. Implementing industrial ecology involves such things as life cycle analysis, closed loop processing, reusing and recycling, design for environment and waste exchange (1).

Need for an integrated complex of sugar-distillery and paper mills has been emphasized, as the sugar industry generates two major by-products; bagasse and molasses, which can be better utilized. At the same time distillery and paper industry are known to have very high pollution potential. Treatment of high organic load containing streams for biogas generation together shall have twin advantage of pollution load reduction at low cost and energy recovery, reducing fuel (coal) consumption in the industries. Hence, development of integrated industrial ecological complex is the need of the hour to achieve sustainable industrial development.

Techno-economic feasibility analysis of a 2500 tcd sugar mill along with a small paper mill and distillery is presented in this paper, with particular emphasis on waste water management in the industrial complex. In India, there are number of distilleries associated with sugar industry, and few paper industries with sugar mill, but a complex integrating these three are in the exploratory stage.

Status of Sugar Industry

Globally, sugar is produced in 121 countries and sugar production in the year 2000-01 was about 128 million tons. Approximately 70% of sugar is produced from sugar cane, which is largely grown in the southern hemisphere tropical countries and the remaining 30% from sugar beet in the temperate zones of the north. The five largest exporters (Brazil, the European Union, Australia, Cuba and Thailand) supply 70% of the world free market exports. India is the largest consumer and second largest producer of sugar in the world. It contributes nearly 15% of the total world's production, and exports only about 5.5% of its total sugar production (2).

With over 450 sugar factories located throughout the country, the sugar industry is amongst the largest agro processing industries, with an annual turnover of Rs. 150 billion. It plays a major role in rural development and its importance for India stretches far beyond the role of a sweetener supplier.

The number of operating sugar mills in the country has increased from 29 in the sugar year 1930-31 to 412 by the year 1996-97. The average capacity of the sugar mills has considerably increased from just 644 TPD in the year 1930-31 to 2656 TPD in the year 1996-97. But the growth in the Indian sugar industry was driven by horizontal growth (increase in number of units) compared to the vertical growth witnessed in other countries (increase in average capacity). The average size of Indian sugar mill is very small as compared to international capacity standards. In India, sugarcane-crushing capacity varies from 400 to 8000 tcd. The distribution of sugar mills according to their capacity range is; up to 1250 tcd: 52%; 1251-2500 tcd: 34% and above 2500 tcd: 14% (3,4).

Indian sugar industries are classified under three major categories; farmers own cooperative sector, private sector and public sector. Currently, out of 453 sugar mills in the country, 252 are in the co-operative sector, 134 are in the private sector and 67 are in the public sector. Besides 136 units in the private sector are in various stages of implementation (3).

Sugar produced in India is mainly of granulated type. The demand for sugar has increased from 3.69 million ton in SY1975-76 to 14.8 million ton for SY1997-98 at a CAGR of 6.48%. The per capita consumption of sugar in the country is only about 25 kg/year (including about 9 kg/year for gur and khandsari), as compared to 54 kg/year in Brazil, 49 kg/year in Cuba, 46 kg/year in Australia and 32 kg/year in USA (2).

Sugar Manufacturing Process

In India, sugar is mainly produced only from sugarcane, as the country's climatic conditions are suitable for cane cultivation. Harvested sugarcane is generally crushed within 8 to 10 hours of cutting, to avoid its deterioration and achieve maximum recovery. The usual sugarcane constituents are: water: 69-75%, sucrose: 8-16%, reducing sugars: 0.5-2%, organic substances: 0.5-1%, nitrogenous bodies: 0.5-1%, ash: 0.3-0.8% and fiber: 10-16% of the total weight. The sugar manufacturing process consists of juice extraction, clarification, evaporation, crystallization and centrifugation. A generalized flow diagram of the sugar manufacturing process is shown in Fig. 1.

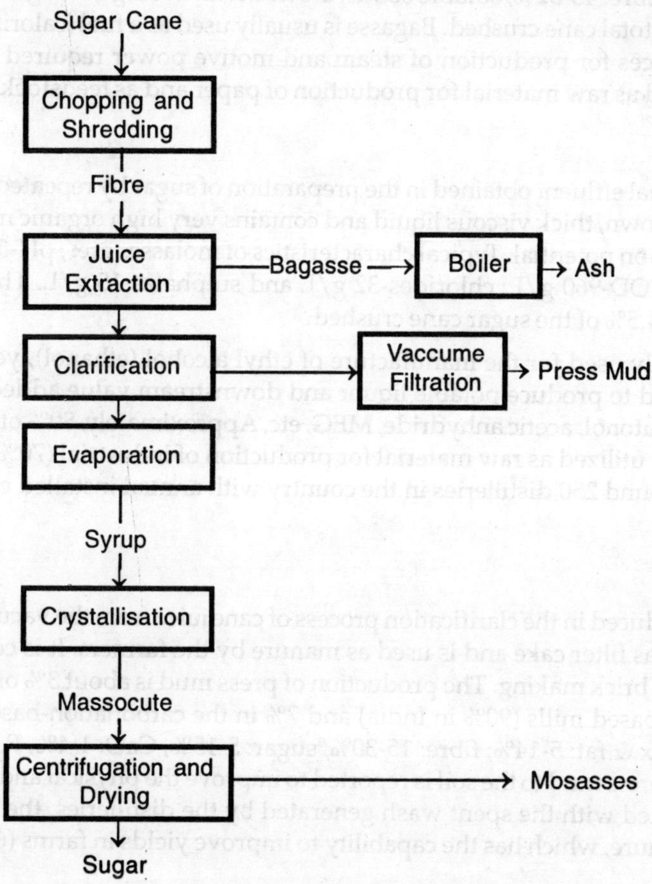

Fig. 1: Sugar Manufacturing Process

The quantum of sugar produced by a mill is determined by the factors like daily crushing capacity, duration of crushing season and percentage of sugar recovery. The crushing season in the country starts from October and reaches its peak in January before finally ending in March or April of the next year; however, it may also vary depending upon the cane availability in different parts of the country. The recovery rate for sugar mills does vary widely across the country, and in the past it has been in the range 9.25 to 10.5% (overall average 10%) (3).

By-products of Sugar Industry

The sugar mill produces many by-products along with sugar. Bagasse molasses and press mud are the major by-products, which if not utilized/disposed properly, have high pollution potential. In fact, these are very valuable products and are used for different purposes as briefly described below (4,5).

Bagasse

Bagasse is a fibrous residue of cane stalk that is obtained after crushing and extraction of juice. It consists of water, fiber and relatively small quantities of soluble solids. The usual bagasse composition is; moisture: 46-52%, fibre: 43-52%, soluble solids: 2-6%. On an average, a sugar mill generates one-third of bagasse of the total cane crushed. Bagasse is usually used as a fuel (calorific value about 1920 Kcal/kg) in the furnaces for production of steam and motive power required (up to 95%) by the industry. It is also used as raw material for production of paper and as feedstock for cattle.

Molasses

Molasses is the final effluent obtained in the preparation of sugar by repeated crystallization. It is a yellowish or dark brown, thick viscous liquid and contains very high organic matters, and as such, it has very high pollution potential. Typical characteristics of molasses are: ; pH-3.5-4.1; TDS-200-320 g/L; BOD-440 g/L; COD-960 g/L; chlorides-32 g/L and sulphates-15 g/L. The yield of molasses varies between 3.5 to 4.5% of the sugar cane crushed.

Molasses is mainly used for the manufacture of ethyl alcohol (ethanol), yeast and cattle feed. Ethanol is in turn used to produce potable liquor and downstream value added chemicals such as acetone, acetic acid, butonol, acetic anhydride, MEG, etc. Approximately, 80% of molasses produced by organized sector is utilized as raw material for production of industrial (70%) and potable (30%) alcohol. There are around 250 distilleries in the country with annual installed capacity of about 2.4 million KL.

Press Mud

Press mud is produced in the clarification process of cane juice from the vacuum filtration unit. It is commonly known as filter cake and is used as manure by the farmers. It is combustible and also used in brick kilns for brick making. The production of press mud is about 3% of cane crushed in the sulphitation process based mills (90% in India) and 7% in the carbonation-based mills. Press mud constitutes; crude wax & fat: 5-14%; fibre: 15-30%, sugar: 5-15%, CaO: 1-4%, P_2O_5: 1-3%, MgO: 0.5-1.5%. Application of press mud to the soil is reported to improve the physical and chemical properties of the soil. When mixed with the spent wash generated by the distilleries, the resultant product is enriched organic manure, which has the capability to improve yields in farms (6).

Development of an Integrated Industrial Complex for Utilization of By-products

During the manufacturing process of sugar from sugarcanes, sugar industry produces many useful by-products. Among which, Bagasse and Molasses are the most important. Most of the molasses is used as raw material in distillery for manufacture of alcohol and various other chemicals, and these industries are generally located near the sugar industry. Bagasse is normally used as fuel in the sugar industry boiler to meet its steam and captive power requirement. Bagasse is also known to have a very good potential for paper manufacture, and several bagasse-based industries are operating in the country along with other non-conventional raw materials.

It is proposed that the concept of industrial ecology or industrial symbiosis may be adopted in case of sugar industry, where the waste of one industry becomes the feedstock to the other industry. An integrated industrial complex (SDP) comprising sugar-distillery and paper industry should be developed in order to achieve maximum utilization of various resources and also for combined treatment of wastewater streams, as these industries are not only water intensive, but also have high pollution potential. The approach has been explained through a case study of 2500 tcd capacity sugar mill, because as per the Govt. of India policy, new sugar factories should have minimum economic capacity of 2500 tcd with no maximum limit on capacity. However in industrially backward areas, co-operative and public sector new units are allowed with an initial capacity of 1750 tcd subject to the condition that the units would expand their capacities to 2500 tcd within a period of 5 years of going into production (3).

A 2500 tcd sugar mill is estimated to produce about 250 MT (@10% recovery) of sugar, and in the process shall generate three major by-products; about 417 MT bone dry bagasse, 112.5 MT (@ 4.5%) molasses and 80 MT (@3.2%) press mud. Accordingly, three more industries utilizing these valuable waste products can be included in the industrial complex. A schematic diagram for the industrial complex along with biogas generation scheme is shown in Fig. 2, and described here briefly.

Utilization of Bagasse-Paper Mill

A huge quantity of bagasse generated can be used partly for paper manufacture and the rest as fuel in boiler for steam and power generation. Depending upon the financial status and availability of other input resources, a small (30 TPD without chemical recovery) or a large (150 TPD with chemical recovery) paper mill can be set up. In order to make self sufficient complex with respect to energy resources, a small paper mill with 30 TPD capacity manufacturing industrial papers and boards is proposed, a major portion of which can to be utilized in the complex itself.

The total demand for paper in the country is about 3.5 million tons per annual, whereas the actual production is around 2.8 million tons. Thus there exists a huge gap between demand and supply, for which additional paper industries are required to be set up in the country. For a 30 TPD pulp and paper mill producing industrial papers from bagasse (70%), waste gunny (10-15%) and waste paper (10-15%), the bagasse requirement would be around 67 MTPD (bone dry), which is about 6% of the total bagasse produced. With a pulp yield of about 30-35%, the annual bagasse requirement would be around 25000 MT (equivalent to 2 months of bagasse production). Since, the sugar mill is operative only for about 6 months in a year, therefore, an equal amount (70-75 MTPD) of bagasse shall need to be stored separately for utilization during non-sugar season. Remaining 275 MT bagasse can be used as fuel in boiler for steam and power generation.

The details of manufacturing process for small paper mills, particularly based on non-conventional raw materials are well established (7). There are two major sections in a paper mill producing industrial

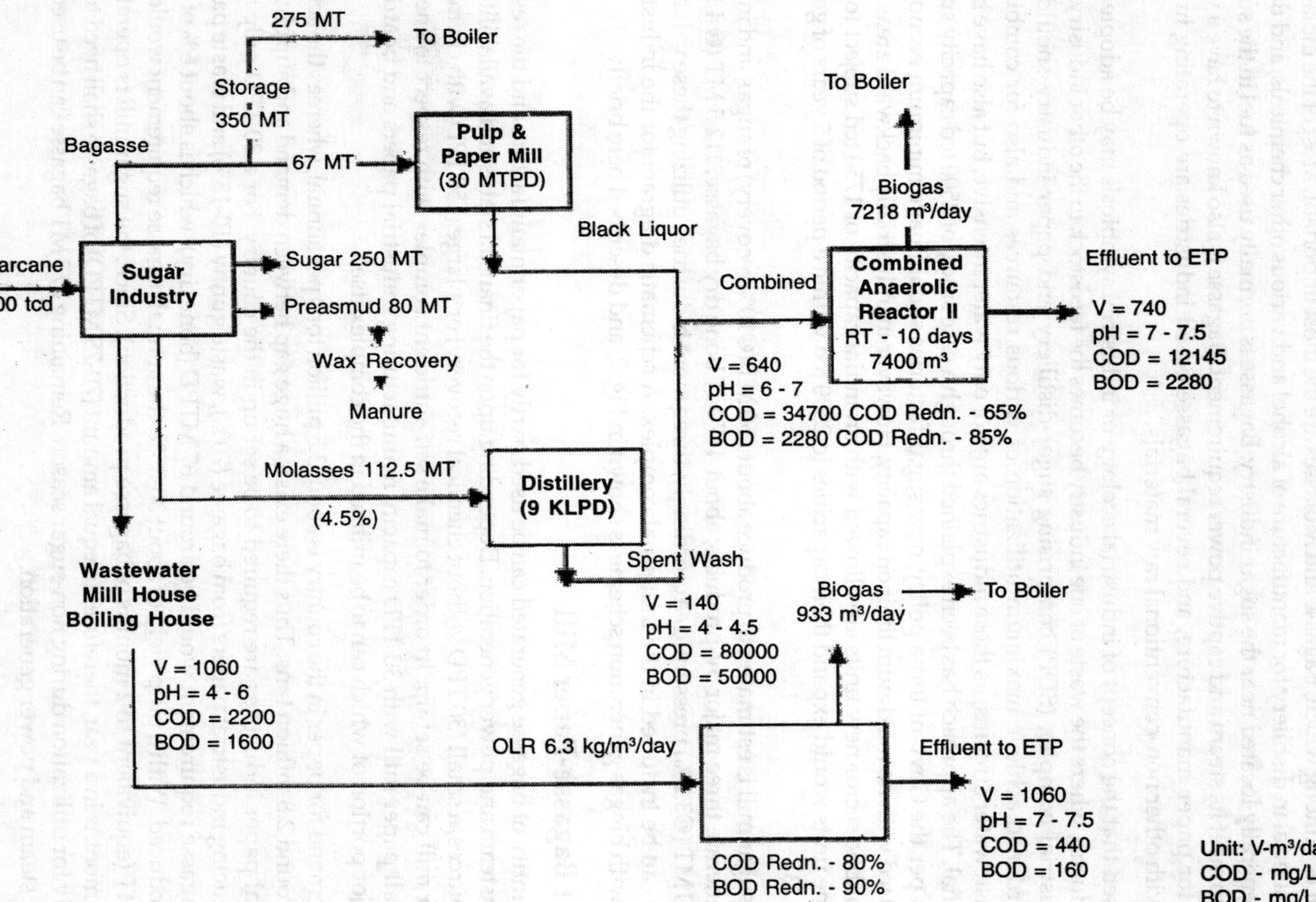

Fig. 2: Scheme of Integrated SDP Complex with Energy Recovery from Potential Wastewater Streams

papers and boards as; pulp mill and paper machine section. Pulp mill section involving chemical digestion of raw materials (bagasse, gunny etc.) and then pulp washing contribute to about 70-80% of the total pollution load from the paper industry. Black liquor containing very high organic load can be separated through screw press to get most concentrated form for the subsequent biogas recovery.

An Indian paper mill (M/s Pravara Pulp & Paper Mill, Distt. Ahmednagar, Maharashtra) has adopted this approach. About 400 m^3/day black liquor is processed through screw press, of which about 100 m^3/day is used for make up volume in the digester and 300 m^3/day black liquor, thus separated is further mixed with brown stock washer filterate to the volume of 500 m^3/day, and is sent to biogas generation (8).

Utilization of Molasses-Distillery

On an average, about 45 MT of molasses is produced for every 100 MT of sugar cane crushed, thus a 2500 tcd sugar mill shall generate about 112 MT molasses daily, to be used for manufacture of alcohol and other products. To produce one litre of alcohol, 3-10 kg (avg. 6 kg) molasses is required. Hence, a total of 18.75 KL alcohol can be produced daily. Since sugar mill operates only for about 6 months in a year, a distillery of 9 KLPD capacity can be set up in the complex, so that it is operated throughout the year. The requisite molasses for the non-sugar mill operation period can be stored properly.

The details of manufacturing of alcohol from molasses are well established (9), and at present about 250 distilleries are operating in the country. For every litre of alcohol produced about 12-15 litre of spent wash, containing very high BOD and COD is generated, which has a very high potential for biogas generation, otherwise leads to high water pollution.

Utilization of Press Mud-Chemical Industry

In India, the average crude wax content of dry sulphitation press mud is approximately 10%. The wax from press mud can be recovered by various solvent extraction methods. Solvents normally used are n-hexane and n-heptane. This wax (crude) is a mixture of hard and soft waxes with other minerals, colouring and resinous matters etc. Hard wax of high melting point representing 70% of crude wax can be separated from crude wax by adopting suitable refining methods. Hard wax can be used for the manufacture of floor polishes, candles, waxed card board, containers and paper cups for electrical insulation etc. Sugar cane wax promises to be very good source for meeting the demand for wax in the country as well as for export purposes. Moreover, the manurial value of dewaxed press mud is reported to be better than that of the ordinary press mud. It is estimated that for every million ton of sugar production, India can produce approximately 9630 MT of crude wax or 6740 MT of hard wax. A 2500 tcd sugar mill can produce 2.4 MT of crude wax or 1.6 MT of hard wax daily.

Utilization of Waste Streams for Biogas Generation

A huge quantity of water is used in sugar as well paper industry, with moderate quantity in distillery. During the manufacturing process of various products, wastewater containing varied amount of pollution loads are generated. The typical characteristics of various streams in sugar mill (10), distillery (11) and agro-residues based small paper mills (12) are summarized in Table 1. Characterization of various waste water streams from these three industries suggest that the following streams containing high organic content can be utilized for biogas generation through anaerobic digestion:

1. Wastewater from mill house and boiling house of sugar industry
2. Black liquor and washer filtrate from pulp mill section of paper mill
3. Spent wash from distillery

Table 1: General Characteristics of Waste waters from Different Sources in Sugar, Distillery and Paper Industry

Parameter	*pH*	*Suspended Solids mg/L)*	*COD (mg/L)*	*BOD (mg/L)*
Sugar Industry				
Mill house	4-6	500-550	1000-1500	700-1000
Boiling house	4-6	400-600	2000-3000	1500-2000
Condensates	6-7	5-10	250-300	100-150
Lime house	8-9	3500-4000	200-250	100-150
Boiler blow-down	6-7	50-100	500-550	30-40
Combined	6-7	150-300	1800-3000	1000-1500
Distillery				
Spentwash	4.0-4.7	8600-13000	62000-85000	30000-42000
Vat cleansing	3.8-4.7	15000-150000	61000-148000	25000-78000
Spentlease	3.4-3.9	10-290	10000-18000	5000-8000
Floor washing	4.5-4.8	600-800	11000-15000	5800-6500
Condensate water	5.7-7.2	16	100	25
Paper Industry				
Black liquor	9-10	2340	53600	10200
Pulp wash	9-9.5	1850	8800	2450
Bleaching	6-7	440	1020	290
Paper Machine	6-7	480	580	180
Combined	7-7.5	850	1100	330

Biogas can be produced through any of the anaerobic digestion process. A fixed bed, fluidized bed, a contact process or UASB system can be employed for anaerobic digestion of these streams. Typical wastewater characteristics used for biogas generation for a 2500 tcd sugar mill (3), 40 KLPD distillery (13) and 30 MTPD paper mill (8) are presented given in Table 2. Since the COD and BOD content of sugar industry stream is comparatively very low, it would be desirable to treat/recover biogas from this stream separately in an anaerobic reactor (AR I).

The other two streams from paper mill and distillery can be mixed together before feeding to another anaerobic reactor (AR II). Another reason for separately treating sugar industry stream is that it operates only for a period of 6 months in a year, whereas the other are operated for the whole year. Further, the hydraulic retention time (HRT) required for AR I and AR II vary considerably, *i.e.* 6 hrs and

Table 2: Typical Characteristics of Segregated Wastewater Streams and Combined Effluent for Biogas Generation

Parameter	*Sugar Industry (Mill house and boiling house)*	*Paper Mill (Pulp Mill Section)*	*Distillery Spentwash)*	*Combined Effluent (Paper Mill + Distillery)*
Flow, m^3/day	1060	500	140	640
pH	4-6	7-8	4-4.5	6-7
COD, mg/L	2200	22000	80000	34700
BOD, mg/L	1600	5500	50000	15200
SS, mg/L	540	700	10000	2700

About 65-80% reduction in COD and 85-90% reduction in BOD can be achieved in the process with biogas yield of about 0.5 m^3/kg COD destroyed. The design parameters and anticipated performance of the UASB anaerobic reactor are summarized in Table 3, and also presented in Fig. 2 along with quality of effluent. The total biogas generation would be around 8151 m^3/day with 65-70% methane content, to be used in boiler. This along with the bagasse combustion during the sugar season shall be equivalent to 140 MT of coal per day, and about 9.0 MTPD during non-sugar season. The details are presented in Table 4. The total energy available during sugar season shall be sufficient to meet the requirements in all the three industries, whereas during non-sugar season, additional arrangement to the tune of about 50% will have to made through coal fired boiler.

Table 3: Design Parameters and Anticipated Performance of UASB Reactors

Parameter	*Sugar*	*Paper & Distillery*
Reactor Capacity, m^3	280	6400
Retention Time	6 hrs	10 days
COD loading rate, Kg/m^3	8.3	3.5
COD Reduction, %	75-85	60-70
BOD Reduction, %	85-90	85-90
Biogas Production, Nm^3/kg COD destroyed	0.5	0.5
Total COD destroyed, kg/day	1866	14435
Total biogas production, NM^3/day	933	7218
Final Effluent Quality		
pH	7-7.5	7-7.5
COD, mg/L	440	12145
BOD, mg/L	160	2280

Integrated Wastewater Treatment

The remaining wastewater streams from all the three industries can be treated together separately before final disposal to water bodies/land for irrigation. The characteristics of remaining wastewater streams are presented in Table 5 along with combined effluent.

Table 4: Energy Generation during Sugar and Non-Sugar Season

Parameter	Value
During Sugar Season	
Bagasse available for combustion	275 MT/day
Calorific Value of bagasse	1900 Kcal/kg
Total heat value of bagasse	52.25×10^7 Kcal/day
Total Biogas Production	8151 m^3/day
Calorific value of biogas	5000 Kcal/kg
Total heat value of biogas	4.08×10^7 Kcal/day
Total heat value	56.33×10^7 Kcal/day
Calorific value of coal	4000 Kcal/kg
Equivalent quantity of Coal	140 MT/day
During Non-sugar Season	
Total Biogas Production	7218 m^3/day
Total heat value of biogas	3.61×10^7 Kcal/day
Equivalent quantity of Coal	9.02 MT/day

Table 5: Estimated Characteristics of Remaining Wastewater Streams and Combined Effluent

Parameter	*Sugar Industry (A)*	*Paper Mill (Paper Machine Section) (B)*	*AR I (Sugar) (C)*	*AR II (Paper + Distillery) (D)*	*Combined Effluent*	
					Sugar Season (A+B+C+D)	*Non-sugar Season (B+D)*
Flow, m^3/day	865	1800	1060	640	4365	2440
pH	6-7	6-6.5	7-7.5	7-7.5	7	7
SS, mg/l	705	750	380	1900	820	1052
COD, mg/l	264	1500	440	12145	2560	4292
BOD, mg/l	57	500	160	2280	590	967

All the remaining streams from sugar mill, paper mill, distillery, and anaerobic reactors can be mixed together in flow equalization tank. The approximate quality of combined effluent has been estimated for the two periods; with and without sugar mill operation.

The wastewater treatment scheme followed shall involve; primary clarifier, anaerobic lagoon, and aerobic lagoon with surface aeration. A schematic diagram of wastewater treatment is shown in Fig. 3, along with anticipated effluent quality at each stage during sugar season. The finally treated effluent conforming to MINAS with respect to pH, SS and BOD can be used for irrigation in the nearby land. The characteristics of treated effluent at every stage of treatment process are presented in Table 6, with corresponding pollution load reductions during sugar and non-sugar seasons.

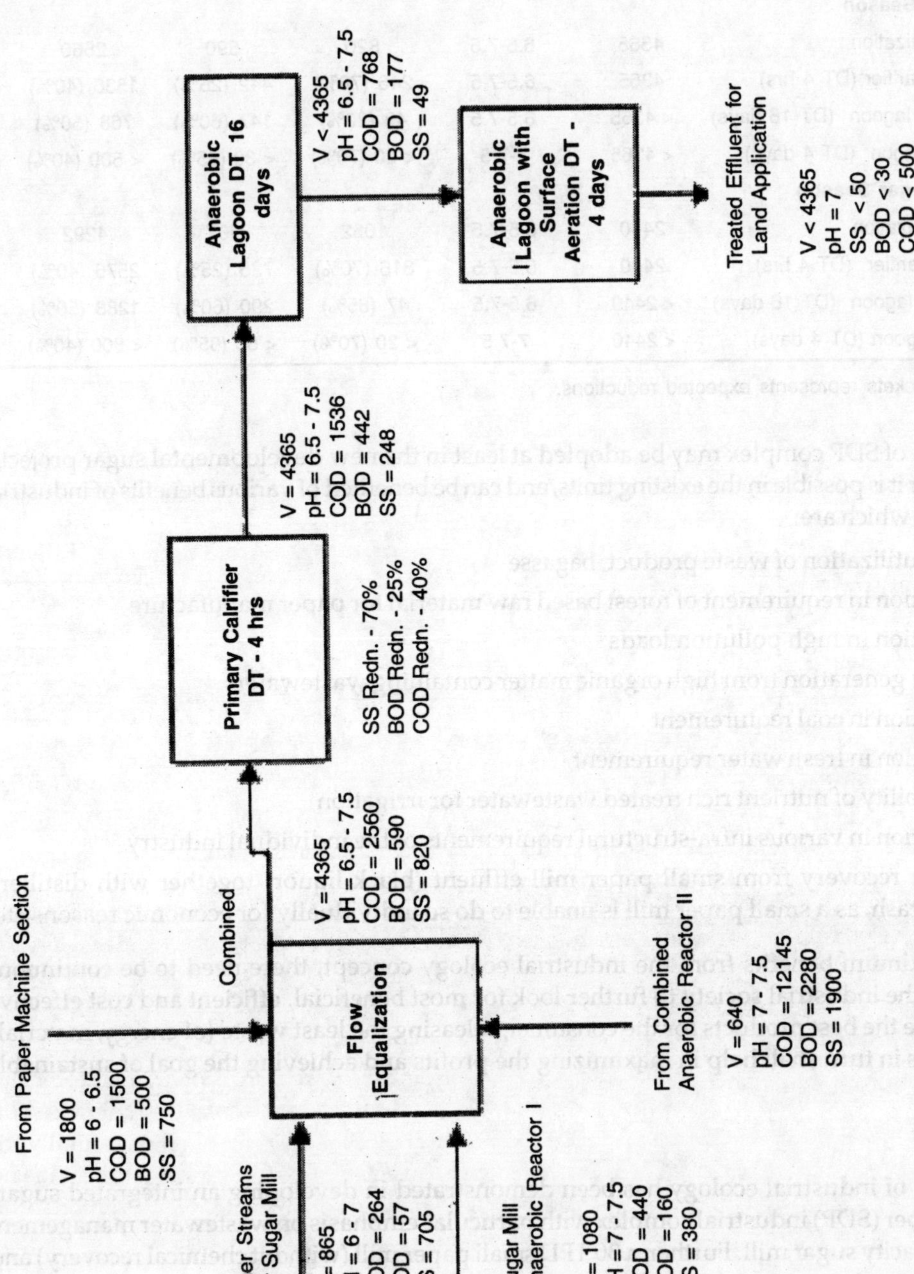

Fig. 3: Scheme of Wastewater Treatment for Remaining Streams from SDP Complex (During sugar season)

Table 6: Estimated Characteristics of Combined Effluent at Each Stage of Treatment

Stage	*Parameters*				
	Flow (m^3/day)	*pH*	*SS (mg/L)*	*BOD (mg/L)*	*COD (mg/L)*
During Sugar Season					
After flow equalization	4365	6.5-7.5	820	590	2560
After primary clarifier (DT 4 hrs)	4365	6.5-7.5	246 (70%)	442 (25%)	1536 (40%)
After anaerobic lagoon (DT 16 days)	< 4365	6.5-7.5	49 (80%)	147 (60%)	768 (50%)
After aerated lagoon (DT 4 days)	< 4365	7-7.5	< 20 (70%)	< 30 (85%)	< 500 (40%)
During Non-sugar Season					
After flow equalization	2440	6.5-7.5	1052	967	4292
After primary clarifier (DT 4 hrs)	2440	6.5-7.5	316 (70%)	725 (25%)	2575 (40%)
After anaerobic lagoon (DT 16 days)	< 2440	6.5-7.5	47 (85%)	290 (60%)	1288 (50%)
After aerated lagoon (DT 4 days)	< 2440	7-7.5	< 20 (70%)	< 50 (85%)	< 800 (40%)

Value in brackets represents expected reductions.

The concept of SDP complex may be adopted at least in the new developmental sugar projects, besides wherever it is possible in the existing units, and can be benefited of various benefits of industrial ecology. Some of which are:

1. Better utilization of waste product; bagasse
2. Reduction in requirement of forest based raw material for paper manufacture
3. Reduction in high pollution loads
4. Energy generation from high organic matter containing wastewaters
5. Reduction in coal requirement
6. Reduction in fresh water requirement
7. Availability of nutrient rich treated wastewater for irrigation
8. Reduction in various infra-structural requirements of the individual industry
9. Energy recovery from small paper mill effluent (black liquor) together with distillery spentwash, as a small paper mill is unable to do so individually for economic reasons etc.

To gain maximum benefits from the industrial ecology concept, there need to be continuous research within the industrial society to further look for most beneficial, efficient and cost effective methods to create the best products for the consumer, releasing the least waste (of energy, materials and capital). This in turn shall help in maximizing the profits and achieving the goal of sustainable development.

Conclusions

The concept of industrial ecology has been demonstrated in developing an integrated sugar-distillery and paper (SDP) industrial complex with particular emphasis on wastewater management for a 2500 tcd capacity sugar mill. Further, a 30 TPD small paper mill (without chemical recovery) and 9 KLPD distillery can be setup in the complex.

Biogas equivalent to about 9 MT coal can be generated from distillery and paper mill wastewaters. During sugar season, the total energy requirement can be met from bagasse combustion and biogas generation (equivalent to 140 MT coal). However, there is a need for further study to look critically for most beneficial, efficient and cost effective methods to create the best products with least waste generation and thus help in achieving the goal of sustainable development.

REFERENCES

1. Garner, Andy and Keoleian, G.A., 1995, Industrial Ecology – An Introduction, National Pollution Prevention for Higher Education, University of Michigan, Ann Arbor, MI, Nov. (Article available on the web).
2. International Facts and Figures about Sugar Industry (Article available on the web).
3. India Infoline Sector Reports – Sugar Industry ((Article available on the web).
4. TIFAC Document, 1996, Techno-market Survey on Pollution Control Technologies for Sugar Mills, prepared by Technology Information, Forecasting and Assessment Council (TIFAC), Department of Science and Technology, Delhi.
5. Comprehensive Industry Document for Sugar Industry, prepared by Central Board for Prevention and Control of Water Pollution (CPCB), New Delhi, COINDS/8/1980-81.
6. Jain, N.C. and Bajpai, A., 1989, Utilization of Filter mud – Present and Future Prospects, Indian Sugar, April.
7. Comprehensive Industry Document for Small Pulp and Paper Mill, prepared by Central Board for Prevention and Control of Water Pollution (CPCB), New Delhi, COINDS/22/1986.
8. Ray, S.S., Subramaniyan, L.R. and Ghogare, B.V., 1998, Treatment of Effluent in Smaller and Medium Capacity Pulp Mill (20-30 TPD) – Agro Pulping – A Case Study of Pravara Pulp and Paper Mill, IPPTA, Vol. 10 (1), March, pp. 29-34.
9. Comprehensive Industry Document for Fermentation Industries, prepared by Central Board for Prevention and Control of Water Pollution (CPCB), New Delhi, COINDS/1981-82.
10. Kaul, S.N., Pathe, P.P. and Nandy, T., 1990, An Overview of Waste Management in Sugar Industry, IAEM, Vol. 17, pp. 55-78.
11. Kaul, S.N., Nandy, T. and Trivedy, R.K., 1999, Book on Pollution Control in Distilleries, published by Enviro-media, Karad, p. 13.
12. Singh, S.P. and Rao, N.J., 1988, A Quick Estimation of Batch Digester Discharge, IPPTA, Vol. 25 (3), pp. 28-33.
13. Goyal, S.K., Seth, R. and Handa, B.K., 1996, Diphasic Fixed Film Bio-methanation of Distillery Spentwash, Bio-resource Technology, Vol. 56, pp. 239-244.

65

Glucose-6 Phosphatase: An Indicator of Pollution Stress for Sulphur Dioxide

V.I. PANDIT, SWATI RAWALGAONKAR & S. HASAN

National Environmental Engineering Research Institute, Nehru Marg, Nagpur - 440 020

ABSTRACT

The physiological and biochemical changes are known to occur in plants as a response to air pollution though the injury is not externally apparent. Photosynthesis, respiration as well as other metabolic processes are affected adversely due to stress conditions caused by air pollutants. Sulphur dioxide is the most important gaseous pollutant present in the ambient atmosphere which damages vegetation. The plant species of beans (Vigna catjang and Dolichos lablab), cotton (Gossypium hirsutum), tomato (Lycopersion esculentum) and wheat (Triticum aestivum) were exposed to different levels of sulphur dioxide. Wheat and cotton plants are sensitive to SO_2 while tomato and bean plants are resistant to SO_2. The plant leaves were analysed for sugar content and the enzyme activity glucose-6-phosphatase while photosynthetic rate and transpiration rates were measured *in situ* by portable photosynthesis system (LiCOR-600 model). The results showed significant reduction in photosynthesis as well as glucose-6-phosphatase enzyme activity. Transpiration, on the other hand, in all the plants increased while sugar content indicated mixed response. The decline in enzyme activity glucose-6-phosphatase was observed with different concentration of sulphur dioxide treatments in all the plants makes it a suitable indicator for sulphur dioxide pollution.

Keywords: *Pollution stress, Static exposure chamber, Bioindicator, Glucose-6 phosphatase, Biochemical alterations, Physiological responses.*

Introduction

Vegetation is the most important natural resource. Sulphur dioxide is one of the major air pollutants in many parts of the world (1). Its effect on vegetation have been well reviewed in terms of foliar injury, physiological and biochemical alternations (2,3). The lower pollution concentration cause injury to plants which are less severe and no loner externally apparent (4). In the absence of visible injury, the changes in the activity of various nzymes due to influence of pollution has been used as an indicator of pollution stress.

Experiments were carried out to study the effect of SO_2 on activity of enzyme glucese-6-phosphatase, transpiration rate, sugar content and photosynthesis in general on different plant species. An attempt was made to find out dose-response relationship.

Methods

To evaluate the responses of plants to a single pollutant demands exposure chamber studies. The studies were carried out using a simple static exposure chamber (Fig. 1). The plants selected include wheat (Triticum aestivum) and cotton (Gossypium hirsutum) sensitive to SO_2 while tomato (Lyopersicon esculenitum) and bean (Vigna catjang and Dolichos lablalb) resistant to SO_2 have been selected. The SO_2 was generated by bubbling air in an aqueous solution of metabisulphite (5). The experiments were carried out with four weeks old plants potted in earthen pots having soil mixture of manure: sand: soil (1: 1: 2). The plants were exposed to 0.25, 0.5 and 1.0 ppm SO_2 for 8 hrs daily while the control pants were not exposed to SO_2.

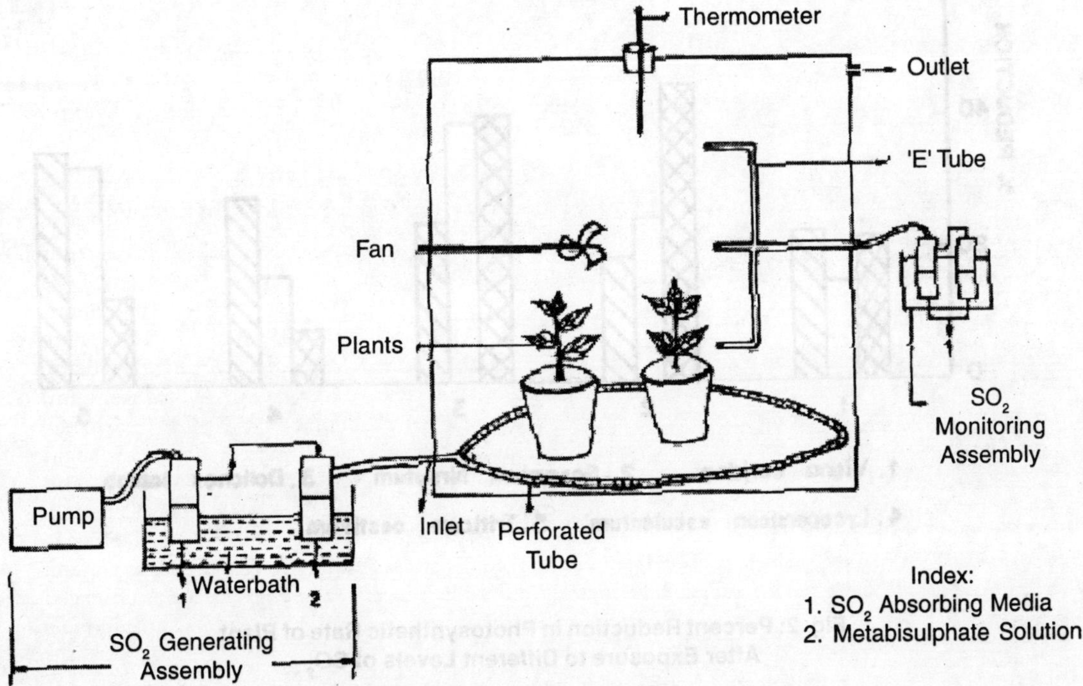

Fig. 1: Schematic Experimental Setup

The position of the potted plants were changed daily so as to avoid positional effect. The leaves of the control as well as exposed plant were harvested after 10 days and analysed for biochemical assay. The activity of enzyme blucose-6 phosphatase was estimated by method of Harper (6) and the sugar content by phenol-sulphuric acid method (7). The photosynthesis and transpiration rates were measured by use of portable photogymetesis system (Licor-6000).

Results

Bean, tomato and cotton plants exposed to 0.5 and 1.0 pm SO_2 exhibited foliar injury. The effect of different doses of SO_2 on photosynthesis and transpiration rate of control and treated plants are presented (Tables 1 & 2). The reduction in photosynthesis rate (8-45%) and increase in transpiration rate (5-50%) was the common response observed in all the treatments for different plants studied (Figures 2 & 3).

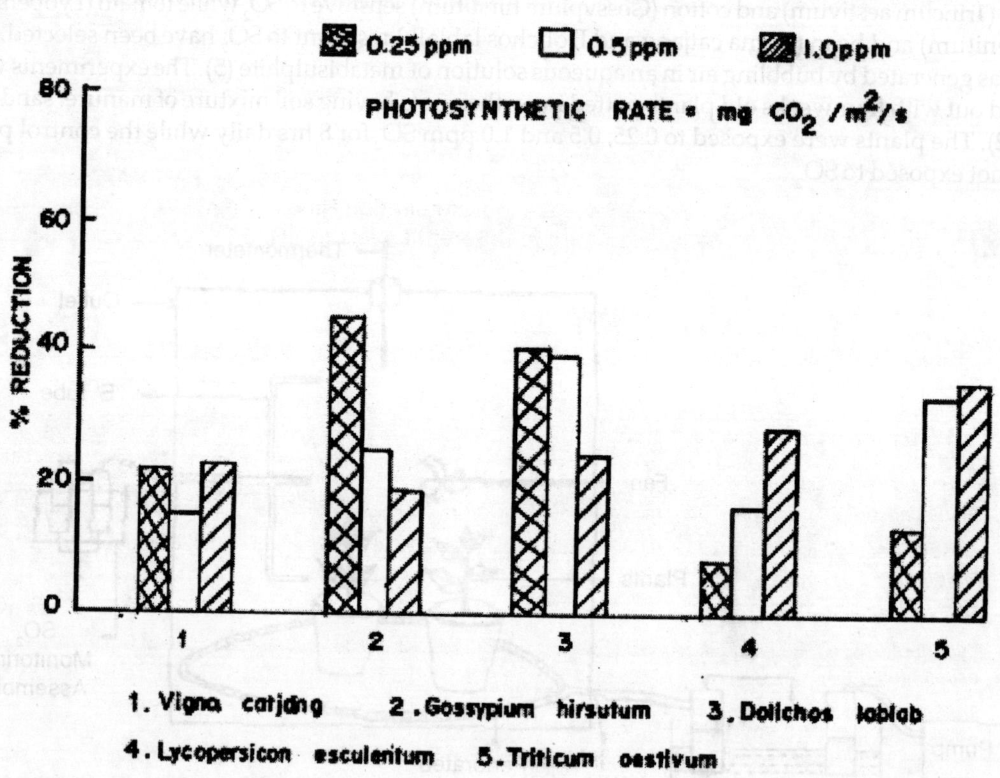

Fig. 2: Percent Reduction in Photosynthetic Rate of Plant After Exposure to Different Levels of SO_2

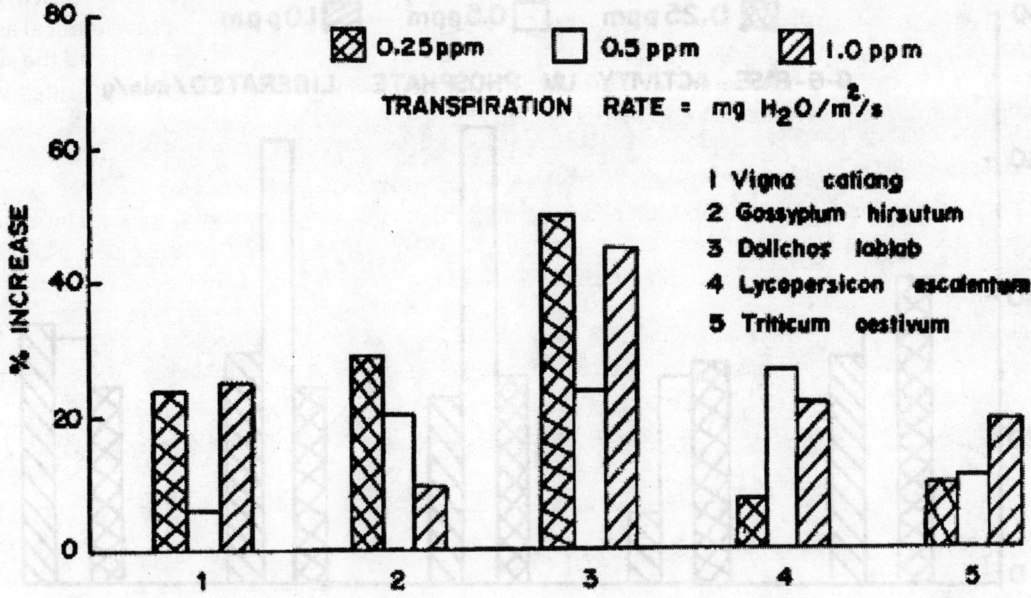

Fig. 3: Percent Increase in Transpiration Rate of Plant After Exposure to Different Levels of SO_2

Table 1: Effects of Different Levels of Sulphur Dioxide on Photosynthetic Rate of Plants

Sl.No.	*Plant Species*	*Photosynthesis Rate* (µg CO²/m²/s)*					
		Control	*Exposed to 0.25 ppm*	*Control*	*Exposed to 0.5 ppm*	*Control*	*Exposed to 1.0 ppm*
1.	*Vigna catjang*	1.2315±0.102	0.9565±0.44	0.3513±0.06	0.2981±0.14	0.3598±0.06	0.2749±0.04
2.	*Gossypium hirsutum*	1.746±0.39	0.9604±0.66	0.2809±0.04	0.2094±0.08	0.3268±0.14	0.2655±0.054
3.	*Dolichos lablab*	0.5697±0.47	0.3424±0.21	0.2933±0.03	0.1766±0.07	0.1834±0.032	0.1395±0.042
4.	*Lycopersicon esculentumum*	3.28±0.388	3.011±0.28	0.8124±0.13	0.6843±0.096	0.3150±0.047	0.2241±0.12
5.	*Triticum aestivum*	2.305±0.09	2.009±0.24	1.2692±0.27	0.937±0.32	1.483±0.32	0.9629±0.11

* Each value is mean of five observations.

Decrease of glucose-6-phosphatase activity was observed in all the treated plants when compared to control (Table 3). Over 60% reduction in enzyme activity was observed in Dolichos lablab and cotton, while in other plants reduction ranged 10-40% (Fig. 4). The sugar content increased with 0.25 ppm exposure while no definate pattern was observed for treatment with 0.5 and 1.0 ppm SO_2 (Fig. 5).

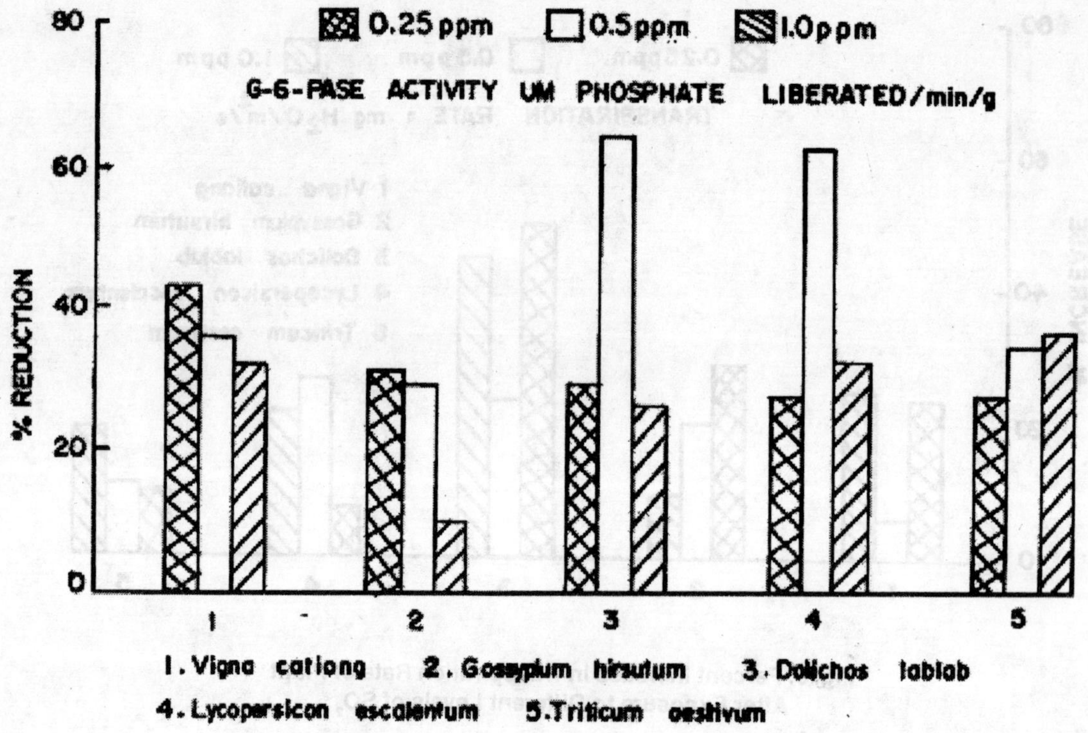

Fig. 4: Percent Reduction in Glucose-6-Photosynthetic Activity After Exposure to Different Levels of SO_2

Table 2: Effects of Different Levels of Sulphur Dioxide on Transpiration Rate of Plants

Sl.No.	Plant Species	Photosynthesis Rate* (μg $CO^2/m^2/s$)					
		Control	Exposed to 0.25 ppm	Control	Exposed to 0.5 ppm	Control	Exposed to 1.0 ppm
1.	*Vigna catjang*	373.5±45.66	464.15±94.20	534.27±32.65	563.26±45.74	512.14±40.31	642.8±78.91
2.	*Gossypium hirsutum*	459.96±109.79	593.3±205.75	489.37±34.36	587.73±61.7	636.31±217.85	695.7±48.98
3.	*Dolichos lablab*	201.85±11.79	302.9±31.04	217.13±48.35	392.05±72.55	262.37±54.22	382.45±15.80
4.	*Lycopersicon esculentumum*	929.15±37.8	995.9±86.23	923.32±53.62	1168.36±92.54	847.62±27.93	1029.52±405.25
5.	*Triticum aestivum*	1550.55±786.93	1694.1±75.36	978.29±62.79	856.27±79.26	598.26±81.49	706.36±42.76

* Each value is mean of five observations.

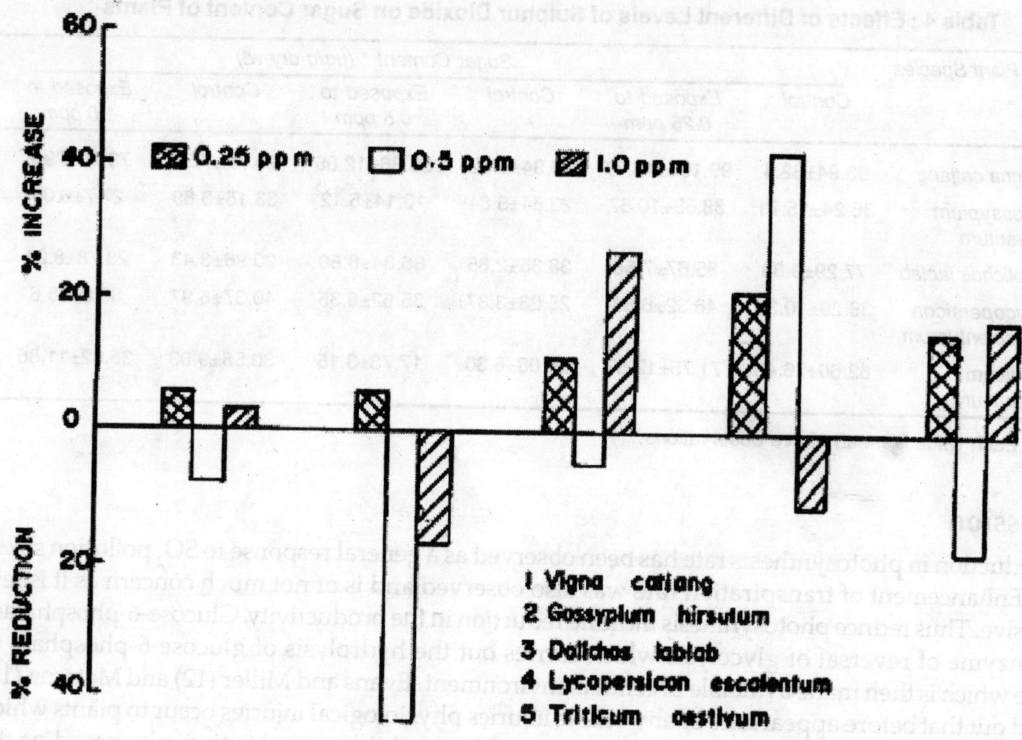

Fig. 5: Percent Change in Sugar Content of Plant After Exposure to Different Levels of SO_2

Table 3: Effects of Different Levels of Sulphur Dioxide on Enzyme Activity Glucose-6-phosphatase of Plants

Sl.No.	Plant Species	Glucose-6-Phosphatase* (µg phosphate liberated/min/g)					
		Control	Exposed to 0.25 ppm	Control	Exposed to 0.5 ppm	Control	Exposed to 1.0 ppm
1.	*Vigna catjang*	2.3942±0.33	1.3627±0.28	3.7595±1.6512	2.3826±1.42	0.92±0.34	0.626±0.04
2.	*Gossypium hirsutum*	1.2473±0.53	0.8624±0.15	1.964±0.77	1.3846±1.02	0.892±0.6	0.8±0.4
3.	*Dolichos lablab*	3.724±0.69	2.632±0.43	4.7895±3.17	1.7253±0.81	1.544±0.07	1.13±0.23
4.	*Lycopersicon esculentumum*	2.6536±0.37	1.926±0.21	6.342±3.91	2.376±0.68	1.372±0.07	0.927±0.27
5.	*Triticum aestivum*	3.2791±0.75	2.379±0.72	3.174±0.35	2.1±0.82	1.696±0.11	1.085±0.04

* Each value is mean of five observations

Table 4 : Effects of Different Levels of Sulphur Dioxide on Sugar Content of Plants

Sl.No.	Plant Species	Sugar Content * (μg/g dry wt)					
		Control	Exposed to 0.25 ppm	Control	Exposed to 0.5 ppm	Control	Exposed to 1.0 ppm
1.	*Vigna catjang*	93.94±32.4	99.105±13.77	36.94±6.62	33.86±12.08	70.78±7.38	73.10±29.2
2.	*Gossypium hirsutum*	36.24±15.71	38.53±10.67	23.64±5.84	13.14±5.12	33.15±3.89	27.7±6.08
3.	*Dolichos lablab*	77.29±9.83	85.67±7.58	38.35±2.65	36.34±8.80	20.96±3.43	28.78±8.24
4.	*Lycopersicon esculentumum*	38.29±10.38	46.32±6.53	25.23±3.37	35.62±6.35	40.37±6.97	35.69±5.6
5.	Triticum aestivum	62.60±26.46	71.75±12.57	23.08±6.30	17.73±3.15	30.56±9.03	35.82±11.86

* Each value is mean of five observations.

Discussion

Reduction in photosynthesis rate has been observed as a general response to SO_2 pollution stress (8-11). Enhancement of transpiration rate was also observed and is of not much concern as it is less responsive. Thus reduce photosynthesis indicate reduction in the productivity. Glucose-6-phosphatase is an enzyme of reversal of glycolysis which carries out the hydrolysis of glucose-6-phosphate to glucose which is then made available in cellular environment. Evans and Miller (12) and Materna (13) pointed out that before appearance of any visible injuries physiological injuries occur to plants which can be used as an indicator of pollution levels. Rabe and Kreeb (14) expresed their views regarding the formulation of physiological or biochemical indicator. The best physiological indicator, according to them, is the one with strongest reaction. They further suggested that for the experiments exposing one species of plant is not enough but comparison of two or more plants exposed simultaneously gives a better picture and this is better if these plants differ in their resistance as studied in present experiments belongs to sensitive class *viz.* wheat and beans as well as relatively resistant species *viz.* cotton and tomato. In all the analysis mentioned a comparison with standard is indispensable. This standard is provided by control plants which are exposed under identical conditions except pollutant supply. It has been reported that in stress condition sugars accumulate as a response to increasing pollution load (15). In the present studies the same was confirmed for 0.25 ppm SO_2 exposure while the depletion in sugars for some plants may be taken as an adjustment towards the tolerance mechanism shown by the plant to combat the stress situation.

Conclusions

In all exposure studies the reduction in photosynthesis rate and increase in the transpiration rate was observed as the common response.

The activity of glucose-6-phosphatase significantly reduced in comparison with the respective control plants. Hence, there is a clear connection between level of pollution the lowering of this enzyme activity may be considered as an indicator of SO_2 pollution. However, further research is desirable for lower exposure concentration normally recorded in polluted urban atmosphere. The lowering of glucose-6-phosphatase activity in all the treatments to different plant species both sensitive and resistant studied make it a suitable indicator of pollution stress.

In all treatments the sugar content increased with 0.25 ppm SO_2 exposure but no defiant pattern was observed for 0.5 and 1.00 ppm SO_2 treatment. These results confirm that at lower exposure the sugars accumulate as a response of pollution load.

Acknowledgement

The authors wish to express sincere thanks to Dr. R.N. Singh, Director, NEERI for kind permission to publish this paper.

REFERENCES

1. Katainen HS, Makinen E Jokinen J, Karjalainen R, Kellomaki S, 1987, Effects of SO_2 on the photosynthetic and respiration rates in Scots pine seelings Environ Pollut., 46; 241-51.
2. Singh SN, Yanus M, Srivastava K, Kulshreshtha K, KJ Ahmed, 1985, Response of Calendula officinalis L. to long term fumigation with SO_2. Environm. Pollut. (Ser A), 39; 17-25.
3. Ewald D, Schlee D, 1983, Biochemical effects of sulphur dioxide on proline metabolism in alga Trebouxiz sp. New Phytol., 94; 235-40.
4. Farooq Mohammed, Akbar Masood, MU Beg, 1985, Effect of acute exposure of SO2 on the metabolism of Holoptelea integrifolia. Environ. Pollt. (Ser A), 39; 197-205.
5. Sharma HC, Thakre RA, 1984, Generation and monitoring of gaseous pollutants. Ind. J.Air Pollut. Contr., 5; 24-27.
6. Harper AE, 1963, Glucose-6-phosphatase in Methods of Enzymatic Analysis Ed by H.U. Bergmeyer. Verlag Chemie, GMBH Weinheim / Bergstr. Academic Press. New York and London, pp 788-792.
7. Plummer DT, 1971, An Introduction to Biochemistry McGraw Hill Book Co., England, p. 223.
8. Olszyk DM, Bytenerowicz A, Fox CA, Kats G, Dawson PJ, Wolf J, 1987, Injury and physiological responses Larrea tridentala (DC) (Coville) exposed in situ to SO_2 Environ. Pollut., 48, 197-211.
9. Lamoraux RJ, Chang WR, 1978, Phosynthesis and transpiration of excised silver maple leaves exposed to cadmium and SO_2. Environ. Pollut., 17, 257-68.
10. Bennet JH, AC Hill, 1974, Acute inhibition apparent photosynthesis by phytotoxic air pollutants. Am. Chem. Soc. Symp., Series No. 3; 115-27.
11. Tanaka *et al.*, 1974, Experimental studies on SO_2 injuries in higher plants. III. Inhibitory effect of sulphite ion on $14CO_2$ fixation. Water, Air, Soil Pollut., 3 (1) ; 11-16.
12. Evans LS, Miller PR, 1972, Ozone damage to pondersoa pine: A histochemical appraisal. Ann.J. Bot., 59; 297-304.
13. Materna J, 1970, EinfluB niedriger Schwefeldioxidk onzentrazionen auf die Fichte. In Wirkungen von Luftveruntereinigungen auf Waldbaume, VII. Int. Tagung Forstl. Rauchschadensachverstandiger Essen., 7-11.
14. Rabe R, Kreeb KH, 1979, Enzyme activities and chlorophyll and protein content in plants as indicators of air pollution. Environ. Pollut., 19, 119-37.
15. Jager HJ, Bender J, Grunhage L, 1985, Metabolic responses of plants differing in SO_2 sensitivity towards SO_2 fumigation. Environ. Pollut. (Ser A), 1985; 39, 317-35.

66

Leachate Generation and Control from Municipal Solid Waste Disposal Sites: Indian Scenario

S.A. GAIKWAD, A.D. PATIL & A.V. SHEKDAR

Solid Waste Management Division, National Environmental Engineering Research Institute, Nehru Marg, Nagpur - 20

ABSTRACT

Open dumping of municipal solid waste is practised in most of the urban centres in India. The landfill sites are not provided with liner and cover material. As a result, the rain water percolates through the landfill and generates the leachate containing higher concentration of COD, BOD, TDS, heavy metals, etc. The leachate pollutes the groundwater sources affecting the public health of adjoining population. The leachate quantity is estimated using the water balance approach for national level. In order to minimise pollution of ground water, the corrective measures need to be provided by installing borewells and pumping out leachate either for recirculation or collection and treatment. The case study of Delhi along with the leachate pollution impact is also described. The procedure for control of leachate for developing new landfill site has been described along with standards for disposal of treated leachate suggested by Ministry of Environment and Forests, Government of India, Delhi.

Keywords: Leachate; Municipal solid waste; Open dumps; Recirculation; Regulations; India.

Introduction

In India, there are nearly 915 class I and class II cities and towns (census 2001) having population around 210 millions and generates solid waste quantity approximately 38.35 million tonnes/year. The waste is commonly disposed on land by open dumping mainly at low-lying areas in and around the cities. For every city there are multiple disposal sites which are operated for 1-20 years. In these open dumps, water is entering through various sources like rains, water contained in the waste and water formed due to the degradation of solid waste. During the degradation process the organic fraction gets disintegrated and leachate is formed which is highly concentrated with various biological and chemical pollutants. As the open dumps are not lined, the leachate finds it's way to the adjoining environments like groundwater, soil, etc. In this paper an attempt has been made to assess this problem and provide probable measures to monitor and control the pollution in order to maintain the public health standards of adjoining population and to prevent degradation of quality of the environment.

Adverse Effects on Public Health Due to Leachate from Solid Waste Dumping Sites

The solid waste consisting of wastes from various sources like residential localities, markets, business, houses, commercial and industrial establishments, contains 40-50% biodegradable fraction and disposed of at low-lying areas known as open dump sites. The silt from open drains and many a times night soil due to open defecation is also transported and dumped in the disposal sites. As cover is not provided, the rain water and moisture in the solid waste percolates through the waste mass, carrying biological and chemical contaminants. During degradation process leachates are produced. When the water holding capacity of waste exceeds field capacity, leachate is generated and finds its way to ground water and contaminate it, as shown in Figure 1. Part of the rainwater which do not percolate through solid waste is evaporated at surface and part of it flows over the waste and known as surface run-off. The pollutants are carried away by the surface water. Mainly the population surrounding the open dumps using water from these sources gets affected which poses threat to human health (1).

Estimation of Leachate from Municipal Landfills

Various methods are available for estimation of quantity of leachate from landfills and most use the water balance approach. The sketch shown in Figure 2 indicates water balance involving summing the amounts of water entering the landfill and subtracting the amounts of water consumed in chemical reactions and quantity leaving as water vapour as well as surface run-off (2).

Tchobanoglous *et al.*, suggested method for generation of leachate by a water balance method, assuming unit area of a landfill site. In this methodology a simple approach such as difference in water entering in the unit area and water lost from the same area is considered (3). The water entering components are moisture in solid waste, moisture in sludge if added, precipitation, moisture in cover material if provided. The factors related to water lost are; water used in the formation of landfill gas, saturated water vapour associated with landfill gas, surface evaporation and water leaving from the bottom of the layer.

Some of the factors such as cover moisture and recovery of landfill gas are not applicable in the prevailing condition in India.

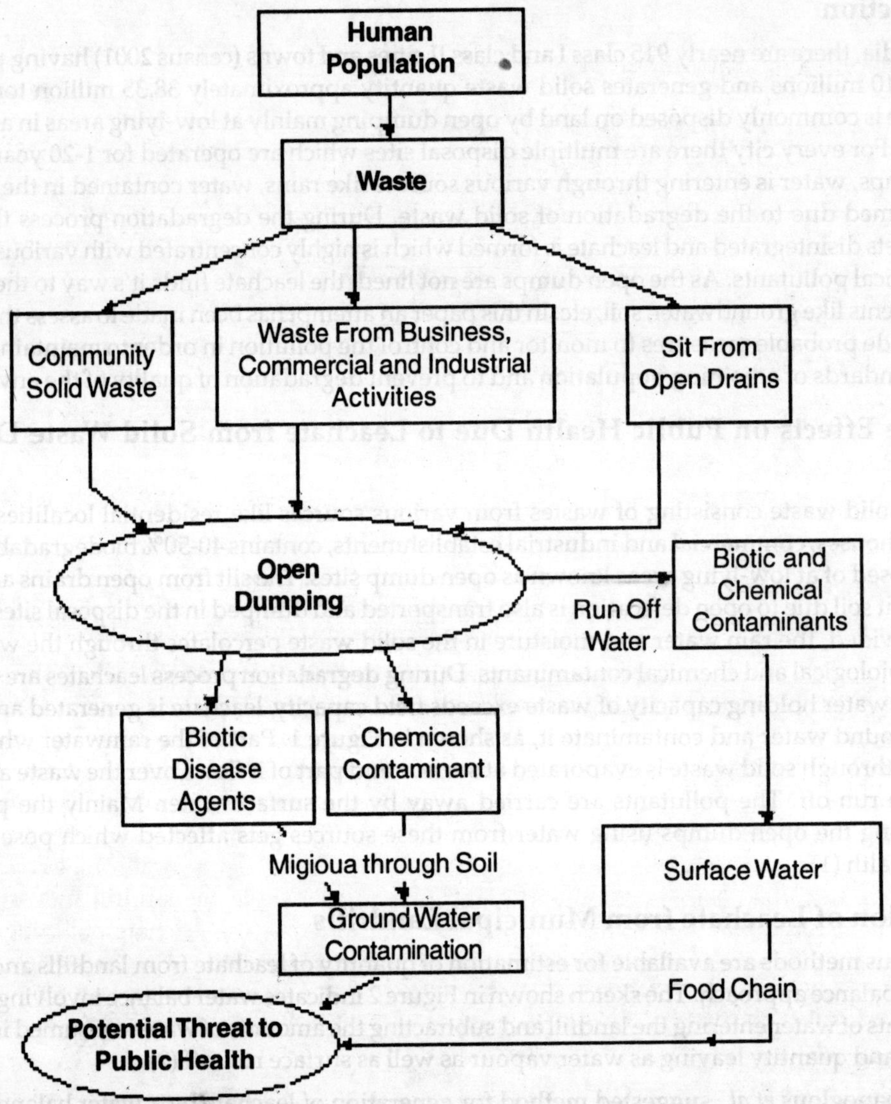

Fig. 1: Adverse Effects on Public Health due to Leachate from Solid Waste Dumping Sites

In Indian condition, the leachate quantity could be estimated by using the simplified water balance approach. In this approach, the water entering the waste mass is the maximum rainfall for a month corrected by using the coefficient of run-off (assumed value as 0.5) and evaporation losses for the month. This information is readily available for all the cities with the meteorological department. The modified approach is presented below:

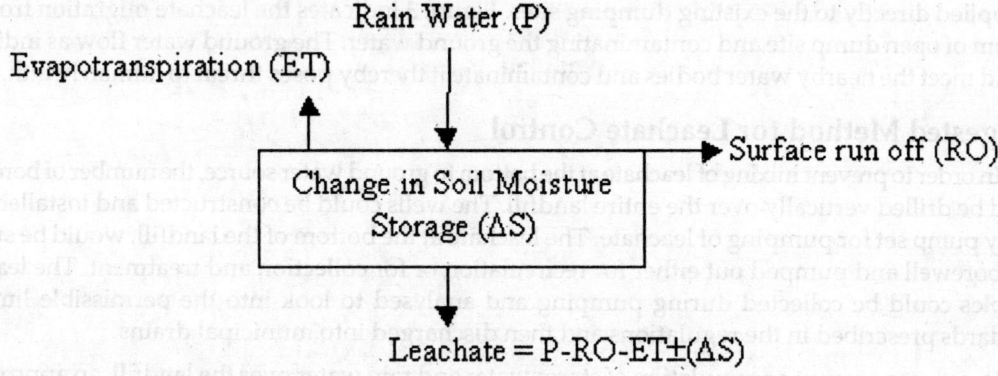

Fig. 2: Water Balance Approach for Calculation of Leachate Quantity

Leachate flow (mm/month) = Maximum rainfall (mm/month) × (Coefficient of run-off) —Evaporation losses (mm/month)

Assuming coefficient of run-off = 0.5

Quantity of leachate for a particular landfill = Surface area (m^2) × leachate flow (m/sec) (m^3/sec)

Based on above approach the leachate quantity is estimated for national level. The solid waste quantity for 915 class I and class II cities having approximate population 210 million (4) is estimated to generate 38.32 millions tonnes/year @0.5 kg/capita/day and would require 1277 hectare land/year for disposal (assuming depth of fill 3 m and in situ density 1000 kg/m^3). Assuming maximum rainfall of 400 mm and average evaporation losses of 150 mm at national level and coefficient of run-off as 0.5, the estimated leachate quantity would be around 7.7 million cum/year. The standard BOD_5 level of 250 mg/l for municipal sewage as suggested by Ministry of Environment and Forests, Government of India, Delhi is compared with BOD of leachate in the landfill sites indicates the pollution level on higher side in the landfills. In order to assess the leachate pollution impact on nearby ground water sources, monitoring of ground water is essential.

Leachate Generation and Control

Developed Countries

In developed countries, various methods are adopted for collection and treatment of leachate to preserve the environmental quality. Also the density of solid waste is lower than the municipal solid waste in Indian condition. Clay and geomembrane liners are provided at bottom and sides of landfill. The bottom of landfill is laid with perforated leachate collection piping, storage sump and treatment facilities. The top surface is provided with cover of composite liners of clay material and geomembrane as a result infiltration of rain water to waste mass is prevented. The appropriate gradient to top surface is provided to prevent accumulation of storm water or rain water. Thus, the leachate generation is minimised.

Leachate Generation and Control in India

In Indian condition, the open dumping of solid waste is practised for most of the centres. These are managed by municipal agencies. The control measures as followed in developed countries cannot be applied directly to the existing dumping sites. Figure 3 indicates the leachate migration from the bottom of open dump site and contaminating the ground water. The ground water flow as indicated would meet the nearby water bodies and contaminate it thereby poses threat to human health.

Suggested Method for Leachate Control

In order to prevent mixing of leachate at the bottom to ground water source, the number of borewells could be drilled vertically over the entire landfill. The wells could be constructed and installed with slurry pump set for pumping of leachate. The leachate at the bottom of the landfill, would be sucked into borewell and pumped out either for recirculation or for collection and treatment. The leachate samples could be collected during pumping and analysed to look into the permissible limits of standards prescribed in the regulations and then discharged into municipal drains.

In order to prevent accumulation of storm water and rain water over the landfill, an appropriate peripheral drain should be provided.

For designing of new disposal site, the bottom and sides of landfill should be provided with liners alongwith leachate collection and treatment facility so that the leachate generated would have minimum impact on water sources.

Studies Carried Out for Leachate Collection in Delhi

Solid waste management at Municipal Corporation of Delhi with a population of 11.48 million (1996-97) generated 4975 tonnes of solid waste per day. The waste disposal was carried out at three disposal sites namely Gazipur, Bhalswa and Okhla sites with an area of 28.2 ha, 16.2 ha and 4.05 ha respectively. In order to assess the leachate pollution impact, the borewells were drilled at disposal site. The leachate samples were collected and analysed (5), the results are shown in Table 1. Based on the water balance approach the leachate generated at Gazipur, Bhalswa and Okhla were estimated as 42.48 m^3/hr, 24.94 m^3/hr and 5.68 m^3/hr respectively.

Table 1: Characteristics of Leachate from Municipal Solid Waste Disposal Sites at Delhi

Sl. No.	*Parameters*	*Values*
1.	pH	6.80
2.	Conductivity, mS/cm at 20°C	2.80
3.	Total Dissolved Solids	3400.00
4.	Total alkalinity as $CaCO_3$	2100.00
5.	Bicarbonate alkalinity as $CaCO_3$	1050.00
6.	Phosphorous as P	6.50
7.	Chloride	1100.00
8.	Sulphate	100.00
9.	Volatile Fatty Acids	2050.00
10.	COD	3100.00
11.	Total Organic Carbon	880.00
12.	BOD_5	2200.00
13.	Organic nitrogen	120.00
14.	Ammoniacal nitrogen	82.00

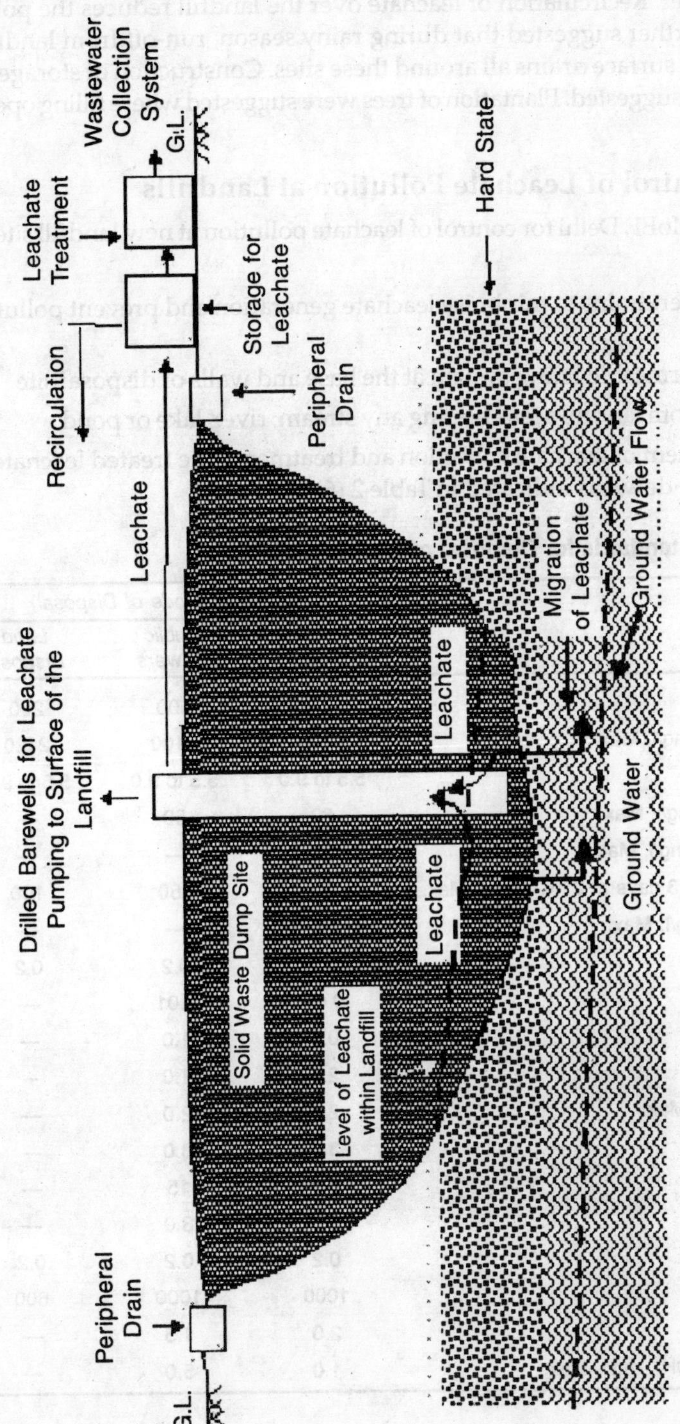

Fig. 3: Sketch Indicating Leachate Pumping Wells at Open Dump Site

It is apparent from the table that the leachate is highly polluted and warrants treatment before discharge into the receiving water. Recirculation of leachate over the landfill reduces the pollution potential substantially. It was further suggested that during rainy season, run-off from landfill site should be removed by providing surface drains all around these sites. Construction of storage sump and pumping out the run-off was suggested. Plantation of trees were suggested where filling operation was completed.

MoEF Regulations for Control of Leachate Pollution at Landfills

Regulations are notified by MoEF, Delhi for control of leachate pollution at new landfill sites. The suggested recommendations are:

1. Diversion of storm water drains to minimise leachate generation and prevent pollution of surface water
2. Construction of non-permeable lining system at the base and walls of disposal site
3. Prevention of run-off from landfill area entering any stream, river, lake or pond
4. Provision for management of leachate collection and treatment. The treated leachate shall meet the standards laid down as indicated in Table 2 (6).

Table 2: Standards for Disposal of Treated Leachates

Sl. No.	*Parameter*	*Standards (Mode of Disposal)*		
		Inland Surface Water	*Public Sewers*	*Land Disposal*
1.	Suspended solids, mg/l, Max.	100	600	200
2.	Dissolved solids (inorganic), mg/l, Max.	2100	2100	2100
3.	pH value	5.5 to 9.0	5.5 to 9.0	5.5 to 9.0
4.	Ammoniacal nitrogen (as N), mg/l, Max.	50	50	—
5.	Total Kjeldahl nitrogen (as N), mg/l, Max.	100	—	—
6.	Biochemical Oxygen Demand (3 days at 27°C), mg/l, Max.	30	350	100
7.	Chemical Oxygen Demand, mg/l, Max.	250	—	—
8.	Arsenic (as As), mg/l, Max.	0.2	0.2	0.2
9.	Mercury (as Hg), mg/l, Max.	0.01	0.01	—
10.	Lead (as Pb), mg/l, Max.	0.1	1.0	—
11.	Cadmium (as Cd), mg/l, Max.	2.0	1.0	—
12.	Total Chromium (as Cr), mg/l, Max.	2.0	2.0	—
13.	Copper (as Cu), mg/l, Max.	3.0	3.0	—
14.	Zinc (as Zn), mg/l, Max.	5.0	15	—
15.	Nickel (as Ni), mg/l, Max.	3.0	3.0	—
16.	Cyanide (as CN), mg/l, Max.	0.2	0.2	0.2
17.	Chloride (as Cl), mg/l, Max.	1000	1000	600
18.	Fluoride (as F), mg/l, Max.	2.0	1.5	—
19.	Phenolic compounds (as C_6H_5OH), mg/l, Max.	1.0	5.0	—

Conclusion

The leachate production in the open dump sites of municipal solid waste in India is a threat to the human health in the nearby residential areas. The leachate quantity from open dumps at national level has been estimated using water balance approach. The extent of pollution has been described through case studies of open dumps at Delhi. In order to control and monitor leachate pollution at open dump sites, some corrective measures are suggested such as drilling bore wells at landfills and pumping of leachate, recirculation, provision of leachate storage and treatment, provision of surface run-off drain all around the landfill. The standards for treated leachate prescribed by MoEF, New Delhi are recommended.

REFERENCES

1. NEERI Report, 2001, "Development of Solid Waste Management System at Port Blair"
2. NEERI Report, 1996, "Manual on Solid Waste Management".
3. Tehobanoglous *et al.*, 1993, "Integrated Solid Waste Management", McGraw Hill.
4. Provisional Population of class I and class II cities, Census of India, 2000.
5. NEERI Report, 1997, Phase II, "Solid Waste Management in MCD Area".
6. Ministry of Environment and Forests, New Delhi, Notification on "Municipal Solid Wastes (Management and Handling) Rules, 2000.

67

Rejuvenation of Clogged Bore Wells to Improve their Water Yield

NAVIN CHANDRA, S.S. AMRITPHALE, SANJEEV BHASIN & NITIN AGNIHOTRI

Regional research laboratory, Habib Ganj Naka, Hoshangabad Road, Bhopal - 462 026

ABSTRACT

The deposition of scale forming inorganic phases in the apertures carrying water from the main water body to the bottom of the bore wells with passage of time leads to a decrease in their water yield. The success rate of simple hydrofracturing process for their rejuvenation is observed to be limited. To achieve higher efficacy and better efficiency, Chemical Hydrofracturing (CHF) involving treatment of the well with a non-toxic chemical composition (to selectively dissolve the clogging materials partially and loosening the remaining clayee materials in the aperture carrying water from the main water body to the bottom of the well) followed by injection of large quantity of high pressure water in the sealed well has been developed in our laboratory. The results of the field trial of the CHF process are discussed in the present paper.

Introduction

Ground water is used extensively in India to meet the demands of irrigation and drinking water. A large number of bore wells have been sunk to extract the water from the ground. Depending upon the water yield, a hand pump or a motor driven pump is fitted in the bore wells. It is observed that with passage of time, in majority of cases the water out put from the bore wells decreases even though the water table is present at the site. The decrease in the water yield is attributed to the clogging or deposition of 'scales' in the apertures, which carry water from the main water body to the bottom of the well. The clogging causes a reduction in permeability (1) and, in extreme cases, may lead to complete drying up of the well. The deposits comprise of phases of water soluble inorganic compounds, which precipitate due to changes in temperature/pressure in apertures (2-8) in association with silicious, clayee materials. The nature and extent of precipitation of these compounds largely depends on the ground water chemistry (9). To overcome the problem of clogging in the bore wells and improve their water yield, a number of methods have been reported in literature. These include blasting of the bore wells using explosives or injection of large quantity of high pressure water in the sealed well (Hydrofracturing). These methods are purely mechanical in nature and thus have limited efficacy in terms of their success rate and efficiency in improving the water yield. To improve efficacy and efficiency of rejuvenation of clogged bore wells, a process has been developed in our laboratory wherein the clogged bore wells are treated with a suitable non toxic chemical composition (evolved by us) followed by sealing of the well and introduction of large quantity of high pressure water. The combination of chemical treatment followed by hydrofracturing of the bore well has been found to be highly effective in rejuvenating the clogged bore wells. The results of the field trials of the process developed by our laboratory in a number of locations in Madhya Pradesh are presented in this paper.

Experimental

The water samples collected from the bore wells, suspected to be suffering from clogging problem were analyzed for their Calcium, Magnesium, Sodium, Potassium, Chloride and iron content following standard methods. The content of total dissolved solids (TDS) was determined by evaporation to dryness. The phases present in the water soluble clogging materials were identified using X-ray diffraction spectrometer (Philips make, Model No. 1710) using Cu-KA radiations. The 'd' values of the experimentally observed peaks were compared with standard 'd' values of the possible phases reported in the JCPDS X-ray diffraction data files (10).

For Field Trials: A Truck Laden Hydrofracturing machine, having facility for pneumatic sealing of the well with the help of rubber packers, High delivery high pressure pump and hydraulically operated facility for removal of pipes, rods, cylinder/motor and pump etc. and their reassembly after Hydro-fracturing operation was used in the present work.

Before carrying out the Chemical Hydrofracturing operation, the water yield per minute of the bore well was measured following standard practice. After removing the metallic parts from the bore well (except casing), the required quantity of non-toxic chemical composition was poured in to the well and allowed to react. After elapse of the reaction time of (4 – 24) hours, the bore well was sealed by inflating the rubber packers of suitable diameter inside the well and then about 2000 liters of water was injected into the well using high pressure pump to flush out the already loosened clogging materials from the apertures. After removing the packer, the well was washed thoroughly. The water yield and draw down were measured again to evaluate the efficacy and efficiency of the chemical hydrofracturing operation for rejuvenation of the well.

Results and Discussion

The results of the field trial of non-toxic chemical hydrofracturing process (CHF) for the clogged borewells located in Matkuli and Pipariya Kala villages in Hoshangabad district, Jaithal and Ghatia villages in Ujjain district and Sulibadi and Padalya villages in Dhar district of Madhya Pradesh are reported in this paper. To diagnose the reasons for low water yield from the bore wells, water samples were collected and analyzed for their chemical constituents. The pH and chemical analysis of the collected water samples is given in Table 1. It is observed that the total dissolved solids are in the range of 250 to 760 mg/l. The Ca + Mg content of the water is observed to be high and iron content is observed to be low. The high values of TDS and Ca + Mg content and low water yield from the bore well, in spite of the presence of water table, indicated that the bore wells may be suffering from the clogging problem.

Table 1: Results of Chemical Analysis of Water Samples Collected from the Clogged Bore Wells before Chemical Hydro-fracturing

Name of Village	*Description of Sample*							
	T.D.S. mg/l	*pH*	*Ca (mg/L)*	*Mg (mg/L)*	*Na (mg/L)*	*K (mg/L)*	*Cl (mg/L)*	*Fe (mg/L)*
Jaithal	758	7.2	142.6	131.3	238	51	120	0.14
Ghattia	254	7	72.28	62.016	55.5	20	40	0.02
Sulibadi	305	6.5	22.088	17.024	270	29	90	0.1
Padalya	250	7.0	84.336	75.392	42.5	15	80	2.8
Matkuli (Near Inder's House)	315	7	65.7	34.8	450	0.6	64.3	0.06
Pipariya Kala (Near Pandit's house)	622	7.4	69.07	66.15	138	1.3	251.6	0.57

Table 2: Co-relation Between Water Improvement Index to Characteristics of Soil-mineral Strata

Sl. No	*Name of Village*	*Before L/min*	*After L/min*	*Mineral Phases*		
				Easy to Clean	*Medium to Clean*	*Hard to clean*
1.	Jaithal	4	15	1. Calcium Magnesium Aluminum Iron Silicate Hydroxide	1. Calcium Silicate Hydrate (L.T) 2. Sodium Magnesium Silicate Hydrate	
2.	Ghattia	3	8	1. Sodium magnesium aluminium silicate hydroxide	1. Calcium Sodium Potassium Aluminum Silicate Hydrate (Jusite) 2. Calcium Silicate Hydrate (L.T)	1.Sodium Magnesium aluminium oxide
3.	Sulibaydi	2	6		1. Calcium Sodium Potassium Aluminum Silicate Hydrate (Jusite) 2. Calcium Silicate Hydrate (L.T) 3. Potassium Aluminium Silicate Hydrate	1. Calcium iron oxide 2. Potassium iron oxide
4.	Padlya	3	10	1. Calcium Magnesium Aluminum Iron Silicate Hydroxide 2. Sodium Iron silicate Hydroxide Hydrate	1. Calcium Sodium Potassium Aluminum Silicate Hydrate 2. Calcium Silicate Hydrate (L.T)	1. Sodium Magnesium aluminium oxide

Contd...

Table 2–Contd...

Sl. No	Name of Village	Before L/min	After L/min	Mineral Phases: Easy to Clean	Medium to Clean	Hard to clean
				3. Calcium Magnesium Aluminum Silicate Hydroxide Hydrate	3. Potassium Aluminum Silicate Hydrate 4. Zinc Magnesium Silicate Hydrate	
5.	Matkuli (Inder Singh's House)	1	6	1. Sodium Magnesium Aluminium Silicate Hydroxide	1. Calcium Sodium Potassium Aluminum Silicate Hydrate 2. Calcium Silicate Hydrate 3. Potassium Aluminum Silicate Hydrate	
6.	Pipariya Kala (Near Pandit's House)	1	12		1. Calcium Sodium Potassium Aluminum Silicate Hydrate 2. Calcium potassium orthosilicate	

To identify the nature of the clogging material, the water sample from respective bore well was evaporated to dryness under slow heating. The X-ray diffractograms for the residue materials are given in Fig.1. The phases identified are observed to generally fall in hydroxide or hydrated mineral categories. In the case of two bore wells, even oxide phases have been observed to be present. The phases have been categorized in easy, medium and hard to clean categories. In the following discussion, the efficiency of rejuvenation of clogged bore wells has been correlated with the nature of the identified phases.

The various steps involved in the Chemical Hydrofracturing of clogged bore wells are (i) removal of metallic parts from the hand pump/bore well (Fig. 2), (ii) pouring of non-toxic chemical composition (Fig. 3), (iii) retention of chemical for specified period for selective dissolution of the clogging material phases (Fig. 4), (iv) sealing of the bore well using inflatable packers (Fig. 5), (v) injection of large – volume, high pressure water in the sealed bore well for hydrofracturing (Fig. 6), (vi) back washing of the well with water after removing the packer (Fig. 7), and (vii) measurement of water yield after assembling the hand pump/tube well assembly (Fig. 8 and 9).

The improvement in water yield from the six bore wells reported in the present study can be seen by comparing the water yield before and after CHF operation (c.f. Fig. 10). It is observed that improvement in the water yield varies from about 3 to 12 times. The water yield from a bore well/hand pump depends on the interplay of a number of complex parameters (11) such as nature of clogging material and extent of clogging, length up to which the clogging has taken place in the aperture, depth of water table from the surface of earth, difference in height between water Table and the position of cylinder of the bore well and the rate of charging (which depends on the diameter of aperture (s) which carry water from main body to the bottom of the well). Keeping these factors and their complex inter play in the mind, the efficiency of rejuvenation of the bore wells can now be analyzed. The water yield before CHF was observed to be between 1-4 L/min. for the six bore wells investigated. After CHF, it is observed to be between 6-15 L./min. In the case of Jaithal, Padlya and Pipariya Kala villages, the water

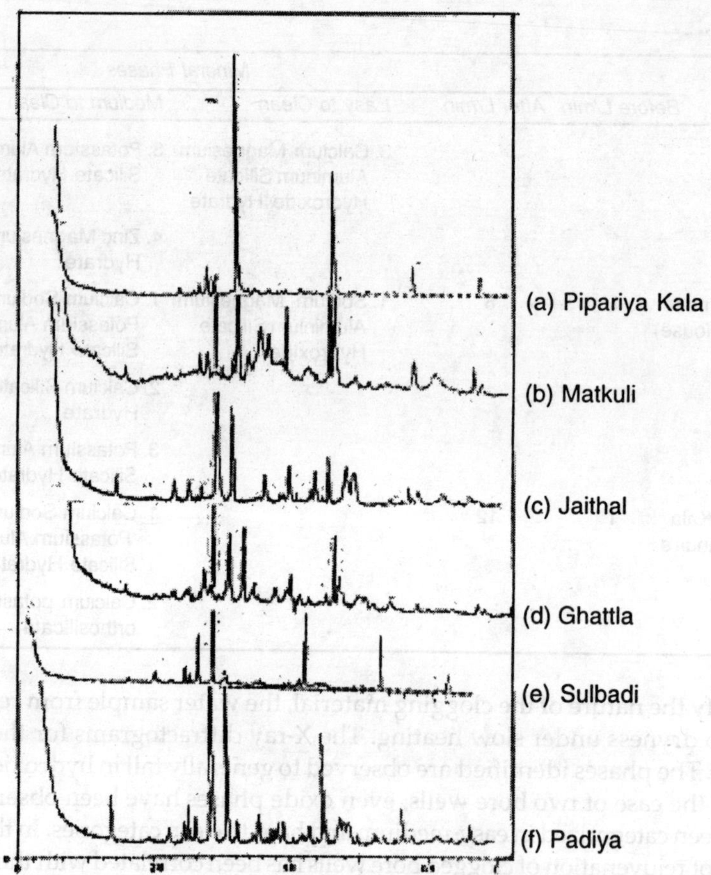

Fig. 1: X-Ray Diffraction Pattern of Clogging Material Prepared Using Water Samples Obtained Before Chemical Hydrofracturing Collected from Hoshangabad, Ujjain and Dhar Districts

yield after the CHF is 15, 10 and 12 L/min. respectively. The high water yield shows the efficiency of CHF for rejuvenation and is attributed to the presence of easy and/or medium to clean phases in the clogging material. In the case of Padalya, the presence of a difficult to clean phase leads to a final water yield of 10 L/Min. only. It is notable that the TDS values in the water of these villages were also high (758, 250 and 622 respectively) leading to severe clogging problems. Hence, due to efficient cleaning, the water yield after CHF became good. In the cases of Ghattia, Sulibadi and matkuli villages, the water yields after the CHF is observed to be 8, 6 and 6 L/Min. respectively. In the case of Ghatia and Sulibadi the presence of a difficult to clean phase and also in Sulibadi, the absence of easy to clean phase in the respective clogging materials may be responsible for less than 10 L/min. water yield after CHF. In the case of Matkuli, however, hard to clean phase (s) are not present and instead, an easy to clean phase is observed to be present. Still, the water yield after the CHF is only 6 L/min. A deeper probe in this apparent anomalous behavior showed that the difference between the water table and position of cylinder in the hand pump was only 20 feet as against 40 – 50 feet, commonly encountered. Thus it is observed that CHF is quite successful in reviving and rejuvenating the clogged bore wells. Further, the water yield after CHF improves substantially if easy to clean phases is present in the clogging materials.

Fig. 2: Removal of Metallic Parts from the Hand Pump/Bore Well

Fig. 3: Pouring of Non-toxic Chemical Composition

Fig. 4: Retention of Chemical for Specified Period for Selective Dissolution of the Clogging Material Phases

Fig. 5: Sealing of the Bore Well Using Inflatable Packers

Fig. 6: Injection of Large-volume, High Pressure Water in the Sealed Bore Well for Hydrofracturing

Fig. 7: Back Washing of the Well with Water after Removing the Packer

Fig. 8: Measurement of the Water Yield of the Hand Pump After Non-toxic Chemical Operation

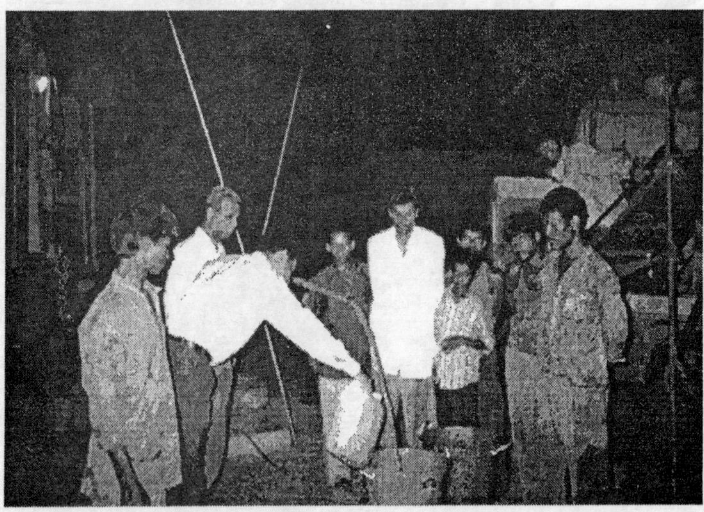

Fig. 9: Measurement of Water Yield of the Bore Well Fitted with Electrical Motor Pump After the Non-toxic Chemical Hydrofracturing

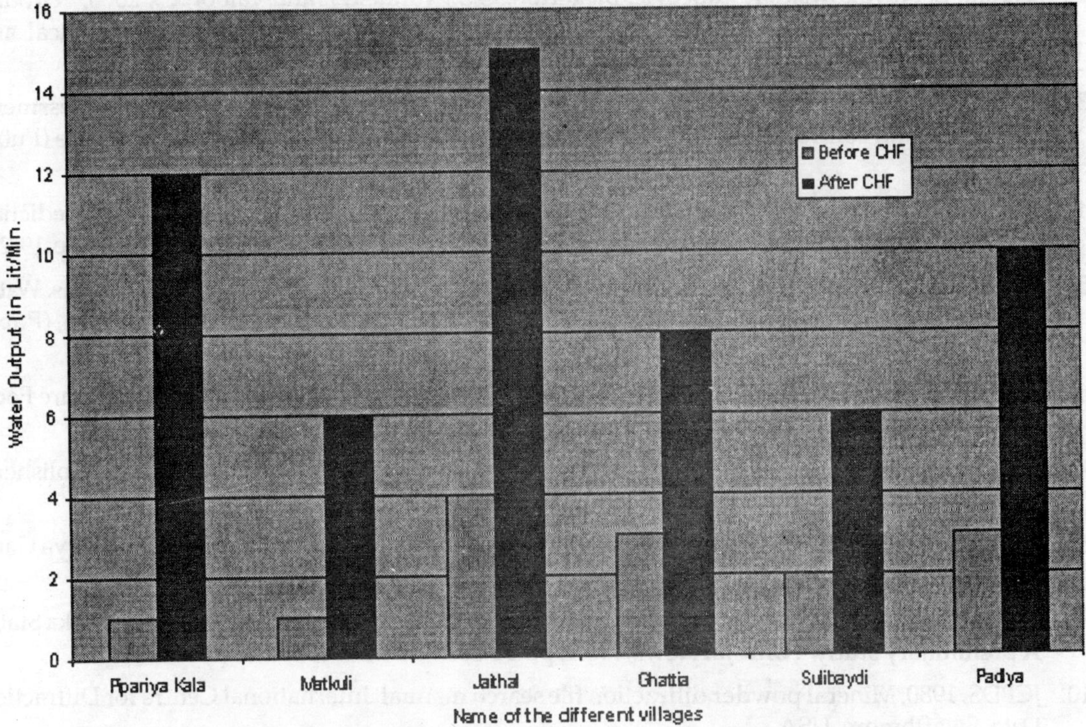

Fig. 10: Bar Diagram Showing Water Yield Results of Before and After Chemical Hydrofracturing Process

Conclusions

From the results of the studies reported above, following conclusions can be drawn:

1. The synergistic effect of treatment of clogged bore wells with non-toxic chemical composition followed by hydro-fracturing is effective in improving the water yield of the bore wells.
2. The improvement in water yield is observed to be several folds due to the CHF process developed in our laboratory.
3. The nature of clogging material influences the efficiency of rejuvenation of the bore wells.

Acknowledgements

The authors are thankful to Dr. S. Das for help in recording the X-ray diffraction work, to Shri B. Kujur for Chemical Analysis, and to Dr. N. Ramakrishnan, Director, RRL, (Bhopal) for his keen interest and permission to publish this paper.

REFERENCES

1. Brown, R.J., 1979, Ground water pollution: Part I. General studies. Avail NTIS. From Govt. Rep. Announce. Index (US) ; V.79, p. 177.

2. Zanini, L, Novakowski, P, Lapcevic, Bickerton, G.S., Voralek, J. and Talbot, C., 2000, "Ground water flow in a fractured carbonate aquifer inferred from combined hydrogeological and geochemical measurements". Ground water vol.38 (3), 350-360.

3. Frank, C.E., 1980, Corrosion and Encrustation in water wells: A field guide for assessment, prediction and control. Food and Agriculture Organisation of the United Nations Rome (Publ), pp.7-19.

4. Frank, 1980, Corrosion and Encrustation in water wells: A field guide for assessment, prediction and control". Food and agriculture Organisation of the United Nations Rome, (Publ), pp.19-22.

5. Ayers, R.S. & Westcot D.W., 1989, Clogging problems in localized (drip) irrigation systems. Water Quality for Agriculture Food and Agriculture Organisation of the United Nations Rome, (Publ), pp 99-104.

6. Ayers, R.S. & Westcot, D.W., 1989, Corrosion & Encrustation. Water Quality or Agriculture Food and Agriculture Organisation of the United Nations, Rome, (Publ), pp.104-106.

7. Walton, W.C., 1991, Principles of ground water engineering. CRC Press., Inc Lewis Publishers, Chapter 6, p. 165.

8. Raghunath, H.M., 1992, Ground water 2nd Edition. Published by Willey Eastern Ltd., Darya Ganj, New Delhi, pp 364.

9. Tamata, S.R., 1983, Mineral-$CaCO_3$ Saturation and stability of ground water in Karanataka State-A preliminary study". Bhu-Jal News, No.1, pp 14-20.

10. JCPDS, 1980, Mineral powder diffraction file search manual. International Centre for Diffraction Data, Swarthmore, USA.

11. Nadim. K, C. and Findikakis A.N., 2000, Quantitative estimates of the uncertainty in the evaluation of ground water remediation schemes. Ground Water 38 129-137.

68

Emerging Scenario in Alcohol Industry Spent Wash and Sludges Treatment and Utilisation for Agro Development: A "Waste to Wealth" Programme with "Recycle-Reuse-Reduce" Concepts

K. SUNDARAMOORTHI

Bannari Amman Sugars Limited (Distillery Division)
Periyapuliyur - 638 455, Erode District, Tamil Nadu

Introduction

India is one of the foremost country growing sugarcane and manufacture Sugar, Alcohol and allied products. Molasses, the sugar rich by-product of the sugar industry forms the raw material for alcohol production and during the process, the spent wash, yeast sludge and spent wash treatment plant sludges are emerging from the industry as outlet streams which are very rich in crop nutrients and organic matters.

Eco-friendly biological treatment of spent wash results in Biogas fuel recovery, a "Waste to Wealth" programme helping the conservation of fossil fuels. The fossil resources like petroleum are depleting at a faster rate and is expected to last for few more decades only. At this juncture, "Recycle-Reuse-Reduce" concepts should be given more thrust, suitably incorporated and practiced to see that the earth resources are conserved and ecology is maintained. Time has riped for the Eco authorities, Scientific community and Technical persons to work in unison to adopt the Eco friendly technologies in production of Alcohol and utilisation of all side streams. Alcohol industry spent wash and sludges with its rich plant origin nutrients are emerging as an important Agro manure input and would certainly play a vital role in the Agro economy.

Alcohol Production and Spent Wash Generation

Molasses, a mother liquor separated after the extraction of sugar crystals comes out as a by-product from the sugar industry and is the raw material for Alcohol production. Molasses is diluted with water, inoculated with yeast micro organism, allowed for fermentation and about 8-10% alcohol formed is distilled out using fractional distillation columns. The liquid stream left out after distillation of alcohol is called as spent wash, stillage, slops, vinasse, dunder, etc., in various regions of the world.

For every litre of alcohol about 10-13 litres of spent wash is produced. Raw spent wash liquid is acidic, brown coloured and contains high percentage of organic and inorganic matter. In a nutshell, spent wash is a NPK liquid mixture with rich organic matter and micro-nutrients, which are inherited from the soil during its growth and left out finally in the molasses and made their way into the spent wash as a result of Alcohol production.

Table 1: General Characteristics of Spent Wash

Colour	:	Reddish brown
Sp. Gravity	:	1.04
B.O.D (mg/L)	:	35,000-45,000
C.O.D. (mg/L)	:	85,000-1,00,000
Total solids (mg/L)	:	80,000-90,000
Volatile solid (mg/L)	:	55,000-60,000
Total Nitrogen as N (mg/L)	:	1,200
Phosphorous as P04 (mg/L)	:	800-1,000
Potassium as K (mg/L)	:	8,000-10,000
Chloride (mg/L)	:	5,000-6,000
Sulphate (mg/L)	:	4,000-7,000

(Composition of Biostill process spent wash would be about 1.5 times higher the value of above contents)

The spent wash rich in organic matter exerts high Biological Oxygen Demand which poses a threat to the water bodies by it oxygen depleting nature and so it cannot be disposed off in water bodies as such. It needs intensive treatment to bring down its B.O.D value to the maximum extent and then may be allowed in the abundantly flowing river water in line with pollution control norms. Even though the environment protection rules permits river water discharge option with stringent norms, its possibilities are very limited as most of the rivers are flowing with less water.

Since the river water source discharge possibility is not a feasible one to most of the distilleries in India and by the virtue of rich plant nutrients content of spent wash enabled the distilleries to work on for Agriculture utilisation as liquid manure input, after necessary pretreatments.

Spent Wash Management Scenario in Decades

In the 1980's, the Anaerobic Lagooning with shallow oxidation ponds prevailed to treat the spent wash to a B.O.D reduction level of about 85%. The Biogas evolved along with traces of H_2S dissipated in the atmosphere and as result the valuable Biogas fuel was lost and the vicinity also experienced some odour. Spent wash after this treatment, allowed to sun dry in pits and the residues were used for Agriculture manure.

In the 1990's, the Anaerobic Bio reactor technology developed at a faster rate. Thanks to the All India Distillers' Association's (AIDA) special delegation to most of the countries in Europe, U.S.A, Japan and other East Asian countries during the year 1984 and the delegation to South America Brazil, the worlds highest alcohol producer based on sugarcane during the year 1986, which paved the way for development of Anaerobic Bio Reactors for treatment of spent wash and harness the Biogas fuel to process boilers and Agro utilisation of spent wash as liquid manure. Most of the distilleries have installed Anaerobic Bio reactors with indigenous and foreign technologies and the B.O.D. reduction of about 85% are achieved through this, simultaneously utilising the evolved Biogas as an alternate fuel in the boiler thereby contributing for fossil fuel resource conservation.

The primary treated spent wash is treated further in the Secondary treatment system based on Aerobic treatment with Surface Aerators etc. The distillery spent wash is not easily aminable to the aerobic treatment as a result of its high total solids content in it. To achieve 10% further reduction in B.O.D. value, the power requirement would be to the extent of 800 H.P. to a distillery of 60KL Alcohol/day, accounting more than the power requirement for production of alcohol and still not meeting out the standards fixed.

The distilleries which are utilising the spent wash for organo mineral supplement of farm lands realize that the organic matter most needed for soil organic increase is volatilised in to CO_2 in the Aerobic spent wash treatment. Today the Agro scientists are vibrating the same thought and opine that the primary treated spent wash (Anaerobic Bio Reactor) could be used for Agro manuring in an eco-friendly and farmer friendly ways.

Some of the distilleries are practicing solar evaporation of spent wash after the treatment and finally utilising the sludge as manure.

Utilisation of Spent Wash for Agricultural Development

The leading Agricultural Research Institutions in India are conducting continuous Research on the utility value of distillery spent wash as Agro manure and found a very positive impact by Recycling the spent wash back to the soil. The mineral nutrients along with organic matter are enriched in the soil, thereby contributes for a sizable reduction in application of fossil mineral fertilizers and needed organic manures. It is estimated that Indian distilleries could contribute to an extent of Rs.1,000 Crores annually by the way of Agro manuring.

Indian Agriculture Research Institution (IARI), New Delhi has done extensive work on land application of spent wash and the Central Pollution Control Board has drawn out the protocols for Agriculture utilization, based on IARI. The distilleries in Tamilnadu studied the recommendations of Central Pollution Control Board and adopted it depending on the land pattern, irrigation water availability, nature of crop and the infrastructure facilities available. In order to handle the spent wash management systems effectively, the distilleries in Tamil Nadu under the consortium of South Indian Sugar Mills Association (SISMA), Tamil Nadu, has sponsored Research and Development schemes.

Ongoing Research and Development Activities on Treatment and Utilisation of Spent Wash

Under this programme the pioneer Environmental Engineering Institution of India, the National Environmental Engineering Research Institute (NEERI), Nagpur is conducting studies on the existing treatment system and suggests methodologies for Improvement. In addition, conducting general impact assessment study on spent wash application methodologies in practice.

The South Indian Sugar Mills Association and the individual distilleries has sponsored Eco-projects with centenary celebrated Agriculture Institution in Tamil Nadu, the Tamil Nadu Agriculture Agricultural University (TNAU), Coimbatore mainly focusing on Agriculture utilization of spent wash as liquid manure and for general Environment improvement.

The Research and Development Advisory Committee formed by the South Indian Sugar Mills Association with 3 reputed scientists in the field of Sugarcane Agronomy, Soil Science and Environmental science along with senior distillery executives as its subcommittee members, work solely on environment management systems monitoring and development. Periodically the R & D team visits the distilleries and suggest measures for improvement.

Under the Research and Development Programme, a 12 member delegation visited U.S. and Brazil during 25th August to 16th Sep-2001 to know the latest developments in Composting Science, Fuel Alcohol Programme and Distillery Spent wash Utilization Schemes to Agriculture crops as Liquid Manure. The U.S. part of the programme was sponsored by United States-Asia Environmental Partnership (USAEP) of USAID under Environmental Exchange Programme (EEP) after witnessing the positive Environmental activities undertaken by Tamilnadu distilleries.

Utilisatoin of Treated Spent Wash as Liquid Manure in Tamil Nadu Distilleries

1. Treated spent wash application along with irrigation water to the crops.
2. One time Pre-application of treated spent wash as liquid manure to the dryland-as basal.

Treated Spent Wash Application Along with Irrigation Water to the Crops

Mostly, spent wash is transported in tanker lorries and applied along with irrigation water. The application rate is about 20-30 m^3/Acre depending on the land needs. For example, when an acre of sugarcane crop is irrigated in 3 hours time, the spent wash from the lorry is mixed uniformly during this period and the required application quantity is maintained by keeping further spent wash transport tanker ready at the site.

The spent wash supply is effected on the specific request of the farmer after inspecting the field conditions. The application of spent wash is done preferably one time at the early stage of crop. The application supplies full requirement of Potash manure, about 20-30% of Phosphate and Nitrogen needs along with micronutrients and organic matter requirements. The organic matter supply ultimately improves the soil humus content and there by build up soil fertility. The application of spent wash once in a season to the sugarcane crops not only improves the yield but also recycles back the organo mineral matters drawn by the crop during its growth and thereby contributes for reducing the use of mineral fertilizers of fossil origins. This method of application is being adopted for crops like tapioca, banana, coconut, etc.

One Time Pre-application of Treated Spent Wash as Liquid Manure to the Dry Land as Basal

Based on the Tamil Nadu Agricultural University Research studies for more than a decade of years and under Research and Development programme, the distilleries supply spent wash as basal application to dry land farming before the raise of crops. Generally dry land rainfed farmer's yield depends on adequate rain during the season and the soil wealth. The dry land farmer's income are limited and are not in a position to apply organic manure and chemical fertilizers. The organo mineral rich distillery spent wash application supplies almost all soil nutrient requirements and builds-up the soil value.

Before the monsoon the dry lands are applied with treated spent wash at the rate of about 40-60 m^3 per acre. The Spent wash is carried in tanker lorries provided with sparger assembly at it back and the vehicles are moved slowly in such a way to apply on the land uniformly. About 4" depth only is wetted and farmers till the land during the first rain and in the second rain period raise the crops. Dry land millet crops are benefited by this programme. The yield of millet and the fodder is good and helps the farmers to maintain milch animals and make their livelihood. The dry land jewar crop millet and fodder yield has increased very well in the applied lands. The application is done once in a year or two, on this scheme.

On going field trials are conducted by the Tamil Nadu Agricultural University in some of the Tamil Nadu distilleries for effective utilization of spent wash as manure, under sponsored research programme by distilleries. At Bannari Amman Sugars Limited-Distillery Division, the Tamil Nadu Agricultural University (TNAU) is conducting a three year research project on "Microcomposting of Crop Residues with Distillery Spent wash and Sludge and Recycling of Spent wash for Agriculture".

The research work interim report reveals the following facts:

1. The grain yield of maize in 60KL/Acre spent wash application registered 1.66 times higher than 100 percent fertilizer recommendation and 2.07 times higher than absolute control. Biomass yield of maize in 60KL/Acre spent wash application was 1.38 times greater when compared with 100% fertilizer recommendation and 2.1 times higher than absolute control.
2. In case of Sunflower, significant increase in grain yield of sunflower with 60KL/Acre spent wash application recorded 1.5 times greater than 100 percent fertilizer recommendation and 2.5 times higher than absolute control. In case of biomass yield 60KL/acre registered 1.39 times higher yield than 100 per cent fertilizer and 2.21 times greater yield in absolute control.
3. In a significant development of R & D studies, Coirpith, a waste material from the coconut coir making industry could be composted using Distillery Yeast Sludge in 8 weeks time and the quality of compost produced is better and further studies are continuing. Based on this findings, Horticultural Research Institutions are now conducting massive trials by drawing yeast sludge from the distillery.

Most of the distilleries own their sugar unit and utilize the spent wash as liquid manure to their nearby sugarcane area. The dry land farmers approach the distillery directly for the supply and is effected after getting their consent in a format and inspecting the field. The spent wash transport tankers carry a transit card containing distance covered, field applied, etc. Representative soil samples, water samples are drawn periodically from the field zones to follow up the programme. The spent wash transport and application is monitored by a special cell containing Agriculture Officers, Environmental Chemists and Field Assistants and the day-to-day datas are recorded. The field applications are reported to the Pollution Control Boards periodically. In recent times the demand for spent wash application to farmlands has increased very much and distilleries are not in a position to meet out the demand.

Research and Development Programme on Raw Spent Wash Utilization for Sodic Soil Reclamation and Calcareous Soil Reclamation

The sodic soil soils with its high pH of above 9.0 not able to sustain the crop development and based on the scientific findings of Tamil Nadu Agricultural University and with government supported programme, Sodic soils reclamation is done using raw spent wash by taking advantage of its pH of 4.0. The land after reclamation becomes productive and the scientists view that sodic soil reclamation using Raw spent wash is better than Zypsum applied reclamation programmes, since the spent wash supplies crop nutrients also. Some of the distilleries partly carryout this programme, as their surrounding area lands are of sodic nature.

Raw spent wash application to high pH calcareous soil also lowers the soil pH to an optimum condition and help to increase the crop productivity. Further research works are continuing on these sectors.

Utilisation of Spent Wash in Biocomposting Process

Sugar industry pressmud and spent wash are the major raw materials for Biocompost manure manufacture. During the manufacture of sugar from sugarcane, a by-product, rich in organic matter and mineral contents called pressmud is produced at the rate of about 3.5 tonnes per every 100 tonne of sugarcane crushed. A sugar unit with an annual crushing of about 7 lakhs tonnes would produce about 25,000 tonnes of pressmud. Generally, farmers used to lift the pressmud and use it on land as basal manure as such or after mixing and keeping it with farmland residues for some months.

The sugar industry pressmud could be effectively composted into Biocompost manure using distillery spent wash. The distilleries with sugar units in their fold have already ventured in composting projects and individual distilleries are on the move of establishing biocompost projects by mobilizing pressmud from their nearby sugar mills.

Aerobic composting is an exothermic process carried out by consortium of microbes and establishes a composting temperature of 55-70 °C which decompose the organic carbon matter into a humus rich compost. The moisture content to be maintained at about 45-50 percent level for the composting operation to proceed and instead of water, spent wash/primary treated spent wash is sprayed to maintain moisture level and enrich the compost. About 2-3 KL of spent wash is utilised for every tonne of pressmud composted and yields about 350-400 Kg. of compost manure.

The Distilleries have established 4 to 10 Acres concreted compost yard inline with the Central Pollution Control Board recommendations. The pressmud is formed into windrows of 1.4 metres to 3.5 metres base depending on the Aerotiller machine employed for turning the windrows to facilitate aeration in the windrows. Aerobic microbial inoculums are used to speed up the composting.

The exothermic temperature of 55-70°C prevails for about 6-8 weeks time and thereafter declines to 40 °C and then allowed for about two weeks curing before marketting. The turning of windrows and spent wash spray is maintained throughout the active composting cycle and the compost ready for marketing is achieved in about 8-10 weeks time. The compost is sold as bulk or in 50 Kg. bags after sieving or pulverizing.

In recent years the Agriculture scientists all over the world are advocating for organic amendments to the soil to buildup farmer friendly soil microfloras. It is viewed that continuous application of chemical fertilizers for the past three decades without sufficient organic amendments has made the soil sick and to rebuild it, the organic amendments to be incorporated sufficiently with reduction in chemical fertilizers. The Biocompost manure produced by the distilleries has become one of the major suppliers of organic manure to the farming community.

Raw Spent Wash Utilisation as Manure in Australia and Brazil

In Australia the distillery spent wash (called as Dunder) is utilised for sugarcane crops as liquid manure without any pretreatment. The distillery members visited during the year 1982 and 1999 witnessed the application to ratoon sugarcane crop. In recent years the Dunder is mixed with additional Nitrogen fertilizer required by the crop and marketed as 'Liquid One Shot' to sugarcane farms and supplied through special tankers with application devices. During the year 2001, an Australian Sugar and Alcohol industry delegation visited Tamil Nadu Sugar and Alcohol units and expressed the valuable use of Dunder and Liquid One Shot for sugarcane development. The Australian delegation appreciated the efforts taken by Tamil Nadu Distilleries in the spent wash management systems.

Brazil, the highest alcohol producing country in the world from sugarcane, utilize the raw spent wash (Vinasse) without any pretreatment generated from their Alcohol industry using cane juice and molasses as raw materials. The application is done through Aspirator Spray Guns once in a season after 10 days of the harvest of the ratoon sugarcane crop. Since most of the cane fields are harvested mechanically, the cane trash bed left out in the field are thus composted effectively with the help of raw spent wash (Vinasse) application, ultimately building up the soil nutrients. The above spent wash application system is in practice for more than two decades successfully and the distillery delegation from India witnessed this during the year 1986 and 2001. The spent wash is supplied to the application zone by pipe line/Lined channel/Twin tanker trucks.

"Recycle-Reuse-Reduce" Concepts Followed by Distilleries

Distilleries in Tamil Nadu are thus marching ahead with "Recycle-Reuse-Reduce" concepts to conserve Energy, Environment, Earth Resources and Nature and emerging as an Eco-green industry.

The word 'Disposal of spent wash' called a few years back has now turned into 'Utilisation of spent wash'. Distilleries which were monitored significantly for water pollution are now looked with New Agendas for composting and Agriculture utilisation schemes (Annexure). Since the spent wash application is done in a controlled manner, with less application frequency in different zones, the ground water pollution is not present.

An estimation says that Indian distilleries could contribute to an extent of 1,000 crores rupees in terms of organo mineral manure by its spent wash utilisation programmes. At this hour, Distilleries needs renewed label and support from Scientists and Eco-control authorities with revised norms for Agro utilisation. The industry which is operating and maintaining the Environmental Management System should be given adequate representation in the Pollution Rules and Regulations framing bodies and also in the technology developing committees of scientific institution to co-ordinate for the development of the industry.

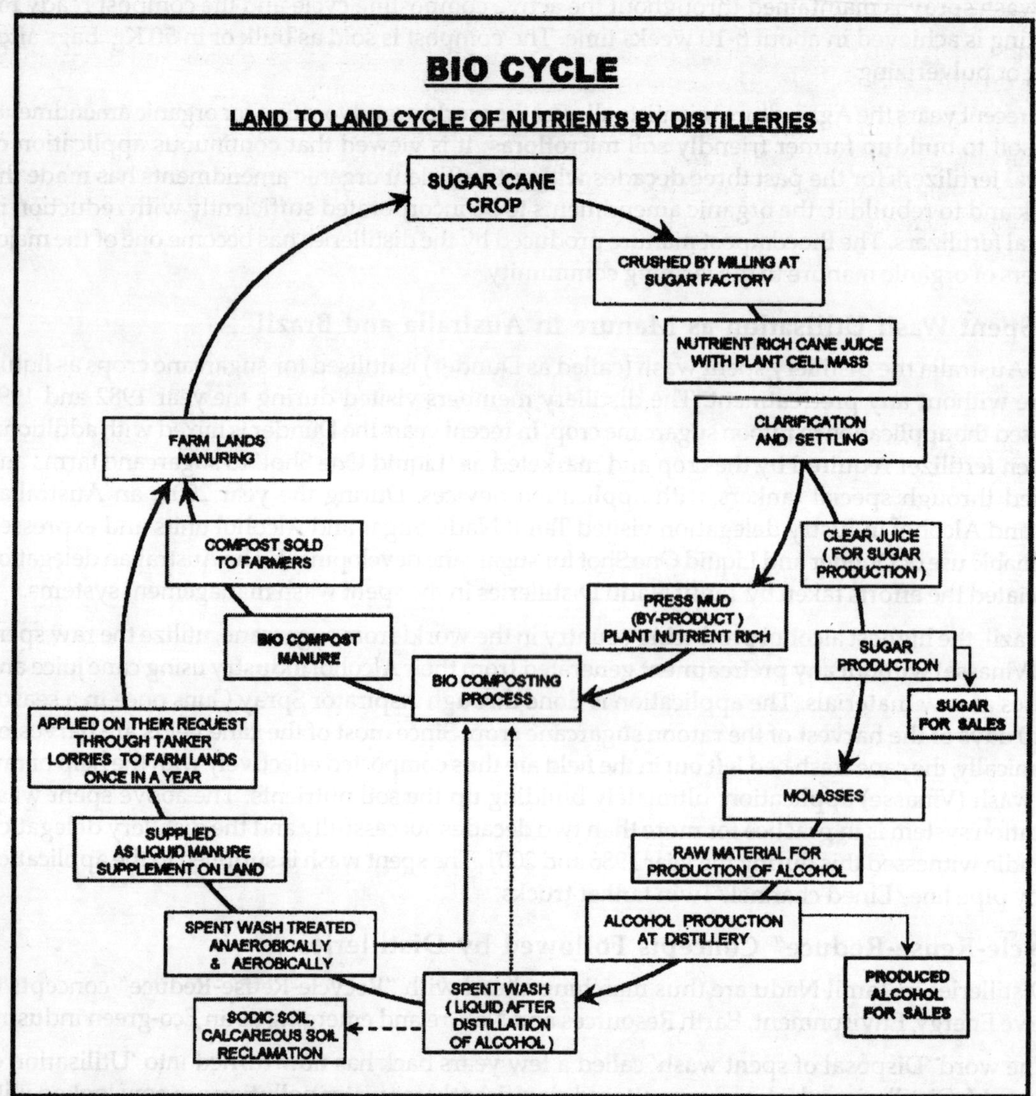

☞ **In a situation of non availability of needed organic manures to the crops and an utmost need to supplement the inorganic fertilisers with organic manures to sustain the soil fertility,**

- **The major Organo Mineral Resources available to the farmers are from Distillery Spentwash – Distillery Sludges – Biocompost Manure.**
- **Organo enriched, Mineral rich distillery spentwash cannot be regarded as an effluent any more. It is a great resource to the Agro development.**
- **More than 1,000 (one thousand) crore rupees manure potential is readily available in Indian Distilleries Spentwash and the farming community could make use of it effectively. By promoting this aspect, Biocycling of farmland nutrients could be achieved.**

The Research Institutions in this country and across the world should look the tremendous potential of Alcohol industry for its contribution to Alcohol, a renewable energy from the Agro source, Renewable Biogas fuel from the spent wash, Agro manure utilisation of spent wash and Biocompost manure production, etc., and address the related issues for its development.

The Indo-Italian workshop on "Emerging Technologies for Industrial Waste water and Environment" organized by NEERI jointly with department of Environmental Sciences, University of Venice, Italy should address this issues in a new prospective and help the industry to move in the Eco-Green direction to conserve energy, conserve earth resources and conserve nature.

69

Missing Links in Technological Advances in Common Effluents Treatment Plant at Industrial Estate

A Case Study of Ankleshwar GIDC Estate Common Effluent Treatment Plant in Gujarat State of India

B.M. JANI

Professor of Economics, Saurashtra University, Rajkot - 360 005

Introduction

Chemicals and Petrochemicals industries consume highest amount of water and hence disposing highest amount of wastewater than any other group of industry in India. The present paper examines technological economics of common effluent treatment plant located at GIDC estate, at Ankleshwar in Gujarat state. It is most necessary for scientists to view advancement of technology and actually utilized by the industry and cost components so as to manage wastewater pollution as per norms of the regulatory agency. This researcher has examined cost regression of treatment of wastewater by Common Effluent Treatment Plant (CEPT), run and manages by Ankleshwar Industries Association

(AIA) as a group effort to manage wastewater treatment for member industries. As such, developing countries require cost reducing technology for treatment of wastewater due to low cost of fresh water available in industrial estate or a park. To stop proliferation of wastewater in open area, it is most necessary to review existing technology and make changes in it periodically. It has been visualized in this study that innovative technology suited to the developing country appear most necessary as in the case chemical and petrochemicals industry, product-wise cost of waste water is different. This plants need innovation in technology pavement and sufficient training for operator of plants. Environment in developing countries is full of technological diffusion.

For this, the present paper is divided into two parts: first one show location of common effluent treatment plants in Gujarat and No. of members and designed capacity, and the second one show techno-economic details of Ankleshwar Industries Association CETP so to know technological cost of treatment of wastewater with present means of technology and the third one analyses to show non-linear regression model analysis of treatment cost.

SECTION I

Status of Common Effluent Treatment Plants (CEPT's) in Gujarat

Before micro level study of Ankleshwar CETP, it is necessary to glance through existing common Effluent Treatment Plants in Gujarat at different GIDC estate levels. The under mentioned tables provides location of CETP, Management agency of it, membership, designated capacity and present status. Out of 14 CETP, four are proposed, whereas, 10 are in working position. Ankleshwar CETP is one of those having initial 96 members (but now 220) and having of designated capacity of 1000 KL.

Table 1: Status of CEPT's in Gujarat

Sl. No.	*Area*	*Name of Management*	*No. of Members*	*Designed Capacity*	*Present Status*
1.	GIDC, Vatava	The Green Environment	450	16000	Already Commissioned
2.	GIDC, Odhav	Odhav Enviro Project Limited	54	1200	Already Commissioned
3.	GIDC, Naroda	Naroda Enviro Project Ltd.	201	3000	Under Constructions
4.	GVMSAV, Odhav	Gujarat Vepari Maha Mandal Sahakar Udyogik Vasahat Ltd.	34	1000	In Operation
5.	Narol Dyestuff Association	Narol Dyestuff Manufacturers Association	38	100	Under Commissioning Stage
6.	GIDC, Nandesari	Nandesari Industries Association	252	5500	In Operation
7.	GIDC, Ankleshwar	Environ Technologies Ltd.	96	1000	In Operation
8.	GIDC, Jhagadia	Jhagadia Industries	24	No Proposal	Under Consideration
9.	GIDC, Panoli	Panoli Industries Association	58	No Proposal	Under Consideration
10.	GIDC, Pandesara	Pandesara Industries Associations	329	No Proposal	Under Construction
11.	GIDC, Sachin	Sachin Industrial Co-op. Society Ltd.	187	No Proposal	Under Construction
12.	GIDC, Sarigam. System Pvt. Ltd.	Project Enviro Control	6	200	In Operation
13.	GIDC, Vapi	Vapi Waste & Effluent Management Co. Ltd.	576	55000	In Operation
14.	Jetpur	Jetpur Dyeing & Printing Association	1200	5400	In Operation

Source: Gujarat Pollution Control Board Publications and GIDC Notes

A Case Study of Ankleshwar GIDC Estate in Gujarat

Ankleshwar Industries Association (AIA) formulated a committee for modus operandi of Rs. 120 crore Pipeline project for Ankleshwar, Panoli and Jhagadia industrial estates in Narmada basin area. Ankleshwar Industrial Estate is located in between Baroda and Surat, at the side of National Highway # 8. Ankleshwar is a major chemical industrial estate in Asia. The Estate houses more than 1500 industries out of which, approximately 500 industries are manufacturing various types of chemicals. Among these chemical industries, about 400 industries are in Small Scale Sector and balance approximately 100 industries are in the medium and large-scale sector. All the medium and large-scale industries are having their own Effluent Treatment Plants. Some of the Small Scale Industries are also having their own treatment facilities. However, there are several Small Scale Industries, who do not have treatment facility and Enviro Technology Limited operates members of the Common Effluent Treatment Plant. The water consumption in Ankleshwar industrial estate is more than 34,000-kL/day and the effluent generation is 22000 kL/day.

Presently wastewater is disposed to estuary of river Narmada through open channel called "AMALA KHADI". Such open disposal of wastewater created problems for people, animals and agriculture considering the extent of hazardous waste water. It is proposed to convey the effluent through a closed system up to sea after polishing treatment. The quantity of effluent is estimated at 60MLD. Total pipeline length will be approximately 40 KMs on land and 14 KMs in sea. For this AIA approached National Environment Engineering Institute Nagpur on order to prepare feasibility report for selection of suitable site for disposal of wastewater through pipeline. Formalities of Economic Impact Assessment studies and cost estimation process are over. Enough rational required for the estimated Rs. 120 crore for the project. Bharuch Eco Aqua Infrastructure Ltd is formed for the execution and operation of the project.

The government of Gujarat has sanctioned subsidy about Rs. 28 crore to encourage the project from environment point of view. The AIA is engaged with experts of GIDC, professional consultants and knowledgeable people to positively execute the project. Bharuch Enviro Aqua Infrastructure Ltd has collected substantial amount from the members as equity and/membership fee to manage the conveyance system. After operation of the project, operation cost is also collected from the members as per volume of wastewater. It is expected that the present project will avoid contamination of ground water and Amalakhadi be available as storm water drain also. Activist Lawyers and Environment Activists are also holding the opinion that effluent need to be conveyed through pipeline project instead of through Amalakhadi.

Advantages of Pipeline

1. Long term solution to the effluent disposal problem from three estates *viz.*, Ankleshwar, Panoli and Jhagadia in Bharuch district.
2. Vacant plots available in estates especially at Jhagadia can be allotted for industries
3. Projects, which are on half way, pending due to uncertainty, can be completed.
4. Can attract new industries in the area.
5. Better acceptance of industries by the communities/farmers in the area
6. Marine discharge Norms can be prescribed by the Gujarat Pollution Control Board.
7. Sustainable growth of Bharuch district, south of river Narmada.

SECTION II

Techno-economic Details of the Plant

A Company *viz.*, Ankleshwar Industries Association (AIA) promotes Environ Technology Ltd. to manage and operate Common Effluent Treatment Plant (CETP) through some of the major industries in the estate. United Phosphorus Ltd is the main promoter holding 51% of the equity. Other major shareholders are-J B Chemicals and Pharmaceuticals Ltd, Link Pharma Chem, SBS Group of Companies, Suyog Group of Companies, Shree Ambeshwar Paper Mills, Sidhi Samrat Dyechem etc. Few members are also equity partners of the Company. The consultants for the project-M/s Associated Environmental Engineers, Baroda-are also equity partners in this Company for initial capital investment in common Effluent Treatment Plant for Ankleshwar GIDC estate. The following table provides equipment details for the operation of the plant.

Table 2: Equipment Details of CETP

Sl.No	Name	Details
1.	Equalization Tank	Units – 2 nos Capacity – 1000 kl each Size – 32 × 12.5 × 2.5 meters Residence time – 24 hours
2.	Lime Storage (Chemical House)	Size – 35 × 25 × 4 meters
3.	Lime Handling System	This system consists of hopper, belt conveyer, bucket elevator and slacker. The capacity of this system is to handle 5 MT of lime/hour.
4.	Neutralizer	Units – 2 nos Capacity – 22.5 kl each Residence time-½ hour
5.	Primary Clarifier (Thickener)	Unit – 1 no Volume – 1200 kl Surface Area – 300 Sq M Size (dia) – 20 meters Side water depth – 4 meters Residence time – 18 hours
6.	Filtration System	Equipment – Rotary Vacuum Drum Filter Units – 3 nos Surface Area – 28 Sq M Capacity – 2.5 MT (wet) /hour
7.	Aeration Tank	Units – 2 nos Capacity – 3725 kl each Size – 70 × 18 × 3 meters Residence time – 120 hours
8.	Secondary Clarifier	Unit – 1 no Capacity – 400 kl Surface Area – 120 Sq M Residence time – 4 hours
9.	Secondary Treated Effluent Sump ($FeSO_4$ + H_2O_2 Reaction)	Unit – 1 no Capacity – 124 kl Residence time – 2 hours
10.	PH Adjustment Tank	Unit – 1 no Capacity – 50 kl Size – 4 × 4 × 3 meters

Contd...

Table 2–Contd...

Sl.No	Name	Details
11.	Tube Settler Clarifiers	Unit – 1 no Capacity – 50 kl Size – 4 × 4 × 3 meters
12.	Discharge Sump	Unit – 1 no Capacity – 50 kl Size – 4 × 4 × 3 meters
13.	Sand Filters	Unit – 2 nos Size – 2 (dia) × 1.7 (height) meters Capacity – 8 kl Sand Bed Volume – 5 kl Residence time – 7 minutes
14.	Activated Carbon Bed	Unit – 2 nos Size – 2.3 (dia) × 3 Sq M Capacity – 16 kl Carbon Bed Volume – 5 kl Residence time – 5 minutes
15.	Guard Pond	Capacity – 750 kl
16.	Hydrochloric Acid Tank	(30% HCl) Capacity – 15 MT × 1 no (10% HCl) Capacity – 10 MT × 2 nos
17.	Polyelectrolyte Solution Tank	Unit – 2 nos Capacity – 10 kl
18.	D G Set	600 KVA
19.	Underground Water Storage Tank	650 kl
20.	Solid Waste Storage Pit with impervious lining	60 × 30 × 3 meters

Table 3: Effluent Quality at Different Stages

Stage of Treatment	*Inlet values (in PPM)*	*Outlet values (in PPM)*
Primary – COD	8000 to 9000	4500 to 5000
Primary – BOD	2900 to 3600	1800 to 2300
Secondary – COD	4500 to 5000	600 to 800
Secondary – BOD	1800 to 2300	60 to 80
Tertiary COD	600 to 800	200 to 250
Tertiary BOD	60 to 80	20 to 30

Discharge Norms: National v/s Gujarat

Minimum National Standards (MINAS) are enacted by the Central Pollution Control Board and Ministry of Environment and Forest in India. These standards are developed

Based on prevailing technology of manufacture and abatement technology available. Such standards for discharge of effluents are prescribed under Environment Protection Rules 1986. These standards are prescribed and based on the type of industries. Also, general standards of effluents are given. If any industry is not falling under the specific category mentioned in the Rules, General standards are applicable. In the standards, authorities have given various norms to be prescribed for the effluent depending up on discharge point *i.e.* in-land surface water, public sewer, land for irrigation and marine coastal areas. The Central Government has prescribed MINAS. However, depending upon the location of industries and assimilative capacity of the receiving body, the State Pollution Control Boards can prescribe more stringent norms. With this addition power with State Boards, they generally prescribing more stringent norms than prescribed in EPA Rules 1989 *e.g.* TDS limits have been removed in the EPA Rules by various amendments. Even then the state Board including Gujarat, are prescribing TDS limit of 2100 mg/1 to the industries.

It may also be pointed here that many norms have been prescribed without considering the prevailing manufacturing technologies and the available effluent treatment technology. Many norms cannot be achieved by the industries by primary and secondary treatment. Industries will have to adopt expensive tertiary treatment system for meeting such stringent norms.

Advantages of Enviro Technology Ltd. Over Other Common Effluent Treatment Plants

Better Collection and Transportation of Effluent

1. Transportation acid proof lining tankers
2. Corrosion problem and choking of pipeline is avoided
3. Each tanker analyzed for COD and acidity and sample is preserved for one month
4. Better control on quality and quantity of effluent
5. Any erring industry, giving shock load, can be identified easily, whereas in other CETPs it is difficult as it is coming through common pipeline

Primary Treatment

1. Primary treatment is included in ETL, whereas other CETPs collect primary/secondary treated effluent.
2. We use huge quantity of lime for neutralization and heavy metal removal. Hence member industry has less botheration for lime and sludge handling
3. RVDF system for sludge dewatering
4. Disposal of sludge to cement industries/landfill through a centralized agency
5. Consistent quality of effluent going for biological treatment

Secondary Treatment

1. Sewage water from residential area is treated along with industrial effluent
2. High residence time of five days to take care of complex effluent
3. Extended aeration system for less secondary sludge generation

Tertiary Treatment

1. Tertiary treatment system for removal of residual COD and BOD are provided whereas in other CETPs it is not included
2. Laboratory Facilities
3. Full fledged laboratory for analysis of all required parameters of effluent

4. Treat ability studies-for all member industries effluent
5. Data bank with respect to products, raw materials, water consumption etc. of all member industries.
5. Round the clock monitoring

Others

1. Auxiliary power supply
2. Computerized billing system
3. Technical committee meeting
4. Pool of experts
5. Support to the member industries, like management training programs
6. Operates like a manufacturing unit, with support of major industries in the area.

Project Cost

Total Project Cost	:	Rs. 599.07 Lacs
Promoters' Equity	:	Rs. 120.00 Lacs
IDBI Loan	:	Rs. 330.00 Lacs (World Bank Line of Credit)

Subsequently Rs 81.23 Lacs additional investment has been incurred by ETL in Project for augmentation of Tertiary Treatment System and other miscellaneous additions. Hence, the total Project Cost as on 31.03.1999 is Rs 680.30 Lacs.The plant has received subsidy worth of Rs. 50 Lakhs.

SECTION III

Economics of Common Effluent Treatment Plant

After examining technical aspects of CETP, it is most necessary to visualize financial and cost aspects of the ETL Table 4 provides financial functioning in terms of profits and financial flow accruing by the ETL It is evident from the table that operational expenses increased from Rs. 45.50 Lacs in 1997-98 to Rs. 630 in 1999-2000 largely due to capacity utilization of the ETL Net Profits going to increase Rs. 70 Lacs in 1999-2000. As a result, the CETP co able to distribute dividend from 10% in 1997-98 to 15% in 1998-99.

Charging System

In addition to the initial membership fees of the CETP, the industries have to pay treatment cost to CETP, based on the volume of effluent given, and the pollution load. The CETP has developed a scientific formula of charging from the beginning and is being revised on a regular basis.

As CETP is taking effluent in rubber linked tankers and each tanker is having 10-KL capacity, all the industries have been asked to give full tanker load of effluent. In case, an industry is giving less quantity, it will be considered as 10-KL of effluent is given, and charging will be based on the capacity of the tanker.

Table 4: Financial Aspects of CETP

Revenue/Year		*Profits and Cash Flow*
Total revenue generated during 1996-97	:	Rs. 39.60 Lacs
Total revenue generated during 1997-98	:	Rs. 604.00 Lacs
Total revenue generated during 1998-99	:	Rs. 716.00 Lacs
Total revenue generated during 1999-00 (projected)	:	Rs. 725.00 Lacs
		Total Operating Expenses
1996-97	:	Rs. 45.50 Lacs
1997-98	:	Rs. 577.00 Lacs
1998-99	:	Rs. 662.00 Lacs
1999-00 (projected)	:	Rs. 630.00 Lacs
		Net Profit
1996-97	:	Nil
1997-98	:	Rs. 24.15 Lacs
1998-99	:	Rs. 43.00 Lacs
1999-00 (projected)	:	Rs. 70.00 Lacs
		Dividend paid to share holders
1997-98	:	10 %
1998-99	:	15 %

As CETP is taking raw effluent from member industries, they have incorporated two essential parameters in the charging formula, one is acidity and other one is organic load. Acidity is measured on actual basis by neutralizing the acidity. The organic load is measured in terms of COD. The operation charges are recovered from member industries based on effluent characteristics-acidity and COD. The charging formula is as follows;

Treatment Charges in paise/liters = Fixed Charges + [Factor for Acidity × Actual Acidity (PPM)] + [(Factor for COD × Actual COD (PPM)]

Initially, the company has kept the fixed charge of Rs. 0.18 paise per liter and as the capacity utilization has increased and due to other factors like reduction in the interest burden, the fixed cost has been reduced substantially over the period of 5 years. Initially, fixed charge was kept which was substantially in the year of 1998 where the fixed charge was also charged at different level rates. Then, the fixed charges are kept in three slabs based on the acidity and COD values. Here, the "polluter Pays" principle comes into picture and the industry having higher pollution load has to pay more fixed treatment charges. The member industry of CETP having less pollution load or having neutral effluent has to pay less.

The fixed charges for effluent having low pollution load has been reduced to the tune of 50% over the years. At the same time, the fixed charges for the industries having very high pollution, has been increased.

Since the cost was worked out based on the pollution load and implemented after acceptance by the member industries, the rates are affordable. Meanwhile, CETP is working out various options to bring down the cost of treatment and make the operation more viable.

The major operation costs are due to:

1. Raw material consumption *e.g.* lime.
2. Power consumption
3. Overhead costs and
4. Sludge disposal cost etc. Lot of measures has been taken to use improved quality of lime so that treatment cost will be reduced.

To reduce power consumption cost, the company has taken various power saving measures. By proper restructuring of the organization and manpower adjustments, CETP could reduced the overhead cost. Another important cost is the sludge disposal cost. CETP has conducted studies and found that sludge can be utilized in Cement industry. By disposal to cement industry, the CETP could save to the tune of Rs. 20-30 lakhs during 2000-01 which other wise could have spent for secured land filling.

Now CETP has conducted studies through Gujarat Agriculture University, Navsari and Industrial Toxicology Institute Lucknow for utilization of the CETP sludge as fertilizer (soil conditioner). The studies have revealed that the sludge can be used as fertilizer. For this, it is necessary to have permission from GPCB to sell such material for the agricultural purpose, and then the CETP can have substantial saving for smooth operation of the plant most economically viable.

Table 5: Cost Economics of Per Tanker Cost of Treatment of Wastewater at CETP, Ankleshwar

Month	*Effluent Quantity Received in No. of tankers*	*Capacity Utilization %*	*Raw Effluent*		*Total Operational Charges in Rs.*	*Per tanker Average Cost in Rs.*
			Average Acidity ppm	*Average COD ppm*		
April-97	585	19.50	42807	8088	1177347.00	2012.59
May-97	718	23.16	27578	8593	2303997.00	3208.92
June-97	1059	35.30	26472	8251	3346915.00	3160.42
July-97	1350	43.54	30056	8317	4078836.00	3021.36
Aug-97	1606	51.80	28100	6100	4696333.00	2924.26
Sept-97	1779	59.30	22716	8109	5672614.00	3188.65
Oct-97	1784	57.54	23306	8498	5864780.00	3287.43
Nov-97	1663	55.40	21343	8076	5058900.00	3042.00
Dece-97	1985	64.03	21971	7911	8190468.00	4126.18
Jan-98	1951	62.93	18300	7987	5531762.00	2835.34
Feb-98	2037	72.75	21718	7408	5906044.00	2899.38
Mar-98	2261	72.93	21031	6599	6889563.00	3047.13
Apr-98	2043	68.10	21112	6595	5770139.00	2824.24
May-98	2004	64.64	23046	7242	6008296.00	2998.15
June-98	1936	64.53	21008	7896	6109650.00	3155.81
July-98	2063	66.54	18807	7036	5856406.00	2838.78
Aug-98	1871	60.35	20425	7053	5678275.00	3034.88
Sept-98	1858	61.93	15845	6205	5120878.00	2756.12
Oct-98	1904	61.47	19310	6781	5350948.00	2810.37
Nov-98	2392	79.73	24969	7371	6919049.00	2892.57

Contd...

Table 5–Contd...

Month	Effluent Quantity Received in No. of tankers	Capacity Utilization %	Raw Effluent Average Acidity ppm	Raw Effluent Average COD ppm	Total Operational Charges in Rs.	Per tanker Average Cost in Rs.
Dec-98	2270	73.22	31249	8011	7115213.30	3134.45
Jan-99	2186	70.51	27475	7694	6593800.39	3016.37
Feb-99	2070	73.92	21051	6965	5374383.41	2596.32
Mar-99	2124	68.51	25519	6583	5826291.56	2743.00
Apr-99	2064	68.80	24297	5754	5516967.00	2672.94
May-99	2128	68.64	22684	5758	5608849.00	2635.73
Jun-99	2079	69.30	22924	5628	5298075.00	2548.37
Jul-99	2229	71.90	24342	6098	5685169.00	2550.54
Aug-99	2480	80.00	26444	5566	6243941.00	2517.71
Sep-99	2560	85.33	20253	5836	5787603.00	2260.78
Oct-99	2836	91.68	21628	5738	6392879.00	2254.18
Nov-99	2628	87.60	22098	5409	6212785.00	2364.07
Dec-99	2860	92.28	23239	6432	5987589.00	2265.00
Jan-00	2736	88.25	20742	6362	6477918.00	2367.66
Feb-00	2584	92.28	20403	6121	5987589.00	2317.17
Mar-00	2805	90.48	21598	6484	6830124.00	2434.98
Apr-00	2591	86.36	21898	6277	6122056.00	2362.81
May-00	2737	88.29	17880	5761	5643286.00	2061.85
Jun-00	2874	95.80	16289	5621	5666215.00	1971.54
Jul-00	2820	90.96	16275	5521	5665650.00	2009.09
Aug.00	2700	87.09	19814	4951	5593913.20	2071.81
Sept-00	2670	89.00	21816	5172	6225244.00	2331.55
Oct-00	2437	78.61	18347	4850	5087764.00	2079.71
Nov-00	2491	83.03	21028	5081	5465793.00	2194.21
Dec-00	2390	77.09	23485	5822	5689222.00	2380.42
Jan-01	2207	71.79	21010	5420	5348461.00	2423.40

Source: Common Effluent Treatment Plant Office, Ankleshwar 23rd March 2001.

The above data are analyzed in a form of Multiple regression with different methods:

Multiple Non-Linear Regression Equation:

$$Y = f\,(A.\ x1\alpha + x2\beta + x3\gamma + x4\delta)$$

where,

Y: Operational cost of waste water (OC)
x1: Effluent Quantity Received (EQ)
x2: Average Acidity ppm (AA)
x3: Chemical Oxygen Dissolved (COD)
x4: Capacity Utilization (CU)

Dependent Variable (s) Entered on Step Number

1. V1: EQ (Effluent Quantity)
2. AA: Average alkality
3. COD: Chemical oxygen dissolved
4. CU: Capacity utilization of the common effluent treatment plant
5. OC: Operational cost (Independent variable)

Here, our Hypothesis is that Cost of wastewater treatment from chemical and allied industry treated by the common effluent treatment plant is substantially affected by volume of wastewater and characteristics of wastewater at point level pollution.

Multiple Regression

List-wise Deletion of Missing Data

	Mean	*Std Deviation*	*Label*
OC	5629956.106	1140594.565	
AA	22689.304	4534.955	
COD	6631.109	1083.524	
CU	71.004	17.302	
V1	2160.978	525.292	EQ

No. of Cases = 46

Correlation, 1-tailed Sig:

	OC	*AA*	*COD*	*CU*	*VI*
OC	1.000	-.478	-.229	.715	.725
	.	.000	.063	.000	.000
AA	-.478	1.000	.438	-.598	-.589
	.000	.	.001	.000	.000
COD	-.229	.438	1.000	-.697	-.702
	.063	.001	.	.000	.000
CU	.715	-.598	-.697	1.000	.993
	.000	.000	.000	.	.000
V1	.725	-.589	-.702	.993	1.000
	.000	.000	.000	.000	.

Multiple Regression

Equation Number 1 Dependent Variable. OC

Descriptive Statistics are printed on earlier Page

Block Number 1. Method: Enter AA COD CU V1

Variable (s) Entered on Step Number

1. V1: EQ (Effluent Quantity)
2. AA: Average alkality

3. COD: Chemical oxygen dissolved
4. CU: Capacity utilization of the common effluent treatment plant

Multiple R	.83052
R Square	.68977
Adjusted R Square	.65950
Standard Error	665561.78160

Analysis of Variance

	DF	*Sum of Squares*	*Mean Square*
Regression	4	40381146366909.3100	10095286591727.3
Residual	41	18161871890268.6100	442972485128.503

F = 22.78987 Signif F =.0000

Var-Covar Matrix of Regression Coefficients (B)

Below Diagonal: Covariance Above: Correlation

	V1	*AA*	*COD*	*CU*
V1	2607158.66	-.05263	.11657	-.98313
AA	-2323.8228	747.73815	-.04289	.13457
COD	24226.9641	-150.96418	16567.8789	-.00484
CU	-77907694	180602.157	-30559.013	2408631226

Multiple Regression

Equation Number 1 Dependent Variable. OC

Variables in the Equation

Variable	*B*	*SE B*	*95%Confdnce*	*Intrvl B*	*Beta*
AA	-26.947898	27.344801	-82.171845	28.276049	-.107144
COD	586.712541	128.716273	326.764752	846.660329	.557356
CU	-29895.34189	49077.80788	-129009.9864	69219.302571	-.453491
V1	3265.386870	1614.669830	4.494971	6526.278770	1.503849
(Constant)	-2582906.749	1583504.394	-5780858.773	615045.27435	

Variables in the Equation

Variable	Tolerance	VIF	T	Sig T
AA	.640131	1.562	-.985	.3302
COD	.506082	1.976	4.558	.0000
CU	.013652	73.249	-.609	.5458
V1	.013683	73.081	2.022	.0497
(Constant)	-1.631		.1105	

Co linearity Diagnostics

Number Eigenval Coned Variance Proportions

Index	Constant	AA	COD	CU		V1
4.84634	1.000	.00016	.00098	.00053	.00003	.00003
.13218	6.055	.00030	.04542	.01909	.00155	.00153
.01815	16.341	.00391	.69248	.32290	.00005	.00014
.00295	40.506	.99563	.25200	.65352	.00690	.00664
.00038	113.321	.00000	.00913	.00395	.99146	.99166

End Block Number 1 All requested variables entered.

Multiple Regression

Equation Number 1 Dependent Variable. OC (Operational cost of treatment)

Residuals Statistics

	Min	Max	Mean	Std Dev	N
*PRED	2336157.750	7144850.5000	5629956.1057	947290.2449	46
*RESID	-1158810.750	2156370.0000	.0000	635292.9664	46
*ZPRED	-3.4771	1.5992	.0000	1.0000	46
*ZRESID	-1.7411	3.2399	.0000	.9545	46

Total Cases = 46
Durbin-Watson Test = 1.331

Conclusion

The technical analysis of the CETP reveals that the pant operation is up to a mark considering inlet and outlet quality of wastewater. Regarding operational financial efficiency, it has distributed divided to its members. It is reasonable also considering paying capacity of the industry. Such common effluent treatment plants are necessary to established by the GIDC and industries association at each estate level in Gujarat. Their periodical evaluation is also necessary by impartial agencies.

Multiple Non-linear regression calculations with different methods reveal that operational cost of wastewater by common effluent treatment plant in our case is highly and significantly co-related with supply of wastewater, value of COD and capacity utilization of the plant. The manager of common effluent treatment plant should constantly monitor operational cost with characteristics of wastewater. Looking to our Model, the results are striking that partial cost elasticity is related with the independent variables to dependent variable. *Magnitude of estimates reveals that adjusted R2 =0.65 show positive sign. Durbin-Watson Test also shows positive sign i.e. 1.331*. This means our hypothesis can be accepted that the operational cost of common effluent treatment plant is highly and substantially affected by the characteristics of wastewater as adjusted R2 is found to be 0. 65, which means in 65% cases society can accept such economic phenomenon. The calculated different regression equations indicate that the hypothesis is fruitfully substantiated.

For cost reducing technology, it is necessary to review technology in-use in CETP as innovative latest technology can treat wastewater better than existing technology. Plant and machinery need to be changed for better management of CETP. Disposal hazardous chemicals in wastewater prohibited by global level agencies are disposed by the industries under this study. There is a case to have product-wise treatment of wastewater by a new technology.

70

Wastewater Management in a Fertilizer Industry: A Case Study

D.G. GAJGHATE, M.Z. HASAN, T. NANDY & S.N. KAUL

National Environmental Engineering Research Institute, Nagpur - 44002

ABSTRACT

The study provides details on wastewater generation, present status of pollution, performance evaluation of existing effluent treatment plants (ETPs) for a fertilizer plant producing diversified products. The implementations of proposed ETP facility achieve the intended objective of pollution control for existing plant and predicted to be effective in maintaining effluent quality to the desired level for future expansion of the plant.

Keywords: Fertilizer industry, ETP, Wastewater treatment

Introduction

Four decades of planning and development of the Fertilizer Industry (FI) have brought India to the frontline of fertilizer producing countries. India today is the third largest producer of nitrogenous fertilizers in the world. There are, at present, 64 fertilizer unit manufacturing a wide range of nitrogenous and complex fertilizers, comprising mainly urea, ammonium sulfate and diammonium phosphate (1).The production capacity of nitrogen has increased from a modest 85,000 tones in 1951-52 to 110,71 lakh tonnes as on 30 Nov. 99 and that of phosphatic fertilizers from 63,000 tonnes to 36,48 lakh tonnes of P_2O_5 during the same period. Despite massive increase in the indigenous capacity and production of fertilizer over the years, substantial imports, either in the form of finished fertilizers or its raw materials and intermediates are required every year to meet the country's growing demand (2&3).

Effective pollution control in a fertilizer industry, necessitates to ascertaining of the present status of wastewater pollution and the control measures practiced by the industry. With this in view, extensive studies have been carried out in a fertilizer industrial complex (4, 5 & 6). The present study pertains to a fertilizer industry which manufacture fertilizers, petrochemicals, industrial plastic and industrial gases and caters to the needs of the several industry. The products manufactured by the industries are as follow:

1.	Fertilizers	:	Urea, ammonium sulphate, diammonium phosphate.
2.	Petrochemicals	:	Cyclohexane, caprolactom, MEK oxime
3.	Plastics	:	Nylon-6, Melamine, Acrylics.
4.	Industrial gases	:	Argon, methane, ammonia, oxo synthesis gas.
5.	Acids	:	Phosphoric acid, sulphuric acid & oleum.

The major raw material used for these products are natural gas, LSHS, naphtha, benzene, sulphur and rock phosphate. The chemicals such as arsenic trioxide, potassium carbonate, caustic soda, hydrochloric acid, catalyst and resin are also required during the manufacturing of fertilizers and petrochemicals. In a fertilizer manufacturing complex, several pollutant are generated They are dissolved solids, suspended solids, urea and ammonia, phosphate, fluoride, oil, grease and organic matter (BOD & COD). The present study present the assessment of the wastewater impact on environment using following attributes:

1. Performance evaluation of the existing effluent treatment facilities
2. Characterization of final effluent being discharged into an estuary of river
3. Preparation of wastewater management plan.

The fertilizer industry consumes large quantity of water for processing fertilizer. The present water requirement under all categories is around 58 MLD (13 MGD). The breakup of the total water requirement for existing as well as proposed expansion is presented in Table 1. A perennial river is a source of water. The artesian radial wells are made in the bed of the river to pump water for its supply to industrial and domestic sector of FI. Increase in water requirement on account of proposed expansion may require some augmentation of existing water source. For this purpose, FI is already equipped with 6 pumps each at the two pumping stations built in the river bed.

Table 1: Breakup of Water Requirement for Existing Units

Sl. No.	*Particulars*	*MGD*
1.	Cooling	6.50
2.	Boiler Feed	2.00
3.	Process	0.70
4.	Domestic (Complex)	0.60
5.	Domestic (Colony)	1.20
6.	Irrigation/Fire/Misc.	0.50
7.	Outside Agencies	1.50
	Total	13.00

Sources of Wastewater and Management

Wastewater in FI at present emanates from different cluster of units such as fertilizer, petrochemicals, plastics & industrial gases. The wastewater from majority of the sources is recycled in the process and the rest after treatment is disposed off in the effluent channel.

The wastewater from these sources are mainly of following types:

1. Ammonia bearing wastewater
2. Urea bearing wastewater
3. Organic bearing wastewater

The effluents generated from ammonia, urea & melamine processes mainly contain ammonical nitrogen and urea nitrogen. For the above plants effluent containing ammonical nitrogen and urea nitrogen are segregated from the respective plant and collected in a separate storage pond. It is reused in process of phosphatic group of plants as and when required. Hence the contaminated effluents from these plants are not discharged.

The arsenic bearing effluent particularly from ammonia-I & II plants are collected in separate pits in the respective plant and this effluent is ejected by steam ejectors to the evaporators. The evaporation takes place by the steam coils and semi-solid residue left out after evaporation is taken out and filled in metallic drum and gunnitted with cement and mortar. The solid waste so collected is stored at FI in large 25 mm shell thickness (MOC-CS with SS-316 Titanium 2 mm thick ladding) vessel installed above the ground level of foundation at an isolated location.

The Phosphoric Acid (PA) is produced by reacting rock phosphate and sulphuric acid and gypsum is a by-product of phosphoric acid plant. Gypsum is mainly used for manufacture of ammonium sulphate and the excess gypsum in the form of slurry is pumped to gypsum pond. The settled gypsum is recovered and is used as soil conditioners by farmers. The supernatant effluent goes to cooling pond. Chalk is a by-product of ammonium sulphate plant. This chalk is pumped in the form of chalk slurry to the chalk pond along with the acidic effluent of PA plant. The chalk settles in the chalk pond while the supernatant is conveyed to cooling pond. This cooling pond water is reused in PA plant for various purposes such as (i) condensers of concentrators, (ii) for scrubbing fumes in fumes scrubber, (iii) for condensers of flash cooling, (iv) washing of pan filter and belt filter.

Effluent from diammonium phosphate (DAP) plant generated is diverted to equalization tank where solid sludge settles and supernatant is pumped to the process requirement. Hence effluent from phosphatic group is not discharged into disposal system. The solid so collected is also reused in the plant for its recovery. Thus there is no discharge from these group of plants.

Effluent from caprolactam processes mainly contains COD & BOD, ammonium nitrogen and nitrate nitrogen. The effluents of these plants are having separate collection system and treatment system. The effluent generated from capro-I & nylon-6 is treated in old BOD plant through activated sludge process for COD reduction and the treated effluent is discharged into the collection pond namely new pond. Similarly effluent generated from new capro plant is also treated by activated sludge process in the new BOD treatment plant for reduction of COD & BOD. For removal of ammonium nitrogen and nitrate nitrogen, ammonia stripper unit and de-nitrification unit is provided. After the treatment the effluent is disposed to new pond from where it gets mixed along with non-contaminated effluents collected from sulfuric acid process and utility plants of the factory and finally discharged in to ECP.

The total wastewater generation from the existing units of FI is around 27320m^3/day. Table 2 gives the quantity of effluent generated from these units. Major portion of wastewater is recycled in the process and only a small portion is taken to the new pump house from where it is discharged in to the effluent channel.

Table 2: Quantity of Effluent Generated in Various Units

Sl. No.	*Source of Effluent*	*M^3/Day*
1.	Old Capro. Treated Effluent	1183.333
2.	Capro. Exp. Treated Effluent	2143.333
3.	Ammonia-I	2120
4.	Ammonia-II	2143.333
5.	Ammonia-III	693.3333
6.	Urea	280
7.	Melamine	1383.333
8.	PGR	1383.333
9.	Bagging	686.6667
10.	Utility	2521.667
11.	Ammonia Storage	333.3333
12.	INOCIL	686.6667
13.	HWP	1688.333
14.	Colony Sewage	4480
15.	SA-IV Cooling Tower	133.3333
16.	SA-III	281.6667
17.	CO-Gen Phase-I	168.6667
18.	CO-Gen Phase-II	2013.333
19.	CO-Gen Phase-III	2013.333
20.	R & D, Central Lab.	686.6667
21.	Miscellaneous	2120
	Total M^3/Day	27320.33

Assessment of Pollution Control Facilities

An evaluation of the existing ETP was carried out by collecting composite sample at six hourly interval over a period of 24 hrs. The samples collected from the following streams:

1. Equalization Tank: (ET-I) of capro expansion plant.
2. Equalization Tank: (ET-II) of capro expansion plant.
3. Equalization Tank: (ET-III) of capro expansion plant.
4. Combined effluent of ET-I, ET-II & ET-III of capro expansion plant.
5. Influent from old BOD capro plant of capro-I.
6. Final effluent from treatment plant of capro-I.
7. Combined effluent

The organic waste is received in Equalization Tank (ET)-III intermittently to non-continuous flow. Effluent from ET-III is received in aeration tank and flows to clarifier. Effluent containing NH_3 and NO_3 is received in ET-I also on intermittent basis which, is send for ammonia striping and then routed through ET-II. Effluent from the panic pond an other stream is also received in ET-II. The effluent from ET-II is denitrified and then sent to aeration tank. The combined effluent from clarifier of ETP-II is discharged in to ETP-I consisting of equalization tank, aeration tank, oxidation ditch and clarifier. The effluent from old capro plant and domestic sewage from colony are also received in equalization Tank. The final treated effluent flow through a disposal line, prior to its discharged into effluent channel.

The various wastewater streams, originating from different section of the complex are received at new ETP. These effluent streams are first discharged in the equalization tank so as to get equalize effluent with respect to flow as well as wastewater characteristics like pH, COD and BOD. The wastewater is then pumped to the treatment plant for removal of these pollutants. A panic pond has been provided which is utilized for storing the process water during any emergency or shutdown etc. This ponds is also used for storing of run off during the rainy season.

Equalization Tank-II received two acid effluents bearing stream. The first one originates from anode plant and contains mixture of formic acid & acetic acid. Second stream is generated from hydroxylamine sulphate plant. Both the streams are collected in equalization tank-II. It is neutralized in neutralization tank-I using 20% caustic solution. Treated effluent of ammonia stripper is also added in neutralization tank-I. The effluent from neutralization tank-I is then taken in Dintrification Reactor (DNR). The reactor is aerobic bio-reactor, where organic carbon present in the effluent is used as a source of carbon by micro-organisms for denitrification process. The overflow of DNR is then fed for aeration tank for COD & BOD reduction.

Equalization Tank-III received the organic bearing effluent streams are mainly generated from Anode plant, lactum plant. These effluent streams are collected in ET-III. The equalized effluents from these streams are neutralized by 20% caustic solution. The neutralized effluent is then fed to aeration tank along with denitrified effluent. The aeration tank is designed as an activated sludge process in extended aeration mode. The biodegradation of organic complex takes place in aeration tank with the help of microorganism. Dissolved oxygen is supplied through fixed surface aerators. Aerated biomass is then fed to final clarifier for separation of sludge and clear effluent. The overflow of clarifier is collected in an intermediate sump and pumped to final polishing stage in post aeration tank.

Floating oil and suspended solids in effluent streams are separated by mechanical device. The effluent free from oil and grease is allowed to flow to equalization tant (ET-I). In equalization tank the effluent detention period is 24 hrs. where the effluent gets self neutralized. Organic load is also equalized before feeding to activated biomass.

Organic matter is degraded by bacteria under aerobic condition in the oxidation ditches. Mixed Liquor Suspended Solids (MLSS) concentration is maintained at 4000 mg/l. The sludge is recirculated to oxidation ditch from primary and secondary clarifiers. The excess sludge is disposed off in the sludge drying beds. Fig. 1 present the flow sheet of effluent treatment plant with wastewater generation. The supernatant is disposed into the disposal system. The characteristics of the different waste streams are given in Table 3.

Disposal of Wastewater

Treated wastewater from FI is discharged into an effluent channel from where it is finally discharged into an estuary. The mixed wastewater disposed off in the effluent disposal channel is used by the farmers for irrigation in the neighboring fields. Treated effluent conforming to the stipulated effluent quality standards is only discharged into the effluent channel. The non-contaminated and/or less contaminated effluent from various processes respect to ammonical nitrogen and total nitrogen are diverted to a collection pond namely new pond along with effluent received from other plants. Thereafter the same is disposed to effluent channel project (ECP). The second part of the non-contaminated water from the cooling tower bleed off of these plants goes to another collection pond, called "old pond" from where the same is also discharged along with other non-contaminated effluent of the factory including domestic waste after its preliminary treatment and or secondary treatment to ECP channel. The characteristics of contaminated and non-contaminated streams is given in the Table 4.

Table 3: Characteristics of Effluent Generation from various Processes and Stage of Treatment

Parameter	*Sample No.*						
	1	*2*	*3*	*4*	*5*	*6*	*7*
Temperature °C	42	38	40	34	35	30	28
Colour	Yellowish	Dark Brown	Colour-Less	Light Yellow	Light Gray	Light Yellow	Colour-less
pH	8.3	8.7	6	7.7	3.6	7.6	6.6
Total Alkalinity as $CaCO_3$,	456	—	120	2830	270	2030	470
Conductivity μS/cm	15398	—	12000	27000	75000	10000	6000
Total Solids	3604	61550	1978	4684	4090	3675	1818
Total Dissolved Solids	3500	61324	1778	4314	4046	3650	1810
Suspended Solids	104	226	200	370	530	25	8
Dissolved Oxygen (DO)	Nil	Nil	Nil	Nil	Nil	1.5	1.7
Biochemical Oxygen Demand	2850	7500	3700	720	1220	418	130
Chemical Oxygen Demand	17600	47600	5299	34836	3880	880	320
Total Nitrogen as N	980	278	30	20	20	7	200
Total Ammonia Nitrogen as NH_3-N	224	3500	241	162	154	15.6	64
Nitrate-Nitrogen as NH_3-N	95.9	180.7	7.4	2.4	0.32	8.7	9.8
Chlorides as Cl	4.2	0.5	0.13	0.16	0.16	0.02	0.09
Sulphate as SO_4	30	1600	40	100	100	70	380
Phosphate PO_4	20	19200	600	700	2500	1000	440
Sodium as Na	52	17200	540	800	740	925	460
Potassium as K	1.6	52	4	7	6	12	11
Calcium as C	44	16	20	20	40	30	36
Magnesium as Mg	24	14	7	10	17	7	31
Oil and Grease	4	54	4	2	2	3	2
Heavy Metals							
Cd	0.011	0.014	0.014	0.016	0.020	0.016	0.016
Cr ND ND	0.033	0.113	0.092	1.840	ND		
Ni	0.016	0.079	0.007	0.059	0.042	0.048	0.040
Fe	2.054	1.587	1.195	1.433	1.213	1.882	0.863
Pb	0.155	0.195	0.200	0.160	0.190	0.160	0.240
Zn	0.047	ND	ND	0.014	0.001	ND	0.008

All values except pH, Temperature, Colour, Conductivity and Turbidity are expressed as mg/L

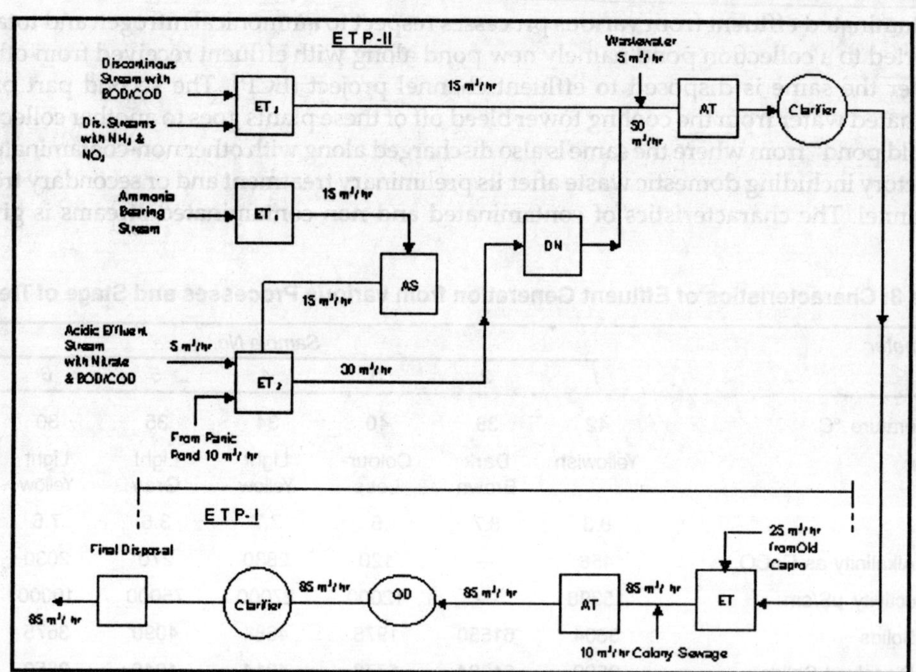

Fig. 1: Flow Sheet of Effluent Treatment Plant

Table 4: Characteristics at Old Pumping station, New Pumping Station and Final Discharge Point

Parameter	*Old Pumping Station*	*New Pumping Station*	*Final Discharge Point*	*Stipulated Standards*
pH	7.4-8.5	6.5-8.5	6.5-8.5	6.5-8.5
Total Dissolved Solids	520-1120	480-2352	494-1800	5000
Suspended Solids	10-64	10-140	6-104	100
Ammonical Nitrogen	6-37	6-50	3-39	50
Total Nitrogen	6-95	6-100	11-112	100
Sulphate as SO_4	5-20	55-360	40-273	1000
Phosphate as PO_4	2-10	1-11	2-11	15
Chemical Oxygen Demand	16-181	38-240	22-152	250
Biochemical Oxygen Demand	—	—	14-30	50
Fluoride as F	—	—	0.25-1.05	1.50

All values are in mg/l except pH

The characteristics of the wastewater from old pond, new pond & combined effluent at effluent channel are given in Table 5. FI is discharging its effluent through two lines to ECP at discharge points via two disposal ponds *viz.* old pond & new pond.

Table 5: Wastewater Quality-Contaminated and Non-Contaminated Streams

Parameter	*Contaminated Stream*	*Non-contaminated Stream*
Temperature, °C	30	31
Colour	Brownish	clear
pH	9.6	7.6
Total Alkalinity as $CaCO_3$	2000	320
Turbidity, NTU	15	1
Conductivity, uS/cm	6255	938
Total Solids	4468	654
Total Dissolved Solids	3988	554
Suspended Solids	480	100
Dissolved Oxygen (DO)	1.2	4.8
Biochemical Oxygen Demand	125	2
Chemical Oxygen Demand	408	100
Total Nitrogen as N	115	7.5
Total Ammonia Nitrogen as NH_3-N	3.8	—
Nitrate-Nitrogen as NO_3-N	1.2	0.65
Chlorides as Cl	150	60
Sulphate as SO_4	1925	28
Phosphate PO_4	455	2.5
Sodium as Na	500	66
Potassium as K	68	3
Calcium as Ca	40	17
Magnesium as Mg	5	62
Oil and Grease	1.2	2

All values except Temperature, pH, Turbidity and Conductivity are expressed as mg/L

Sewage Treatment Plant

The sewage treatment plant is designed for a flow rate of 200m³/hr and comprises of the following processes:

1. Activated sludge process followed by clarification and disinfection
2. Pressure filtration, Softening and Clarification (Tertiary treatment)

A separate collection system has been provided for collection of domestic wastewater which is treated in the sewage treatment plant. Performance of the sewage treatment plant is shown in Table 6. The reclaimed treated waste is used for cooling towers and gardening. A portion of the sewage is used for seeding purpose in biological treatment of the wastewater in the BOD reduction plant.

Table 6: Performance Evaluation of Sewage Treatment Plant

Parameter	*Influent Wastewater*	*Treated Effluent*
Temperature, °C	31	30
Colour	grey	clear
pH	7.0	7.5
Total Hardness as $CaCO_3$	230	200
Calcium Hardness as $CaCO_3$	115	110
Total Dissolved Solids	460	430
Suspended Solids	90	30
Biochemical Oxygen Demand	100	3
Chemical Oxygen Demand	140	8
Ammonical Nitrogen as NH_3-N	16	0.3
Chlorides as Cl	100	100
Sulphate as SO_4	11	11

All values except Temperature and pH are expressed as mg/L

Revamping of Pollution Control Facilities

The Effluent Treatment Plant was installed and commissioned alongwith caprolactam expansion plant to remove the major pollutants present in the effluent generated from different sections of caprolactam expansion plant. The major pollutants present in the wastewater streams are; ammonical nitrogen, BOD, COD, and nitrate nitrogen. In the effluent treatment plant, ammonical nitrogen and nitrate nitrogen are removed by stripping and denitrification. BOD & COD are removed by activated sludge process through aeration tank having capacity of 4800 M^3 and six aerators of 40 HP are fitted. The Effluent Treatment Plant is designed to treat 50 M^3/hr of effluent.

In order to meet the statutory effluent norms the effluent treatment plant is revamped by modifying the existing facility/addition of new equipment's and addition of tertiary treatment for removal of BOD & COD from the effluent so as to make the effluent treatment plant having two stage biological treatment with sludge handling facilities and tertiary treatment. The details of revamping designed parameter is shown in the Table 7.

The effluent from various streams such as DNRC, ET, III & PANIC pond is collected in the Biotower feed Sump. The feed sump is designed for 30 minutes of retention time for flow capacity of 80 M^3/hr. The acid and alkali is added for pH correction in the feed sump. The nutrient is added in this sump from existing facility. The effluent from the feed sump is pumped to Biotower with the help of Biotower feed pumps. The Biotower is designed for 2.65 kg/m^3/day of BOD loading rate and 1.5m^3/m^2/day of wetting rate. The treated effluent from Biotower bottom is collected in the Biotower recirculation sump from where it is pumped back at the inlet of Biotower. The normal flow of 80 M^3/hr is to be taken from recirculation line to existing aeration tanks and new aeration tanks in ratio of 65 % and 35 % approx. for further treatment for parallel operation or series operation.

Table 7: Revamp of Effluent Treatment Plant Designed for Input and Output Parameters

Parameter		Value
Inlet Parameter		
Flow	:	80 M^3/hr.
pH	:	6.5 to 8.5
Design BOD Load	:	300 Kg/hr.
BOD	:	4000 mg/l (Normal)
	:	6000 mg/l (Maximum)
Design COD Load	:	600 Kg/hr,
COD	:	8000 mg/l (Normal)
	:	12000 mg/l (Maximum)
Suspended Solids	:	200 mg/l
Ammonical Nitrogen	:	100 mg/l
Outlet Parameter		
pH	:	6.5 to 8.5
BOD	:	50 mg/l (max.)
COD	:	250 mg/l (max.)
Suspended Solids	:	10-20 mg/l
Ammonical Nitrogen	:	50 mg/l

The effluent from first stage Biotower is subjected to extended aeration type activated sludge process for further removal of BOD & COD. The effluent is divided in two streams and two separate sets of extended aeration type activated sludge process are operated in parallel. One set of existing system which includes aeration tank, secondary clarifier and sludge recirculation facilities is used for the same, whereas new set consisting of aeration tank, secondary clarifier and sludge recirculation facilities are provided. Provision is also made to operate both the systems in series, and hydraulically both the sets of above systems are designed for minimum 80 M^3/hr flow capacity.

The treated effluent from the existing intermediate sump after biological treatment is pumped to clarified water sump/new aeration tank through pipeline. Two numbers of new effluent transfer pumps of suitable capacity are provided for the same. The new sludge drying beds are constructed considering total sludge to be generated in Biotower and two sets of extended aeration type activated sludge processes. The filtrate from the sludge beds is recirculated back to new aeration tank with the help of two filtrate transfer pumps.

The treated wastewater after both the secondary clarifiers is collected in the clarified water sump from where the same shall be subjected to tertiary treatment. The treated effluent from secondary clarifiers after biological treatment is collected in clarified water sump from where it is pumped to the tertiary units. The clarified water transfer is designed for 30 minutes of retention time based on 80 M^3/hr flow. Tertiary treatment consists of two number pressure sand filters. The suspended solids as well as COD are removed with the help of pressure sand filters and activated carbon filter. The pressure sand filters are provided with air and water backwash whereas activated carbon filters are provided with only water backwash. The water for backwash is drawn from discharge line of activated carbon filters. The backwash water is collected in filtrate sump from where it is pumped to new aeration tank.

The treated effluent water from the tertiary treatment facility is taken to existing pH correction sump in which acid/alkali is added for final pH correction before sending to existing new effluent pond.

Conclusions

In the fertilizer industry the effluent from the treatment plants are monitored for incoming effluent parameters, in process effluent treatment parameters and final effluent discharge from the effluent treatment facility. The proposed revamping of the pollution control facilities by NEERI will improve its final effluent characteristic and will also take care of any proposed expansion meeting the statutory Standard. The final treated effluent conforming to prescribed Standards are used for green belt development. Maximum emphasis will be given to wastewater recycling after commissioning of the proposed expansion. This will help to maintain total quantum of pollutant discharge well below the stipulated limits. The wastewater generated from water treatment plant and the boiler blow-down water shall not be allowed to mix up with the ammonia and urea plant effluents. Proper segregation of effluent will be made for ensuring adequate treatment when the concentration of the pollutants are high in the final effluent. Corrective measures will be taken at source and the effluent shall be discharged at a regulated rate after treatment to conform to the prescribed limits of statutory authority.

Acknowledgement

Authors are grateful to Dr. R.N. Singh, Director, NEERI, Nagpur for permission to present the paper in the workshop.

REFERENCES

1. India, 2001, A Reference Annual, Publication Division, Ministry of Information and Broadcasting, Govt. of India. Patiala House, New Delhi, 110001.
2. Economic Survey, 1998-99, Govt. of India, Ministry of Finance, Economic Division, Delhi.
3. CPCB, New Delhi, Comprehensive Industry Document: Fertilizer Industry: COINDS/50/1994-95.
4. NEERI Report, 1998, Rapid environmental Impact and Risk Assessment Studies for the Proposed Urea Project, Vadodara.
5. APHA, AWWA, WPCF, 1989, Standard Methods for Examination of Water and Wastewater. 16th Edition.
6. D.G. Gajghate, M.Z. Hasan, T. Nandy and S.N. Kaul, 2001, Performance evaluation of a full scale ETP for a fertilizer industry with recourse to capacity expansion, Jr. of Industrial Pollution Control 17 (2), 245 – 255.

71

Wastewater Management in Antibiotic Industries

SHANTA SATYANARAYAN, RAMAKANT, ATUL P. VANERKAR, RITA S. DHODAPKAR & S.N. KAUL

National Environmental Engineering Research Institute, Nehru Marg, Nagpur

ABSTRACT

Antibiotics are one of the largest groups of drugs manufactured in India. At present there are more than 16000 bulk drug industries, manufacturing antibiotic formulations. Two major sources of wastewater generation are from the manufacture of basic drugs and another from formulation units. Waste generation from the basic drug manufacturing houses are higher in quantity than the formulating units. Strength of wastewater generated in an antibiotic industry depends on manufacturing process, nature of products, raw materials used and purification process of the finished products. Treatment required for different wastewaters depend upon their characteristics. A common method of treatment for many antibiotic wastewaters are equalization, neutralization, sedimentation followed by biological methods such as trickling filter, activated sludge system, oxidation alone or in combination. Better housekeeping and segregation of high strength wastewater reduces the pollution load on the effluent treatment plant for better removal efficiency. Wastewater management using cleaner technologies reuse/ recycling, are being tried now a days to treat the wastewater efficiently and also conserve the water usage thereby reducing wastewater generation. An attempt has been made in this paper to discuss the feasibility of different treatment technologies used in different antibiotic industries and the quality of the resultant effluent quality.

Kewwords: Antibiotics, wastewater, physico-chemical, aerobic, anaerobic.

Introduction

Antibiotics are chemical compounds derived from microorganisms which have the capacity of inhibiting the growth of and even destroying other microorganisms. In the production of these antibiotics, huge quantities of water are required and wastewater generation is also large. Antibiotic industry is one of the most polluting industries in India. Spent broth and column wash from solvent extraction unit, generated by antibiotic industry has dark brown colour and has high COD/BOD and TDS. Once the antibiotic industry effluent is discharged without treatment it creates odour nuisance as it purifies faster in nature. Of late there has been a tremendous growth in the Indian antibiotic/pharmaceuticals and bulk drug industries. According to the latest information there are more than 16000 units in India, which manufacture and formulate various drugs and pharmaceutical products. This phenomenal growth has created problems of environmental protection from various emissions and discharges from these industries.

Production in antibiotic industries varies quite frequently based on the market demands. The wastewaters resulting from these units contain considerable amount of pollutants due to which it is not possible to discharge the untreated wastewater into surface water bodies and also into the Municipal sewers without creating great damage to the environment. There are so many industries which manufacture antibiotics/pharmaceuticals, but no effluent from one industry is similar to another and same effluent from one industry also changes its characteristics very frequently as production process are of batch wise nature. This non-uniformity in the effluent characteristics makes it problematic during treatment process. Due to these problems, antibiotic wastewater needs a treatment plant which is a combination of two or more different treatment systems.

Wastewater from pharmaceutical plants producing antibiotics, penicillin and similar antibiotics are strong in nature (high COD, BOD and SS) and generally should not be treated with domestic sewage, unless the entire load is considered in the design and operation of the treatment plant. Effluent from the manufacturing of antibiotics presents new and difficult disposal problems because of the presence of antibiotic substances along with large amounts of organic matter.

Antibiotic industry effluent frequently results in the loss of needed/desirable microorganisms/ flora in the biological treatment systems but other organism also develop. Hence proper acclimatization along with modifications in the treatment systems is the need of the hour. Hence an attempt has been made to review the wastewater management practices and problems encountered in the antibiotic industries.

Antibiotic Manufacturing Processes

Antibiotics are chemical compounds derived from microorganisms, which have the capacity of inhibiting the growth of and even destroying other microorganisms. Antibiotics and vitamins are produced by the fermentation of fairly complex nutrient solutions of organic and inorganic salts by molds, fungi and bacteria. The antibiotics are produced by certain fungi, bacteria, algae and lichens are presented in Table 1. Major raw materials used in the production of antibiotics are presented in Table 2.

Process Details

Majorities of the antibiotics are produced by fermentation followed by solvent extraction. Depending on the antibiotic to be produced the manufacturing process varies accordingly. But basic fermentation step remains more or less same for all the antibiotics. According to the growth characteristics of the mold/fungus/bacteria, they are cultured in the sterile conditions in the laboratory. From the main culture or as it is called the mother culture, large volumes of culture is prepared. For preparing inoculant media, peptone, yeast extract, beef extract and dextrose are used, once the culture is ready, it passes through the fermentation stage. Fermentation process consists of inoculation, seeding and fermentation,

Table 1: Details of Different Molds/Fungi/Bacteria Used for the Production of Antibiotics

Srl No.	*Antibiotics*	*Molds/Fungi/Bacteria*
1.	Cephlosporin	Cephalosporium – acremonium
2.	Penicillin	Penicilium notatum/chrysogenum
3.	Streptomycin	Griseus
4.	Erythromycin	Steptomyces erytherus
5.	Hamycin	Streptomycin primprina
6.	Griseoflulvin	Penicillium griesolfulvro
7.	Gentamycin	Micromonospora purpurea

Table 2: Major Raw Materials Required for Various Antibiotic Production

Sl. No.	*Description*	*Sl. No.*	*Description*
1.	Ammonium Acetate	13.	Dextrose
2.	Ammonium Sulphate	14.	De-emulsifier
3.	Ammonium Solution	15.	Ethyl Acetate
4.	Acetic Acid	16.	Ferrous Sulphate
5.	Acetone	17.	Formaldehyde
6.	Antifoam	18.	N-Hexane
7.	Activated Carbon	17.	Formaldehyde
8.	Butanol	19.	Potassium Carbonate
9.	Sucrose	20.	Sulphuric Acid
10.	Corn steep liquor	21.	Sodium Sulphate
11	Calcium Carbonate	22.	Sodium Chloride
12.	Caustic Soda	23.	Urea

Inoculation stage is initiated by transferring the inoculum under aseptic conditions into large inoculating tank, having sterile medium consisting of corn steep liquor, soyabean meal, ammonium sulphate and corn starch. The agitation, aeration and the maintenance of temperature continue in inoculation for at least 24-30 hours till sufficient growth of fungus/mold/bacteria are produced. Then the entire content from the inoculum tank is transferred to the seed tank. After transferring the entire growth of the first inoculation stage into the seed tank stage, the same process is continued for obtaining abundant growth. At the fermentation stage, production of antibiotics like Penicillin, Cephalosporin, Hamycin etc takes place and it takes 154-208 hours after which the culture is harvested. Harvested material is sent for recovery of antibiotics to recovery section. Recovery stage includes filtration, absorption through resin column, elusion, ion exchange, concentration and precipitation. Further extraction of antibiotics varies for different antibiotic products. Few important antibiotic's manufacturing process details are discussed below:

Cephalosporin

After fermentation, the crude cephalosporin powder is mixed with acetone and water in a mixing vessel in which activated carbon is added. The mixture is sterilized and the product is separated out and dried. The generated mother liquor is distilled to recover solvent, which is recycled, and the organic matter with little solvent is incinerated. Process flow diagram with wastewater generation sources for the production of Cephalosporin is shown in Fig. 1.

Penicillin

Penicillin is produced by the aerobic fermentation using the mold penicillin chrysogynum and penicillin notatum. The fermented broth is then subjected to the filtration through rotary vacuum filter. A solid waste in the form of mycelia is generated which also exerts organic load during washings. But major mycelial mass is sold for the preparation of protein meal. The filtrate is subjected to penicillin extraction with organic solvents (butanol) so that entire penicillin is transferred from aqueous phase to solvent phase and spent broth is discarded as wastewater. The extracted solvent containing penicillin is further refined in solvent phase is further extracted with buffer (potassium carbonate) at higher pH and solvent is sent to solvent recovery section. Penicillin contained in buffer is finally subjected to azeotropic distillation followed by crystallization, filtration and drying through carbon treatment and sparkler filter. Process flow diagram with wastewater generation sources for the production of Penicillin is shown in Fig. 2.

Streptomycin

Streptomycin is produced in a similar way as cephalosporin, using a streptomyces griseus culture on a medium consisting of glucose, corn steep liquor etc. The fermentation broth is filtered and the filtrate is adsorbed on charcoal or a resin. The streptomycin is eluted from the charcoal or resin with dilute acids. The elute is then neutralized and concentrated. The crude streptomycin is then precipitated by the addition of acetone and further purified. Process flow diagram with wastewater generation sources for the production of Streptomycin is shown in Fig. 3.

Erythromycin

It is very important to maintain proper environmental conditions like pH, temperature, nutrients and dissolved oxygen for effective and optimized production of erythromycin. The fermented broth is extracted in the ethylene dichloride solvent to extract erythromycin. Moreover at this stage ammonium thiocyanate is also added to convert erythromycin into erythromycin thiocyanate. The extract is then concentrated through evaporator cum crystallizer to obtain crystals of erythromycin. The crystallized slurry is then filtered in a pressure filter cum dryier when the crystals are dried. Process flow diagram with wastewater generation sources for the production of Erythromycin is shown in Fig. 4.

Aureofungin and Hamycin

After fermentation the fermented broth is filtered and extracted. Extracted liquid is evaporated and chilled. After chilling it is centrifuged and washed several times with hexane and ethyl acetate. The washed wet cake is put through drier followed by ball mill, sieving and finally Aureofungin and Hamycin powder are packed. Process flow diagram with wastewater generation sources for the production of Aureofungin and Hamycin are shown in Figs. 5 & 6.

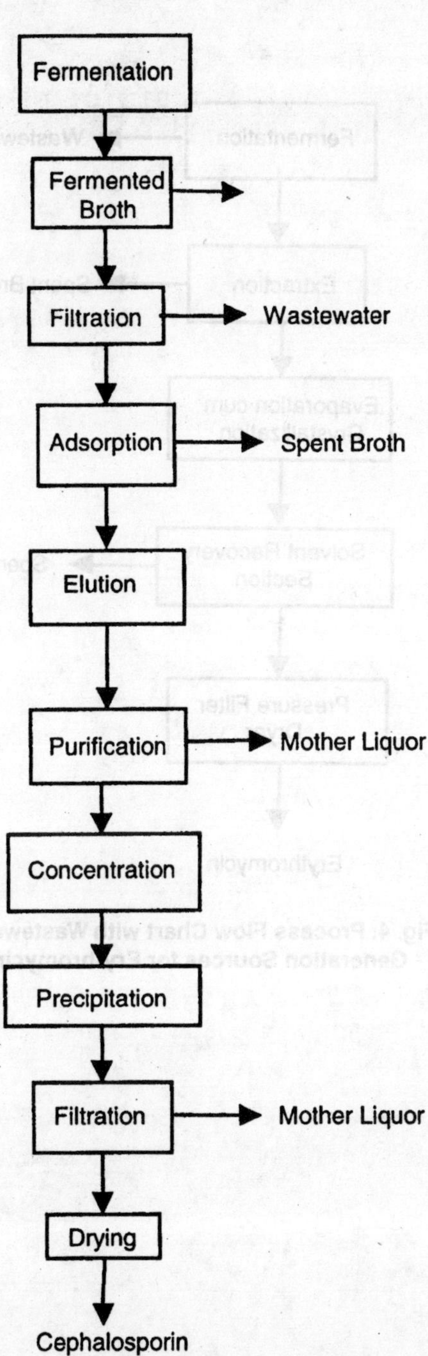

Fig. 1: Process Flow Chart with Wastewater Generation Sources for Cephalosporin

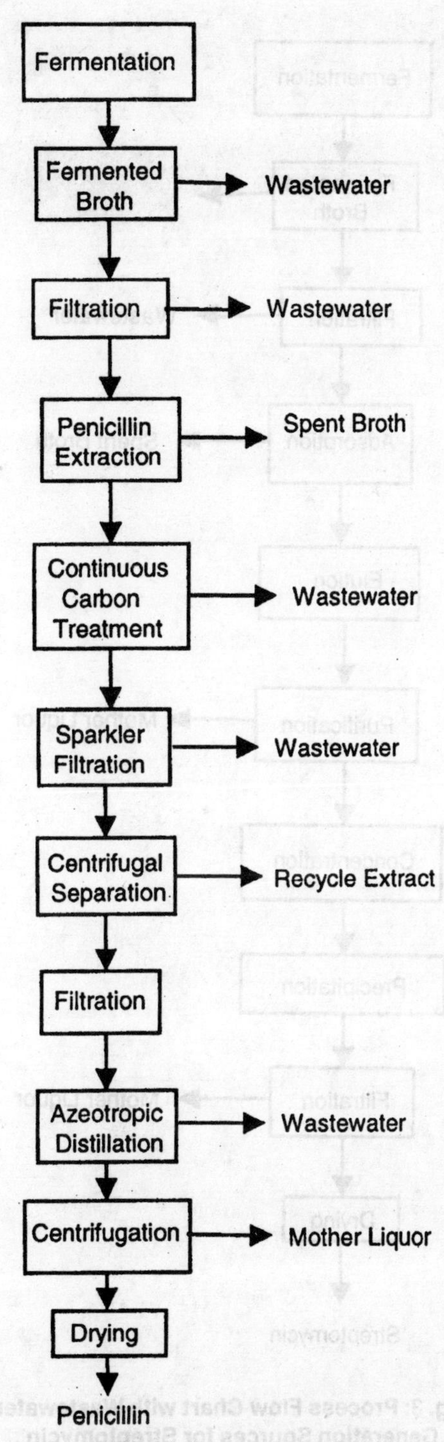

Fig. 2: Process Flow Chart with Wastewater Generation Sources for Penicillin

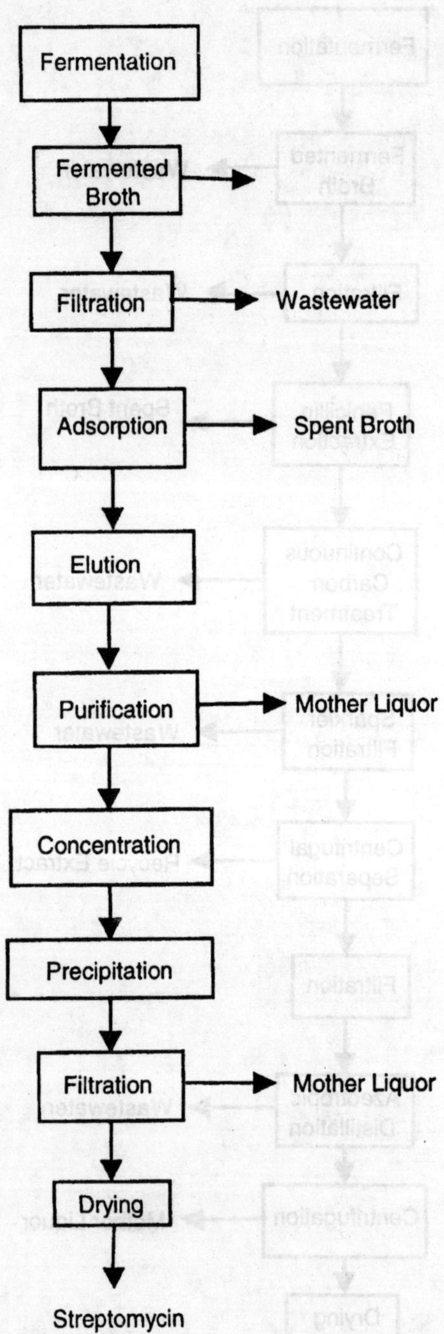

Fig. 3: Process Flow Chart with Wastewater Generation Sources for Streptomycin

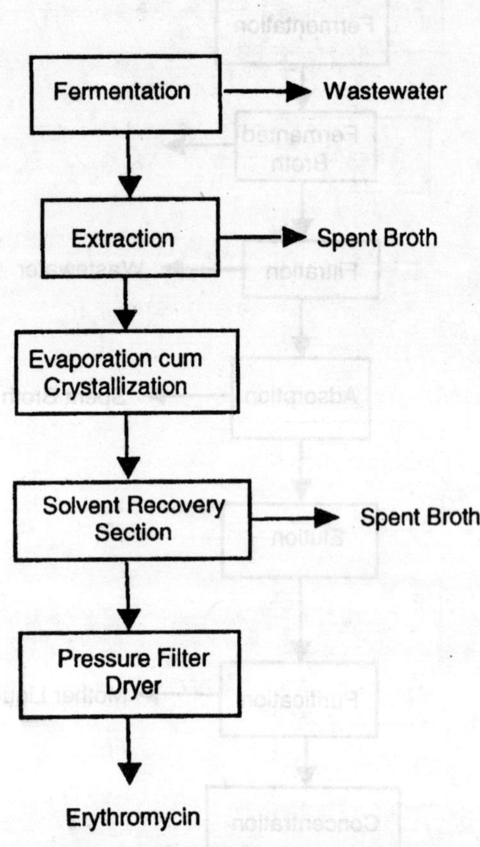

Fig. 4: Process Flow Chart with Wastewater Generation Sources for Erythromycin

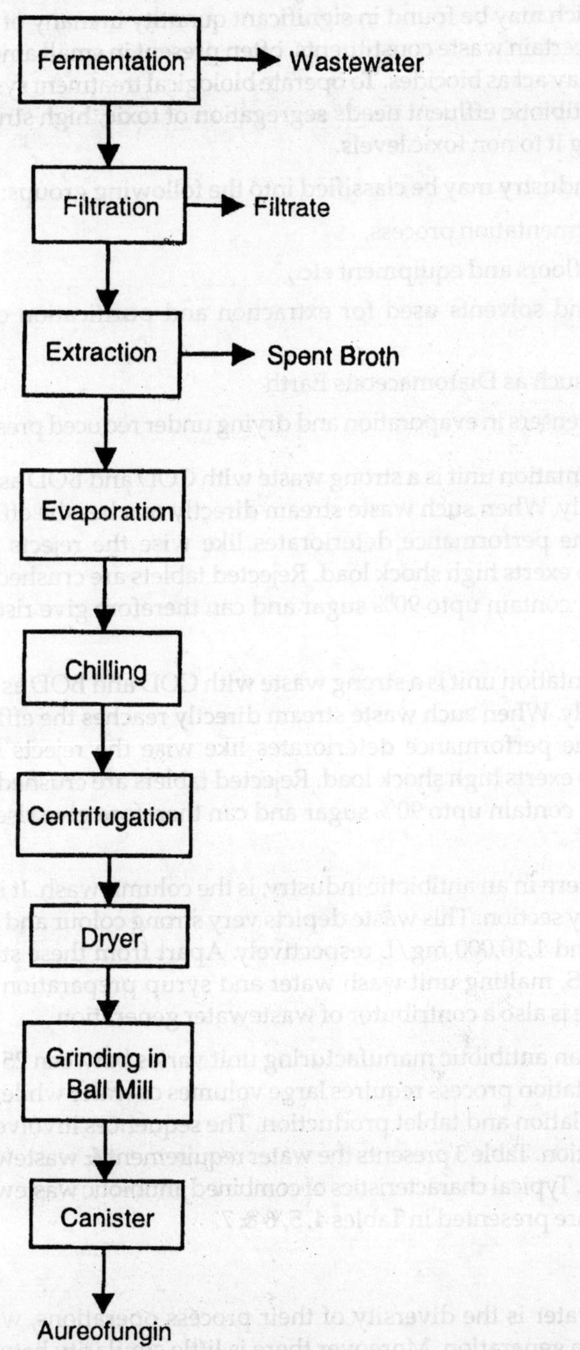

Fig. 5: Process Flow Chart with Wastewater Generation Sources for Aureofungin

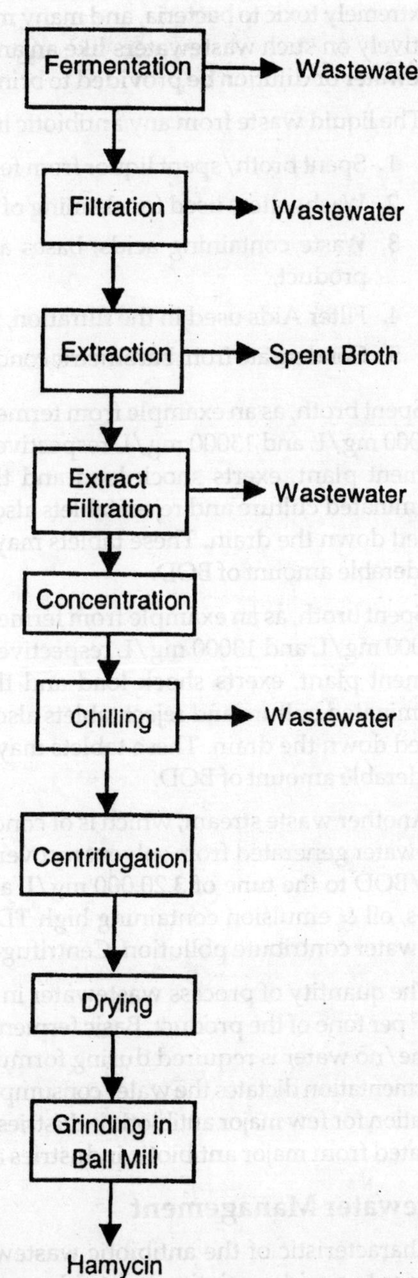

Fig. 6: Process Flow Chart with Wastewater Generation Sources for Hamycin

Wastewater Generation and Characterization

The majority of organic chemicals which may be found in significant quantity in many of these process wastes are readily biodegradable, certain waste constituents, often present in small amounts are extremely toxic to bacteria, and many may act as biocides. To operate biological treatment systems effectively on such wastewaters like an antibiotic effluent needs segregation of toxic, high strength wastewater or dilution be provided to bring it to non toxic.levels.

The liquid waste from any antibiotic industry may be classified into the following groups:

1. Spent broth/spent liquor from fermentation process,
2. Washwaters used for cleaning of floors and equipment etc.,
3. Waste containing acids, bases and solvents used for extraction and purification of the product,
4. Filter Aids used in the filtration, such as Diatomaceous Earth
5. Condensate from barometric condensers in evaporation and drying under reduced pressure.

Spent broth, as an example from fermentation unit is a strong waste with COD and BOD as high as 26000 mg/L and 13000 mg/L respectively. When such waste stream directly reaches the effluent treatment plant, exerts shock load and the performance deteriorates like wise the rejects from contaminated culture and reject tablets also exerts high shock load. Rejected tablets are crushed and washed down the drain. These tablets may contain upto 90% sugar and can therefore give rise to a considerable amount of BOD.

Spent broth, as an example from fermentation unit is a strong waste with COD and BOD as high as 26000 mg/L and 13000 mg/L respectively. When such waste stream directly reaches the effluent treatment plant, exerts shock load and the performance deteriorates like wise the rejects from contaminated culture and reject tablets also exerts high shock load. Rejected tablets are crushed and washed down the drain. These tablets may contain upto 90% sugar and can therefore give rise to a considerable amount of BOD.

Another waste stream, which is of concern in an antibiotic industry, is the column wash. It is the wastewater generated from solvent recovery section. This waste depicts very strong colour and high COD/BOD to the tune of 3,20,000 mg/L and 1,10,000 mg/L respectively. Apart from these strong wastes, oil & emulsion containing high TDS, malting unit wash water and syrup preparation unit wash water contribute pollution. Centrifuge is also a contributor of wastewater generation.

The quantity of process wastewater in an antibiotic manufacturing unit varies between 25 m^3 – 975 m^3 per tone of the product. Basic fermentation process requires large volumes of water while, less volume/no water is required during formulation and tablet production. The sequences involved in the fermentation dictates the water consumption. Table 3 presents the water requirement & wastewater generation for few major antibiotic industries. Typical characteristics of combined antibiotic wastewater generated from major antibiotic industries are presented in Tables 4, 5, 6 & 7.

Wastewater Management

Characteristic of the antibiotic wastewater is the diversity of their process operations, which gives rise to a wide variation in liquid waste generation. Moreover there is little similarity between effluents from different antibiotic units and individual effluents may alter continually as a result of process charges. This lack of homogeneity leads to difficulty in categorizing waste of this type and prevents standardization of any treatment methods and necessitating detailed and individual approach to effluent assessment. In many instances, careful segregation and collection of particularly toxic or high strength wastewater solve the more critical problems.

Table 3: Details of Water Requirement and Wastewater Generation per tone of the Product

Sl. No.	*Antibiotics*	*Water Requirement*	*Wastewater Generation*
1.	Cephalosprin	775.0 m^3	612.0 m^3
2.	Penicillin	590 m^3/mmu*	855 m^3/mmu
3.	Penicillin G	84.25 m^3	56.2 m^3
4.	Aureofungin	70 m^3/kg	90 m^3/kg
5.	Cephalexin	137.0 m^3	8.5 m^3
6.	Ranitidine	147.0 m^3	6.5 m^3
7.	Griseofulvin	727.0 m^3	25.0 m^3

*mmu: mega million unit

Table 4: Physico-chemical Characteristics of Different Streams from an Antibiotic Industry*

Sl. No.	*Parameters*	*Extraction Unit Wastewater*	*Wastewater from Penicillin G Unit (Spent Broth)*	*Combined Wastewater*
1.	Flow, m^3/d	20	150	190
2.	Temperature,°C	37	97	42
3.	pH	5.81	6.27	4.45
4.	Oil & Grease	9.5	15.0	12.0
5.	Total Suspended Solids	145	19720	5732
6.	Total Dissolved Solids	1048	27453	19300
7.	BOD	369	14130	8429
8.	COD	1155	35335	25545
9.	Sulfide	1.0	233	181
10.	Chloride	84.0	500	3519
11.	Phosphate	10.0	306	330
12.	Sulfate	490	17000	3625
13.	Ammonia Nitrogen	278	927	1025
14.	Phenols as C_2H_5OH	0.60	3.5	1.2

*Source: Torrent Gujarat Biotech Ltd., Vadodara, Gujarat
All values are expressed in mg/L except flow, temperature and Ph

Wastewater Management

Characteristic of the antibiotic wastewater is the diversity of their process operations, which gives rise to a wide variation in liquid waste generation. Moreover there is little similarity between effluents from different antibiotic units and individual effluents may alter continually as a result of process charges. This lack of homogeneity leads to difficulty in categorizing waste of this type and prevents standardization of any treatment methods and necessitating detailed and individual approach to effluent assessment. In many instances, careful segregation and collection of particularly toxic or high strength wastewater solve the more critical problems.

Table 5: Physico-chemical Characteristics of Combined Wastewater from an Antibiotic Industry*

Sl. No.	*Parameters*	*Value*
1.	Flow, m^3/d	160
2.	pH	10.6
3.	Oil & Grease	26.8
4.	Suspended Solids	454
5.	Total Dissolved Solids	3280
6.	BOD	4330
7.	COD	9680
8.	Ammonia Nitrogen	33.6

*Source: Glaxo India Ltd. Ankleshwar, Gujarat.
All values are expressed in mg/L except flow and pH

Table 6: Physico-chemical Characteristics of Combined Wastewater from an Antibiotic Industry*

Sl. No.	*Parameters*	*Value*
1.	pH	7.2-7.5
2.	Suspended Solids	100-200
3.	Total Solids	2000-2500
4.	BOD	600-800
5.	COD	1250-1500
6.	Alkalinity	350-550
7.	Total Nitrogen	110-130
8.	Total Phosphorous	30-40
9.	Sulphate	1100-1200

*Source: Ranbaxy Laboratories Ltd., Paonta Sahib, HP.
All values are expressed in mg/L except pH

Table 7: Physico-chemical Characteristics of Different Streams from an Antibiotic Industry*

Sl. No.	*Parameters*	*Formulation Unit and Domestic*	*Spent Broth*	*Column Wash*
1.	pH	4.6-7.0	5.1-6.8	6.6-7.0
2.	Suspended Solids	200-600	680-2700	1730
3.	Total Dissolved Solids	1900-3000	26300-28600	360000
4.	BOD	1100-3908	10680-11100	—
5.	COD	2100-6400	20900-22100	100000
6.	Total Kjeldhal Nitrogen	50-100	280-388	6000
7.	Phosphate	10-40	30-40	—

*Source: Hindustan Antibiotics Ltd., Pune, MS; All values are expressed in mg/L except pH

Treatment of Antibiotic Wastewater

Due to the complexity of the antibiotic wastewater the resultant treatment plants may be multi stage and in some cases may comprise of two or three completely separate plants. Serious problems may occur in the operation of installed treatment plants when process change takes place. This can be overcome by considering future product planning at the time of design of the treatment plant.

Treatment of the liquid wastes formed during the production of antibiotics start from the plant itself. For economic reasons, most of the organic solvents used in the process, like butanol in case of penicillin, methanol in case of cephalosporin, acetone for ampicillin trihydrate are recovered in the plant. The recovery of alcohols reduce the organic load like BOD of the wastewater to a great extent. Different treatment technologies adopted for the treatment of antibiotic wastewaters have been cited in literature. General treatment processes tried are physical, physico-chemical, biological, land disposal, soil injection technique etc. Treatments suggested by various authors are presented in Table 8.

Table 8: Treatment Suggested for Various Types of Antibiotics Wastewater

Authors	*Type of Wastes*	*Treatment Suggested*
Koziorowski B	Pharmaceutical wastewater	Anaerobic treatment
NEERI	Antibiotic wastewater	Physico-chemical, Complete mixed Activated sludge system
Deshmukh SB	Synthetic drug plant wastewater	Complete mixed activated sludge system
Muss DL	Penicillin wastes	Two steps trickling filters
Brown JM	Pharmaceutical wastes	High rate trickling filter
Horne WR	Pharmaceutical wastes	Completely mixed activated sludge system
Otto R	Pharmaceutical wastes	Laboratory testing for biological control
Howe RHL	Antibiotic production wastes	High-rate two-step aeration composes the first stage of biological treatment
Eckenfelder WW Jr.	Pharmaceutical wastes	Completely mixed activated sludge system
McKenny RE	Antibiotic wastes	Completely mixed activated sludge system
Herion RW Jr.	Penicillin waste broth and miscellaneous plant wastes	Completely mixed activated sludge system
Mayabhate SP	Pharmaceutical wastewater	Physico-chemical and activated sludge system
Jennett JC	Pharmaceutical wastes	Anaerobic filter
Venkatamohan S	Pharmaceutical wastewater	Anaerobic suspended film contact reactor
Qi Peshi	Antibiotic wastewater	Integrated two stage anaerobic reactor
Gulmez B	Pharmaceutical and Baker's yeast common wastewater	Upflow anaerobic sludge blanket reactor system
Kadam RV	Penicillin, Erythromycin and Distillery combined wastewater	Upflow anaerobic fluidized bed system
Saravanane R	Cephalexin drug based pharmaceutical effluent	Upflow anaerobic fluidized bed system
Sachs EF	Pharmaceutical wastes	Anaerobic filter
Yamagiwa	Highly foaming pharmaceutical wastewater	Activated sludge with modified bubble columns equipped with a mechanical foam-breaker (MFRD)
Melcher RR	Pharmaceutical wastes	Soil injection
Colovos GC	Pharmaceutical wastes	Land disposal
Anderson	Pharmaceutical wastes	Extended aeration plant followed by flow splitter box

Pharmaceutical wastes may consist of spent fermentation broth, chemical wastes from fine chemical production or a combination of the two. The spent broth usually vary in BOD from 3000 mg/L to 15000 mg/L and contains nutrient salts, organic residues from fermentation, small quantities of solvents, defoaming oils and autolyzed organisms. The chemical wastes will contain a variety of organic compounds and may vary in BOD from several hundred to several thousand mg/L.

Physico-chemical Treatment

It is reported in literature that the antibiotic waste can neither be clarified in settling tanks nor can be chemically coagulated to reduce BOD (1). The poor response of waste to the coagulation is due to the fact that the most of the substances contributing to BOD appear in solutions. But contradictory to this, reports are also available where in spent broth has been treated by coagulation and two stage clarification.

Physico-chemical treatment of penicillin producing industry wastewater yielded no results. Detail laboratory scale experiments (NEERI) using even higher doses of hypochlorite, lime, alum and combination of conventional coagulants along with polyelectrolyte resulted in no significant removals (2). Naturally occurring polyelectrolyte as coagulant and coagulant aid is not satisfactory for pharmaceutical wastewater treatment. It is reported that ferric chloride resulted in good removals followed by ferrous sulphate and alum in case of synthetic bulk drug wastewater (3).

Biological Treatment

Biological treatment systems are comparatively cheaper, and also cost effective than other organic compound processes. It is well documented that the dilute organic compounds produced by the chemical and pharmaceutical industries are amenable to biological treatments. But the toxic compounds present in the antibiotic effluent lead to problems during biological treatment, hence proper segregation, acclimatization and treatment scheme has to be selected.

Aerobic Treatment

The penicillin wastewater readily responds to almost all methods of treatment. Due to their high strength, special treatment scheme is needed which may be anaerobic-aerobic or aerobic-aerobic with special operating procedures. Aerobic system may be two stage aeration or it may be two stage aeration followed by third stage aerobic system like trickling filter depending on the effluent quality requirement for disposal. A pilot plant consisting of two major units of aeration tanks and two shallow trickling filters and one deep trickling filter has been applied for the treatment of streptomycin wastes alone and in combination with penicillin has been reported way back in 1951 (4). This was followed by many more reports on the treatment of penicillin wastewater through aerobic means using high rate trickling filter (5). It has been observed that the trickling filters are efficient in treating wastewaters from antibiotics, vitamins and sulpha drugs manufacture. A combined waste of sewage and pharmaceutical effluent has been treated in an activated sludge treatment plant followed by two trickling filters. It has been reported that the performance of the aerobic system was quite satisfactory (6). The fermentation and fine chemical waste have been treated at bench scale level using shaking flask test and studies are also reported on the effect of temperature in activated sludge (7).

Detailed studies on complete biological processes for the treatment of antibiotic production effluent is also cited in the literature (8). In this report it has been quoted that the antibiotic liquid waste is preneutralized with lime/caustic soda and the pH of 8.0–8.5 is maintained prior its further treatment in the biological system (high rate aeration tanks). This high rate two step aeration comprises the first stage biological treatment and in the first stage of treatment the average BOD reduction of 80% and the average antibiotic concentration reduction of 95% is reported have been achieved. This treatment is followed by two biological filters as second stage biological treatment.

Biological treatment facilities and the variables involved in the design and operation of the system is also reported (9). The disposal of spent fermentation broth from the manufacture of antibiotics has created many environmental problems because of their high organic concentration and low pH. Spent broth from penicillin manufacture is readily treated by biological methods, but conventional biological treatment processes were not able to handle these wastewaters satisfactorily. Modifications of the conventional biological treatment systems have been used with varying degrees of success. One such system using a completely stirred activated sludge plant has been reported for the treatment of fermentation broth (10). Followed on the heals of earlier workers, many permutation and combination of biological treatment systems have been tried for the treatment of antibiotic/pharmaceutical wastewaters (11,12). Schematics of effluent treatment plant (ETP) for the treatment of different antibiotic wastewaters are shown in Fig. 7 & 8.

In late nineties a combined physico-chemical and aerobic system was tried. The physico-chemical study using different coagulants revealed that all the coagulants used are not effective and their doses required were also very high for COD reductions (13). Antibiotic/Pharmaceutical wastewater treatment efficiency has been studied along with effect of temperature on COD removals and bacterial community development. Antibiotic industry wastewater is reported to be treated by an aerobic activated sludge system but the performance was very poor which may be due to toxic pollutant from penicillin production getting mixed with combined wastewater.

Anaerobic Treatment

Anaerobic treatment serves as a primary treatment facilities it can not be the ultimate treatment process. Anaerobic treatment generally is used for stabilizing 80% of the organics thus reducing the load on the secondary aerobic treatment. In this process cost of the treatment reduces considerably during aerobic treatment. Recent years, efforts have been made to combine different industrial wastewaters along with antibiotic wastewater and treat it to achieve efficient removal in organic pollutants. One such work was the treatment of combined pharmaceutical wastewater in combination with yeast industry wastewater.

Combined wastewater from a pharmaceutical factory has been successfully treated in an anaerobic filter with steady state removal efficiency of 93.7–97.8% (14). The pharmaceutical industry effluent comprising organic solvents have been treated by anaerobic suspended film contact reactor. Organic loading rates ranging between 0.25 and 2.5 kgCOD/m^3.d with the COD reductions of 60-80% is reported and optimum loading of 1.25 kg/m^3.d has been achieved (15). An integrated two stage anaerobic reactor has been used to treat the antibiotic wastewater with the influent COD as high as 26347 mg/L and the maximum volumetric loading rate was 8.57 kgCOD/m^3.d. The absolute concentration of SO_4^{-2} of 1325 mg/L and COD/SO_4^{-2} ratio 3.0 was obtained. The results showed that this anaerobic two stage reactor is capable of resisting impacts of organism inhibiting agents and shock loadings (16).

Feasibility of anaerobic treatment of pharmaceutical and baker's yeast industry effluent has also been tried. Lab scale anaerobic treatability studies on UASBR with influent COD concentration of 10000 mg/L was studied. It is reported that the pharmaceutical wastewater has 5% inhibition on the COD removals at the dilution rate of 1: 100 or more in the combined treatment of pharmaceutical effluent with baker's yeast effluent (17). Likewise upflow anaerobic fluidized bed system for treatment

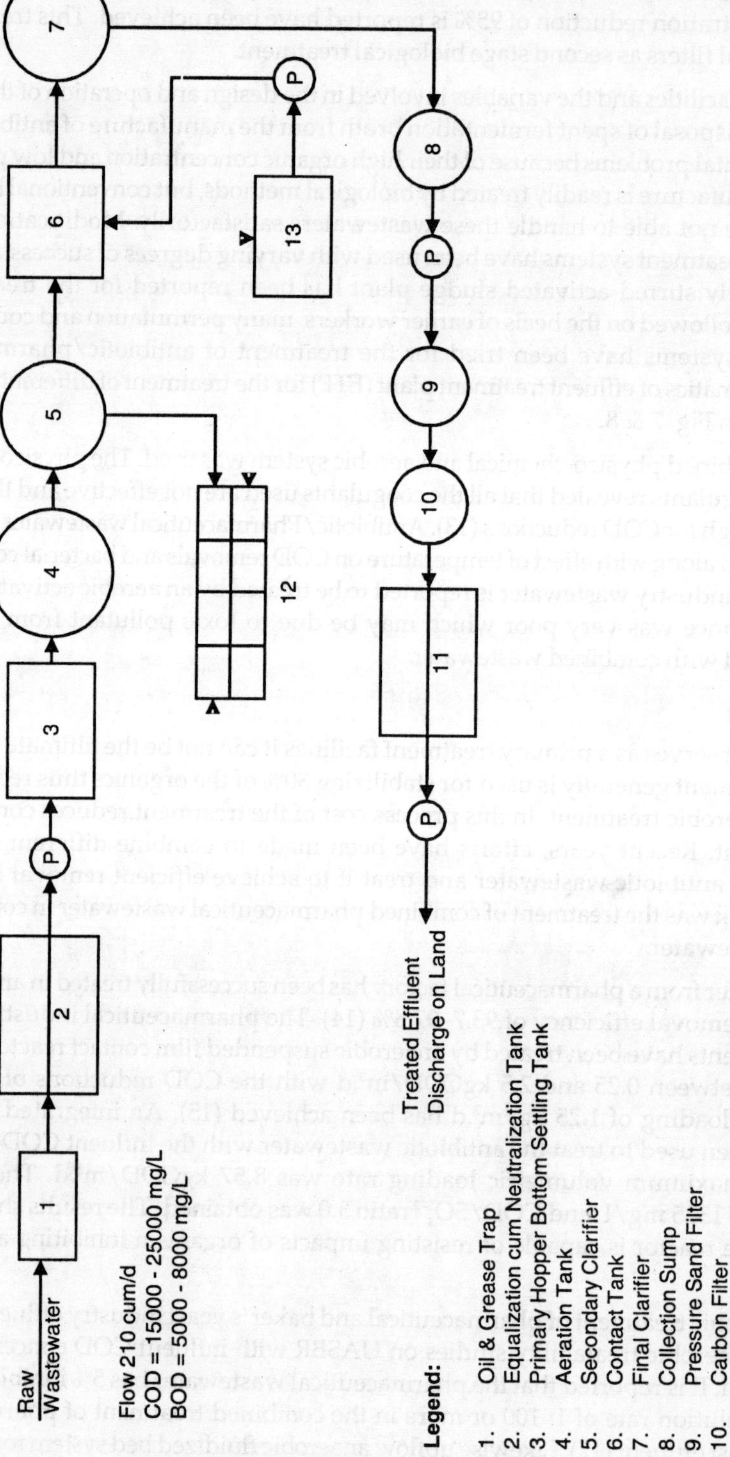

Legend:

1. Oil & Grease Trap
2. Equalization cum Neutralization Tank
3. Primary Hopper Bottom Settling Tank
4. Aeration Tank
5. Secondary Clarifier
6. Contact Tank
7. Final Clarifier
8. Collection Sump
9. Pressure Sand Filter
10. Carbon Filter
11. Equalization cum Neutralization Tank
12. Sludge Drying Beds
13. Stabilization Tank

Fig. 7: Schematics of Effluent Treatment Plant at M/s Glaxo India Limited

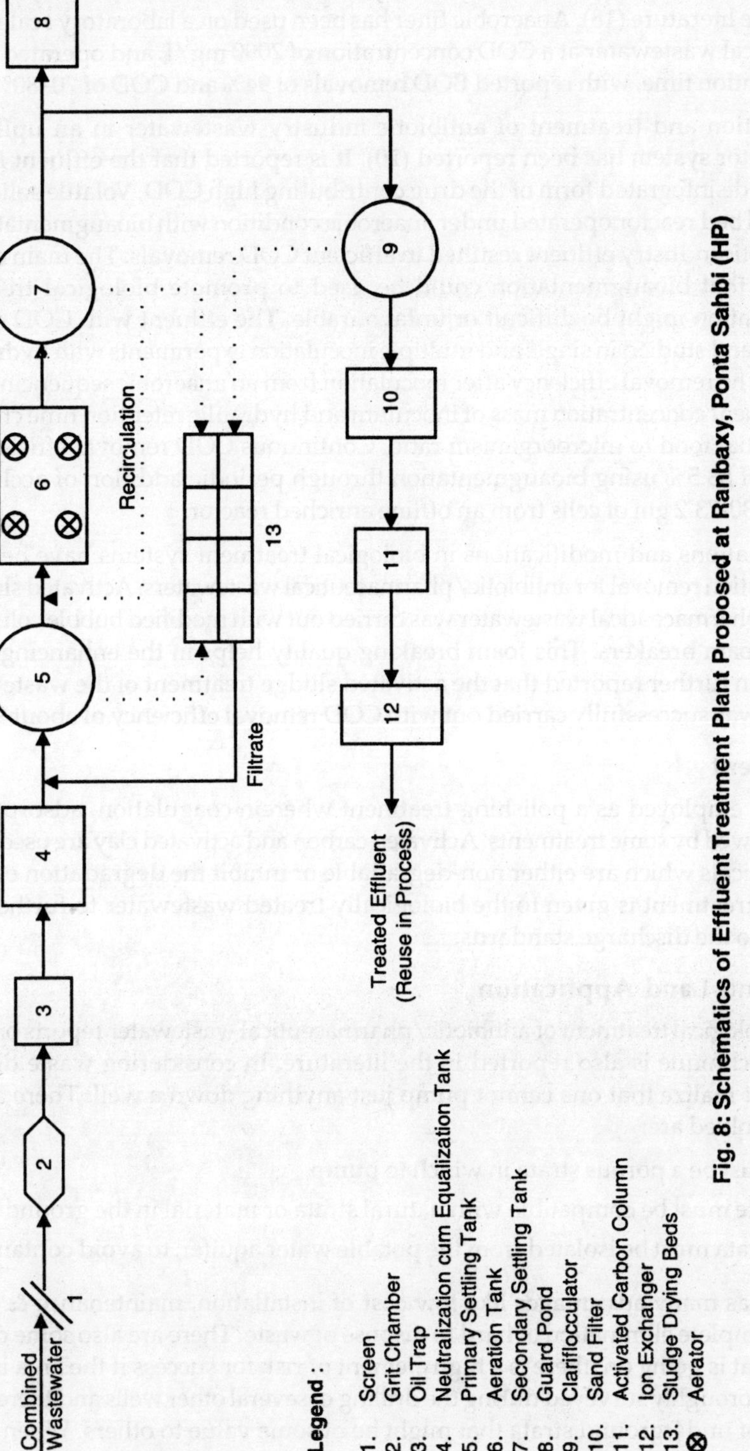

Fig. 8: Schematics of Effluent Treatment Plant Proposed at Ranbaxy, Ponta Sahbi (HP)

of combined wastewater from penicillin, erythromycin and molasses based distillery wastewater is also reported in the literature (18). Anaerobic filter has been used on a laboratory scale experiment to treat pharmaceutical wastewater at a COD concentration of 2000 mg/L and operated at 35°C and 36 hrs hydraulic retention time, with reported BOD removals of 94% and COD of 70–80% (19).

Bioaugmentation and treatment of antibiotic industry wastewater in an upflow anaerobic fluidized bed reactor system has been reported (20). It is reported that the effluent from antibiotic industries contain disintegrated form of the drug contributing high COD, Volatile solids and organic solvents. Fluidized bed reactor operated under anaerobic condition with bioaugmentation to treat the cephalexin antibiotic industry effluent resulted in efficient COD removals. The main objective of the study is to show that bioaugmentation could be used to promote biological treatment where conventional operation might be difficult or unfavourable. The effluent with COD of 12000-15000 mg/L was diluted and studied in single and multiple inoculation experiments with hydraulic retention times of 3-12 Hrs. The removal efficiency after inoculation from an anaerobic sequencing batch reactor was related to influent concentration mass of inoculum and hydraulic retention time characterized by calculating the initial food to microorganism ratio. Continuous COD removal efficiency attained a maximum value of 88.5% using bioaugmentation through periodic addition of acclimated cells at every 2 days with 30-73.2 gm of cells from an offline enriched reactor.

Many combinations and modifications in biological treatment systems have been reported to achieve better pollution removal for antibiotic/pharmaceutical wastewaters. Activated sludge treatment of highly foaming pharmaceutical wastewater was carried out with modified bubble columns equipped with mechanical foam breakers. This foam breaking quality helps in the enhancing the treatment capacity. It has been further reported that the activated sludge treatment of the wastewater with the modified column was successfully carried out with COD removal efficiency of about 90% (21).

Tertiary Treatment

It is generally employed as a polishing treatment wherein coagulation, adsorption and other treatments are followed by some treatments. Activated carbon and activated clay are used for adsorption of refractory chemicals which are either non-degradable or inhibit the degradation of other organic compounds. This treatment is given to the biologically treated wastewater to further improve the quality to bring it to the discharge standards.

Soil Injection and Land Application

Apart from biological treatment of antibiotic/pharmaceutical wastewater reports on their disposal by soil injection technique is also reported in the literature. In considering waste disposal by soil injection, one must realize that one cannot pump just anything down a well. There are three main considerations involved are:

1. There must be a porous strata in which to pump,
2. The waste must be compatible with natural strata or material in the ground and
3. These strata must be isolated from the potable water aquifer, to avoid contamination (22).

This system has many advantages like low cost of installation, maintenance & operation and most important complete elimination of hard to dispose of waste. There are also some disadvantages, one cannot see what is going on, there is a high element of risk for success if the area in question has not already been thoroughly surveyed during the drilling of several other wells and there is a possibility of contamination of underground strata that might be of some value to others. When the volume of wastes is not excessive, the underground strata are of the proper nature, the chances of contamination are negligible, the waste are of a nature that can be properly assimilated into the strata and finally no surface method of treatment appears possible or adequate to handle the problem.

In the pharmaceutical/antibiotic manufacturing units, there is a large volume of varied solid and liquid wastes that needs removal. Solids are readily disposed by simple incineration or dumped in secured landfill area. Liquids on the other hand, are treated by biological methods. But an attempt has been made to see the feasibility of land application of pharmaceutical wastewater. Land application by way of spray irrigation has been tried and reported. Orchard irrigation equipment with aluminum pipe and rotating sprinklers were used. Since the liquid effluent was highly aromatic, a pretreatment with chlorine was applied which reduced the odour to an appreciably level. Chlorinated wastewater was sprayed on land until equivalent of 1½ inch of rainfall was reached. This method of disposal of high BOD fermentation spent water was proved very satisfactory (23).

Antibiotic Wastewater Treatment: A Case Study

An antibiotic industry producing Penicillin, Aureofungin, Gentamycin and Hamycin were studied. This unit was manufacturing 3000 MMU/year of penicillin, 0.208 MT/month of Aureofungin, 0.37 MT/month of Gentamycin, 0.021 MT/month of Hamycin and 11.25 MT/month of Carbnedazim. Apart from above antibiotics, vials, IV bottles, tablets, capsules and dry syrup were also being produced.

Water requirement was around 8000 m^3/d including industry processes, and colony drinking purposes and the wastewater generation was 6550 m^3/d at the antibiotic industry. Physico-chemical characteristics of different wastewater streams are presented in Table 7. The wastewater generated in the factory was being treated by the existing effluent treatment plant, shown in Fig. 9. But the performance evaluation studies of effluent treatment plant clearly indicated poor reductions in polluting parameters. Samples were collected on hourly basis and composited for 24 hours and combined wastewater was prepared on flow basis. Detail bench scale treatability studies were carried out using combined wastewater.

Physico-chemical treatment method, *viz.* coagulation and flocculation was tried on various wastewater streams (spent broth, column wash, etc) to achieve BOD and COD removals. Various coagulants, *viz.* lime, alum, ferrous sulphate etc. and polyelectrolytes were tried. The studies revealed no significant removals of BOD and COD.

Anaerobic fixed film fixed bed reactor made of glass and aggregates was used as a support media, was used for the treatment of wastewater. The studies were carried out using only spent broth and also combined spent broth and column wash, by mixing the two streams in the ratio of their flow. Reduction in BOD and COD was found in the range of 50 to 54.2% and 33 to 34.6% respectively at HRT of 5 days and 65.4 to 68.3% and 52.9 to 53.5% respectively at HRT of 10 days in case of combined waste.

A bench scale model, made of perspex sheet, of activated sludge system was used for aerobic treatability studies. The studies were carried out in two stage aeration system. First stage treatment studies indicated BOD and COD removals of 81.9 to 86.1% and 61 to 67.4% respectively for HRT of 1 day and 54.5 to 56.6% and 36.4 to 37.6% respectively at the same HRT at second stage of treatment.

Likewise a bench scale trickling filter was also used for the treatment of combined wastewater. Removal efficiency of BOD and COD were in the range of 78 and 82% and 60 and 63.2% respectively at 1.5 $kgBOD/m^3.d$ organic loading rate. The studies revealed that the removal efficiency reduced as the organic loading increased. Based on the above studies, treatment options were arrived at. Schematics of effluent treatment plant suggested for the treatment of high strength antibiotic wastewater are shown in Fig. 10.

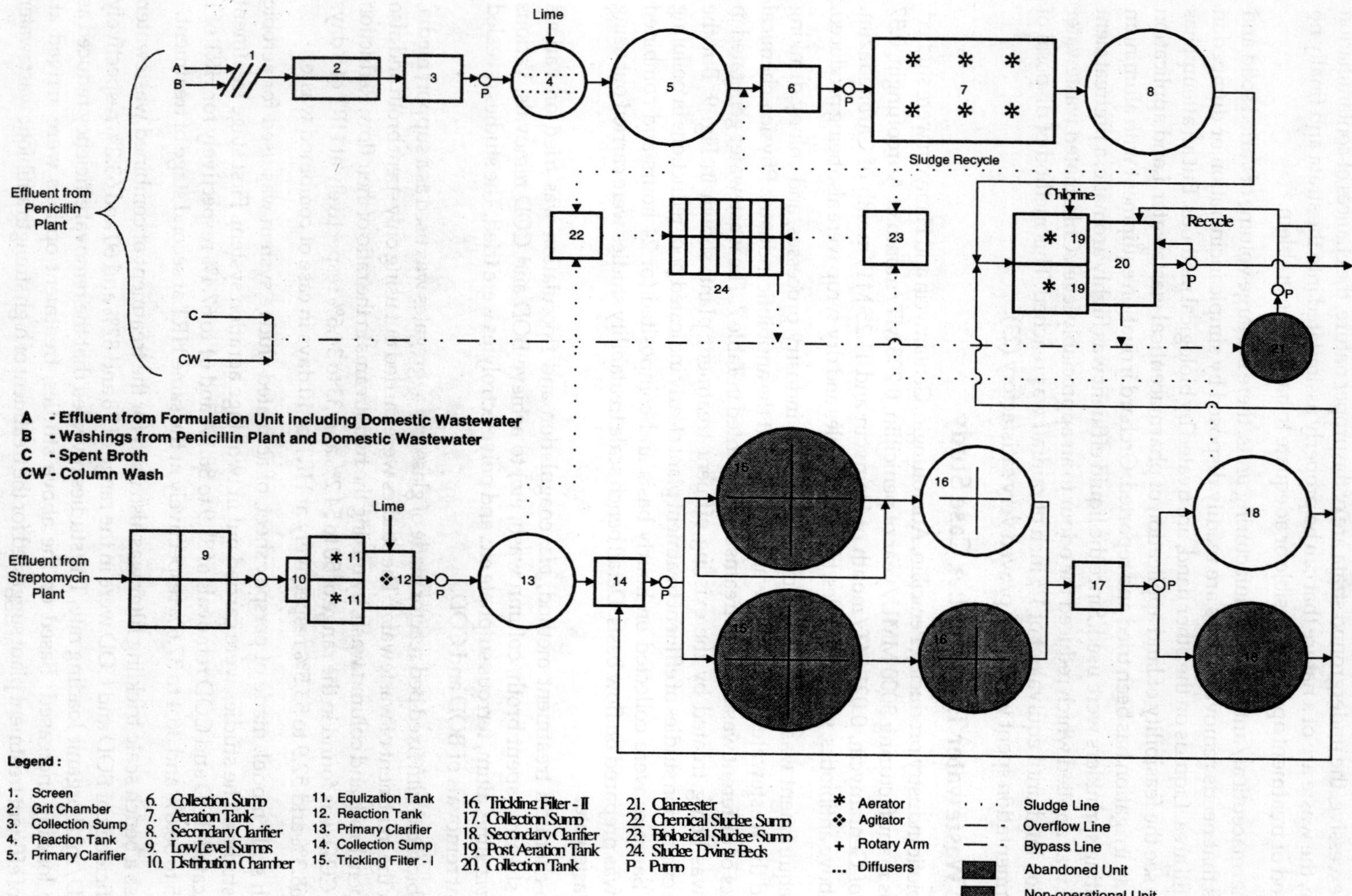

Fig. 9: Schematics of Existing Effluent Treatment Plant

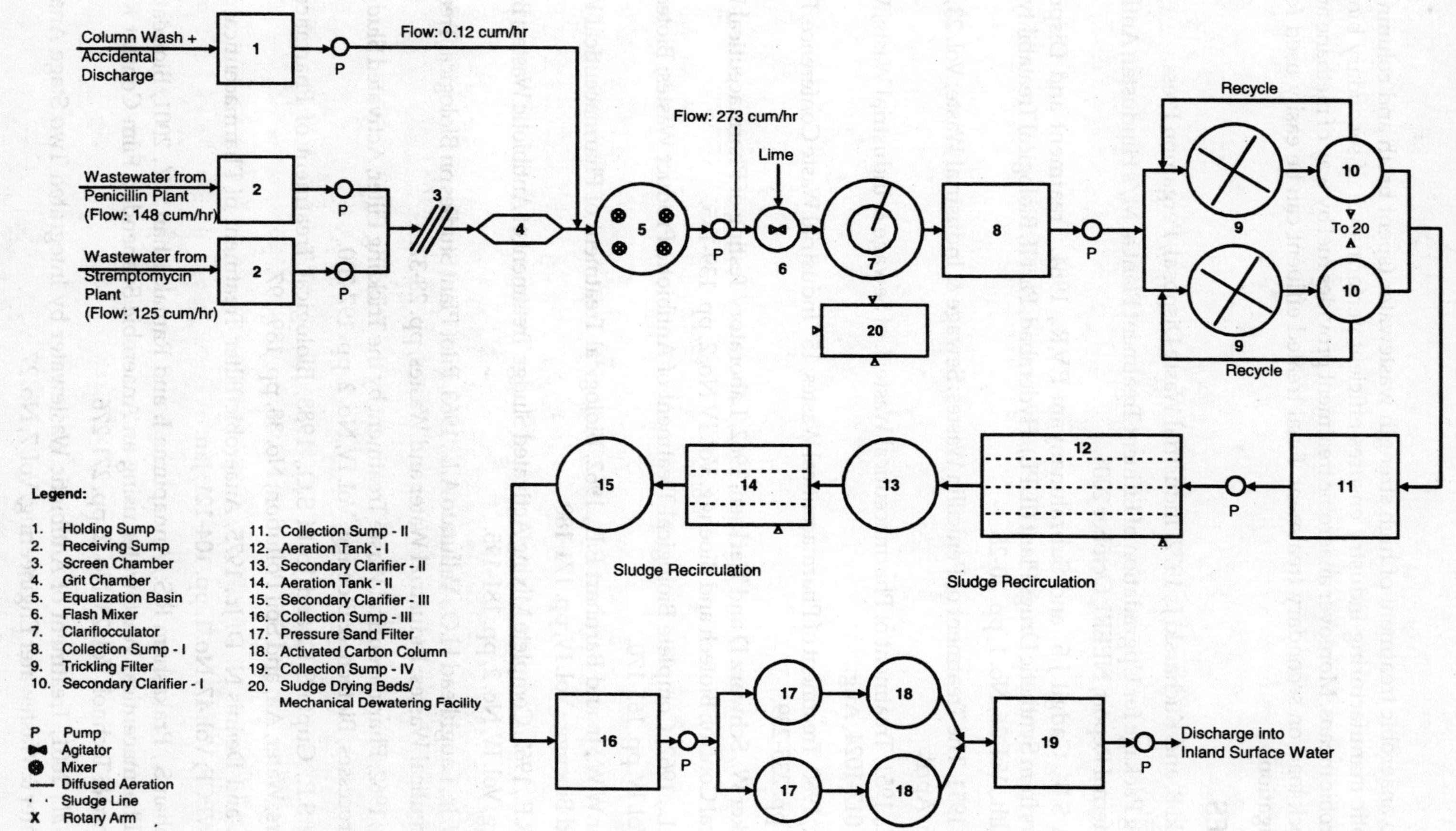

Fig. 10: Schematics of Proposed Effluent Treatment Plant

Conclusion

Recourse to anaerobic treatment of high strength wastewater (spent broth and column wash) from an antibiotic manufacturing industry ensures efficient performance of secondary biological treatment by aerobic means. Moreover anaerobic treatment provides fuel by way of methane and also reduces the shock load on secondary treatment. Final treated effluent can be easily used for land application/irrigation.

REFERENCES

1. Koziorowski B. and Kucharski J., 1972, Industrial Waste Disposal, Pergamon Press.
2. Engineering Package for Upgradation of Effluent Treatment Plant at M/s Hindustan Antibiotics Limited, Internal Report, NEERI, October 2001.
3. Deshmukh S.B., Gadgil J.S. and Subrahmanyam P.V.R., 1984, Treatment and Disposal of Wastewaters from Synthetic Drugs Plant (IDPL), Hyderabad, Part II, Biological Treatability, Ind. J. Environ. Hlth., Vol. 26, No. 1, pp. 20-28.
4. Muss D.L., 1951, The Treatment of Penicillin Wastes; Sewage & Industrial Waste; Vol. 23, No. 4, pp. 486-496, April.
5. Brown J. M., 1951, Treatment of Pharmaceutical Waste; J. of Sewage & Industrial Waste, Vol. 23, No. 8, pp. 1017-1024, Aug.
6. Horne W.R., 1960, Treatment of Pharmaceutical Wastes; 15th Industrial Waste Conference, Purdue University, pp. 235-239.
7. Otto R, Barker W., Schwarz D and Tjarksen, 1962, Laboratory Testing of Pharmaceutical Wastes for Biological Control; Biotech and Bioengg. Vol. IV No. 2, pp. 139-145.
8. Howe R.H.L., 1962, Complete Biological Treatment of Antibiotic Product Wastes; Biotech and Bioengg., Vol. IV, pp. 161-170.
9. Eckenfelder W.W., Jr. and Barnhart E.L., 1962, Biological Treatment of Pharmaceutical Wastes, Biotech. and Bioengg., Vol. IV, pp. 171-180.
10. Mckinney R.E., 1962, Complete Mixing Activated Sludge Treatment of Antibiotic Wastes; Biotech. and Bioengg. Vol. IV, No. 2, pp. 181-195.
11. Herion R.W. Jr., Laughhead H.O., Vellutato A.L., 1963, Pilot Plant Studies on Biological Treatment of Pharmaceutical Wastes; Industrial Water and Wastes, pp. 28-32.
12. Molof A.H., 1962, Pharmaceutical Waste Treatment by the Trickling Filter, Activated Sludge and Compost Processes; Biotech. & Bioengg. Vol. IV, No. 2, pp. 197-200.
13. Mayabhate S.P., Gupta S.K. and Joshi S.G., 1988, Biological Treatment of Pharmaceutical Wastewaters; Water, Air, and Soil Pollution; No. 38, pp. 189-197.
14. Jennett J.C. and Dennis N. D. Jr., 1975, Anaerobic Filter Treatment of Pharmaceutical Waste; Journal of WPCF; Vol. 47, No. 1, pp. 104-121; Jan.
15. Venkatamohan S., Prakasham R.S., Annapurna J. and Ramakrishan S.V., 2001, Biotreatability Studies of Pharmaceutical Wastewater using an Anaerobic Suspended Film Contact Reactor, Water Science & Technol. Vol. 23, No. 2, pp. 271-276.
16. Qi Peshi *et al.*, 2001, Treatment of Antibiotic Wastewater by Integrated Two Stage Anaerobic Reactor, Water & Wastewater Engineering, Vol. 7, No. 27.

17. Gulmez B., Qzturk J., Aip K., Arikan O.A., 1998, Common Anaerobic Treatability of Pharmaceutical and Yeast Industry Wastewater, Water Sci. Technol., Vol. 4-5, No. 38, pp. 37-44.

18. Kadam R.V., Saxena G.S. and Shukla G.L., 2000, Anaerobic Treatment of Combined Wastewater from Penicillin, Erythromycin and Distillery Industry – A Case Study; Bio Energy News., Vol. 4, No. 2, pp. 12-15, June.

19. Sachs E.F. and Jennett N.C., 1982, Pharmaceutical Waste Treatment by Anaerobic Filter, ASCE, J. of Environmental Engineering Division, No. EE2, pp. 108.

20. Sarvanane R., Murthy D.V.S., Krishnaiah K., 2001, Bioagumentation and Treatment of Cephalexin Drug based Pharmaceutical Effluent in an Upflow Anaerobic Fluidized Bed System, Biores. Technol. Vol. 3, No. 76.

21. Yamagiwa K., Yoshida M., Ohkawa A and Takesono S., 2000, Biological Treatment of Highly Foaming Pharmaceutical Wastewater by Modified Bubble-Column under Mechanical Foam Control, Water Science & Technology, Vol. 42, No. 3-4, pp. 331-337.

22. Melcher R.R., 1962, Pharmaceutical Waste Disposal by Soil Injection; Biotech and Bioengg.; Vol. IV, pp. 147-151.

23. Colovos G.C. and Tinklenberg N., 1962, Land Disposal of Pharmaceutical Manufacturing Wastes; Biotech. and Bioengg. Vol. IV, pp. 153.

72

Common Effluent Treatment Plant (PETL) at Patancheru, Andhra Pradesh

P. VENKATA RAO, V.S.N. RAJU, P. SIVACHANDER & V.R. AKELLA

Patancheru Enviro-Tech Limited, Patancheru - 502 319 (Andhra Pradesh)

ABSTRACT

The performance of the Common Effluent Treatment Plant is studied when the input T.D.S is of the order 10000-12000 mg/L and the C.O.D is around 6000 mg/L. The study parameters are flow through, MLSS/MLVSS, power input to aeration system in terms of Watts per Cu.M., volume of Aeration Tank, removal efficiency of B.O.D and C.O.D. It is observed that maximum efficiencies are obtained at 5500-6500 mg/L of MLVSS and 40 Watts per Cu.M. of Aeration Tank Volume

Keywords: Aeration System, MLVSS, F/M ratio, Efficiency.

Introduction

The concept of Common Effluent Treatment Plant emerged as a necessity consequent upon industrial development in Patancheru Industrial belt with no. of small and medium scale units. The various incentives offered by the Government of Andhra Pradesh has created a hub of activity in this area during 1970-80. Thus, Patancheru is reckoned on the industrial map. The establishment of Chemical, Pharmaceutical industries brought a distinction to this area and has become an address for bulk-drug manufacturers. It is a known fact that with this development, pollution has become an issue a part.

To overcome the pollution problem, and maintain the ecological balance in Patancheru, the industrial units, in particular, pharmaceutical and bulk drug units with the encouragement from Andhra Pradesh Industrial Infrastructure Corporation Limited (APIIC) floated a company in 1989 – M/s PATTANCHERU ENVIRO TECH LIMITED under the provisions of Indian Companies Act, 1956. The objective being establishment of a common effluent treatment plant for treating the industrial effluents emanating from this industrial belt for Pollution Control.

M/s Associated Industrial Consultants Limited, a reputed and experienced consultants in environment and pollution control prepared the project report and a Common Effluent Treatment Plant (CETP) was established with treatment capacity of 7500 m^3 per day.

The cost of the project was Rs. 5.70 Crores. The Plant started its operations in 1994 and the management of the plant was taken over from APIIC by the directors representing Industry. An additional amount of Rs. 3.00 Crones was spent subsequently for upgrading the treatment facilities in the CETP.

Liquid Oxygen injection technology was adopted for the first time in the country for biological treatment in a CETP of this magnitude.

The development of industries in Patancheru industrial belt in the last three decades resulted in high economic growth evidencing marked raise in standard of life and living environment. The socioeconomic growth created ample employment opportunities for the local people.

Unit Operation and Facilities at CETP

Quality Control Facilities at the Entry to CETP

The pretreated effluents from various units reach CETP by road tankers with a 6 copy manifest system. Samples from each tanker are taken from both top and bottom of the tank, mixed in equal quantity and tested for pH and TDS/conductivity. The Chemist permits tanker for unloading in receiving sump only after confirming to the admissible limits.

Receiving Sump and Terminal Pumping Station

Pre-treated effluents received by road tankers are tested/analysed for inlet standards and unloaded in the sump. The influents are pumped to Equalisation Tanks, to equalize the effluents.

Equalization Tanks

4 Nos. of Equalisation Tanks are provided to store the effluents for about 2-3 days with aeration grid and air blower to equalize the influents and to strip off any organic volatile matter. The effluents from equalization tanks are being pumped to Dissolved Air Floatation Unit.

Dissolved Air Floatation Unit

This unit is used to remove suspended solids from the influents. The treatment chemicals Alum and Ployelectrolyte are used for coagulating the S.S. and these solids are separated with dissolved air in the form of fine bubbles. The suspended solids as collected at the top of the unit are removed and collected in a sludge storage tank. The sludge/slurry separated is pumped to Decanter – 1 (centrifuge) for further separation of liquids (centrate) and solids in the form of cake. The clarified effluents from DAF flow to Buffer Tank by gravity.

Decanter No. 1 (Centrifuge)

The suspended solids removed in DAF with 1-2% concentration are pumped to Primary Decanter. Polyelectrolyte solution is added to improve separation efficiency. In this unit sludge is separated into clear liquid and solid cake. The clarified water (Centrate) goes to Buffer Tanks, for further treatment. The sludge cake is discharged to sludge drying beds / or secured land fill unit for storage.

Sludge Drying Beds

The sludge cake from Decanter – 1 (Primary decanter) is taken to secured land fill after drying in these drying beds.

Secured Land Fill Unit

The dried sludge from Decanter / Sludge Drying Beds is stored in Secured Land Fill Unit. This unit is constructed with well consolidated and cement grouted base, over which gravel and Bentonite layers are provided. The tank is covered with a layer of 1.8 mm thick imported HDPE sheet. Drainage system is provided with media to collect the leachate. The filtered water from leachate collection tank is pumped back to Equalization Tank.

Buffer Tanks

The clarified effluents from DAF unit are collected in these tanks. The raw sewage from BHEL sewage treatment and disposal plant, is added to the clarified effluents in these tanks. The sewage is added to improve the biological treatment efficiency in the next units, and as a constant seed in Aeration Tanks.

Aeration Tanks

There are 2 nos. of Aeration Tanks, each capacity of 4,300 cum. 4 nos. of 50 HP Surface Aerators are provided to supply oxygen from atmospheric air. In addition to the surface aerators, there is Liquid Oxygen Storage Tank and Mixflow Oxygen pumps to supply pure oxygen up to 10000 Kg / Day. The flow rate of oxygen is regulated from control panel and by checking the D.O. levels in Aeration Tanks at regular intervals.

The influents of Aeration Tank are biodegraded by micro – organisms (MLSS / MLVSS) present in the Aeration Tank, there by reducing the pollution loads (BOD & COD). As a result, the micro-organisms multiply and MLSS levels are maintained in the Aeration Tank by recirculation of sludge from next unit (SST) and also wastage of excess sludge.

Secondary Settling Tanks

The outlet of the Aeration Tank is connected to Secondary Settling Tanks for separation of Biological sludge along with Suspended Solids. The MLSS settles at the bottom in the form of sludge in Secondary Settling Tank. Settled sludge is recirculated back to Aeration Tanks to maintain MLSS levels in the Aeration Tanks and the excess sludge is wasted through Decanter – 2. The clarified water from Secondary Settling Tanks is presently discharged into Isakavagu.

Decanter-2 (Centrifuge)

The excess sludge produced in Aeration Tanks as collected in Secondary Settling Tanks at the bottom is sent to Sludge pit of Decanter, for wastage. The sludge is separated as sludge cake and clarified water in this Decanter. The clarified water / centrate goes back to Aeration Tank and the sludge cake is transferred to Secured Land Fill Unit.

Quality Control Laboratory

The laboratory is having facilities for analysis of the following:

1. pH
2. Oil & Grease, mg / L

3. BOD, mg/L (3 days, 27° C or 5 days, 20° C)
4. COD, mg/L (by titration)
5. TDS, mg/L (by gravimetric method) / Conductivity method
6. N-NH_3, mg/L.
7. Suspended Solids, mg/L
8. Chlorides, mg/L
9. Phosphates, mg/L
10. Sulphates, mg/L
11. Metal Ion concentrations by AAS.

Special Features at this CETP

The wastewater treatment facility was designed as a common facility to member industries with certain design load to the system. Due to various reasons the characteristics of the wastewaters subsequently reaching the treatment plant have increased substantially. Then, the CETP management has upgraded the process units by introducing more unit operations. The source industries/member units also have taken additional measures to contain pollution.

Upgradation by CETP, Patancheru

1. Adopted liquid oxygen injection technology for the first time in the country for biological treatment at a CETP with 3.75 MLD capacity.
2. Adopted protective measures for storage of solids generated in the form of sludge at plant site in a Secured Land Fill Unit with HDPE lining and with Leachate system for ultimate disposal to TSDF.
3. Created lined land fill arrangement for storage of decanted primary and secondary sludge.
4. Has become a member of TSDF (Treatment, Storage and Disposal Facility) located in Dindigal for ultimate disposal of solid sludge.
5. Secured higher capacity Decanter for processing more volume of effluent by wastage of excess sludge for improvement in the treatment.
6. Non-acceptance of effluents with heavy metals beyond specified norms.

Further, the CETP has taken several measures in the positive direction resulting in appreciable reduction in pollution load in this area. Also as a self disciplinary measure, progressively CETP has fixed inlet standards for accepting effluent tankers at CETP.

1.	Earlier to taking over in 1996	Inlet TDS	> 40000 mg/l
2.	After taking over the management by industries in 1997	Inlet TDS	< 20000 mg/l
3	Subsequently reduced to	Inlet TDS	< 15000 mg/l
4	Subsequently, as per the Hon'ble Supreme Court directions during 1998	Inlet TDS Inlet COD pH Suspended Solids	< 15000 mg/l < 15000 mg/l 6.5 – 8.5 < 1000 mg/l
5	Subsequently, identification and non-acceptance of effluents with limits	At Inlet	Heavy metals beyond specified norms or with Arsenic

Additional Measures Taken by Individual Industry

Member industries have taken the following steps as pretreatment before sending the effluent to Common Effluent Treatment Plant (CETP).

1. Adopted segregation of organic and inorganic effluents and High, Medium and Low TDS effluents at source.
2. Have setup secondary treatment facilities by individual units generating more than 40-KL effluents per day.
3. Adopted forced evaporation and multiple effect evaporation technologies for reducing the hydraulic load and disposal of solids generated to TSDF facility directly.

Results and Discussion

The treatment process is illustrated in the Flow Diagram attached and are self explanatory.

The plant performance results for a period of one year are presented in the following tables for evaluation, analysis and to arrive at best possible values for F/M ratio and MLVSS to achieve best performance results when the T.D.S. values in the effluents are in the range 9500 to 14500 mg/L and the elevation of Patancheru is about 540 M above Sea level.

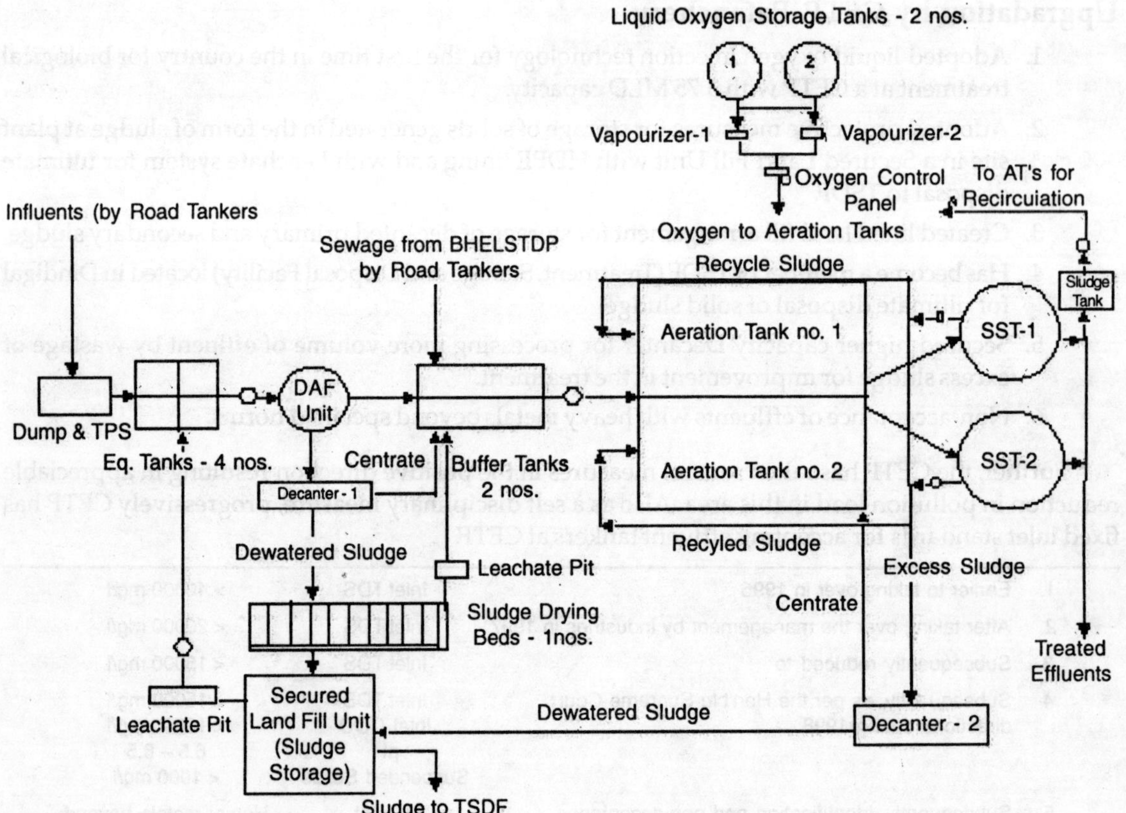

Fig. 1: Flow Diagram of Treatment Process

Table 1: Range of Values for Raw Effluent and Treated Outlet

	May, 2001	June, 2001	July, 2001	Aug, 2001	Sept, 2001	Oct, 2001	Nov, 2001	Dec, 2001	Jan, 2002	Feb, 2002	Mar, 2002	Apr, 2002
Inlet												
COD mg/L Min.	5200	5200	5000	5000	5200	5200	5200	5200	5200	5280	5000	5000
Max.	6720	7800	7000	6600	7600	6600	7200	8800	7600	7800	7000	6800
Avg.	5939	6180	5983	5858	6176	5858	6107	6439	6155	6410	5813	5620
TDS mg/L Min.	11250	10700	10850	10360	11100	10100	10455	10350	9930	10690	9495	10030
Max.	14500	14350	13505	14060	14970	13320	13885	13220	13590	14125	13510	11920
Avg.	12740	12836	12060	12054	12426	11867	11691	11829	11420	12117	11324	11043
BOD mg/L Min.	1800	2000	1800	1600	1677	1677	1800	1875	1750	2000	1875	2000
Max.	2750	2500	2800	2333	2500	2333	2500	2625	2500	2500	2625	2500
Avg.	2195	2180	2214	1878	2008	2012	2151	2252	2142	2281	2144	2117
Outlet												
COD mg/L Min.	1640	1920	1520	1560	1720	1520	1240	1480	1440	1400	1200	1120
Max.	2520	2760	2240	2080	3120	2120	2480	2360	1970	1800	2000	1520
Avg.	1990	2401	1933	1855	2167	1806	1788	1964	1677	1611	1506	1303
TDS mg/L Min.	9200	9200	7790	8150	8350	8685	8575	7310	7125	7630	7445	8015
Max.	11700	12600	10310	10990	11000	10820	11230	10775	9800	10665	10110	9330
Avg.	9944	10712	9380	8305	9472	9522	9401	9128	8454	9336	8739	8521
BOD mg/L Min.	80	130	100	60	200	110	88	53	45	30	20	20
Max.	214	325	285	290	333	248	212	266	113	90	50	50
Avg.	134	236	222	164	260	165	142	153	78	60.3	38.1	35

Table 2: Range of Values for Aeration System and Treated Outlet

	May, 2001	June, 2001	July, 2001	Aug, 2001	Sept, 2001	Oct, 2001	Nov, 2001	Dec, 2001	Jan, 2002	Feb, 2002	Mar, 2002	Apr, 2002
Inlet to A.T.												
COD mg/L Min.	4160	3840	3360	3840	3360	3520	4000	4160	4160	3840	3520	3840
Max.	5440	5760	5600	5120	5120	5440	6560	6760	5720	6080	5440	5120
Avg.	4759	4863	4273	4485	4400	4644	4667	5103	4964	4970	4444	4299
TDS mg/L Min.	7550	7900	7890	8180	7650	8080	10455	10350	6950	10690	9495	10030
Max.	11800	12750	10850	10700	11810	11465	13885	13220	10125	14125	13510	11920
Avg.	9716	10036	9380	9243	9289	9818	11691	11829	8454	12117	11324	11043
BOD mg/L Min.	1250	1375	1333	1250	1200	1125	1400	1400	1300	1300	1400	1400
Max.	2000	2000	2167	1625	1900	1875	1900	2100	2000	2000	2000	1800
Avg.	1644	1685	1599	1399	1465	1529	1629	1810	1687	1740	1621	1593

Contd...

Table 2–Contd...

	May, 2001	June, 2001	July, 2001	Aug, 2001	Sept, 2001	Oct, 2001	Nov, 2001	Dec, 2001	Jan, 2002	Feb, 2002	Mar, 2002	Apr, 2002
Outlet												
COD mg/L Min.	1640	1920	1520	1560	1720	1520	1240	1480	1440	1400	1200	1120
Max.	2520	2760	2240	2080	3120	2120	2480	2360	1970	1800	2000	1520
Avg.	1990	2401	1933	1855	2167	1806	1788	1964	1677	1611	1506	1303
TDS mg/L Min.	9200	9200	7790	8150	8350	8685	8575	7310	7125	7630	7445	8015
Max.	11700	12600	10310	10990	11000	10820	11230	10775	9800	10665	10110	9330
Avg.	9944	10712	9380	8305	9472	9522	9401	9128	8454	9336	8739	8521
BOD mg/L Min.	80	130	100	60	200	110	88	53	45	30	20	20
Max.	214	325	285	290	333	248	212	266	113	90	50	50
Avg.	134	236	222	164	260	165	142	153	78	60	38	35
MLSS Mg/L Min.	4200	1950	1292	2191	1500	2175	2800	2900	4125	6005	7235	7565
Max.	7850	4210	5350	6066	7925	6375	6041	5200	8725	8892	9285	9697
Avg.	3192	3065	2966	4200	5078	4165	4533	3982	5655	7463	8267	8324
MLVSS Mg/L Min.	2200	1200	775	1166	800	1375	1925	1300	3175	4780	5547	5667
Max.	5750	3245	4266	4366	5900	4650	4571	4075	6207	6777	7920	7700
Avg.	2267	2173	2119	3087	3685	3106	3506	2954	4337	5819	6434	6507
Oxygen Cons. (Kg.) Min.	5615	6673	5752	6446	3322	5016	4042	6695	7146	7021	7037	6397
Max.	10522	11495	10572	10770	7920	8933	8888	10385	13780	11056	10499	10599
Avg.	8175	7995	8282	7622	6385	6824	6024	8168	9264	8592	8731	8686

Composite samples of the effluents from member industries sending more than 30 Tankers per month are analyzed at regular intervals on monthly basis for metal ion (Cd, Cr, Cu, Fe, Mn, Ni, Pb, Zn) concentrations. The results are found always within the permissible limits for " Inlet effluent quality for CETP " as per the CPCB Manual (1993).

Further, the characteristics of the treated outlet waters are analyzed regularly at the Quality Control Laboratory of C.E.T.P. and also thrice in a week by APPCB and the results are found within the permissible values as specified by APPCB, CPCB, and the Hon'ble Supreme Court of India directions.

Table 3: Plant Performance – Efficiencies

	May, 2001	June, 2001	July, 2001	Aug, 2001	Sept, 2001	Oct, 2001	Nov, 2001	Dec, 2001	Jan, 2002	Feb, 2002	Mar, 2002	Apr, 2002
(a) Aeration system												
COD mg/L Min.	43.75	39.50	40.40	51.90	33.30	52.27	51.72	54.77	54.33	62.00	52.55	64.00
Max.	67.19	61.36	67.86	66.40	62.93	68.33	71.15	71.60	72.14	76.91	72.06	78.15
Avg.	57.91	50.11	54.12	58.45	50.29	60.94	61.53	61.29	65.89	67.18	65.87	69.57
BOD mg/L Min.	85.60	80.90	79.37	78.24	72.22	83.47	87.69	84.35	92.53	96.47	96.67	96.67
Max.	95.20	90.68	93.30	95.47	86.46	94.29	94.36	97.06	97.35	98.13	98.69	98.89
Avg.	91.40	86.10	85.85	88.26	82.04	89.05	91.31	91.52	95.33	96.57	97.34	97.79

Contd...

Table 3–Contd...

	May, 2001	June, 2001	July, 2001	Aug, 2001	Sept, 2001	Oct, 2001	Nov, 2001	Dec, 2001	Jan, 2002	Feb, 2002	Mar, 2002	Apr, 2002
(b) Over All Plant Performance												
COD mg/L Min.	59.35	54.00		60.00	54.12	64.00	63.08	59.26	65.00	65.90	70.50	69.30
Max.	73.75	70.26	72.73	72.69	73.33	74.60	77.86	78.86	77.84	80.56	78.29	82.50
Avg.	66.32	60.92	67.50	68.27	64.78	69.14	70.66	69.19	72.54	74.61	74.10	76.60
BOD mg/L Min.	90.00	84.00	83.60	83.10	81.00	87.60	89.50	88.20	93.60	96.00	95.90	97.50
Max.	96.40	93.50	93.70	97.00	90.40	95.40	96.10	97.60	97.80	98.60	99.00	99.10
Avg.	93.84	89.18	89.87	91.15	86.95	91.75	93.43	93.19	96.30	97.35	98.19	98.34
Eff Flow Min. m^3/d	1500	1570	1570	1810	1670	1180	1710	1710	1780	1820	1750	1680
Max.	2000	2030	2160	2170	2120	2010	2220	2110	2230	2260	2150	2190
Avg.	1850	1840	1960	1980	1880	1740	1870	1960	1950	2090	1930	1990

Conclusions

The following conclusions are arrived from the results:

1. High MLVSS concentrations of 5500-6500 mg/L in aeration tanks with extended aeration system has resulted in 95 % and above efficiencies in removal of B.O.D and 65 –70 % in removal of C.O.D.
2. High mixing in the aeration tank in terms of watts per Cubic Meter of Aeration Tank Volume due to extensive contact between Food and MLVSS also resulted in good performance.

The Aeration tanks in general are provided with 15 to 26 w/M3 of tank volume, Manual on Sewerage and Sewage Treatment (First and Second Edition,1993). In this case of CETP receiving industrial wastes waters with high TDS and C.O.D concentrations required 35 to 40 watts per Cubic Meter of the tank volume to achieve best results:

1. High Oxygen requirements of about 2.5 Kg – 3.0 Kg per Kg of B.O.D. removed are observed. This may be due to (a) altitude 540 M above Sea Level and (b) T.D.S values (around 12000 mg/L)
2. Detention time in Aeration tank is also in the range 72 – 96 Hrs. as this is a common effluent treatment plant with inlet flow fluctuations and T.D.S. concentrations of around 10000 mg/L.

All these are possible because this CETP has the advantage of adding liquid Oxygen (as required either increase or decrease the quantity) based on the inlet organic load.

Acknowledgements

The authors take this opportunity to thank all the staff at the common effluent treatment plant especially the operating staff including the Shift-In-Charges for their sincere efforts in maintenance of various components, both static and rotating equipment and the electrical units. They thank the Dy. Manager and his team of chemists in Quality Control Laboratory for their constant monitoring and analysis of various samples. The authors thank particularly Dr.D.R.Rao, Chairman of C.E.T.P.-P.E.T.L,

PATANCHERU for his keen interest and encouragement and free hand in improvements in performance of the plant. The plant performance is also being monitored by the Andhra Pradesh Pollution Control Board, Regional office at Sangareddy and the Zonal Office at Ramachendrapuram by collecting the samples and constant monitoring.

REFERENCES

1. Pollution Control Series: PCL/2/1992 (VOLUME –I), CPCB, September 1997. Clause 55 page 297.
2. Manual on Sewerage and Sewage Treatment (Second Edition, 1993), page233, Ministry of Urban Development, New Delhi.
3. Manual on Sewerage and Sewage Treatment (First Edition 1980), Table 13.1,page 82 Ministry of Works and Housing, New Delhi.

73

Hazardous Waste Effluent Treatment

P.T. RAGHURAM

Department of Chemical Engineering, Indian Institute of Science, Bangalore - 560 012

ABSTRACT

Radioactive isotopes are used for different applications in industries and research laboratories and during its usage, it also contaminates the environment. A knowledge of the theoretical background and handling techniques will reduce the chances of pollution and help clear the atmosphere better. Details for the reasons of contamination, methods of decontamination, handling techniques of radioisotopes, instruments used, personal and area monitoring and disposal methods are given this paper. It is equally important to be award of the user's responsibilities of the radioactive isotope (RAI) to safeguard the environment from pollution while hazardous wastes are generated.

Introduction

In this paper, effluent aspects of radiation in environment are considered. Radioactive isotopes re being used in industries, hospitals and research and development instates for different purposes and benefits are being derived. At the same time, there is sustained chances of polluting the environment, due to continuous disposals. This pollution may in atmosphere, via air pollution and water pollution.

A thorough knowledge of the radioactive isotope being used is very much essential for the person in order to utilize it safely and properly. At the same time, one should use the minimum quantity of the isotope to succeed the job. Any little extra material, may be unnecessary and its disposal to the atmosphere would further aggravate. Hence complete knowledge regarding the reasons of contamination, decontamination methods etc are necessary prior to its usage and the following topics discuss on the same, in brief.

Selection of Material

It is very important to select the right isotope for the required experimental/application job. It should as minimum in quantity as required and should have good half life period, availability and its energy level should be easily detectable with the available instruments. Handling procedures for such isotopes should also be easy and complete scientific information about the material must be known well in advance.

Contamination

During the application of RAI, surroundings may be contaminated either directly or indirectly. The reasons for contamination can be numerous. It can be due to the following:

1. Leakage while handling the material
2. Spillage during transfer of material
3. During operation itself
5. Leakage from the glove box/fumehood/storage points, waste boxes.
6. Negligence.

All these must be carefully controlled to minimise/avoid contamination.

Decontamination

It is our primary responsibility that the atmosphere is not contaminated. If contaminated we have to take remedial methods to decontaminate the surroundings quickly. Contamination can occur to the personal and as well as to the area. Hence we look into both situations. Decontamination is the process of reducing the level or removing the contamination. This must be done and continued until the level reaches below the safety limits.

Personal Decontamination

This can be achieved by the following methods, in case person is contaminated: cleaning with good amount of tap water.: cleaning with dilute about 10 per cent detergent solution. cleaning with 5 per cent EDTA solution in 10 per cent teepol. Unfortunately, if the radioisotope is taken internally, it must be seen that it has very minimum residence time in the body. Hence it has to be expelled as quickly as possible. For this laxatives are recommended and also physicians consultation required.

Instrumentation

Instrumentation plays very important role in the handling of radioactive isotopes. It is very difficult to know the level and degree of damages that may occur, without instrumentations. Literature says wide range of instruments are available for detection and as well as for measurement. In this, paper it is considered to highlight few important instruments that detect the isotope.

Instrumentation can be categorized on its utilization to the Personal and as well as to the Area. Personal monitoring and area monitoring are discussed hereunder:

Personal Monitoring

Personal monitoring is the assessment and controlling of radiation level for individual. Person who uses/handles their RAI must be monitored periodically. For this purpose certain devices are available. Film badge is one among them. The film badge should be worn during handling operation, even during transpiration RAI from one place to another. The radiation quanta received by the individual is assessed by this. The level received by the operator is checked periodically, generally on a monthly basis.

The film badge mainly contains six windows, encased in a small frame arrangement, each window containing metallic/non-metallic film. These films absorb the free radiations around the human being, and analyzing it gives the type and level of radiation received by the individual, which can be estimated. Different type of badges are available, *viz.* wrist badge, chest badge, head bade etc. with the usage of film badges, the level received by the individual can be monitored and also checked that one never exceeds the prescribed limits legislated. Other devices used for personal monitoring are thermo-luminiscent detector, pocket dosimeter etc.

Area Monitoring

Area monitoring refers to the regular checking and monitoring of the radiation level of the laboratory or working place or storeroom etc. For this purpose, instrument like Ionizing chamber, Geiger-Muller counter, Pie counter are used. These help to study the safe limits for the area protection.

When a premise get contaminated, which is detected by instruments then immediate decontamination of the area has to be undertaken. While decontaminating an area, the procedure followed is like this: first cordon the place with barricades, put show-barriers, clean from top (roof) to wall to floor by wet mopping method only. Brooming is not recommended as the dust may fly all over. Check the level with monitor and repeat the process till it reaches the safe low level.

The same instruments can also be used for detection of leaks etc in a pipeline, using radio tracer technique. Consider the example in a long pipeline (say 100 km) where petroleum is pumped and there a leak somewhere, in between. To detect such leak, radiotracer is added in the feed itself and the liquid is pumped in the usual way. The place (underground) where the leak exists, will have little radioactive material resided. Later the instruments like GM counter is run along the pipeline and it will detect the radiotracer and signal is produced. Hence the leak location found. The instruments are generally portable and compact in size for the jobs it is intended. Interval half technique will reduce the labor. Also suspected area can be checked first.

Disposal Method

Disposal of RAI is an important point and care has to be taken to see that the material is not disposed as an ordinary waste of laboratory discharge. Even the waste can have live radioactive property and has to be treated very carefully. It should be put in PVC bags, tied and buried in pits. If the waste is liquid in nature, large amount of dilution to be done and permitted out. Log-books have to be maintained for the purchase, usage and disposal data etc. Burial pits are recommended for disposal of RAI, and they have to be fenced and proper sign-boards kept. Also placard symbols must be used at all places wherever the radioactive operation is going on. (at storage, usage and disposal sites). After 10 half life period of RAI, it can be considered as normal waste and discharged.

Conclusion

Effluent aspects of radiation in environment is a wide subject and in this paper, a little has been discussed. Very special care has to be taken while handling, storing, using and disposing of radioactive material. During these operations, extreme care has to be taken that it is not misutilised, spilled or polluted. The methods of area and personal monitoring is little highlighted. When the subject is wide, the author has touched little points only, and the readers are recommended to look for more details and hoping that these will remain with them.

74

Innovative Systems for Achieving Zero Discharge of Pulp and Paper Mill Coloured Wastewaters: A Unique Experiment at Orient Paper Mills, Amlai

SANTOSH KUMAR SINGH, M.C. PANDE & N.K. THUSU

Forest Officer, Orient Paper Mills, Amlai (M.P.)

Introduction

In the process of pulp and paper manufacturing, large quantity of water is required. At Orient Paper Mills, Amlai, on an average about 190 M^3 water is required to produce 1 MT of paper. Nearly eighty per cent of this water comes out as industrial wastewater. For a daily production of 220 MT of paper at the Mills, about 39,000 M^3 wastewater comes out from the Mills, out of which about 18,000 M^3 is coloured wastewater, which has sodium ligno compounds that do not degrade easily. The wastewaters undergo elaborate treatment before being discharged into the environment and confirm to all the statutory parameters prescribed by the regulatory authorities. The treated wastewater is free of most of the objectionable parameters except the colour for which scientists all over the world are trying to find an economically feasible solution. Some methods have been experimented at pilot scale, but have not been found commercially viable.

Orient Paper Mills, Amlai and National Environmental Engineering Research Institute (NEERI), Nagpur have, in the past, undertaken two field research projects to find alternative uses of this coloured effluent so as to reduce the volume of coloured wastewater going into the surface water environment. Research works conducted by the Mills in collaboration with NEERI has shown that the treated wastewaters can be successfully utilised as irrigation water to grow many agricultural and forestry crops in light textured soils. It was during early nineties that NEERI developed "High Rate Transpiration System (HRTS) " as a unique land treatment technology for utilisation of pulp and paper mills wastewater, which has been demonstrated successfully for the first time at Orient Paper Mills, Amlai (M.P.) at full-scale level.

Nature of Wastewaters and their Treatment

In an integrated pulp and paper mill, practically, all the unit operations such as bamboo/wood washing and chipping, digester house and pulp washing, bleaching, paper machine and chemical recovery contribute towards generation of liquid wastes. At Orient Paper Mills, Amlai there is a well laid out drainage system to carry out different wastewaters individually from each section and also has built in provisions for segregating uncontaminated waters. Depending upon the nature of wastewaters, they are designated as Grade-I, Grade-II and Grade-III.

Grade-I water is the clean wastewater consisting of condensates from Paper Machine, Steam Turbines and Evaporators and cooling water from spray ponds, which is fully recycled in the mill for various processes. Grade-II wastewater comprises of white water from stock preparation, paper machines, outflow from chlorination tower and hypo section of bleach plant, wash water from chipper house and grit collectors, drain water from sludge filter vacuum pumps and supernatant of sludge ponds. Grade-III wastewater is the coloured wastewater, which originates from pulp mill, brown stock washers, spillages from digester house and caustic extraction effluent from bleach plant.

Total wastewater generated at the Mills is about 39,000 M^3 per day out of which about 18,000 M^3 per day is the coloured wastewater. The entire wastewater coloured and non-coloured receives elaborate treatment in suitably designed effluent treatment plant installed with a capital investment of Rs. 6.0 crores and spread over an area of 300 acres. The Mills have been spending about Rs. 2.00 crores annually in operation and maintenance of the above effluent treatment plant, which works out to nearly 1% of its annual turnover. A schematic diagram of the effluent treatment plant is shown in Fig. 1.

The characteristics of Grade-III wastewater before and after treatment are shown in the Table 1.

It is the Grade-III wastewater, which despite elaborate treatment has a pale brownish yellow colour that causes aesthetic pollution of the river water after its discharge. Orient Paper Mills, Amlai was conscious of the problem of coloured wastewater and its impact right from its start up stage in 1965. The Mills made many pioneering studies for pollution control and was the first in the country to have installed a most comprehensive effluent treatment system in collaboration with NEERI in early seventies. The elaborate treatment system significantly helped to reduce the pollutants to meet the standards prescribed by MoEF and MPPCB. To stop the discharge of about 18,000 M^3 of treated coloured wastewater per day for nine non-monsoon months in a year itself is a very challenging task that the Mills have taken up in addition to the regular treatment operations. The irrigation water quality of Grade-III wastewater is shown in the Table 2.

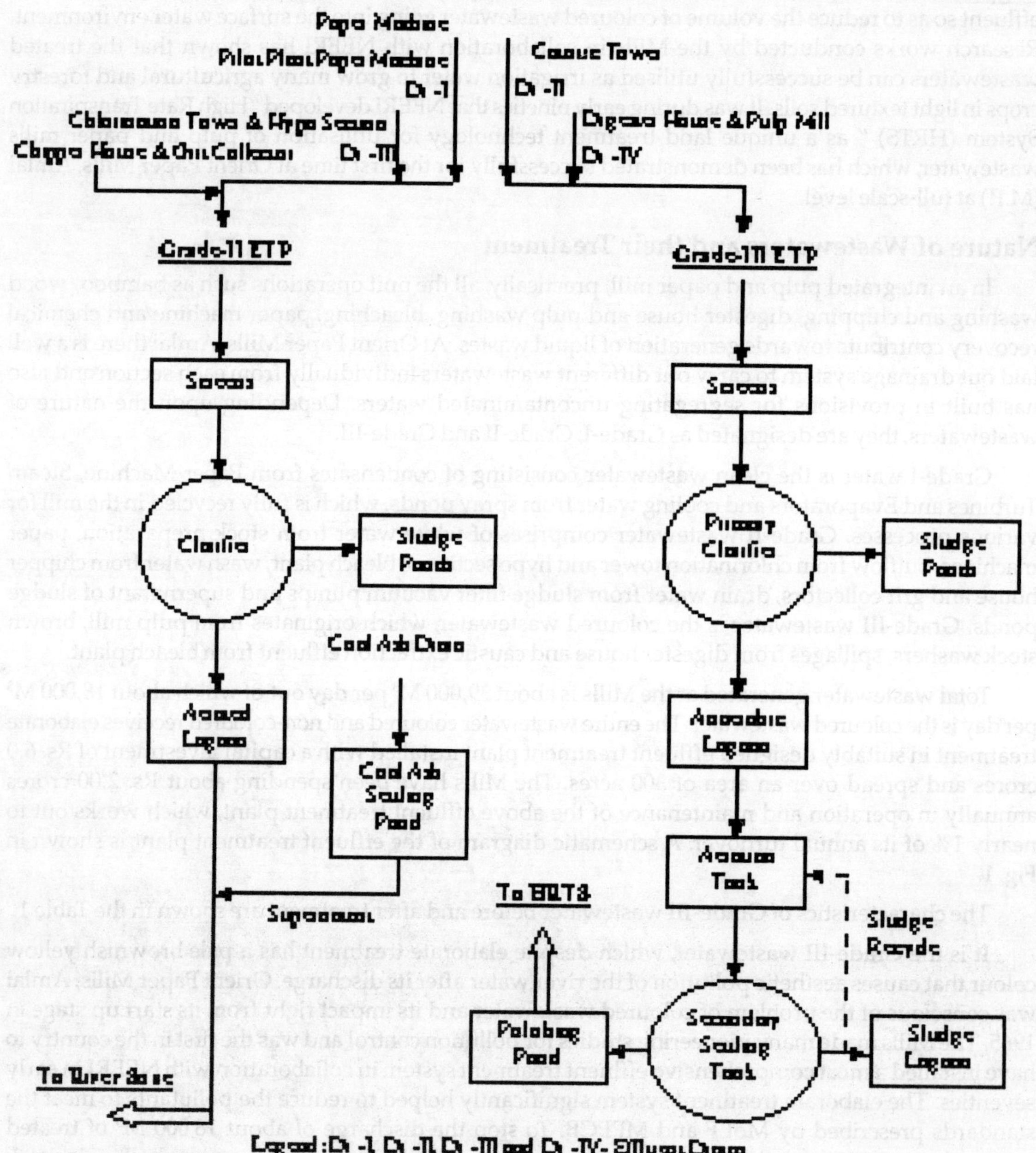

Fig. 1: Schematic Diagram of the Effluent Treatment Plant

Table 1: Characteristics of Grade-III Wastewater Before and After Treatment

Parameters	*Before Treatment*	*After Treatment*	*MoEF Standards*
pH	7.7 – 9.4	7.3 – 7.7	5.5 – 9.0
EC_w, mS/cm	1.4 – 3.1	2.5 – 3.8	—
SS, mg/l	282.0 – 596.0	26.0 – 59.0	100
COD, mg/l	525.5 – 1351.8	171.36 – 300.3	250
BOD_5, mg/l	80.0 – 280.0	20.0 – 30.0	30
Ca^{++}, mg/l	186.5 – 369.6	293.0 – 364.2	—
Mg^{++}, mg/l	7.99 – 69.24	16.1 – 63.9	—
Cl^-, mg/l	303.33 – 800.28	658.29 – 684.10	1000
HCO_3^-, mg/l	146.6 – 536.8	122.0 – 463.6	—

Table 2: Irrigation Quality of Grade-III Wastewater

Parameters	*Observed Values*	*MoEF Standards for Discharge on Land for Irrigation*
pH	6.70 – 7.90	5.5 – 9.0
EC_w, mS/cm	2.50 – 3.80	NA
Na^+, meq/l	6.87 – 13.10	NA
K^+, meq/l	0.32 – 0.95	NA
Ca^{++}, meq/l	9.30 – 15.50	NA
Mg^{++}, meq/l	0.40 – 8.08	NA
HCO_3^-, meq/l	1.50 – 15.16	NA
Cl^-, meq/l	15.69 – 23.49	600
SAR	2.78 – 4.56	NA
BOD, mg/l	20 – 30	100
COD, mg/l	171 – 300	NA
SS, mg/l	26 – 59	200

The irrigation water quality of the coloured effluent show that the wastewater is neutral to lightly alkaline in reaction and its pH varies from 6.7-7.9. Soluble salt content in the wastewater, expressed as EC (mS/cm), ranges from 2.26-3.50. The sodium adsorption ratio of wastewater is low and ranges from 3.10-4.56 indicating that it will not impose sodicity problems after its disposal on land. Wastewater contains appreciable quantities of chlorides, which range from 15.69-23.49 meq/l. Although the wastewater belongs to the irrigation water class of 'Increasing Problem to Severe Problem' with respect to salinity and chlorides, it is not likely to cause any serious problem in the long run because soils in the applicant area are mostly loamy sand to sandy loam in texture (coarse texture). Further the species grown in the HRTS are salt tolerant in nature and not likely to get affected due to wastewater quality.

Genesis of HRTS at Orient Paper Mills

As mentioned earlier, Orient Paper Mills, Amlai have installed a comprehensive effluent treatment plant for treatment of wastewaters after segregating them according to their characteristics. The coloured wastewater Grade-III effluent undergoes three-stage treatment, primary clariflocculation, anaerobic treatment in an anaerobic cum equalisation lagoon, activated sludge treatment and post treatment polishing, resulting in the treated wastewater meeting all the statutory parameters for discharge in surface water environment, as specified by Ministry of Forest and Environment and the state Pollution Control Board. It may be mentioned that even though there has not been any standard specified for the colour in the statutory regulations, the Mills, in order to stop aesthetic pollution of the river Sone, sponsored National Environmental Engineering Research Institute (NEERI), Nagpur in June 1994 for development and implementation of a land treatment system for full utilisation of Grade-III wastewaters.

The conventional physico-chemical methods to remove colour are not economically viable at plant scale. Therefore, to tackle the colour problem, an environmentally friendly technology, *i.e.* High Rate Transpiration System, developed by NEERI was installed at Amlai in 1994 under the sponsorship of Orient Paper Mills, Amlai. The System was earlier designed by NEERI during a project sponsored by United Nations Environment Programme.

Concept and Basic Approach of HRTS

The treatment and disposal of wastewater through High Rate Transpiration System (HRTS) provide cost effective and environmentally acceptable solution to manage the problem of coloured wastewater. The system also promotes development of environmental forests and green belt and enables creation of much required sink potential for absorption of green house gases.

The High Rate Transpiration System envisages the use of dynamic, multicomponent soil system as a live filtration device to renovate the wastewater through adsorption, ion exchange, precipitation and stabilisation of pollutants through microbial degradation. The high transpiration capacity of plants grown on soil matrix enables the system to serve as biopump. Various plant species like Acacia, Bamboo, Neem, Shisham and Eucalyptus transpire water equivalent to 5 to 13 times that of potential evapo-transpiration from the soil matrix alone. The soil system works in close conjunction with plants on it that provide a bio-pump through their high transpiration capacity. The filter media provided in the furrows mainly consists of bamboo/wood dust, fly ash and gypsum at the rate of 2-5 tons per hectare. Thickness of filter media varies from 5-15 cm depending on the colour intensity of the wastewater, soil texture and permeability of the soil. In addition to this, artificial media and leaf fall from standing plants provides thick mat and forms a filter bed that is also responsible for retention and assimilation of colour bodies.

Initially, both semi-treated as well as fully treated wastewaters were used in HRTS. This system has, however, been modified now to use only treated wastewaters. Land application of wastewaters is maintained at 250 m^3 per hectare per day to avoid wastewater logging and undesirable downward percolation. This application rate, however, remains effective only during peak summer months and necessary adjustments became unavoidable during winter months when evapo-transpiration rates slow down.

The system works on following main principles:

1. Transpiration of large amount of water through stomatal network.
2. Waste renovation using soil as a physico-bio-chemical reactor (living treatment filter).
3. Assimilation of colour in filter media.

The Cross section of HRTS layout is shown in Fig. 2.

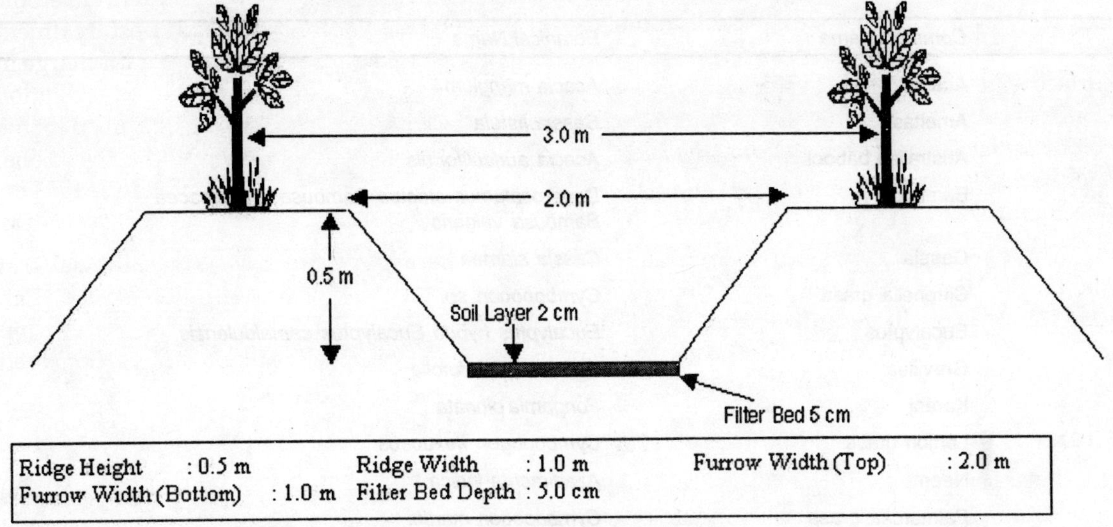

Fig. 2: Cross Section of HRTS Layout

Development of HRTS at Orient Paper Mills, Amlai

Orient Paper Mills, Amlai have gone beyond just statutory compliance of meeting the standards for its wastewaters to be discharged and are in the process of minimising the discharge of coloured wastewater to prevent any aesthetic pollution of the river water. It started with the study of possible utilisation of this wastewater for irrigation. Once it was found that that it can be used for irrigation in light textured soils, the Mills had started wastewater-irrigated agriculture on pilot scale within its premises. After establishment of the HRTS technology by NEERI, the Mills have implemented it on plant scale for its beneficial utilisation for growing pulpwood plantation during 1994-97. The Orient Paper Mills have been the first industrial unit in India to have adopted and installed HRTS. After successfully implementation and experience of the HRTS on 80 Ha, which is able to utilise 30-75 % of the coloured wastewater generated at the Mills depending upon atmospheric evapo-transpiration conditions, the Mills have further expended the HRTS network in 70 ha of its own land for complete utilisation of its coloured wastewater during the period of lean flow in the river. The Mills achieved zero discharge of its coloured wastewater into the river on 31st March 2002 and became the first integrated pulp and paper mill in the country to have achieved so. The discharge of coloured wastewater continued to be zero throughout the summer of 2002.

The HRTS has been developed in the existing plantation areas of the Mills as well as new areas that have also been subsequently brought under it. The existing plantations mainly consisted of Eucalyptus and Bamboo. In the new areas brought under HRTS, acacia, cassia, grevillea, neem and shisham have been grown. Lemon, citronella, palmarosa and vetiver grasses have also been planted on the ridges on experimental basis. Typha and phragmatis which are wetland plant species have also been utilised.

Details of species of tree and grasses planted in HRTS area is given in Table 3.

Table 3: Tree and Grass Species Planted in HRTS Area

Common Name	*Botanical Name*
Acacia	*Acacia mangium*
Amaltash	*Cassia fistula*
Australian babool	*Acacia auriculiformis*
Bamboo	*Dendrocalamus strictus Bambusa arundinacea Bambusa vulgaris*
Cassia	*Cassia siamea*
Citronella grass	*Cymbopogon sp.*
Eucalyptus	*Eucalyptus hybrid Eucalyptus camaldulensis*
Grevillea	*Grevillea pteridofolia*
Karanj	*Pongamia pinnata*
Lemon grass	*Cymbopogon flexuosus*
Neem	*Azadirachta indica*
Palmarosa grass	*Cymbopogon martini*
Siras	*Albizia lebbeck*
Vetiver grass	*Vetiveria zizanioides*

Agricultural Crop Cultivation with Effluent Irrigation

Along with the development of HRTS, the Mills have been promoting the use of treated wastewater as irrigation water amongst farmers of the adjacent villages. Initial studies on agricultural utilisation

Fig. 3: Bamboo Plantation Under HRTS

of treated Grade-III wastewaters were carried out by NEERI in mid seventies at the Mills. Based on the studies, spanning three years, standard crop agronomic practices have been developed which have since been followed for successful development of non-traditional agriculture in the vicinity of the Mills. The farmers in about 30 acres of their land have been utilising the wastewater for irrigating paddy and wheat crops with significant improvement in their yields.

To make the proposition more attractive to the user farmers the Mills have introduced incentives like free seeds of improved crop varieties, free supply of NPK fertilizers and technical guidance.

In an area where 90 per cent of agriculture is rain fed only, the treated Grade-III effluent has proved to be a boon for the farmers in the vicinity of the Mills. Cultivation of non-traditional crops like wheat has now become possible in the neighbourhood of the Mills due to ready availability of irrigation water aided by input support from the Mills.

Fig. 4: Eucalyptus Plantation Under HRTS

Scientific Research and Technological Innovation

For developing the HRTS technology the initial research was conducted with lysimeters in a controlled environment to assess the potential evapo-transpiration from the soil and to characterise the leachate generated due to the wastewater loading. In the next phase research was conducted to identify the tree species that could tolerate wastewater loading and at the same time increase transpiration from their stomatal network. Research has also been conducted to assess the change in wood characteristic in HRTS network.

The above research has helped in planning the HRTS for Orient Paper Mills, Amlai so that the discharge of coloured wastewater into the surface water environment can be minimised. The lysimeter study has helped in assessing the total area required for complete utilisation of the coloured wastewater. However, there have been observed deviation in the actual transpiration rates in the field due to different evapo-transpiration conditions which are being duly incorporated in the second phase of HRTS expansion. The identification of suitable species that could be planted and having higher transpiration was also an important finding of the research and the tree and grass species for planting have been selected accordingly.

Fig. 5: Effluent Irrigated Wheat Crop

Pollution Prevention Land Significant Environmental Benefits

The "High Rate Transpiration System" has been designed in such a way that it serves not only as a means of irrigation of plantations by treated wastewater but acts as a land treatment system. The irrigation system has been improved upon by developing ridges and furrows and planting tree species on the ridges only. These species are salt tolerant in nature and not likely to get affected due to wastewater quality. The furrows have been laid with low cost filter media consisting of gypsum, fly ash and sawdust. The fall of leaf litter further adds up to the filter media and bacterial activity in them acts as a live filter for the treated wastewater. The wastewater loading is done only in furrows which ensure controlled dosing of wastewater to prevent any groundwater contamination. Regular monitoring of water samples from piezometric holes developed in the HRTS area is conducted to detect any seepage of wastewater at an early stage. Thus the irrigation system has been modified to minimise any chances of groundwater pollution.

The High Rate Transpiration System has helped the Mills in reducing the aesthetic pollution of the surface water environment by reduction of wastewater being discharged directly into the surface water environment. In addition, the HRTS has helped in greening the barren wastelands and has provided effective sink for Mills wastewaters and air pollutants. The HRTS has also helped in resource generation in terms of ground water recharge, generating industrial and domestic wood and fodder. Along with creation of employment opportunities for local villagers the land treatment of coloured wastewater has helped in conservation of natural resources.

The environmental issue (s) that the programme addresses are:

1. Generating green cover on barren lands arresting soil erosion.
2. Providing effective sink for Mills wastewaters.
3. Providing effective sink of air pollutants by the planted trees.
4. Carbon sequestration by the plants.
5. Addition to the biodiversity of erstwhile barren land with the help of multiple tree species and grasses.

Cost Implications of HRTS

The total cost of HRTS development at Orient Paper Mills, Amlai has been about Rs. 1.00 Lacs per hectare. The cost includes earthwork for developing ridges and furrows, cost of filter media and its laying, cost of power arrangement at the site, cost towards necessary infrastructure such as motors, pumps etc., cost of irrigation system development through laying of distribution piping network with control valves and the cost of required civil constructions. The operations and maintenance costs amount to about Rs. 5.00 Lacs per month. The cost of land application thus comes to about Rs. 0.90 per m^3 of effluent.

Performance So Far

The HRTS has largely been successful. In a span of seven years from 1994, the discharge of coloured wastewater has been brought down to Zero during the periods of lean flow in the river. Regular monitoring of ground water is being done through piezemetric holes and other ground water sources such as wells and tube wells to assess the quality of ground water in the area. There has not been any adverse effect on the ground water. The monitoring results of the water samples collected from the piezometric holes is summarised below in Table 4.

Table 4: Monitoring Results of Piezometric Holes

Parameters	*Observed Values*
pH	7.2-7.9
BOD_5, mg/l	2-14
COD, mg/l	4-78
SS, mg/l	4-52
Total Solids, mg/l	410-1886
Chlorides, mg/l	60-240
Colour, Pt Co	Nil
Hardness as $CaCO_3$, Ca-H	160-330
Hardness as $CaCO_3$, Mg-H	20-70
Total Hardness as $CaCO_3$	180-390

Periodic monitoring of soil samples is also carried out to assess any impact on them due to wastewater utilisation. The monitoring results have been encouraging as there has not been any adverse impact on them. The monitoring has established the sustainability of the HRTS. In spite of its success the regular monitoring will be continued in the future also to establish its efficiency and efficacy in the long term.

The programme is replicable very easily for similar quality of wastewater elsewhere. The technology and the installation of HRTS is simple and requires construction of ridges and furrows, laying of filter media in furrows, planting of high transpirational ability and tolerant plants and development of irrigation infrastructure. At the same time regular monitoring is required for proper wastewater loading and assessing the quality of ground water and soils.

The biomass growth observed has been phenomenal. Studies have been made to assess the wood growth in Eucalyptus species planted in the HRTS and it has been found that in similar rotation period, the wood mass per hectare increased by about 40% when compared to the plantations in rain fed conditions. Similarly the yield of paddy crop per unit area increased to threefold compared to that in the rain fed conditions.

Constraints

Being a new technology and having been implemented at such a large scale for the first time by an industry in India, the HRTS at Amlai has led to some important observations and conclusions, *e.g.*:

1. Evapo-transpiration factor varies greatly in different seasons so much so that peak winter performance gets reduced to thirty per cent of the summer performance. Area requirement calculations must, therefore, keep the region's climatic condition well in mind before arriving at any conclusion in respect of area to be brought under HRTS.
2. Geology of the area is equally important. Since lateral and downward percolation of the irrigant is a contributing factor, the nature of rock formation and soil characteristics must be studied carefully before implementing HRTS. In light textured soils with rocky pan at a depth of 2 meters the lateral movement of the irrigant is inevitable leading to water logging in certain pockets even at generally recommended dosing of irrigant.

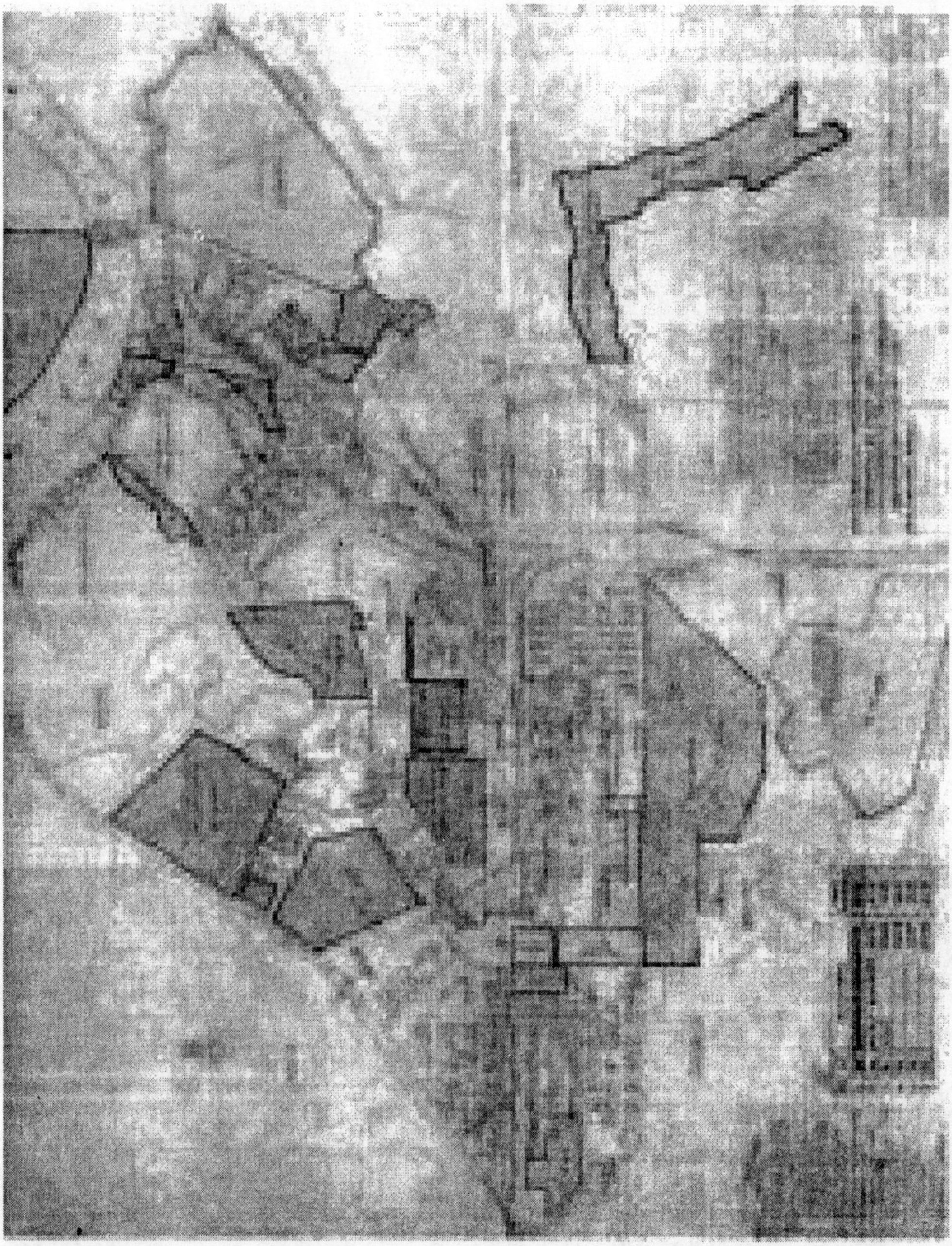

Fig. 6: A Plan of HRTS Layout at Orient Paper Mills, Amlai

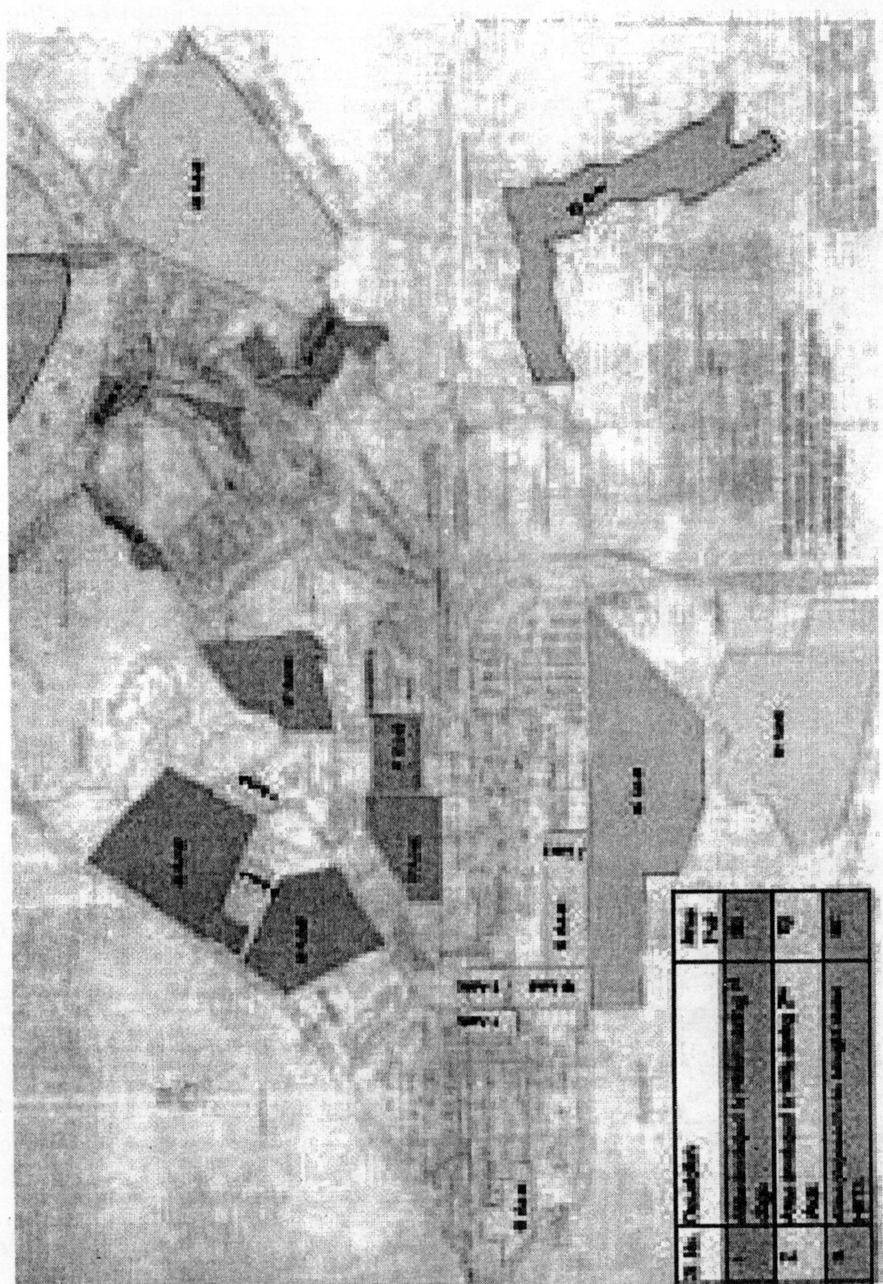

Fig. 7: A Plan of HRTS Layout at Orient Paper Mills, Amlai

3. More studies are required to widen the choice of plant species to be planted in HRTS plantations and evaluate long-term efficiency of plants in terms of transpiration factor. It will also be necessary to study the effects of chemical constituents of the irrigant effluent (*e.g.* chlorine derivatives) on plant metabolism and functional physiology like stomatal efficiency. Being the first technology implementation of this kind and scale in Indian pulp and paper industry, this aspect needs prolonged and intensive investigations to arrive at a cogent conclusion.

Future Perspective

The High Rate Transpiration System is not only a simple technology system but has been very successful in the tropical climate of our country having lot of sunshine, high temperature and availability of broad leafed tree species with high transpiration ability. Its unique suitability in our geo-climatic conditions has appealed to many organisations for utilisation of wastewater generated by them. The system is also economic in long run although the initial investment are high due to large tracts of land requirement, developing them as HRTS, planting and irrigation facility development. Subsequently only application and monitoring costs are involved. The system is unique from other wastewater treatment systems in the sense that it helps in resource generation also along with other environmental benefits. However, the abundant availability of revenue wastelands in our country may facilitate development of this system if suitable policy decision is obtained from the state and central governments. The development of revenue and forest wastelands into HRTS network elsewhere will also help in much needed re-greening and rehabilitation of the wastelands. The minimisation of wastewater discharge into the surface water environment may rejuvenate many rivers by restricting their aesthetic pollution.

Acknowledgements

Development of HRTS, its monitoring and expansion has been a teamwork involving a cross section of Mills personnel. We gratefully acknowledge the Mills management for permission to publish this paper. Authors are deeply indebted to Dr. S. N. Kaul, Dy. Director, National Environmental Engineering and Research Institute, Nagpur for his encouragement and keen interest in preparing this paper.

REFERENCES

1. Anon., 2000, *Descriptive Law on Pollution and Environment*, Akalank Publications.
2. CPPRI, 2001, *A Final Report on Evaluation of Performance of Existing Effluent Treatment Plant at M/s Orient Paper Mills, Amlai (M.P.)*, prepared by Central Pulp and Paper Research Institute, Saharanpur.
3. NEERI, 1979, *Utilisation of Pulp & Paper Mill Wastewater for Agriculture*, sponsored by Orient Paper & Industries Ltd., Amlai.
4. NEERI, 1998, *Treatment and Disposal of Wastewater through High Rate Transpiration System at Orient Paper Mills, Amlai*, by National Environmental Engineering Research Institute, Nagpur.
5. RRL, 2000, *Re-evaluation of Existing Effluent Treatment Plant of Orient Paper Mills, Amlai (M.P.)*, by Regional Research Laboratory (CSIR), Bhopal.
6. RRL, 2001-2002, *Monthly Analysis of various wastes generated at Orient Paper Mills, Amlai (M.P.)*, by Regional Research Laboratory (CSIR), Bhopal.

76

A New Approach Towards Effluent Handling at Sterlite Copper (CCR Division)

B.P. KAMATH, A.K. MITRA, K.P. SHETTY & MEENAKSHI SABHARWAL

Sterlite Industries (India) Limited, Silvassa

ABSTRACT

The Sterlite Copper produces continuous cast copper wire rods at Silvassa, India. The plant was commissioned in 1997 and the present capacity of the plant is 1,75,000 MTA. The plant receives copper anodes from Tuticorin as its raw material. The plant is certified for both ISO-9002 and ISO-14001.

This paper discusses the various changes, modification made and innovative ideas implemented in this plant to achieve the status of "zero discharge" and "environment friendly" plant. The paper also discusses the strategies adopted with the purpose of revamping and modernizing the treatment system to treat the effluents and wastes generated and convert them in to saleable products in a safe manner.

The paper also describes the sources of various effluents, the processes involved in treating of these effluents/wastes and the also the safe disposal/usage of these effluents. Also an attempt has been made in this paper to outline the steps taken to conserve the natural resources. The path towards achieving ISO-14001 and compliance towards the environmental management systems is also described in brief.

Introduction

Raw material for Plant located at Silvassa is impure copper anode. These anodes are fire refined from blister, which in turn is produced from copper concentrates. These anodes having a purity of 99.6% copper are Electro-refined to produce pure copper cathodes of purity 99.99%. These cathodes are in turn melted, cast and rolled in a controlled atmosphere to produce wire rods of the required size. These wire rods find their uses in electrical and cabling industries.

The challenge at Sterlite CC rod Plant is to maintain the purity of copper cathodes at a much higher level as specified in LME grade – A to meet the requirements of customers as well as to produce useful products from the byproducts and to ensure zero discharge of effluents from the plant.

This paper describes the method developed and implemented which ensures the *"ZERO discharge"* status. The waste produced are converted into useful products and disposed off.

Process Description

The plant at Silvassa has the following divisions.

Refinery Plant

Converts impure copper anodes to pure copper cathodes.

Continuous Copper Rod Plant

Converts copper cathodes to CC Rods.

Anode Casting Plant

Converts spent anode to fresh anode for use in refinery.

Each plant operation is discussed in brief. The Fig. 1 gives the flow chart of material.

Refinery

The main purpose of refinery is to convert impure copper anodes received from Tuticorin, which contain many impure elements to pure cathode copper of 99.99% purity by using *electrolysis process.* The technology adopted by copper refinery is ISA-PROCESS developed by MIM-Australia.

The first commercial production of cathode started in November 1996. The installed capacity of refinery was 60,000MT. The refinery was expanded in steps over last 5 years and the present capacity of the refinery is 1,75,000 MT. Another expansion is on road to enhance the capacity to 1,80,000 MT.

The 99.99 % purity copper is necessary to obtain the required electrical conductivity for the electrical applications and mechanical properties for further processes. Purity is achieved by the near elimination of the impurities such as lead, arsenic, antimony, bismuth, nickel, iron, cobalt, sulfur, gold, silver etc present in anode copper. Electrolytic refining also serves a purpose that these impurities can be recovered as by-products, particularly precious metals silver and gold.

The anodes are first dressed in Anode Preparation Machine before placing them in the Electro-refining tank-house. Prepared anodes are then taken to electrolytic cells. A cell in turn consists of 56 anodes and 55 cathodes placed alternately, *i.e.* one cathode between two anodes. Cathodes are stainless steel plates on which copper gets deposited during the process.

The electrolysis cells are filled with electrolyte which consists of copper sulfate, sulfuric acid and anode impurities in solution. The electrolyte flows through the cell, is reheated and recirculated. The electrolyte composition and temperature are controlled to a particular level to optimize the process.

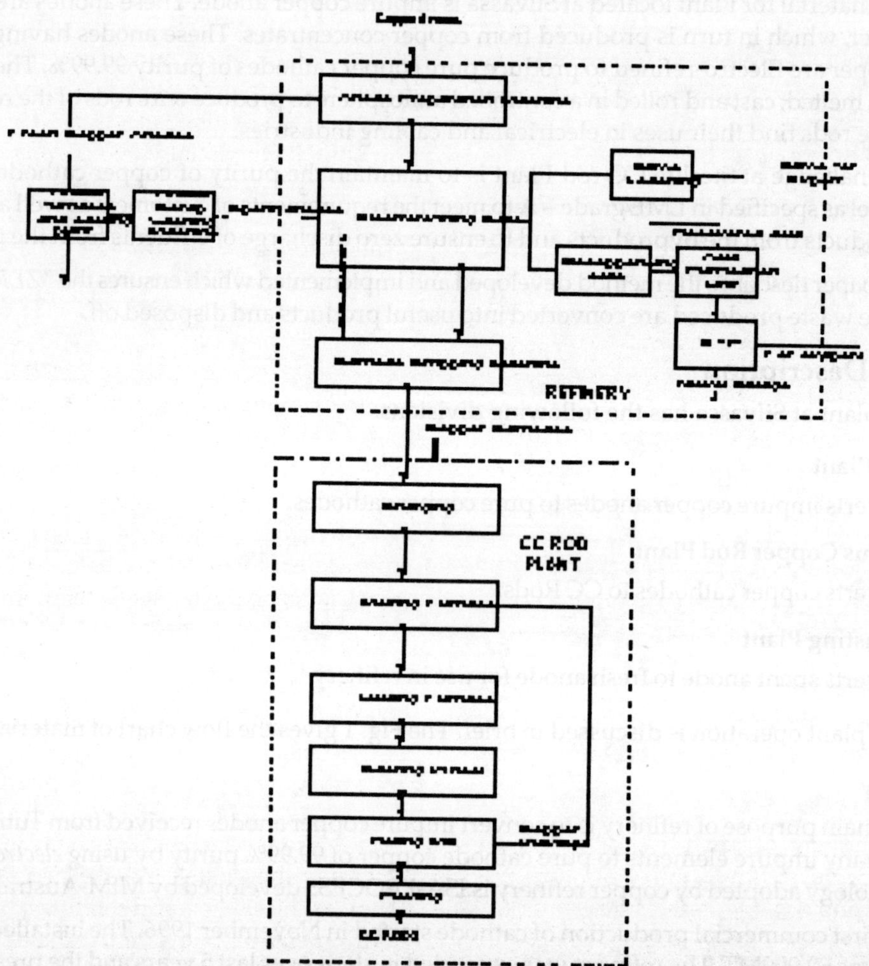

Fig. 1: Process Flow Chart

Some addition reagents are also added to the electrolyte to achieve a good quality cathode with smooth fine grained deposits.

During electrolysis, when DC current is passed through the cells, the copper is dissolved from the anode and deposited on the cathode plate, at a rate dependent on the strength of the current. Electrolysis is carried on for seven days, called a *crop*. Anodes once charged are used to yield three cathode crops of 7 days each *i.e.* for 21 days. This constitutes one *anode cycle*.

Cathodes, which are steel plates with deposited copper are removed from the process and passed through the Cathode Stripping Machine after each crop. Here cathodes are washed and the copper deposit is stripped off from the plate. Copper is then bundled, weighed and strapped, ready for use in Rod plant. The empty cathode plates are returned to the process for further deposition.

Once the anode cycle is completed, the un-dissolved part of the anode (12-15%), called *spent anode*, is washed off from electrolyte and slime. It is then melted and recast as anode in our Anode Casting Plant for reuse.

The insoluble impurities (anode slime) settle down at the bottom of the cell. Whenever anodes are replaced, the cell is emptied of electrolyte and slime is washed through a plug hole in the bottom of the cell to a collection system. The slime is then leached of copper in leaching tank, filtered and dried. This slime contains precious metals such as gold, silver and is of commercial value and is sold

Continuous Cast Copper Rod Plant (CCR)

The cathodes received from the refinery are then taken to CCR plant and charged in furnace. The CCR plant has following operations.

Furnace Operation

The melting furnace having a capacity of 40 MT, is initially charged up to the silo of the charge section. At start-up, all burners are turned on and a low firing rate is maintained until brickwork is hot and melting starts. The firing rate is then increased until the desired melting rate is reached. Molten metal drips and flows down the sides of the charge, which stands solid on the furnace bottom. The charge descends downward as its head-softened bottom supporting the main charge column melts away. Charge level is maintained depending on melting rate and amount of solid charge added each time by additions made at intervals. The melting furnace is followed by a holding furnace and a casting unit. The holding furnace helps in averaging of variation in molten metal temperature, melting rate and charge composition.

Liquid Metal Collecting

Copper cathodes must be melted and through any transfer system must be poured into a collecting system before been cast. This is called TUNDISH. It is a pre-heated ladle placed on the casting machine.

Casting Process

The molten copper from holding furnace is poured through a transfer system called Tundish using a nozzle. A flow control pin throttles the delivery of the molten metal. Then it flows into a lubricated and cooled rotating circular mold called "The Casting Wheel" and solidifies. While rotating at the proper speed, the casting wheel fills with liquid metal on one side. At the same time, the solidified cast bar comes out from the other side.

Cast Bar Preparation

Before threading the cast bar into the rolling mill, it is processed in the following sequence:

1. A roller arm senses and guides the cast bar loop by varying the mill speed around a narrow range. The roller support table called roller way is used to straighten and cut the castbar (at the beginning of the casting process and whenever necessary the bar is cut manually or automatically).
2. Before reaching the rolling mill entry, the bar is straightened by a series of driven rolls. It is also shaved on the top corners by a set of blades and the bar is cleaned with two sets of brushes. The whole set-up is known as "Bar Preparation Unit".
3. At the brushing machine outlet, the bar is held straight by a second roller table. The temperature of the cast bar is measured by optical pyrometer. At the rolling mill entry, pressurized emulsion sprays remove the cast bar oxide coating.

Rolling Process

Once the bar is prepared, it is pushed to the first rolling mill stand by a pinch roll unit. The bar rolling starts in the two roll stands, and finishes at the specified rod diameter in the finishing three roll stands. A soluble oil emulsion cools and lubricates both the rolled bar and the rolling rolls during the rolling process.

Rod Quenching and Pickling Process

After the rod is rolled it passes through a series of water diffusing head, which quench and pickle it.

Rod Drying Process

This step dries the cooled rod using compressed air, preparing it for the waxing process.

Rod Waxing Process

The wax spraying process protects the rod surface against environmental agents (air, moisture, etc.) for long storage periods.

Rod Coiling Process

In this process, first a pinch-roll pulls the rod. This keeps it in tension after leaving the last stand of the rolling mill. The pinch roll then pushes the rod into the orbital-coiling head, which forms the rod loops. Finally the loops drop inside a rotating basket system.

The CC ROD thus coiled is packed and now ready for sale.

Anode Casting Plant (ACP): Spent Anode Recycling Unit

In anode casting plant, the spent anodes received from refinery is melted and cast as fresh copper anodes for reuse in refinery.

The spent anodes received from refinery are inspected and then charged in melting reverberatory furnaces. Melting is controlled manually by furnace operator. Once the charge is melted completely, slag is skimmed out of the furnace. Green wood lags are used for reduction of oxygen content of the melt. Melt parameters are analyzed for oxygen and temperature.

Once melt is prepared for casting, it is cast in the preheated copper moulds through the red-hot launder, intermediate ladle and weigh ladle. Weight of anode and movement of cast conveyors are controlled by PLC in semiautomatic mode.

Cast anodes are ejected out of the mould and kept in the bath for cooling. Cooled anodes are then inspected for irregularities and then sent to refinery for reuse after dressing.

Sources of Effluent Generation

The flowchart with the point of effluent generation is given in Fig. 2.

The impure copper anode (99.6 % copper) contains various impurities like selenium, bismuth, antimony, nickel, iron, gold, silver etc. These impurities can be divided in to three groups based on their behavior during Electro-refining:

Metals More Electropositive than Copper

These metals does not dissolve in electrolyte and almost completely falls to the bottom of the cells and forms a part of anode mud (slime). *e.g.*: Gold, Silver

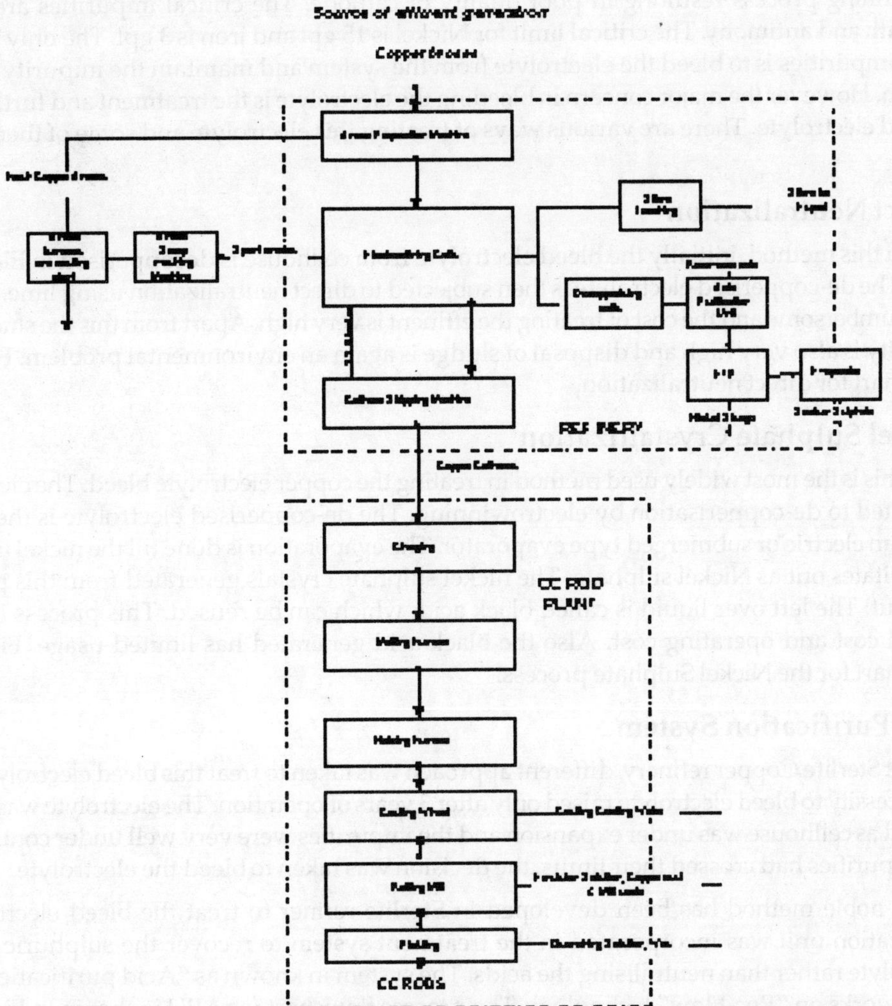

Fig. 2: Sources of Effluent Generator

Metals More Electronegative than the Copper

These metal dissolve and pass almost completely in to the solution, thereby gradually accumulating in the electrolyte. *e.g.*: Nickel, Iron

Metals with Potentials Very Near to the Copper

These metals are most harmful, as they readily pass on to the solution and remains in the solution and co deposits with the copper under conditions of the low copper concentration, high concentration of these metals in the electrolyte and high current density. Also these metals goes to the anode slime.

Metals dissolving in electrolyte are major concern for any refiner, as over a period of time, they accumulate in the electrolyte. If the level of these impurities reaches the critical levels, then it may harm the refining process resulting in poor quality of cathode. The critical impurities are Nickel, Iron, Bismuth and antimony. The critical limit for Nickel is 15 gpl and iron is 3 gpl. The only way to control these impurities is to bleed the electrolyte from the system and maintain the impurity balance in the system. However the major concern in bleeding the electrolyte is the treatment and further disposal of treated electrolyte. There are various ways of treating this electrolyte and some of them are outlined here.

Direct Neutralization

In this method, initially the bleed electrolyte from cellhouse is decopperised in Electro-winning cells. The de-copperised electrolyte is then subjected to direct neutralization using lime. The process is very cumbersome and the cost of treating the effluent is very high. Apart from this the sludge-generated quantity is also very high and disposal of sludge is again an environmental problem. Fig. 3 gives the flowchart for direct neutralization.

Nickel Sulphate Crystallization

This is the most widely used method in treating the copper electrolyte bleed. The electrolyte is first subjected to de-copperisation by electrowinning. The de-copperised electrolyte is then evaporated using an electric or submerged type evaporator. The evaporation is done till the nickel in electrolyte is precipitates out as Nickel sulphate. The nickel sulphate crystals generated from this process can be sold out. The left over liquid is called black acid, which can be reused. This process involves lot of capital cost and operating cost. Also the black acid generated has limited usage. Fig. 4 gives the flowchart for the Nickel Sulphate process.

Acid Purification System

At Sterlite Copper refinery, different approach was taken to treat this bleed electrolyte. At Sterlite, the necessity to bleed electrolyte raised only after 3 years of operation. The electrolyte was continuously diluted as cellhouse was under expansion and the impurities were very well under control. Only once the impurities had crossed their limits, the decision was taken to bleed the electrolyte.

A noble method has been developed in Sterlite refiner to treat the bleed electrolyte. A acid purification unit was incorporated in the treatment system to recover the sulphuric acid in bleed electrolyte rather than neutralising the acids. The system in known as "Acid purification Unit (APU) " and works on "Rec-Flow" technology. The process flowchart for APU is shown in Fig. 5.

The Bleed electrolyte is first taken to decopperisation cells, where the copper is recovered from the electrolyte by electrwinning process. The decopperised electrolyte is then taken to APU.

APU is an ion-exchange column. This process works on the principle that the certain ion exchange resins have the capability to sorb strong acids from the solution. While excluding metallic salts of those acids. The process is reversible, in that acid can be readily desorbed from the resin with water. It is thus possible, by alternatively passing contaminated acid and water through a bed of this resin, to separate the free acid from the metal. Fig. 6 gives the operational sequence of APU.

In copper refining process the electrolyte after decopperising in purification cells, is passed through this resin bed. The resin adsorbs the sulphuric acid in the electrolyte, whereas the major part of impurities such as Nickel, iron, bismuth, antimony goes in waste liquor. In the next step, water is passed through the resin, which elutes acid from the resin and the resulting product containing recovered sulphuric acid is sent back to the electrolyte system. The waste liquor generated is further treated with sodium sulphide and neutralized with caustic up to pH of 8-9 to precipitate impurities as

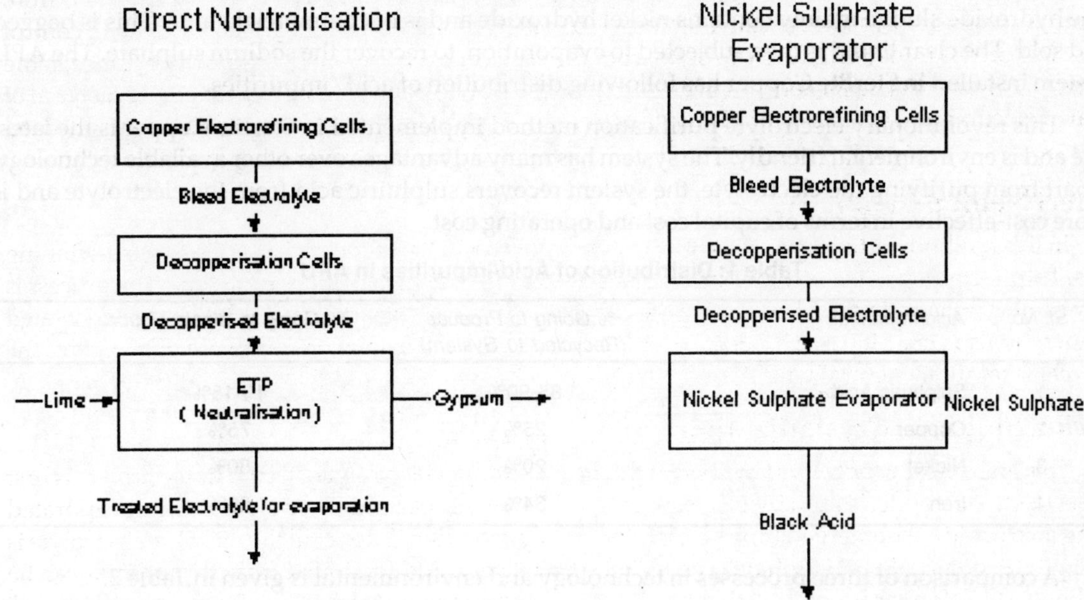

Fig. 3: Direct Neutralisation

Fig. 4: Nickel Sulphate Evaporator

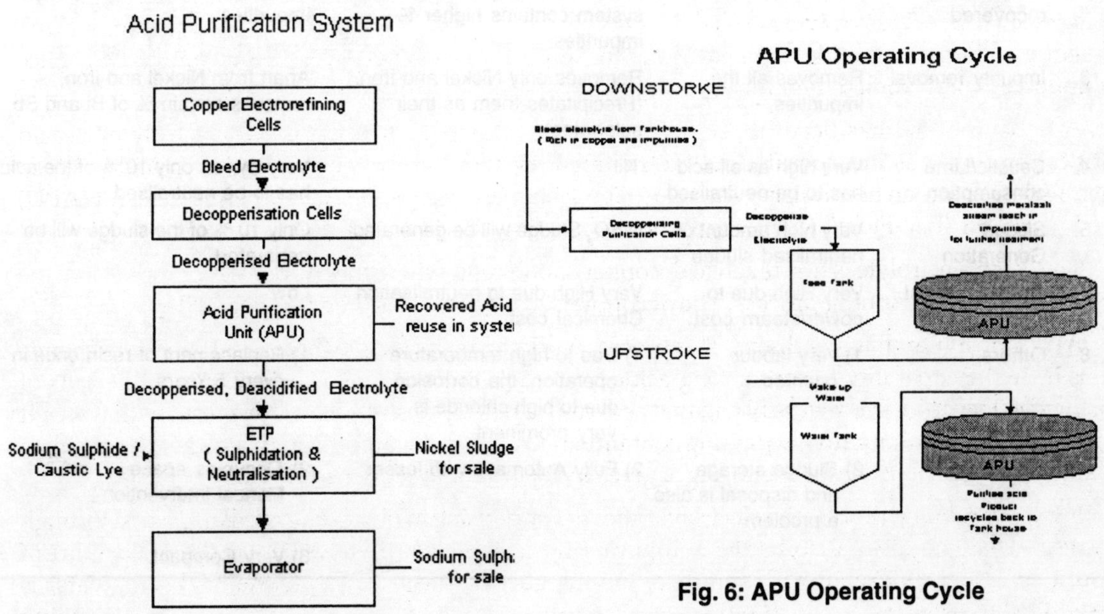

Fig. 5: Acid Purification Unit

Fig. 6: APU Operating Cycle

their sulphides and hydroxides respectively. This is further filtered to separate the hydroxide sludge. The hydroxide sludge mainly contains nickel hydroxide and is of commercial value. This is bagged and sold. The clear liquid is then subjected to evaporation, to recover the sodium sulphate. The APU system installed in Sterlite Copper has following distribution of acid / impurities.

This revolutionary electrolyte purification method implemented in Sterlite Copper is the latest one and is environmental friendly. The system has many advantages over other available technology. Apart from purifying the electrolyte, the system recovers sulphuric acid from the electrolyte and is more cost-effective in terms of capital cost and operating cost.

Table 1: Distribution of Acid/Impurities in APU

Sl. No.	*Acid/Impurities*	*% Going to Product (Recycled to System)*	*% Going to Waste Liquor*
1.	Sulphuric Acid	85-90%	10-15%
2.	Copper	25%	75%
3.	Nickel	20%	80%
4.	Iron	34%	66%

A comparison of three processes in technology and environmental is given in Table 2.

Table 2: Comparison of Various Electrolyte Treatment Processes

Sl. No.	*Parameter*	*Direct Neutralization*	*$NiSO_4$ Evaporator*	*Acid Purification Unit*
1.	Acid recovery	No	Yes (70 %)	Yes (90 %)
2.	Quality of acid recovered	——	Recovered acid sent back to the system contains higher % impurities	Recovered acid with very less impurities
3.	Impurity removal	Removes all the impurities.	Removes only Nickel and iron. (Precipitates them as their sulphates)	Apart from Nickel and iron, removes certain % of Bi and Sb
4.	Caustic/Lime consumption	Very high as all acid has to be neutralised	Nil	Very low, as only 10 % of the acid has to be neutralised
5.	Sludge Generation	Very high amount of neutralised sludge	$NiSO_4$ Sludge will be generated	Only 10 % of the sludge will be generated
7.	Operating Cost	Very High due to power/steam cost	Very High due to neutralisation Chemical cost	Low
8.	Others	1) Very labour oriented. 2) Sludge storage and disposal is also a problem	1) Due to high temperature operation, the corrosion due to high chloride is very prominent 2) Fully Automatic with lesser	1) Replacement of resin once in every 5 Years. 2) Occupies space Manual intervention 3) Very Compact.

Waste Generated at CCR Plant

Table 3 gives the effluents/waste generated at CCR plant:

Table 3: Effluent/Waste Generated at CCR

Sl. No.	*Source of Generation*	*Waste Product*
1.	Rolling mill	Emulsion
2.	Pickling Line	Quenching solution
3.	Waxing unit	Wax
4.	Casting Wheel	Casting cooling water
5	Rolling mill	Copper dust
6.	Rolling mill	Mill scale

Emulsion

Emulsion used in rolling mill acts as a lubricant as well as a cooling agent. As this waste emulsion contains oil, water and fine copper dust, the direct disposal of this material is of high environmental concern. So a innovative process was implemented to convert this waste emulsion in to useful saleable product.

In the first step, the waste emulsion is filtered through a filtration system to remove the fine copper dust. Copper dust is collected in drums and recharged back in anode casting furnace. This ensures maximum copper recovery in the system. Further the filtrate is subjected to evaporation in a steam evaporator at around 100 °C temperature. The water evaporates, leaving behind the oil in evaporator. Oil is collected in barrels and sold to lubricant manufacturers

Wax Solution

In the same evaporator, wax solution is also treated where water gets evaporated and the residual wax is again reused in process.

Quenching Solution

Quenching solution is used for cooling the CC Rod. It also acts as a reducing agent to prevent the oxidation of copper. This solution comprises of water, 1-3 % Iso-propyl alcohol (IPA). This waste quenching solution taken to anode casting plant and is used to cool the anodes.

Casting Cooling Water

Casting cooling water is used in casting wheel for cooling purpose. It contains water along with carbon soot. So direct usage of this waste casting cooling water is not possible for gardening purpose. Before using this water in gardens, we are treating it with a suitable coagulant/flocculent and pass it through Lamella clarifier. Prior to use in gardening a sample is sent to laboratory for pH and heavy metals testing. Once it is cleared by the laboratory, it can be used for gardening.

Copper Dust and Mill Scale

Copper dust and mill scale generated at rolling mill is collected in drums and sold to outside interested parties.

Conservation of Natural Resources

Water Conservation and Recycle

Water conservation is minimizing the use of water for each specific application. Some of the water conserving measures taken by Sterlite:

1. Reduction and control of usage of water for different purposes by installation of flow-controllers, etc.
2. Minimization of leakage's and water wastage.
3. Reduction of evaporation loss by covering Cell covers in cell house and open tanks.
4. Increasing the output between regeneration in water treatment plant and thus reducing the water required for regeneration.
5. Work is in progress for rainwater harvesting.

Water recycle is the use of water for a succession of downgraded uses before it is finally treated. Recycling is carried out by two ways:

On-line Recycle

In this case, wastewater from particular process section (*e.g.* casting cooling blow down water, boiler blow down water.) is collected and recycled back to the same section as input.

Off-line Recycle

This involves combining wastewater from all the sections, equalizing them and then recycling after proper treatment.

The recycled water is used for non-potable purposes like toilet flushing, cooling water make-up, irrigation, gardening etc.

Some of the major water recoil steps being taken are:

1. Using water treatment regeneration water, boiler blow down water for cooling in anode casting plant.
2. A sewage treatment plant is being installed to convert the sewage in to clean water for use in gardening.

Conservation of Fuel

Following steps are taken in order to reduce the fuel consumption in various furnaces and boilers and thus reducing the air pollution.

1. Installation of modulating burner and air –preheater in boiler.
2. Installation of recuperator and waste heat recovery system in anode furnaces.

ISO 14001 Environmental Management System implementation

Sterlite has successfully implemented and practicing the ISO-14001 environmental management systems. The implementation of the ISO-14000 has helped in documenting the waste disposal procedures, identifying and mapping out the different environmental aspects. Based on these aspects different Environmental Management Plans are taken and implemented successfully.

Conclusion

Sterlite has adopted a different, more positive and proactive approach to the challenges posed by environmental issues. At Sterlite Copper the waste generation is minimized in processes by adopting the latest and innovative technologies and to a maximum extent the resulting products are either recycled back to the process till its life time or are reconditioned for reuse in processes

All other effluents and wastes generated in CCR plant is converted into saleable/useful products in a safe manner and are sold or used internally The main emphasis is also given in conserving the natural resources like water and fuel. This has helped in reducing the effluent generated at source itself. The concept of "Reduce the waste at source" is successfully implemented at Sterlite by implementing the above measures. The implementation of ISO-14001 system has played a key role in regularising the environmental systems with proper procedures.

With all these measures the Sterlite is able to achieve the status of "*zero discharge*" and "*environment friendly*" plant. With its continued efforts in conserving the environment, the Sterlite has a vision of becoming a model plant in protecting the mother earth.

Acknowledgement

The authors express their deep sense of gratitude to the management of Sterlite Industries (I) Limited, for giving total support to them, in preparation of this paper with our endeavor to contribute to the protection of environment and mother earth. We also grateful to them for granting the permission to attend this seminar and present the paper.

REFERENCES

1. A.K.Biswas and W.G. Davenport, 1994, Extractive Metallurgy of Copper. 3rd edition, Pergamon Press, Oxford, England.
2. B.P. Kamath, B. Latchayya, K. P. Shetty, 1999, "Quality Control of Copper Cathodes: An Overview", ICDC, Seminar, Mumbai, "Testing & Quality Control of Copper & Copper Alloy Products" pp. 1 – 14.
3. B.E. Langer and P. Stantke, 1995, EPD Congress, The Minerals, Metals & Materials Society, pp, 559-569.
4. Michael Sheedy, 1998, JOM, Vol.50, No.10, October, pp, 66-69.

76

Sludge Management: Indian Perspective

J.K. BHATTACHARYYA & A.V. SHEKDAR

Solid Waste Management Division, NEERI, Nehru Marg, Nagpur - 440 020

ABSTRACT

A substantial amount of sludge is generated from the industries like refineries, distilleries paper mill, sugar mill, etc. Production units and effluent treatment plants are the major sources of sludge generation. Oily sludge, bio sludge, chemical sludge, press mud, yeast sludge and lime mud are some of the major sludges generated in the industries. Heavy metals like chromium, cadmium, copper, zinc, lead, nickel, etc. are present in majority of the sludges. Recycling of sludges in an environment-friendly manner is a one of the appropriate routes of sludge management. Secure landfill system should be adopted if industries have no alternatives other than land disposal. Major components of secure landfill system are liner, cover, leachate management and gas collection. Appropriate design and operation of secure landfill system prevent environmental damages.

Keywords: Sludge management; Leachate; Ground water pollution; Secure landfill; Land farming

Introduction

Sludge is generated in sizeable amount in the industries like refineries, distilleries, paper mills, sugar mill, etc. These sludges are generated from production units as well as effluent treatment plants. The sludges are often contaminated with organic as well as inorganic pollutants. Uncontrolled handling of these sludges often leads to environmental pollution and also affects the aesthetic quality.

Recent legislation desires environment-friendly sludge management systems in the industries. This paper has highlighted the problems and lacuna in existing sludge handling system and mitigation measures need to be adopted on the basis of field studies carried out at some of the industries.

Generation of Sludge

Various types of raw materials are used during manufacture of different products in the industries. During this conversion process sludges are generated which are treated as waste. Industries are often equipped with effluent treatment facilities. Various types of treatment offered to the wastewater are physical, chemical and biological. During the treatment of effluent, sludges are generated from various units of effluent treatment plant. Sludges generated from various sources have been presented in Table 1.

Table 1: Sludge Generation from Various Industries

Sl. No.	*Waste*	*Source*
1.	Oily sludge	Refinery
2.	Bio sludge	Refinery
3.	Chemical sludge	Refinery
4.	Yeast sludge	Distillery
5.	Press mud	Sugar mill
6.	Lime mud	Paper mill

Oily Sludge from Refinery

Oily sludge is generated in significant amount in the refineries. Crude oil is usually stored in storage tank. Impurities present in the oil are deposited at the bottom of the tank. During cleaning of the tanks, the sludge is recovered and is treated as waste. Oily sludge is also generated from the treatment plant of oily wastewater. The sources of oily sludge are API separator, TPI unit, etc.

Bio Sludge from Refineries

Bio sludge is generated during biological treatment of wastewater in the refinery. It is usually generated from trickling filter which is often called biotower.

Chemical Sludge from Refineries

Caustic washing and mercaptans oxidation facility are provided for removal of mercaptans and sulphide from lighter products obtained from crude oil. Chemicals sludge is generated from the flocculation tank during mixing of caustic treated effluent with ferric chloride and a poly-eletrolyte.

Yeast Sludge from Distilleries

Fermentation of molasses is conducted at the fermentator in presence of yeast to produce alcohol. The yeast sludge is recovered from the settling tank.

Press Mud from Sugar Mill

Cane juice is subjected to filtration for removal of impurities during manufacture of sugar. Press mud is generated as residue during filtration of cane juice in a rotary filter.

Lime Mud from Paper Mill

Lime is initially calcined to improve the quality before use in paper mill. Lime mud is generated during recovery of caustic soda in pulp and paper making process.

Quantity of Sludge

The sludges contribute significantly in total waste generation in the industries. The magnitude of the sludge management problem is decided by the quantity of sludge generation. The quantum of sludge generation per annum from various sources are presented in Table 2.

Table 2: Quantity of Sludges

Sl. No.	Source	Capacity	Type of sludge	Sludge quantity (TPA)
1.	Refinery	3 MTPA of crude oil processed	Oily sludge	430
2.	Refinery	3 MTPA of crude oil processed	Bio sludge	830
3.	Refinery	3 MTPA of crude oil processed	Chemical sludge	140
4.	Distillery	24000 KLPA of alcohol produced	Yeast sludge	300
5.	Sugar industries	7,50,000 TPA of cane sugar processed	Press mud	22,500
6.	Paper mill	90,000 TPA of paper produced	Lime mud	1,56,000

TPA: Tonnes Per Annum; KLPA: Kiloliter Per Annum; MTPA: Million Tonnes Per Annum
Source: NEERI Reports on
1. Solid Waste Management in Numaligarh Refinery,
2. Carrying Capacity based Development Planning for Doon Valley Region
3. Environmental Impact Assessment Cacher and Nagaon Paper Mill
4. Wastewater Management in Tamil Nadu Distillery

Characteristics of the Sludges

Sludge characteristics differ from product to product depending upon raw material used and manufacturing/treatment process involved. Oily sludge, which is generated in massive quantity from refineries, is highly viscous in consistency. Sludge contains phenol (90-100 mg/kg) and sufficient amount of grease and waxy material. The calorific value of the sludge is also high (4000-6000 kcal/kg). Calcium carbonate is the major constituent in lime mud. The other constituents include aluminium oxide and iron oxide. Press mud contains silica, alumina, calcium oxide, magnesium oxide, potassium oxide, phosphorus pentoxide, etc. It has been observed that heavy metals like chromium, cadmium, copper, nickel, lead, zinc, etc. are commonly present in majority of the sludges generated in industries. Heavy metals in the sludges are usually contributed by raw materials and chemicals used in the process. Wastewater enriched with heavy metal is often treated in effluent treatment plant. Thus the sludges become contaminated with the heavy metals during treatment of wastewater in effluent treatment plant. Heavy metal content in oily sludge, bio sludge, chemical sludge and yeast sludge is presented in Table 3.

Table 3: Heavy Metal Content in Some of the Sludges

Type of Sludge	Chromium	Cadmium	Zinc	Lead	Copper	Nickel
Oily sludge	80.00	2.00	80.00	40.00	120.00	25.00
Bio sludge	20.00	0.60	58.00	29.66	120.83	24.66
Chemical sludge	35.50	ND	500.00	32.00	220.00	26.00
Yeast sludge	ND	0.20	36.15	ND	43.90	45.50

All values are expressed in mg/kg and on dry weight basis

Existing Sludge Handling System and its Impact on the Environment

It has been observed that industries pay more attention to wastewater management and air quality monitoring. Sludge management aspect is given low priority. Uncontrolled management of sludge often leads to serious environmental pollution. Various management practices adopted by the industries have been presented in Table 4.

Table 4: Existing Sludge Management Handling System at Different Industries

Sl. No.	*Type of Sludge*	*Existing System*
1.	Oily sludge	Stored in lagoon
2.	Bio sludge	Stored in sludge drying bed
3.	Chemical sludge	Stored in lagoon
4.	Yeast sludge	Agricultural purpose
5.	Lime mud	Land disposal
6.	Press mud	Agricultural purpose

It has been observed that certain sludges are often disposed of on lagoon or ponds, which are not provided with appropriate liners. During rainfall, water becomes contaminated with the pollutants during its contact with the sludges. This water carrying pollutants often finds access to the groundwater through soil strata causing groundwater pollution. The sludges are often directly exposed to environment. Thus the volatile and suspended particles are emitted polluting the air environment. They often affect the aesthetic quality. Sludges containing high concentration of pollutants are directly disposed of on land without any precautionary measures. These practices also cause adverse impact on land, water and air environment. Aesthetic quality is also degraded.

Recycling of Sludges

Bio sludge obtained from effluent treatment plant is dried in sludge drying bed. The dried sludge can be used as a soil conditioner for agricultural and horticultural purpose. Nitrogen, phosphorus and potassium present in the sludge contribute in enrichment of nutrients in the soil. However, it should be noted that the heavy metal in the sludges should not be present in excess amount if these are to be used for agricultural purpose. Excessive amount of heavy metals in the sludges may cause adverse impact on soil quality and enrichment of heavy metal in the leachate. Application rate of bio sludge should be decided on the basis of the amount of heavy metal required for plant and vegetation growth.

Yeast sludge and press mud are often directly applied on land by the farmers for agricultural purpose. However, it is better to mix this sludge with press mud and convert to compost prior land application. The compost prepared from yeast sludge and press mud should be monitored for heavy metals. In some industries untreated or partially treated water is used in composting process for enrichment of nutrients. This is not a good practice from environmental point of view.

Current research on land farming using oily sludge is expected to open a pathway for better utilisation of oily sludge. Land farming involves the decomposition of oily sludge by microbial action in cultivated soil. Though this process is tried in a few countries on experiment basis, more R&D is needed for large-scale exploitation. Limiting application rate of the sludge in the soil and groundwater contamination due to migration of leachate are the two major aspects, which need thorough study. Nitrogen, Phosphorus and Potassium content in different sludges have been presented in Table 5. The values reveal the potential of the sludges for use in agricultural purposes.

Table 5: Nitrogen, Phosphorus and Potassium Content in the Sludges

Sl. No.	*Type of sludge*	*Nitrogen*	*Phosphorus as P_2O_5*	*Potassium*
1.	Oily sludge, %	0.35	0.22	0.10
2.	Bio sludge, %	3.86	2.30	1.08
3.	Press mud, mg/kg	55[a]	4500	4000
4.	Yeast sludge, %	1.82	0.98	2.20

All values are on dry weight basis
[a]Available nitrogen

Lime mud from paper mill is basically a calcium carbonate sludge. It can be upgraded by calcining at high temperature. The calcined lime mud can be used for upgrading subgrade lime stone. This calcined lime has also the potential of raw material for portland cement manufacture.

Incineration of Sludges

Incineration systems are usually designed to destroy only organic compound of waste. In India sludge incineration is not popular. Though oily sludge possesses sufficient calorific value it is not thermally treated. Various constraints in sludge incineration are as follows:

1. Emission of toxic gases and vapours
2. Non-availability of properly designed incinerator with appropriate emission treatment facility
3. Problems of disposal of ash contaminated with heavy metals
4. High operating and capital cost

Disposal of Sludges

When waste cannot be recycled and incinerated, the only alternative left is secure landfilling. In secure landfilling special measures are adopted to protect groundwater from environmental pollution and also prevent air pollution. Salient features of secure landfilling are as follows:

Liner System

The liner acts as an impermeable barrier and prevents leachate from entering the subsoil water body. In secure landfill system a combination of clay/clayey soil and synthetic liner is applied. It is desirable that the permeability of the soil liner should be equal to or less than 1×10^{-7} cm/sec. Various types of synthetic liner used include polyvinyl chloride, high density polyethylene, poly propylene, butyl rubber, neoprene, etc. During selection of liner, it is necessary that the liner material should be compatible with the particular sludge. It has been observed that High Density Poly Ethylene (HDPE) liner is compatible with maximum industrial waste.

Cover System

Landfill cover is applied for minimizing leachate generation. It also ensures good aesthetics and minimised exposure of the waste to the atmosphere. Major components of cover system are vegetable cover, filter layer, drainage layer and clay layer. The upper most layer is vegetation layer. Major functions of vegetation layer are as follows:

1. Reduces erosion.
2. Reduces infiltration of precipitation
3. Enhances evapotranspiration
4. Enhances aesthetics quality

The function of drainage layer is to enhance lateral drainage of any precipitation that infiltrates through vegetable cover. A geotextile filter serves to maintain separation between the layers and act as a filter to minimise migration of material.

Leachate Management

The secure landfill system is usually equipped with leachate collection and treatment facility. Leachate is collected in perforated Poly Vinyl Chloride (PVC) pipe and finally stored in sump. From the sump it is sent to treatment facility.

Gas Collection System

In a secure landfill, gas pressure may build up due to emission of volatiles and biodegradation of waste material. Hence gas collection system is necessary to collect generated gas migrating from the landfill for subsequent venting to the atmosphere. The system is typically composed of porous sand, gravels and gas collection piping.

Conclusion

Sludges are generated from manufacturing units as well as wastewater treatment facilities located at the industries. Sludges often contain organic and inorganic pollutants. Uncontrolled handling and disposal of sludges lead to environmental pollution. Some of the sludges which are generated in substantial amount are Oily sludge, bio sludge, chemical sludge, press mud, lime mud and yeast sludge. Recycling of sludges in an environment-friendly manner is one of the appropriate solutions of sludge management problem. When sludges cannot be recycled or incinerated, the only option left is secure landfilling. Secure landfill system is adopted so that sludge disposal does not cause adverse impact on the environment. Important components of secure landfill system are liner, cover, leachate management and gas collection. Proper design and operation of secure landfill are necessary for environmentally safe disposal of sludges.

REFERENCES

1. Shekdar, A.V., Bhattacharyya, J.K., 2000, Design of Environmentally Solid Waste Management System, Seminar 'Chem Vision-2000'.
2. Bhattacharyya, J.K., Shekdar. AV., 2001, Waste management in refinery-Search for low or non-waste technology-Con. Petrotech-2001.
3. Lagrega, M.D., Buckingham, P.L., Evans, J.C., 1994, Hazardous waste management. McGraw Hill.
4. NEERI Report, 1999, Solid Waste Management in Numaligarh Refinery and Numaligarh Refinery Marketing Terminal.
5. NEERI Report, 1989, Environmental Impact Assessment Studies for Cacher Paper Mill. HPCL.
6. NEERI Report, 1995, Carrying Capacity based Developmental Planning for Doon Valley Region.

77

Industrial Hazardous Waste Treatment and Management

B. CHAKRADHAR

Regional Research Laboratory (CSIR), Bhopal

ABSTRACT

The generation of solid and hazardous waste increasing day by day with the rapid development of industrial growth in India. Knowledge of technologies and regulation for hazardous waste generation, storage, transportation, treatment and ultimate disposal or utilization is essential for sound hazardous waste management. The hazardous waste generated from industrial operations is not a single material, but an incredibly diverse collection of chemical waste. The waste characteristics shows a great variability in the toxic parameters. Improper disposal of the hazardous waste poses a threat to public health and causes serious environmental damage. The Government of India has promulgated the Hazardous Waste (Management & Handling) Rules (HW & MH) in 1989 through MOEF, New Delhi under the aegis of Environment (Protection) Act (EP Act) 1986. Based on the further suggestions received and considering the various new methodologies GOI has notified the new amendments as the HW (M&H) rules, 2000 and suggested modifications in Schedule-1 with list of process generating hazardous wastes and Schedule-2 with list of waste substances with concentration limits.

Most often hazardous wastes are treated by physico-chemical treatment technologies. Knowledge of the various technologies and regulation for hazardous waste generation, storage, transportation, treatment and ultimate disposal is essential for sound hazardous waste management. Treatment and disposal of hazardous waste by land filling and stabilisation techniques are of recent development.

Studies were carried out to develop a proper landfill site for containing zinc extraction sludge generated during the recovery of electrolytic zinc. The hydraulic conductivity of the soil is determined in the laboratory and found to be in the range of 1.95×10^{-7}cm/sec. The Toxicity Characterisation Leachate Procedure (TCLP) test was carried out for the metal extraction and various physical and thermodynamic properties were estimated. Different binders like polymers, cement and lime were used to solidify the waste and leachate extraction test reveals that there is no toxic metal present in the leachate from hazardous waste. The materials development can be suitable for making usable building materials.

Keywords: Hazardous waste, Landfill site, TCLP, Stabilisation, Hydraulic conductivity

Introduction

Production of various chemical substances on a large scale during the past three decades in India has led to the release of toxic and hazardous wastes into the environment. Hazardous wastes means wastes other than radioactive wastes which by reason of their chemical reactivity or toxicity, explosive or corrosive or other characteristics causing danger or likely to cause danger to health or the environment, whether alone or when coming into contact with other wastes, are legally defined as hazardous waste in the state in which they are generated or in which they are disposed of or through which they are transported.

The Government of India has promulgated the Hazardous Waste (Management & Handling) Rules ((HW & MH)) in 1989 through MOEF, New Delhi under the aegis of Environment (Protection) Act (EP Act) 1986, based on the further suggestions received and considering the various new methodologies GOI has notified the new amendments as the HW (M&H) rules, 2000 and suggested modifications in Schedule-1 with list of processes generating hazardous wastes and Schedule-2 with list of waste substances with concentration limits.

Characteristics of Hazardous Wastes

There are four main characteristics of hazardous waste (1) *viz.* Ignitability, corrosivity, reactivity and extraction potential toxicity.

Ignitability

Ignitable wastes are liquids with a flashpoint below 600C, or solids capable of causing fire under standard temperature and pressure.

Corrosivity

Corrosive wastes are aqueous wastes with a pH below 2 or above 12.5, or which corrode steel at a rate in excess of 0.25 inches per year.

Reactivity

Reactive wastes are normally unstable, react violently with air or water, or form potentially explosive mixtures with water. This category also includes wastes that emit toxic fumes when mixed with water and materials capable of detonation.

Toxicity

The fourth characteristic is Extraction Potential (EP) toxicity. The characteristic of toxicity is more difficult to define. The objective of this parameter is to determine whether toxic constituents in a solid waste sample will leach into ground water if the waste is placed in a municipal solid waste landfill. If this is the case, then the waste will be declared hazardous.

The EPA has developed two tests for determining toxicity. The initial test referred to as the EP Toxicity Test, required grinding of solid samples, followes by rotary extraction in acetic acid for 24 hours. The material would then be filtered and the filtrate analysed and compared to published standards (generally these were one hundred times the concentration levels acceptable in drinking water for specified metals and organic compounds). If the concentration in the exceeded any one of the specified levels, the original solid waste was considered hazardous.

A number of shortcomings with the EP Toxicity Test led the EPA to develop a modified test referred to as the Toxicity Characteristic Leaching Procedure (TCLP) (2). The latter test was more specific in many operational parameters, required a zero head-space extractor so that volatile organics would not be lost, and had more stringent quality control measures. Primary extration is a measure of immediate leachability while secondary extraction determines long term leachability of contaminants to the ground water. If even one compound exceeds the specified level then the original waste is a hazardous waste.

Management of Hazardous Waste

The potential damage for public health and to the environment from the mismanagement of hazardous waste justify the neet for implementation of an effective Hazardous Waste management (3) programme. A Hazardous Waste Managment Programme include minimisation of waste, treatment of waste and secured land disposal (4) of residues that remain hazardous after treatment.

Treatment of Hazardous Waste

Waste may be made less hazardous by physical, chemical, or biological treatment. Treatment of hazardous waste may be costly, but it can serve to prepare the material for recycling or for ultimate disposal in a manner safer than disposal without treatment.

Physical Treatment

Physical treatment methods are conducted in order to reduce the volume of the wastes and facilitate the solid-liquid separation. Several physical processes including sedimentation, clarification, centrifugation, floatation, filtration, evaporation, distillation, reverses osmosis etc. are used in hazardous waste management. The various physical treatment technologies available for different applications are given in Table 1.

Chemical Treatment

Chemical treatment methods involve the use of reactions to transform hazardous waste streams into less hazardous substance. The various chemical treatment technologies available for different applications are given in Table 2.

Biological Treatment

Biological or biochemical biodegradation may be a practical detoxification method for some hazardous organic wastes. It is the degradation of organic wastes by the action of microorganisms. If biological treatment of a hazardous waste is contemplated, care is required to ensure that the other components in the waste neither poison the organisms nor render the residue unfit for landfill disposal.

Disposal of Hazardous Wastes

Landfill is currently the most widely practiced method for disposing of hazardous wastes (5). A landfill is defined as that system designed and constructed to contain discarded waste so as to minimise releases of contaminants to the environment. Landfills are necessary because hazardous waste minimisation technologies cannot totally eliminate the waste generated and treatment technologies produce residues. Thus the generation of hazardous waste can not be reduced to zero in the foreseeable future.

Table 1: Physical Treatment Technologies

Technology	*Application*	*Comments*
Carbon Adsorption	Organic contaminated liquids and vapours. Activated carbon can be customised to meet specific needs	Expensive; require regeneration or disposal. Only suitable for dilute organics and performance may be difficult to predict
Air Stripping	Dilute wastewater or ground water	VOC emissions a problem
Filtration	High volume aqueous wastes	Pretreatment or multiple screening may be required
Centrifuging	Sludge dewatering and volume reduction	Uniform feed requirements usually necessitate pretreatment
Distillation	Solvent recovery, glycols, amines and waste oils	Energy intensive; may require prefiltration; residues usually hazardous
Evaporation	Recovery of inorganic salts from organic or inorganic solvents	VOC emissions can be a problem. Energy intensive
Solidification	Slurries and sludges but not suitable for organics and heavy metals	Dewatering may be required. Volume increases. May require landfill
Encapsulation	Mercury bearing wastes, asbestos	Waste not converted or destroyed. May require special landfill

Table 2: Chemical Treatment Technologies

Technology	*Application*	*Comments*
Hydrolysis	Sludges, slurries and liquids contaminated with organics. Successfully used for chemical warfare agents	Require high temperatures and pressures; may produce toxic by-products
Neutralisation1	All acids and caustics	Simple, and disposal easy. Not suitable for sludges and solids
Oxidation/Reduction	Cyanides, chrome vi, arsenic, silver	Expensive and requires dilute solutions. May require special equipment and produce off gases that need to be controlled
Precipitation	Metal bearing solutions	Post-treatment may be required if toxic.
Fixation	Liquids, slurries and sludges particularly those containing heavy metals. Boiler and incinerator ash are also prime candidates	Increases volume. May require landfill.
Ion Exchange	Wide variety of metals/salts streams	Expensive and species specific; require treatment/ disposal of residue
Coal Agglomeration	Contaminated soils	Pilot scale at this point but has potential for site clean-ups especially wood preservative and heavy oils

Hazardous Waste Management (Secured Landfill Site) for Zinc Extraction Industry

We are reporting here about hazardous waste management (Secured Landfill Site) for zinc extraction industry, in which electrolytic zinc and copper are extracted from zinc ash/skimming and Brass dross. The operations are broadly divided into three sections:

1. Leaching of zinc and copper bearing ash
2. Solvent extraction of zinc and copper separately
3. Electrowining of zinc and copper in separate cell houses.

As a result of these operations the solid waste generated consisting of metallic impurities which needs proper treatment before disposal. The objectives of the Hazardous Waste Management studies are as follows:

1. Quantification of the waste generated
2. Identification and characterisation
3. Characterisation of waste using TCLP testing
4. Selection of Hazardous waste disposal site for the safe disposal and storage of the waste generated from various activities.

Characterisation of the Hazardous Waste

The solid waste sample has been characterised as follows:

1. Analysis of total heavy metals
2. Characterisation of leachate
3. Corrosivity
4. Reactivity

The total heavy metals concentration and leachability results of various metals present in solid waste collected from zinc extraction industry are listed in Table 3.

Table 3: Total Heavy Metals Concentration and Leachability Results of Various Metals Present in Hazardous Waste Sample

Metals	*Total Con.of Metals (mg g^{-1})*	*Leachability (ppm)*
Iron	42.5	0.032
Copper	12.5	1.72
Aluminium	42.5	NIL
Zinc	21.9	0.136
Chromium (Total)	0.60	0.016
Chromium (Hexavalent)	0.25	NIL
Cobalt	0.31	0.056
Nickel	1.5	0.06
Lead	0.25	0.011

Based on the presence of the heavy metals in the waste sludge, this waste has been categorised under category '3' of Hazardous Waste (Management & Handling) Rules, 1989.

The pH of the hazardous waste is neutral and hence it is not considered as corrosive and in the absence of sulphide it is not likely to react with water or air. The moisture content of the waste cake is around 35%.

Design of Secured Landfill

The industry has identified a site for establishing hazardous waste disposal site for wastes generated in the factory premises. It is proposed to construct a secured landfill as per laid down standards so that the impounding of wastes do not affect the land and environment.

1. The total quantity of hazardous waste to be disposed is 11.1 MT/day (on wet basis).
2. The column required for impounding the generated waste would be 1797 m^3.
3. The proposed size of the secured landfill (after lining etc) would be 2500 m^3
4. One pit would thus be sufficient for one and half years.
5. This can be viewed from the site plan that there lies enough space adjacent to our set-up for the creation of further pits, if required.

The details of the secured landfill are as follows:

The soil samples from proposed site have been analysed for heavy metals. The heavy metal concentration present in the soil samples are almost negligible and shown in Table 4.

Table 4: Analysis of Soil Samples

Sl. No.	Heavy Metals	Soil Sample*
1.	Cu	0.005
2.	Fe	0.095
3.	Mn	0.600
4.	Zn	0.035
5.	Ni	0.041
6.	Co	0.040
7.	Cr^{+6}	ND
8.	Cr^{+3}	0.005
9.	Al	0.045
10.	Pb	ND

* All values is mg/l obtained in one gram.
ND: Non-Detectable

The drinking water from the source nearby the proposed hazardous waste disposal site is collected and analysed for heavy metal contents. Most of the metals were found to be in non-detectable range. The results are shown in Table 5.

The engineering design of the proposed landfill will comprise of:

1. Site suitability, Compatibility of native soil, Characteristics of Solid Waste.

2. Side Slope, Liner System (s), Specification for liner and other materials used for secured landfill, Drainage, Sump.

Table 5: Ground Water Quality Analysis

Sl.No.	*Parameters*	*Concentration*
	Physical Parameters	
1	pH	7.85
2.	Turbidity (NTU)	1.3
3.	Conductivity (mS/cm)	0.556
4.	Total Solid (mg/l)	312
5.	Suspended Solid (mg/l)	2
6.	Total Dissolved Solid (mg/l)	310
Sl.No.	*Parameters*	*Concentration*
	Inorganic Parameters	
1.	Alkalinity	130
2.	Total Hardness as $CaCO_3$	170
	Calcium Hardness as $CaCO_3$	130
	Magnesium Hardness as $CaCO_3$	40
3.	Sulphate (SO_4^{2-})	30
4.	Chloride (Cl^-)	42
5.	Sodium (Na^+)	19
6.	Potassium (K^+)	30
7.	Nitrate (NO_3^-)	19
8.	Fluoride (F^-)	0.62
9.	Phosphate (PO_4^{3-})	23.7
Sl.No.	*Metals*	*Concentration (mg/l)*
	Heavy Metals	
1.	Iron (Fe)	0.04
2.	Nickel (Ni)	ND
3.	Zinc (Zn)	0.10
4.	Copper (Cu)	ND
5.	Cobalt (Co)	ND
6.	Manganese (Mn)	0.01
7.	Chromium (Cr)	ND
8.	Cadmium (Cd)	ND
9.	Lead (Pb)	ND

ND: Non Detectable

Suitability for Secure Landfill

All potential Solid Waste disposal sites have been analysed and evaluated based on a set of factors as outlined in the Guidelines for Management & Handling of Hazardous Wastes published by Ministry of Environment & Forests, Government of India, 1991. These are:

1. Landfills should be located at elevation which are lower than nearby residential areas.
2. Landfills should be located in an area that is clean prior to construction. If the site contains refuse from previous dumping or layers of organic material, mud, etc., it should be checked for safety and overload conditions.
3. Landfills should be placed in areas not likely to be used for any residential, commercial or public building construction in the future.
4. Landfills should be located in areas with as many hydrologic barriers as possible surroundings the site. These barriers include a fault, ground water divide, confining bed, discharge area, drainage, etc.
5. Landfills should be put in areas as remote as practicable with prevailing winds blowing away from residences.
6. Landfills should not be located in a flood plain, wetlands or potable water supplies.
7. Landfills must not be situated over limestone, dolomite or marble beds, over past or present subsurface mining activities or in gravel pits.
8. Dumps should not be in areas where leachate could enter a free flowing water body and then be carried towards residences.
9. Landfills should be provided with an impermeable natural or manmade bed
10. Landfills should not be placed on a ridge because their leachate will pollute not only one, but several water sheds because of their elevation and location.
11. Landfills should not be located in the upstream portion of a watershed draining to an on-stream reservoir, including areas that drain directly into off-stream reservoirs.

Compatibility Test

In addition to determination of hydraulic conductivity, tests were carried out to determine the compatibility of the soil with the solid waste. This was carried out to explore the possibility of utilising the locally available soil as liner material in the landfill.

From the tests it can be concluded that the clay soil present in the proposed landfill site can be used as liner material alongwith Bentonite powder so as to give a permeability rate of less than 1×10^{-7} cm/sec.

Liners

The function of a liner system is to minimise and control the migration of polluting constituents in the waste or liquid being contained and prevent them from entering the environment either through the air or through the ground.

The liners are sometimes composites, constructed of different materials, some of which are polymeric. Each component of these system is designed to fulfill as specific function while it is at the same time exposed to compressive, tensile, and multiaxial stresses.

Interaction between waste constituents and the liner system becomes important in terms of the long-term functioning of the system because of the combined effects of mechanical stresses and interaction with the wastes constituents.

Drainage Layer

The function of the drainage layer is to act Leachate Collection and Removal System (LCRS) proposed to be constructed above the clay foundation is to detect and collect any liquid that has entered the system, *i.e.* leaked through the top layers, throughout the lifetime of the unit including the post-closure care period.

Protective Layer

It is suggested to provide a soil layer of 20 cm thickness above the drainage so that the waste can be placed not directly above gravel. In the absence of soil liner, the waste material may enter the drainage layer and close the pipes and the very purpose of drainage thus gets defeated.

The protective cover on the slope is also to be constructed maintaining the slope 2%. A coating of bentonite or silicon sealant would strengthen the stability of the slope and reduce erosion.

Waste Layer

The Solid Waste from the plant will be transported to the landfill site by conveyors/Loaders. The waste volume per day is above 1.75 m^3/day with a moisture content of about 35 %. The design depth of the waste layer is 1.85 mt.

Further it is proposed to have no of cells with each cell having a waste volume of about one month. At the end of each month, it must be covered with soil. The soil thickness will be about 10 cms.

Sump

A sump is to be constructed outside the secure landfill to collect the leachate generate from the landfill. Even though the landfill is designed to avoid leachate generation, there is likelihood of leachate generation. The leachate that is generated has to be collected and treated if necessary. The leachate from the drainage layer has to be taken to the sump for collection and subsequent treatment if warranted.

The main function of the sump is to collect leachate. The capacity of the sump is equivalent to one day storage capacity of the expected leachate volume.

The sump has to be constructed below the ground level so that the leachate can travel by gravity from the landfill.

Final Cover

A final cover ha to be provided at the end of the active life of a landfill *i.e.* when the landfill volume gets exhausted.

The final cover system has been designed so that it can function with minimum maintenance, promote drainage, minimise erosion, accommodate settlement and subsidence, and have a transmission rate less than or equal to that of the bottom liner system.

Final cover systems are multilayer structures constructed in layers on top of a mass of waste that may settle unevenly. The barrier layer is the most important layer because it prevents water from infiltrating the fill.

Once the landfill becomes full, the landfill should be closed. This is to prevent entry of rain water and surface run off into the fill.

Conclusion

The zinc extraction industry produces hazardous waste sludge in the form of hard cake consisting of 35% moisture. The waste mainly comprises of toxic elements *viz.* iron, copper, aluminium, zinc, chromium, cobalt and nickel and falls under category '3' of hazardous waste (Management & Handling) rules. TCLP testing reveals the need for proper treatment and disposal method. Secured landfill site has been considered with leachate collection system, proper liner and final cover system. The volume of the landfill is 2500 m^3 with a depth of 1.85m.

Acknowledgement

The authors are grateful to the Director, Regional Research Laboratory, Bhopal for giving permission for the submission of this paper.

REFERENCES

1. La Grega M.D, Buckingham P.L, Evans J.C., 1994, Hazardous Waste Management. New York: McGraw-Hill Inc..
2. Lichtenberg J.J, Winter J.A, Weber C.I, Fredkin L., 1988, Am. Soc. Test. Matls. Phildelphia, Pennsylvania, pp.129-140.
3. Dawson, G.W. and Marcer B.W., 1986, Hazardous Waste Management. New York: John Wiley and Sons.
4. Freeman, H.M., 1988, Standard Handbook of Hazardous Waste Treatment and Disposal. New York: McGraw-Hill, Inc..
5. Ham R.K., 1989, New Approaches to Technology –Landfills. Fourth National Conference on Environmental Issues, Academy of Natural Sciences, Philadelphia, PA.

78

Control of Sulphide Inhibition in Anaerobic Treatment

V.B. MANILAL & AJIT HARIDAS

Wastewater Technology, Regional Research Laboratory, CSIR, Trivandrum - 695 019

ABSTRACT

Sulphide is a serious problem in the anaerobic treatment of sulphate containing wastewater. The available methods are cumbersome and expensive for controlling and managing sulphide inhibition in the anaerobic treatment. A cost-effective process developed for the anaerobic treatment of high-sulphate effluent is described in this paper. In the process the digester liquor from a UASB reactor is circulated through an external stripper where air is introduced to remove the H_2S, and recycled to the digester to absorb more H_2S. The device can be incorporated in high-rate anaerobic reactors such as fluidised bed, expanded bed, or fixed film digester also.

The crucial innovation is the use of air as stripping medium, while avoiding oxygen toxicity, which makes the process exceptionally simple, robust and economical. In some cases, a buffer tank is used to prevent oxygen toxicity. The stripped air containing H_2S is treated in a proven air pollution control equipment.

Introduction

Anaerobic digestion is widely used for the treatment of biodegradable wastewaters. It is energy efficient because of the conversion of organic matter to methane and environmentally friendly as it produces small amount of biomass. Currently, anaerobic digestion is the best technology available for the treatment of industrial, domestic and urban wastewaters, and for the stabilisation of biological sludge[13,24]. This has largely been due to the advancements in anaerobic processes and reactor technology, which enable high organic loading of the reactors with short hydraulic residence times and good process stability.

Many streams of wastewater containing biodegradable matters have significant amount of sulphate and COD/SO_4 ratio of effluent may be as low as 3. This is very undesirable for the best performance of anaerobic treatment that results in the inhibition of the biological process. There have been substantial efforts to develop suitable method to control both the sulphide generation and sulphide control in anaerobic process, but little progress could be made to practice in actual fields because of technical non-feasibility and high cost.

This paper discusses a cost-effective technology for the control of sulphide inhibition in anaerobic process, which enables the anaerobic treatment of high-sulphate wastewater treatment. It has also shown success in effluent treatment plants at industrial scale.

Methanogenesis and Sulphate Reduction

Industrial Effluents Containing Sulphates

There are many industries discharging high-BOD effluents contain inorganic sulphates or other sulphur containing compounds. The origin of sulphur compounds in industrial wastewaters is usually from the use of sulphuric acid, the most common mineral acid solvent, or such chemicals as sodium sulphate and sulphites applied in pulping, sulphur dioxide used for clarification and bleaching, bisulphites used as preservatives in food processing, alum used as coagulants, etc.

Some of the industries that discharge effluents containing substantial amounts of sulphates are:

Latex Centrifuging Factories (Malaysia, India)

Here sulphuric acid is used for neutralisation of ammonia preserved latex for coagulation. The effluent from latex centrifuging factory has a medium to high COD of 12000mg/l possessing a COD/SO_4 ratio of effluent may be as low as 3. Usual anaerobic treatment systems are unable to handle this effluent[6,16].

Molasses-based Distilleries and Yeast Factories (India, Brazil)

Molasses is a by-product of sugar manufacture, where sulphur dioxide is used for clarification and bleaching. The effluent produced by distilleries has very high COD (80000 to 100,000mg/l). The sulphate concentration of effluents that use cane-molasses is in the range of 5000 mg/l and biogas produced in commercial anaerobic effluent treatment plants usually has 1 to 2% hydrogen sulphide[12]. High corrosion rates are reported for ETP steel structures and for boiler stacks.

Many Bulk Drug and Pharmaceutical Industries Use Molasses as Substrate for Fermentation (India, Japan)

The effluents discharged from such industries have similar characteristics as that of distillery effluents. For example, citric acid is produced by fermentation of molasses with *Aspergillus niger.* The process discharges 18-20 tonnes of wastewater per tonne of citric acid, with a COD of 20,000 to 35,000 mg/l in media containing sulphuric. The spent broth effluent discharges may have COD/SO_4^{2-} ratio less than 3[22]. In addition, there are several other synthetic drugs manufacture units require the use of sulphuric acid, which are discharged along with biodegradable organic solvent.

Vinasses from Wine and Cognac Manufacture

Tartaric acid is recovered lees vinasses by recuperation with calcium sulphate. The effluents produced contain sulphates in the range of 1000 to 4000 mg/l and COD in the range of 15000 mg/l[9].

Pulp and Paper Mills

The sodium sulphate based "Chemical-Thermo Mechanical Pulping" (Canada, Finland, India) industries discharge a medium to low strength effluent with exceptionally high COD/SO_4^{2-} ratio (between 5 and 2)[37]. The problems caused by sulphate in anaerobic treatment of pulp and paper process and waste streams are tremendous[32]. The thermophyllic anaerobic treatment of sulphate rich thermomechanical pulping process white water has been recommended to reduce the white water load[32].

Food, Fruit and Vegetable Processing Industries

Many of these industries discharge effluents containing sulphates or sulphites or bisulphites. For example: raisin finishing industry (Greece) discharges an effluent with COD of 20000 mg/l and sulphite of 1000 to 3000 mg/l[2]. Anaerobic treatment of the wastewater results in unionised H_2S of 200 mg/l. although failure of the process has not been reported, the biogas obtained from the process has more than 5% H_2S which severely restrict its use.

Effect of Sulphate Reduction on Methanogenesis

Thermodynamic Considerations

Sulphate reduction is an energetically favoured reaction compared to methanogensis for both decarboxylation of acetate as well as reduction using hydrogen as shown in the following reactions[5].

Methanogenesis

$$CO_2 + 4H_2 = CH_4 + 2H_2O \quad G = -135\ KJ/M$$

$$CH_3COOH = CH_4 + CO_2 \quad G = -28.8\ KJ/M$$

Sulphidogensis

$$SO_4^{2-} + 4H_2 = H_2S + 2H_2O + 2OH^- \quad G = -154\ KJ/M$$

$$SO_4^{2-} + CH_3COOH = H_2S + 2HCO_3^- \quad G = -43\ KJ/M$$

Because of this SRB is favoured energetically over MPB under standard conditions, however, the actual free energy changes are dependent on the activities of the reactants and the products of each reaction in the reactor[26].

Kinetic Considerations

In anaerobic digestion about 70% of the methane is produced from decarboxylation of acetate and the rest from CO_2 and hydrogen. While sulphate reduction uses varied substrates such as acetate, lactate, ethanol, propionate, and hydrogen. So there is a competition between SRB and MPB for the substrates, acetate and hydrogen.

The growth of MPB is slower than that of SRB when acetate or hydrogen is provided as sole carbon source[35], although in actual digesters the situation is unclear. In terms of Monod kinetics of bacterial growth the maximum growth rate of SRB is approximately the same as that of MPB, but the half saturation constant *Ks* is lower for SRB. It has a value of 1 to 2 ìM H_2 while for MPB the value is 6 to 20 ìM H_2[23]. The lower *Ks* value for SRB favours faster growth rate at lower H_2 partial pressure.

Effect of Waste Composition

Certain synthetic wastewaters tend to promote the growth of MPB as opposed to SRBs[26]. In mesophilic digesters, SRB commonly out-compete methanogens for hydrogen, but competition for acetate is unpredictable. It is rare to find methanogens out-competing SRB in industrial anaerobic digesters. Therefore, it is reasonable to expect the domination of SRB over MPB when there is adequate availability of sulphate. In practice, such situations arise in effluents with a low COD/SO_4 ratio. It is also seen that the sulphate in raw effluents are almost invariably converted 100 % to sulphides during the treatment of common industrial effluents.

However, this does not mean that growth of SRB is always undesirable under all conditions. The affinity of SRB for hydrogen usually lowers the hydrogen partial pressures and contribute to the breakdown of some of the volatile acids such as propionic acids[10,11]. Thus SRB may contribute favourably to the overall digestion of materials if adverse effects of hydrogen sulphide generation can be avoided.

Hydrogen Sulphide Equilibrium in Anaerobic Digesters

Hydrogen sulphide exists as following three main species in the pH range of interest in anaerobic digesters.

$$H_2S\,(aq) \Longleftrightarrow H^+ + HS^- \Longleftrightarrow {}^2H^+ + S^{2-}$$

The first equilibrium has more relevance in the pH range of 6.5 to 7.5 at which anaerobic reactors operate. The equilibrium constant has the value 1.49×10^{-7} at 35°C. At pH 7.1, 50% of the total sulphides are present as dissolved H_2S gas. The equilibrium between dissolved hydrogen sulphide gas and hydrogen sulphide in the gas phase is expressed in terms of a Henry's Law constant.

$$^{39}\,(H_2S)_{solution} = 1.83 \times (H_2S)_{gas} \text{ at } 35^\circ C.$$

It has also been shown that inter-phase mass transfer of H_2S from solution to rising as gas bubbles in the digester liquor is sufficiently fast, so that equilibrium conditions are a valid approximation, meaning local variation in hydrogen sulphide concentration may not be significant. In the case of other gases in the digester, namely methane, H_2 are not in equilibrium with the gas phases as extremely high concentration in the liquid phase (as high as 70 times the gas-phase equilibrium value) while, CO_2 is in equilibrium[22].

Adverse Effects on Anaerobic Treatment Technology

When anaerobic technology is applied to effluents containing sulphates, the following undesirable effects are observed.

Toxicity of H_2S from Sulphate Reduction to MPB

The toxic effect of H_2S to living systems including microbes is well documented[25]. The toxicity of various species of sulphur compounds is rated as H_2S > total sulphides> sulphite> thiosulphate> sulphate[21]. Hydrogen sulphide can readily permeate cell membranes and denature cell proteins or interfere with assimilatory metabolic pathways of sulphur. The unionized species is highly toxic by at least a factor of 10 as compared to ionic species. It implies that the gas phase H_2S concentration is a valid indicator of the toxic concentration, taking into account the gas-transfer consideration as stated elsewhere. It has been reported that the inhibition of methane production occurs when the gas-phase H_2S reaches 5%. Free hydrogen sulphide at about 50 mg/l is inhibitory to acetotrophic methanogenesis. There are reports that 50% reduction in MPB has occurred at a hydrogen sulphide concentration of 250 mg/l in the liquid phase. Beyond that level, methane production is affected drastically and complete cessation of digestion process.

Intrinsic Toxicity H_2S to SRB

Interestingly, SRB is also inhibited by high concentration of H_2S because of the intrinsic toxicity of the gas to living systems. A concentration of H_2S gas between 500 and 600 mg/l can completely stop the growth of SRB[31]. This is a very important observation as for the biological treatment of effluents with very low COD to sulphate ratio.

Deterioration of Biogas Quality

Presence of H_2S in biogas is a nuisance. Gas utilization machinery such as gas engines usually specify H_2S to be less than 100 ppm vol%. Even when biogas containing H_2S is utilized in such equipment as boilers, the effects of corrosion are quickly noticed in such accessories as stacks. It is reported that in distillery effluent treatment plants in India where cane molasses spent wash is used for methane generation, the boiler stacks have the life span from 6 to 12 months. The resultant flue gases need further treatment to meet the SO_2 emission standards.

Gas Handling Equipment

Being highly toxic to human as respiratory irritant and systemic poison leading to respiratory paralysis and death it is necessary to ensure gas handling equipment leak proof when H_2S containing biogas is transported and stored. Gas compressors are particularly prone to seal leakage hazard. In some cases, the problems are so severe that pressurized reactors are constructed to avoid the need for compression gas and it is an expensive option.

Control of Sulphide in Anaerobic Systems

Use of Specific Sulphate Reducing Bacteria Inhibitors

Bacteriostatic or batericidal inhibitors of SRB could prevent H_2S formation from sulphate. Among the inhibitors identified are molybdate salts, the theory being that molybdate is a stereochemical analogue of sulphate. However, mixed results are reported in the literature for inhibiting SRB in high sulphate effluent. This may be due to the inhibition of useful methane bacteria in addition of SRB[30]. The high chemical cost and molybdenum discharge to the receiving waters are drawbacks of this technique.

The addition of various chemicals such as surfactants, antibiotics, pigments and growth inhibitors have been claimed in a series of patents[17], to selectively inhibit the growth of SRB. There is little evidence either to the efficiency or cost-effectiveness of the claimed methods. Organic halogen compounds have also been claimed to have similar effects[29].

Higher Temperature Digester Operation

It has been mentioned that thermophilic digestion may tolerate higher sulphate concentration and produce gas which has lower hydrogen sulphide[21].Palm oil mill effluent treatment digesters operating at 55°C produce only 0.1 to 0.2% H_2S. However, there are also reports that SRB out-compete acetoclastic methanogens at 55°C when treating a synthetic volatile acid mixture containing sulphate[40]. A study of thermomechnical pulp effluent at mesophilic and thermophilic treatment conditions found lower COD and SO_4^{2-}removal efficiency at thermophilic conditions[27].

Fixed Film Reactors

Fixed-film reactors are believed to tolerate higher than usual sulphate loading rates before inhibition. Theoretically, the higher substrate utilisation rates of SRB, will result in thinner SRB biofilms,because of substrate (SO_4^{2-}) diffusion mass transfer limitation. Thus methane producing bacteria finds a niche in the inner layers of biofilms, but to what extent it will be shielded from the toxic effects of H_2S is unclear[28]. While this allows the treatment of sulphate containing effluents to some extent, the H_2S content of the biogas will be correspondingly higher.

Two-stage Digester Operation with H_2S Removal

A patent has been taken on the process of stripping of H_2S after the first stage in two-stage digestion systems. This is an interesting option but the following disadvantages are noted: (1) The acidogenic reactor operates optimally at $pH<5$, while sulphate reduction requires a $pH>6.5$[7] in order to maintain sufficiently high efficiency of sulphate removal. This requires high alkalinity consumption. (2) Stripping of H_2S is efficiently carried out at the acidogenic pH, but the moment pH control is optimised for sulphate reduction, the stripping efficiency is lost. (3) The process obviously looses the simplicity and economy of the single stage digester. Air has been suggested as the stripping gas in such systems[26].

H_2S Removal with Inert Gas

Stripping of H_2S from the reactor by introduction of a neutral gas such as nitrogen into the bottom of an anaerobic reactor has been proposed as a method of treatment of wastewater containing sulphates and apparatus for the same have been patented[8]. The neutral gas is recycled back into the reactor after absorbing H_2S. Mixing of liquid in the reactor has been proposed by the stripper gas.

Such schemes have also been studied from the point of view of biological removal and resource recovery for sulphate-rich industrial effluents or waste gypsum using cheap carbon sources for sulphate reduction. It has also been tried at laboratory level for biological treatment of dilute sulphuric acid[38].

H_2S Removal with Biogas

The removal of H_2S by gas stripping has been shown to be an effective method of control of inhibition. In laboratory studies H_2S has been removed from the biogas using an absorbent solution and used to strip the digester liquor in an external stripper. The improved performance of such scheme for both COD and SO_4^{2-} removal has been demonstrated in laboratory investigations. Anaerobic treatment apparatus for such systems have been designed and covered by patents[19].

Precipitation of Sulphides

Precipitation of sulphide from solution as insoluble iron sulphide complex is the most widely practised method of decontamination of biogas as well as to prevent inhibition. Various technological innovations on the addition of iron salts have been covered by patents in Brazil[4], South Africa[41] and Japan[18]. The use of insoluble ferric phosphate in a thermophilic process has been studied[27].

But there are several problems with this method of control. These include high consumption and cost of ferric salt, accumulation of precipitates inside the reactor, corrosion and black colour of the effluent. Thus use of iron salts is not attractive as a viable long-term option for H_2S control, although it is the only popular method as of now[34].

Materials and Methods

Experimental Set-up in the Laboratory

A glass upflow anaerobic sludge bed (UASB) reactor having 9 litre volume was used as a model system for the studies in the laboratory. The scheme of experimental set-up used is shown in Figure 1.

Two important modifications were made in the experimental set-up during sulphate loading: 1) introduced a stripping device in the recycle line of UASB and 2) an open buffer tank of 30 litre after the stripping device, before the liquor was being pumped back into the reactor (scheme shown in Fig.2.).

The stripper facilitated air contact with the reactor liquor, which were of two types: 1) packed column and 2) venturi contactor.

The granular sludge in the UASB reactor was developed from seed sludge of earlier UASB using sucrose and starch initially. A typical influent composition was ammonia nitrogen content of

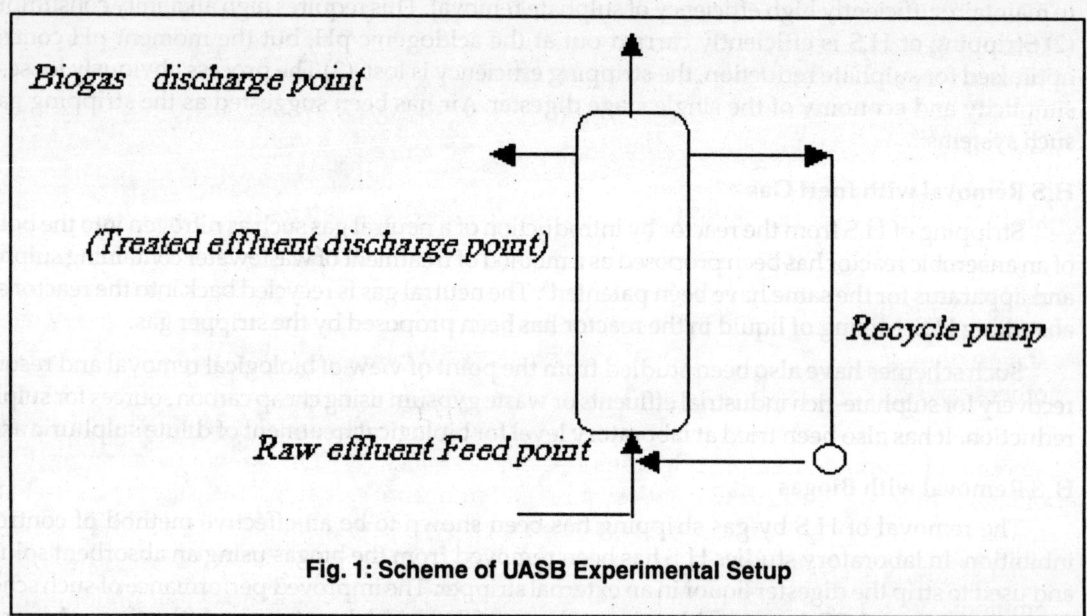

Fig. 1: Scheme of UASB Experimental Setup

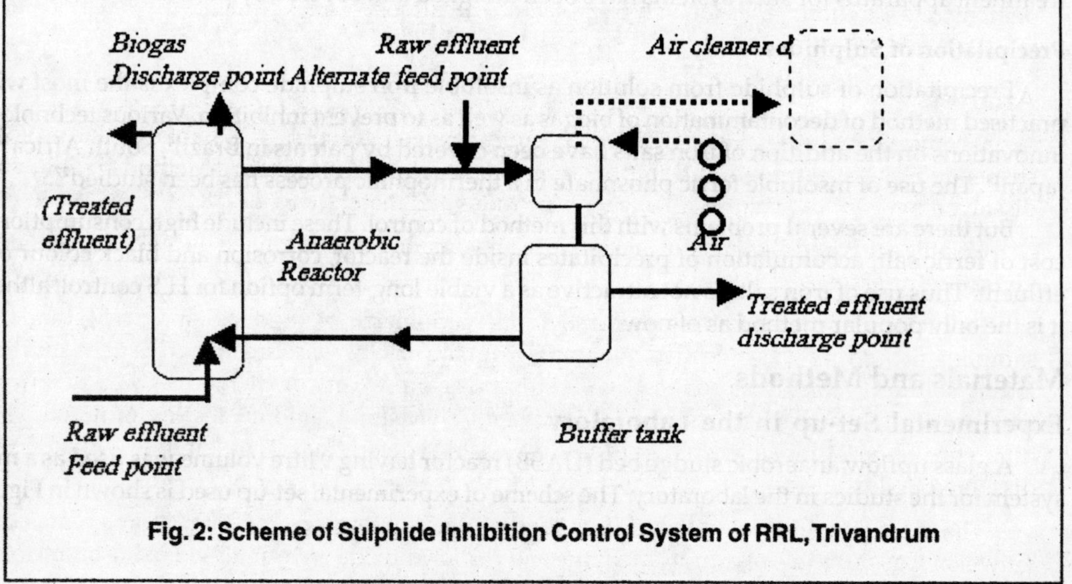

Fig. 2: Scheme of Sulphide Inhibition Control System of RRL, Trivandrum

900 mg/l, and COD of 6000 to 8000mg/l and trace elements[26]. Sodium bicarbonate was used to support alkalinity and maintain pH in the reactor. The sulphate content in the feed was increased by gradually replacing ammonium chloride with ammonium sulphate to test the use of the reactor for high sulphate effluents by using a synthetic feed ammonium sulphate, sodium sulphate, sulphuric acid and sucrose as the carbon source.

Experimental Procedure

All experiments conducted in the laboratory were carried out with a single reactor set-up and values reported are for steady state conditions with varying operational parameters and feed conditions. Initially acclimatization of the sludge was done in every case of changing feed characteristics. The various phases of the experiments are given below:

Operation of UASB with synthetic feed prepared from sucrose and ammonium chloride so as to simulate the COD and ammonia concentration of latex centrifuge effluent. The purpose of these sets of experiments was to acclimatise the anaerobic sludge to high ammonia concentration and to COD, alkalinity and salt concentration expected when feeding sulphate free latex effluent.

Operation of UASB with sulphate free latex effluent. Skim latex was stripped without pH adjustment to remove ammonia. It was then coagulated with HCl and skim rubber was removed by straining through a filter cloth. The serum obtained was diluted to the required concentration. This stage of the experiments should show whether, latex effluent contains any other inhibitory substance other than sulphuric acid.

The experimental UASB was also run with synthetic sucrose solution containing increasing amounts of ammonium sulphate and sulphuric acid: simulating COD, ammonia and sulphate content of latex. These experiments had the objective of demonstrating control measures so as to enable the treatment of high-sulphate containing effluent inhibition of methanogenesis.

Operation of UASB with latex effluent. Skim latex was stripped without pH adjustment to remove ammonia. It was then coagulated with H_2SO_4 and skim rubber was removed by straining through a filter cloth. The serum collected was diluted to the required concentration. These experiments would demonstrate the effectiveness of high-rate anaerobic treatment of latex effluent when sulphide inhibition is controlled.

Analysis

Chemical oxygen demand (COD), biochemical oxygen demand (BOD), Total dissolved solids and suspended solids, ammonia and sulphate were measured as per Standard Methods[36]. The analysis of dissolved oxygen was made with oxygen electrodes and meter (Jenway, UK).

Volatile acids and alkalinity of anaerobic reactor liquor was determined by following the procedure given by Anderson *et al.* (1992)[1].

Sulphide in the liquid phase was analysed by iodometric method in Standard Methods, but the sampling procedure was modified. A sample from the reactor was directly introduced into acidified iodine solution which was reduced with sulphides and excess of free iodine was determined by titration with standardised sodium thiosulphate. The sample amount was determined by weight difference of the iodine solution. The procedure enabled determination of sulphide without loss of H_2S dissolved in the liquid.

The H_2S content in biogas was determined by determining the amount of iodine reduced using a Tutweiler Gas Burette Apparatus. Methane and carbon dioxide were analysed using online NDIR gas analysers.

Results and Discussion

The granular sludge developed in the laboratory was easily acclimated to sucrose feed containing ammonium chloride. Granular sludge was formed at a COD loading rate of 4 kg/m^3/d. There after, a mixture of sucrose, ammonium chloride and HCl coagulated latex serum (10 ml per litre of skim latex) was fed to the reactor and the COD loading rate was increased. The ammonia concentration after anaerobic treatment was observed to increase above 1000 mg/l. This is expected from the contribution of degraded nitrogenous compounds in latex serum. However, the anaerobic reactor should not be affected with the increased ammonia[3].

The reactor was in operation over a period of 325 days. Most of the time the reactor was operated with high sulphate effluent (COD/SO_4^{2-} ratio between 3 and 10 and few days the feed was medium sulphate effluent (COD/SO_4^{2-} = 24). The COD loading rate varied from 10 to 30 kg/m^3/day and the profile is given in Figure 3.

Hydraulic residence time in the reactor was maintained at 18 hours during the early period, which was reduced to 12 hours from day 300 to 325. At steady state which was possible to achieve consistently, the COD removal was greater than 90% for each of the compositions (Figure 4). A basic parameter which is a measure of the volumetric productivity of the reactors system is the specific gas production or the volume of gas produced per day per unit volume of the reactor (Figure 5). The specific gas production remained consistently above 5d^{-1}. It shows that the COD load on the reactor was effectively converted to gas. The reactor volatile acids were measured every day, and except during an upset lasting 1 day, it never exceeded 10 mM (as acetic acid). During most of the operation, the volatile acids was below 5 mM (acetic acid) which shows the fast utilization of acids produced in the reactor (Figure 5).

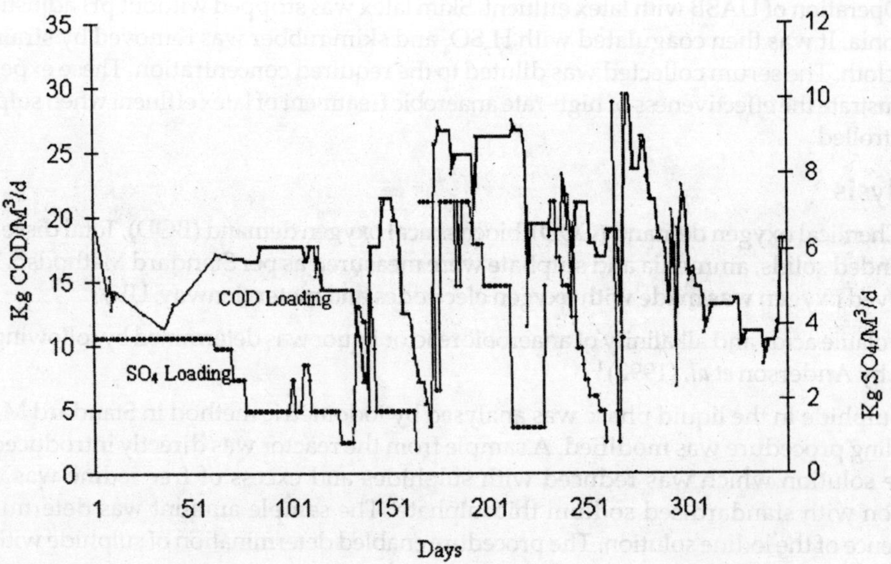

Fig. 3: COD and SO_4 Loading Rates in the Laboratory UASB Reactor

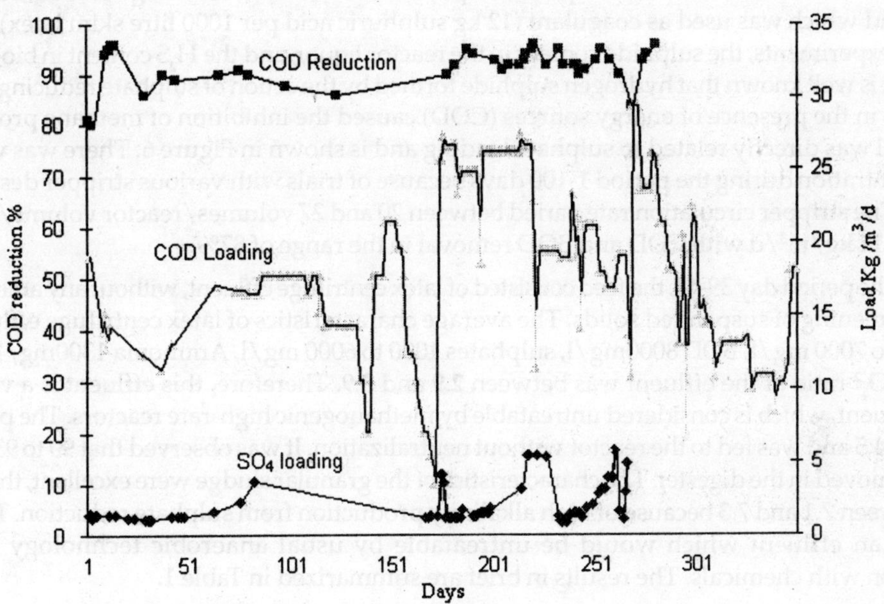

Fig. 4: Reactor Performance: COD Removal

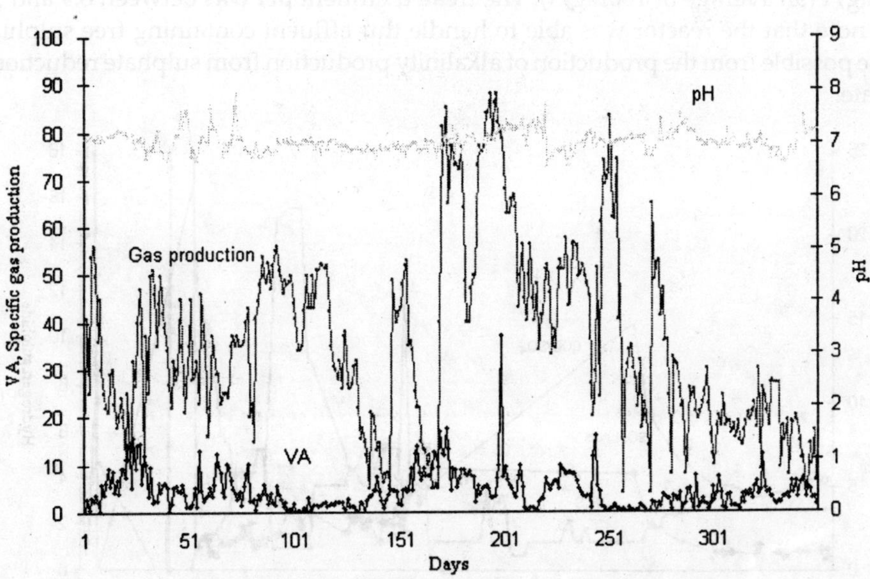

Fig. 5: Reactor Performance: Specific Gas Production

The sulphate loading of the reactor was varied in the range of 1-7 kg/m³/d, by the addition of different combinations of ammonium sulphate, sulphuric acid and latex centrifuge effluent containing sulphuric acid which was used as coagulant (12 kg sulphuric acid per 1000 litre skim latex). During this stage of experiments, the sulphide content in the reactor liquor, and the H_2S content in biogas were monitored. It is well known that hydrogen sulphide formed by the action of sulphate reducing bacteria on sulphates in the presence of energy sources (COD) caused the inhibition of methane production. The H_2S level was directly related to sulphate loading and is shown in Figure 6. There was variation in H_2S concentration during the period 1-100 days because of trials with various stripper designs and circulation. The stripper circulation rate varied between 20 and 27 volumes/reactor volume/day was increased to 12 kg/m³/d with COD and BOD removal in the range of 87%.

During the period day 39-44, the feed consisted of latex centrifuge effluent, without any amendment, except for screening of suspended solids. The average characteristics of latex centrifuge effluent are: COD 13000 to 7000 mg/l, BOD 8000mg/l, sulphates 4000 to 6000 mg/l, Ammonia 1300mg/l, pH 4.5. The COD/SO_4^{2-}ratio of the effluent was between 2.9 and 3.9. Therefore, this effluent is a very high sulphate effluent, which is considered untreatable by methanogenic high-rate reactors. The pH of this effluent was 4.5 and was fed to the reactor without neutralization. It was observed that 90 to 93% of the COD was removed in the digester. The characteristics of the granular sludge were excellent, the reactor pH was between 7.1 and 7.3 because of high alkalinity production from sulphate reduction. This is an example of an effluent which would be untreatable by usual anaerobic technology without neutralization with chemicals. The results in brief are summarized in Table 1.

During the period, day 92 to 124, the feed had 33 g/l sucrose, 5 g/l ammonium sulphate, 3g/l sulphuric acid, corresponding to an average feed COD 18500 mg/l, COD/SO_4^{2-}ratio of 3.5 the pH was between 1.9 and 2.2. The COD loading rate during operation was 26 kg/m³/d. The treatment efficiency in terms of COD removal was between 90-95%. The total sulphide in the treated effluent was between 140 and 240 mg/l (an average of 160mg/l). The treated effluent pH was between 6.9 and 7.05. It is interesting to note that the reactor was able to handle this effluent containing free sulphuric acid, which could be possible from the production of alkalinity production from sulphate reduction and the high recycle rate.

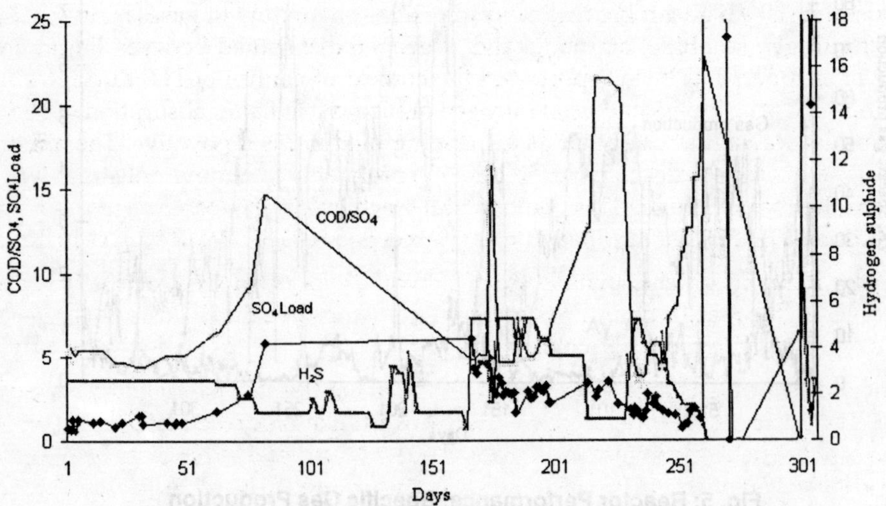

Fig. 6: Reactor Performance: H_2S Generation with SO_4 Loading

Table 1. Results of Feeding Natural Ratex Serum with High Sulphate in UASB Reactor

Parameter	Raw Effluent	Treated Effluent	Removal %
COD mg/l	14,500	1,300	91
BOD mg/l	9,100	570	91
TDS (mg/l)	16,500	2,600	84
Ammonia nitrogen mg/l	1,360	1,370	-
Sulphate mg/l as SO4	4,000	0	100
pH	4.5	7.2	-
Sulphide mg/l	nil	80	-

During the period, day 273 to 300, stripping was accomplished in a loop independent from the recirculation to the reactor bottom. The stripped liquor was injected into the digester a point just below the gas –liquid-solid separator. The reactor was fed with medium sulphate effluent (COD/SO_4^{2-}= 23.6) at a COD loading rate of 19 kg/m^3/d. The reactor worked well, with average H_2S content in biogas of 0.17%. The reactor was fed with a high sulphate effluent also with COD/SO_4^{2-}ratio of 3.9 at a COD loading of 14.5 Kg/m^3/d. The COD removal averaged 95% with a biogas H_2S at an average of 0.7% (Fig. 6). The nature of the sludge granules were totally unaffected by the stripper incorporation. A notable point about this method is the very high removal of total dissolved solids. This is because no chemicals were added for neutralization and whatever inorganic sulphate present in the influent was removed during anaerobic treatment. The process performance was tested to see if the system could be loaded with more sulphate so as to use the same for other industrial effluents also. We were able to show that even with very high concentration of sulphate (5800 mg/l as sulphate and 16000mg/l COD), the reactor was able to maintain more than 90% removal of COD, while the sulphide concentration in the reactor was 150 mg/l. The reactor was also able to run using a synthetic feed with a pH of 2.1 because of free sulphuric acid. The treated effluent had a pH of 7.1, indicating the transformation of sulphuric acid to the H_2S which was continuously removed in the stripper. The sulphate load in the reactor under these condition was more than 7 kg SO_4/m^3/d, which is higher than the previous reports. An anaerobic reactor system has as a mixture of gases mainly, methane, carbon dioxide, hydrogen sulphide, ammonia and hydrogen distributed between liquid and gas phases. The use of strippers has been reported in the context of control of H_2S during anaerobic treatment. Previous use of strippers had used nitrogen or biogas itself after absorption of H_2S as the stripping medium. However, the use of air as a stripping medium is innovative. The air stripper devices were designed in the studies to have the function preferably to remove only toxic hydrogen sulphide gas from the reactor liquid. This could actually achieve in the experimental set up with realistic flow conditions. In this case stripper efficiency exceeded 80% of total sulphides in the liquid.

Conventionally, it was considered that the use of air would expectedly lead to toxicity because of absorption of oxygen in the reactor liquid. However, we could observe that by using a large enough buffer tank, the oxygen toxicity could be avoided.

The efficiency of stripping is highly depended on the pH of the stripped liquid because of the ionisation of H_2S. Thus under usual operating conditions of an anaerobic reactor, the pH at which the stripper operates is in the neutral range and only about 50% of the dissolved H_2S can be removed. In order to remove sufficient quantity of H_2S via the stripper, it is required to increase the flow through the stripper ie., to increase the recycle ratio.

The construction of the stripper has important implications on the performance of the process. A countercurrent packed column is expected to have a higher efficiency than a venturi type stripper. However, in our experience, the packed column results in lower overall efficiency because of expected unusual phenomenon-the growth of sulphur bacteria, which oxidizes sulphide to sulphur in the presence of air. This phenomenon is noticed by the difference in sulphide content of immediately after stripping.

Case Study: Sulphide Inhibition Control for Anaerobic Treatment of Penicillin Factory Effluent

M/s ABC, is one of the country's largest manufacturer of Penicillin-G using sucrose solution as the culture medium. It discharges 800 cubic meter of effluent per day. The penicillin produced by fermentation is extracted from the fermentation broth in a series of operations that include acidification using sulphuric acid. The spent broth forms the main component of plant effluent. The effluent has COD in the range of 14000 to 18000mg/l. The effluent contains besides sugars, sulphuric acid, ammonium sulphate, and mycelia of the penicillin culture. The effluent treatment system of the plant consists of primary anaerobic treatment and secondary aerobic treatment. The anaerobic treatment system is a slow rate bulk volume fermenter (BVF) which has a high residence time of 30 days. In this BVF COD reduction was 70-75% initially, but it was declined to around 65%.

The factory has been facing the following problems in the operation of their ETP:

1. High sulphate in raw effluent-2700 mg/l; which may increase to 4500 mg/l in future with modification in the production processes and penicillin cultures.
2. Low COD to sulphate ratio in raw effluent-COD 14500 mg/l to 18000 mg/l.
3. Therefore, generation of sulphide causing inhibition and failure of methanogenesis in BVF.
4. Necessitating addition of ferric chloride and lime to control sulphide inhibition by chemical precipitation of sulphide as iron sulphide.
5. Incurring high chemical costs for operation.
6. Build up of iron sulphide and calcium sludge inside anaerobic reactor resulting in slow deterioration in the performance of BVF in terms of COD removal.

Initially the company was adding nearly 1 tonne ferric chloride per day and 700kg/d lime and 500 kg/d soda ash to provide alkalinity as per the recommendations of the system supplier. As lime addition was leading to an increase in non-volatile suspended matter in the digester sludge, a certain fraction of alkalinity was supplied as soda ash. Over a period the company started reducing the chemical consumption in an effort to cut costs. At the time of contact, the consumption of ferric chloride as 40% solution was 0.8 t/d and lime addition was 0.5 t/d.

The hydrogen sulphide concentration in the BVF effluent was consistently higher than 7% corresponding to total sulphide of 350 mg/in liquor. This situation was inhibitory to methanogenesis severely as per all prevalent literature[14,15]. Surprisingly, the performance of the digester was showing up to 70% removal of COD. With time the company was definitely detecting a gradual decline in COD removal. It was postulated that the high residence time permitted the retention and adaptation of a methanogenic population that is tolerant to high sulphide levels. The slow deterioration of performance was perhaps a result of the decline in sludge methanogenic activity as a result of build up of iron sulphide precipitates.

However, the company felt the adverse effect of increase in dissolved sulphide in the downstream secondary treatment operation. The secondary treatment operation is an activated sludge unit designed for carbon oxidation and nitrification. The increased oxygen demand from sulphide totally reduced the oxygen availability for nitrification. The resulting high ammonia concentration in the treated effluent, was detected and noted as non-compliance by the regulatory agency. In addition, the TDS level in the treated effluent increased to reflect the oxidised sulphate.

The company decided to retrofit the BVF with RRL process for the control of sulphide inhibition because it would improve the nitrification capacity of the ETP, while totally eliminating chemical consumption for BVF operation, thus reducing the costs significantly.

The BVF recirculation rate was sufficient to reduce the concentration of sulphide to non-inhibitory levels when an air stripper was installed by RRL. A baffle plate stripper was chosen for the stripping operation. Although it was a low efficiency mass-transfer device, there was no clogging with sludge solids. The baffle plate stripper required high flow rate of air for the mass transfer of hydrogen sulphide. The off-gas from stripper has hydrogen sulphide in the range of 1500 ppm v/v. The stripper removes 600 kg sulphur per day; ie., 60% of the total sulphide generated in the anaerobic reactor. The stripping system is designed to reduce total sulphide concentration in the reactor to 220 mg/l from 450 mg/l. There is no need for chemical addition and the biogas H_2S is < 3% as compared with 7.2% prior to modification.

The odorous off-gas from the stripper is treated in a biofilter to remove hydrogen sulphide concentration less than 10 ppm v/v before releasing to the atmosphere.

Estimated payback as a result of chemical savings at manufacturer recommended dosage is less than 4 months. At the actual dosage used by the operator, the payback period is between 24 and 30 months. In addition, compliance with regulations with regard to ammonia removal as a result of downstream nitrification is expected as a result of oxygen demand in the secondary treatment system being lower by 368 kg oxygen per day, which was estimated as equivalent to augmentation of the diffused aeration system by the addition of a 22 KW blower. A further benefit is reduction of TDS in the treated effluent.

REFERENCES

1. Anderson, G.K., Yang, Y., 1992, Determination of bicarbonate alkalinity and total volatile acid concentration in anaerobic digesters using a simple titration, Water Environ. Res. 64: 53-59.

2. Athanassopoulos, N., Koujnis, J., Papdimitriou, A. and Koutinas, A.A., 1989, Effect of high sulphite contents on anaerobic digestion of raisin-finishing wastewater, Biological Wastes, 30: 53-60.

3. Bhattacharya, S.K. and Parkin, G.F., 1989, The effect of ammonia on methane fermentation process. J.WPCF, 61: 55-59.

4. BR 8901150 dated January 1990.

5. Bruser,T.,Lens, P. and Truper, H., 2000, The biological sulfur cycle. In: Environmental technologies to treat sulfur pollution-Principles and Engineering, Lens, P.N.L. and Hulshoff Pol,L. Ed, International Water Association, London, pp 47-85.

6. Colleir, H.M. and Chick, W.H., 1977, Problems and potential in the treatment of rubber factory and palm oil mill effluents, Planter Kaula Lumpur, 53: 439-488.

7. EP 418121 dated March 1991, Assignee: Degremont SA.

8. EP241999 dated August 1991 and EP 241602 dated October 1987; Assignee: Gist-Brocades NV.

9. Ehlinger, F., Gueler, J., Ball, F.X. and Prevot, C., 1992, Treatment of lees vinasses of red wine by methanogenic fermentation in presence of tannin and sulphide, Water Sci. Tech, 25: 275-284.

10. Fdz-Polanco, F., Fdz-Polanco, F., Fernandez, N., Uruena, M.A., Garcia, P.A. and Villaverde, S., 2001, Combining the biological nitrogen and sulfur cycle in anaerobic conditions. Water Sci. Technol. 44: 77-84.

11. Fdz-Polanco, F., Fdz-Polanco, F., Fernandez, N., Uruena, M. and Villaverde, S., 2000, Simultaneous organic nitrogen and sulfate removal in anaerobic GAC fluidised bed reactor. In proceedings of Latin American workshop-seminar on wastewater anaerobic treatment, Nov. Pacife, Brazil, 36-43.

12. Fernandez,N., Fdz-Polanco, F., Motalvo, S.J., Toledano, D., 2001, Use of activated carbon and natural zeolite as support materials in anaerobic fluidised bed reactor for vinasse treatment, Wat. Sci. Technol., 44: 4,1-16.

13. Franklin, R.J., 2001, Full-scale experiences with anaerobic treatment of industrial wastewater, Wat. Sci. Technol., 44: 1-6.

14. Hilton, B.L. and Oleszkiewicz, J.A., 1987, A comparison of anaerobic reactors operating with and without the addition of sulfates, Wat. Poll. Res. J. Canada, 22: 444-455.

15. Hulshoff Pol, L.W. Leus, P.N., Weijma, J. and Stam, A.J.M. (2001, New developments in reactor and process technology for sulfate reduction. Wat. Sci. Technol., 44: 67-76.

16. Ibrahim, A. and John C.K., 1985, Aeration system for the treatment of effluent from latex concentrate factories, Proc. International Rubber Conf. Kuala Lumpur 202-208.

17. JP 01249196 to JP 01249199 dated October, 1989: Assignee: aqua Renaissance GI.

18. JP 3131196 dated June 1990: Assignee: Kubota Corporation.

19. JP 3278892 dated December 1991 and JP 3270792 dated December 1991; Assignee: Toshiba KK.

20. JP 428296 dated 10.92: Assignee: Ishikawajima Harima Heavy Industry.

21. Khan, A.W., Trottier, T.M., 1978, Applied. Environ. Microbiol., 35: 1027.

22. Koepp H-J. *et al.*, 1985, Anaerobic treatment of wastewater from citric acid production, Conservation and recycling, 8: 211-219.

23. Kristjansson, J.K., Schonheit, P. and Thauer, R.K., 1982, Different K, values for hydrogen of methanogenic and sulfate reducing bacteria: An explanation of the apparent inhibition of methanogenesis by sulfate, Arch. Microbiol. 131: 278-282.

24. Lettinga, G. de mam, A., van der Last, A.R.M., Wiegant, W., vam Knippenberg, K. Frijns, J. and vam Buuren, J.C.L., 1993, Anaerobic treatment of domestic sewage and urban wastewater. Water Sci. Technol., 27: 67-73.

25. Maillacherruvu, K.Y., Parkin, G.F., Peng, C.Y., Kuo, W.C., Oonge, Z.I., and Lebduschka, V., 1993, Sulfide toxicity in anaerobic systems fed sulfate and various organics. Water Environ. Res. 65: 100-109.

26. McCartney, D.M., and Oleszkiewicz, J.A., 1993, Competition between methanogens and sulfate reducers: effect of COD/SO_4^{2-}ratio and acclimation, Water Environ. Res. 65: 655-664.

27. McFarland, M.J. and Jewell, W.J., 1989, In situ control of sulfide emission during thermophilic anaerobic digestion process. Water Res. 23: 1571-1577.

28. Neilsen, PH., 1987, Biofilm dynamics and kinetics during high-sulfate reduction under anaerobic condition, Appl. Environ. Microbiol. 53: 27-32.

29. NL 9100060 dated August 1992, Assignee: Paques BV, AB Balk, The Netherlands.

30. Puhakka J.A., Fergusan, J.F., and Vuoriranta, P., 1990, Effect of molybdate ions on methanation of simulated and natural wastewaters. Appl. Microbiol. Biotechnol., 32: 494-498.

31. Reis, M.A.M., Lenros, P.C., Almeida, J.S. and Carrondo, M.J.T., 1991, Evidence for the intrinsic toxicity of H2S to sulphate –reducing bacteria. Appl. Microbiol. Biotechnology, 36: 145-147.

32. Rintala, J., Sanz, M.J.L. and Lettinga, G., 1991, Thermophilic anaerobic treatment of sulfate rich pulp and paper integrate process water, Wat. Sci. Technol. 24: 149-160.

33. Rinzema A., Lettinga, G., 1988, Anaerobic treatment of sulphate containing wastewater In Biotreatment Systems Vol. III p107-121; Ed: Wise, D.L., CRC Press Inc., Boca Raton, USA.

34. Sarner, E., *et al.*, 1988, Anaerobic treatment using treatment for controlling H_2S toxicity, TAPPI J. 41-45.

35. Schonheit, P., Kristjansson, K. and Thauer, R.K., 1982, Kinetic mechanism for the ability of sulphate reducers to out-compete methanogens for acetate. Arch. Microbiol. 132: 285-288.

36. Standard methods for the Examination of Water and Wastewater (1995), 19th Ed. American Public Health Association/American Water Works Association/Water Environment Federation, Washington DC, USA.

37. Stephenson, R.J., Branion, R.M.R., and Pinder, K.L., 1994, Anaerobic 35 °C and 55°C treatment of a BCTMP/TMP effluent: sulphur management strategies, Water Sci. Tech. 29: 433-445.

38. Stucki, G., Hanselman, K.W. and Hrzeler, A., 1993, Biological sulfuric acid transformation: reactor design and process optimization, Biotech. Bioeng. 41: 303-315.

39. Tursman, J F. and Cork, D.J., 1989, Influence of sulfate and SRB on anaerobic digestion technology, In biological Waste Treatment, 273-285, Alan R. Liss Inc.

40. Visser A, Gao, Y. and Lettinga, G., 1992, Anaerobic treatment of synthetic sulphate –containing wastewater under thermophilic condition. Water Sci. Technol. 25: 193-202.

41. ZA 8907731 dated July 1990; Assignee: Gencor Ltd.

Index

A

D

E

F

G

H

I

J

K

L

M

P

R

S

W

X

Y

Z